THERMODYNAMICS OF NUCLEAR MATERIALS 1974

The following States are Members of the International Atomic Energy Agency:

AFGHANISTAN
ALBANIA
ALGERIA
ARGENTINA
AUSTRALIA
AUSTRIA
BANGLADESH
BELGIUM
BOLIVIA
BRAZIL
BULGARIA
BURMA
BYELORUSSIAN SOVIET SOCIALIST REPUBLIC
CANADA
CHILE
COLOMBIA
COSTA RICA
CUBA
CYPRUS
CZECHOSLOVAKIA
DEMOCRATIC PEOPLE'S REPUBLIC OF KOREA
DENMARK
DOMINICAN REPUBLIC
ECUADOR
EGYPT
EL SALVADOR
ETHIOPIA
FINLAND
FRANCE
GABON
GERMAN DEMOCRATIC REPUBLIC
GERMANY, FEDERAL REPUBLIC OF
GHANA
GREECE
GUATEMALA
HAITI
HOLY SEE
HUNGARY
ICELAND
INDIA
INDONESIA
IRAN
IRAQ
IRELAND
ISRAEL
ITALY
IVORY COAST
JAMAICA
JAPAN
JORDAN
KENYA
KHMER REPUBLIC
KOREA, REPUBLIC OF
KUWAIT
LEBANON
LIBERIA
LIBYAN ARAB REPUBLIC
LIECHTENSTEIN
LUXEMBOURG
MADAGASCAR
MALAYSIA
MALI
MAURITIUS
MEXICO
MONACO
MONGOLIA
MOROCCO
NETHERLANDS
NEW ZEALAND
NIGER
NIGERIA
NORWAY
PAKISTAN
PANAMA
PARAGUAY
PERU
PHILIPPINES
POLAND
PORTUGAL
ROMANIA
SAUDI ARABIA
SENEGAL
SIERRA LEONE
SINGAPORE
SOUTH AFRICA
SPAIN
SRI LANKA
SUDAN
SWEDEN
SWITZERLAND
SYRIAN ARAB REPUBLIC
THAILAND
TUNISIA
TURKEY
UGANDA
UKRAINIAN SOVIET SOCIALIST REPUBLIC
UNION OF SOVIET SOCIALIST REPUBLICS
UNITED KINGDOM OF GREAT BRITAIN AND NORTHERN IRELAND
UNITED REPUBLIC OF CAMEROON
UNITED STATES OF AMERICA
URUGUAY
VENEZUELA
VIET-NAM
YUGOSLAVIA
ZAIRE
ZAMBIA

The Agency's Statute was approved on 23 October 1956 by the Conference on the Statute of the IAEA held at United Nations Headquarters, New York; it entered into force on 29 July 1957. The Headquarters of the Agency are situated in Vienna. Its principal objective is "to accelerate and enlarge the contribution of atomic energy to peace, health and prosperity throughout the world".

Printed by the IAEA in Austria
June 1975

PROCEEDINGS SERIES

THERMODYNAMICS OF NUCLEAR MATERIALS 1974

PROCEEDINGS OF A SYMPOSIUM ON
THE THERMODYNAMICS OF NUCLEAR MATERIALS
HELD BY THE
INTERNATIONAL ATOMIC ENERGY AGENCY
AT VIENNA, 21-25 OCTOBER 1974

In two volumes

VOL.II

INTERNATIONAL ATOMIC ENERGY AGENCY
VIENNA, 1975

THERMODYNAMICS OF NUCLEAR MATERIALS 1974,
IAEA, VIENNA, 1975
STI/PUB/380
ISBN 92-0-040175-9

FOREWORD

These Proceedings form the record of the International Atomic Energy Agency's fourth Symposium on the Thermodynamics of Nuclear Materials. The first meeting in 1962 provided a means for thermodynamic experts from many countries to assess the available thermodynamic information on nuclear materials, which was very limited at that time. The second symposium held in 1965 was focussed on diffusion and emf methods while still emphasizing the importance of basic thermodynamic data. The third symposium was concerned primarily with the thermodynamics of solution systems, alloys, semi-metal compounds, transition metal and actinide carbides, solid solutions, and molten-salt systems. Again the importance of basic thermodynamic data was stressed, as the symposium served as a means for the international experts to review the existing status.

The technological theme of the fourth symposium was the application of thermodynamics to the understanding of the chemistry of irradiated nuclear fuels and to safety assessments for hypothetical accident conditions in reactors. Sessions were also held on the more basic thermodynamics, phase-diagrams and the thermodynamic properties of a wide range of nuclear materials.

The main technological aspects of this meeting included the transport behaviour and diffusional properties of nuclear materials of immediate importance in reactor systems. The chemical and physical interactions resulting from the formation of fission products in nuclear fuel systems were also presented. It is clear that our detailed understanding of fuel/clad reactions is still far from complete despite considerable out-of-pile and in-pile experiments. Significant advances have been made in the theoretical prediction of creep rates and in the formulation of equations of state. New experimental methods were described involving the use of laser bombardment to obtain extremely high temperatures.

This symposium showed that while considerable progress has been made in the thermodynamic analysis of nuclear fuels, there is still a great need for fundamental data in this field, especially for the advanced fuels such as carbides, oxycarbides and nitrides, where computer calculation of multi-component systems is likely to be of growing importance.

EDITORIAL NOTE

The papers and discussions have been edited by the editorial staff of the International Atomic Energy Agency to the extent considered necessary for the reader's assistance. The views expressed and the general style adopted remain, however, the responsibility of the named authors or participants. In addition, the views are not necessarily those of the governments of the nominating Member States or of the nominating organizations.

Where papers have been incorporated into these Proceedings without resetting by the Agency, this has been done with the knowledge of the authors and their government authorities, and their cooperation is gratefully acknowledged. The Proceedings have been printed by composition typing and photo-offset lithography. Within the limitations imposed by this method, every effort has been made to maintain a high editorial standard, in particular to achieve, wherever practicable, consistency of units and symbols and conformity to the standards recommended by competent international bodies.

The use in these Proceedings of particular designations of countries or territories does not imply any judgement by the publisher, the IAEA, as to the legal status of such countries or territories, of their authorities and institutions or of the delimitation of their boundaries.

The mention of specific companies or of their products or brand names does not imply any endorsement or recommendation on the part of the IAEA.

Authors are themselves responsible for obtaining the necessary permission to reproduce copyright material from other sources.

CONTENTS TO VOL. II

THEORETICAL MODELS OF NUCLEAR FUELS (Session VI)

BASIC PHASE EQUILIBRIA OF NUCLEAR FUELS (Session VII)

BASIC THERMODYNAMIC PROPERTIES (Session VIII)

Session VI
THEORETICAL MODELS OF NUCLEAR FUELS

Chairman: A. PATTORET
FRANCE

SYSTEMATIC THERMODYNAMIC PROPERTIES OF ACTINIDE METAL-OXYGEN SYSTEMS AT HIGH TEMPERATURES

Emphasis on lower valence states

R.J. ACKERMANN
Argonne National Laboratory,
Argonne, Ill.,
United States of America

M.S. CHANDRASEKHARAIAH
Bhabha Atomic Research Centre,
Trombay, Bombay,
India

Abstract

SYSTEMATIC THERMODYNAMIC PROPERTIES OF ACTINIDE METAL-OXYGEN SYSTEMS AT HIGH TEMPERATURES: EMPHASIS ON LOWER VALENCE STATES.

The thermodynamic data for the actinide metals and oxides (thorium to curium) have been assessed, examined for consistency, and compared with the lanthanides. Correlations relating the enthalpies of formation of the solid oxides with the corresponding aquo ions make possible the estimation of the thermodynamic properties of $AmO_2(s)$ and $Am_2O_3(s)$ which are in accordance with vaporization data. The known thermodynamic properties of the substoichiometric dioxides $MO_{2-x}(s)$ at high temperatures demonstrate the relative stabilities of valence states less than 4+ and lead to the examination of stability requirements for the sesquioxides $M_2O_3(s)$ and the monoxides MO(s). Sequential trends in the gaseous metals, monoxides and dioxides are examined, compared, and contrasted with the lanthanides.

The production of the transuranium elements via successive neutron capture occurs along with the formation of fission fragments in nuclear fuel burnup. From the rather substantial amount of information about the fuel materials, principally oxides, and a growing amount dealing with thorium, neptunium, americium, and curium compounds it is instructive to examine the available information for consistency, reliability and trends, and to extend this knowledge to the estimation of the thermodynamic properties of lower-valent oxides including stable and metastable phases that have not been experimentally measured.

Critical reviews and assessments of the phase behavior, chemical and thermodynamic properties of uranium, and plutonium compounds have been published by the International Atomic Energy Agency [1,2], and a variety of reviews of the actinides are available [3-10].

The high temperature chemistry of these systems has not been examined thoroughly but should be emphasized because it is based largely on the chemical bonding in relatively simple gaseous molecules, and furthermore, it elucidates the relative stabilities of a variety of valence states that co-exist in non-stoichiometric solid phases that are thermodynamically stable and experimentally accessible only at high temperatures. In general the lower valence states, not stable at room temperature, become relatively more stable at higher temperatures.

The chemistry of the actinides is challenging because a great deal of the systematization appears to be a synthesis of transition metal [(n-1)d] and lanthanide [(n-2)f] chemistry. The lack of general similarity in behavior of analogues of the (4f) lanthanides and (5f) actinides will become apparent in the course of the present examination although certain more specific intergroup trends appear to persist. In order to ascertain the role of f-orbitals in chemical bonding of the lanthanides and extend it to the more complex actinides let us examine some of the thermodynamic properties of lanthanide metals, ions, and monoxides with increasing nuclear sequence, i.e., with increasing occupation of f-orbitals by electrons. The results are given in Table I. Both the heats of sublimation of the metals, $\Delta H^0_{298}(M)$, and dissociation energies of the monoxides, $D_0(MO)$, show distinct parallel trends of decreasing strength in bonding with increasing occupation of f-orbitals. A sharp increase in stability is reflected in both quantites at gadolinium and lutecium

TABLE I. SOME THERMODYNAMIC PROPERTIES OF LANTHANIDE METALS AND OXIDES IN VAPOR, SOLID AND AQUEOUS PHASES

	Electronic configuration [11] 4f	5d	6s	$\Delta H^0_{298}(M)$ [12] (kcal)	D_0 (MO) [15] (kcal)	$\Delta E(M)$ $4f^n \rightarrow 4f^{n-1}5d^1$ sol'n [17] (kcal)	gas [9] (kcal)	$-\Delta H^0_f[M^{3+}, aq.]$ [18] (kcal)	$-\frac{1}{2}\Delta H^0_f$ <M_2O_3, 298 K> [19] (kcal)
La	0	1	2	102.3 [13]	189.9 [13]	-45.7	-43.1	168.8	214.4
Ce	1	1	2	99.3 [14]	190.5 [16]	-17.1	-13.6	166.1	214.6
Pr	3	0	2	85.0	179	5.8	11.4	167.7	217.9
Nd	4	0	2	78.3	167	11.5	19.3	166.4	216.1
Pm	5	0	2	(79)	-	-	22.9	(166)	-
Sm	6	0	2	49.4	133	39.9	44.3	164.1	218.0
Eu	7	0	2	41.9	129	65.7	71.8	144.6	197.4
Gd	7	1	2	95.0	161	-22.8	-31.3	163.0	217.0
Tb	9	0	2	92.9	164	8.5	8.2	164.4	222.9
Dy	10	0	2	69.4	145	31.4	21.6	(166.7)	222.6
Ho	11	0	2	71.9	148	31.4	22.0	169.4	224.8
Er	12	0	2	75.8	146	28.6	20.5	169.5	226.8
Tm	13	0	2	55.5	121	51.2	37.5	167.9	225.7
Yb	14	0	2	36.4	97	83.0	66.4	161.2	216.8
Lu	14	1	2	102.2	158	-	-	166.7	224.4

($4f^{14}$) at which point the f-orbitals are respectively half-filled ($4f^7$) and completely filled ($4f^{14}$), and each atom contains one d-electron. Hence, one observes the greater strength of chemical bonding of the d-orbital, i.e., the recurrence of transition metal behavior. The repetitive behavior in chemical bonding associated with these trends in the lanthanides has been called the "double periodicity" [20].

However, a closer inspection of the thermodynamic data suggests that the latter half of the trend is not a simple repeat of the first half. For instance, the heat of sublimation for erbium is either grossly incorrect, or a substantial increase in chemical bonding in the metal occurs. The energy necessary to transform the $4f^n$ configuration to a $4f^{n-1}5d^1$ based on spectroscopic measurements in solution [17] and in the gas phase [9] also shows this departure from a monotonic trend. Further evidence of this departure from the simple double-periodicity is given by the heats of formation of the M(+3) aquo ions [18] and the solid sesquioxides [19] that are shown in Figure 1. In addition, the half-filled $4f^7$ of europium and the filled $4f^{14}$ of ytterbium reflect the considerably lower stabilities of the trivalent metal in solution and in the respective sesquioxides.

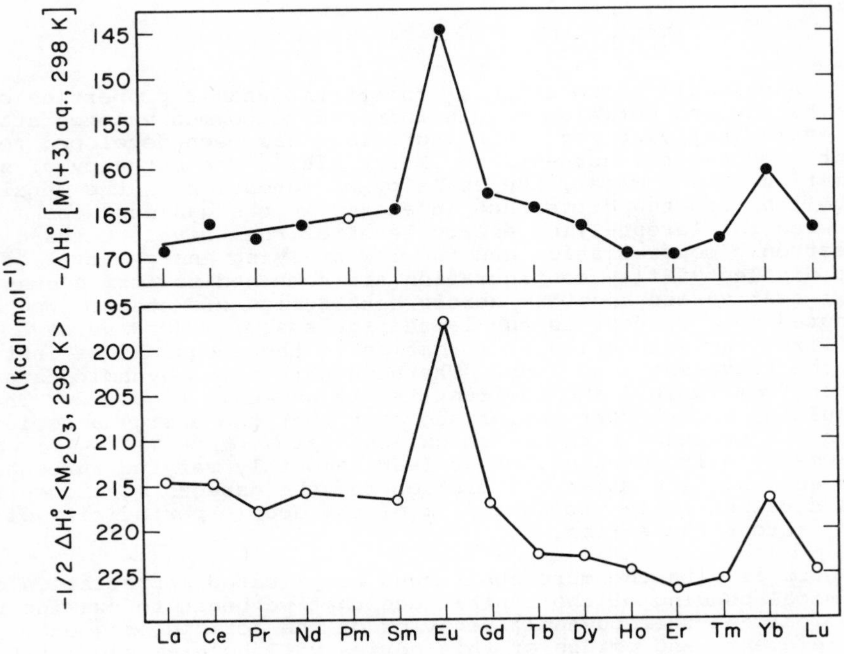

FIG.1. Sequential trends of the enthalpies of formation of the M(+3) aquo ions and of the solid sesquioxides at 298 K for the lanthanides.

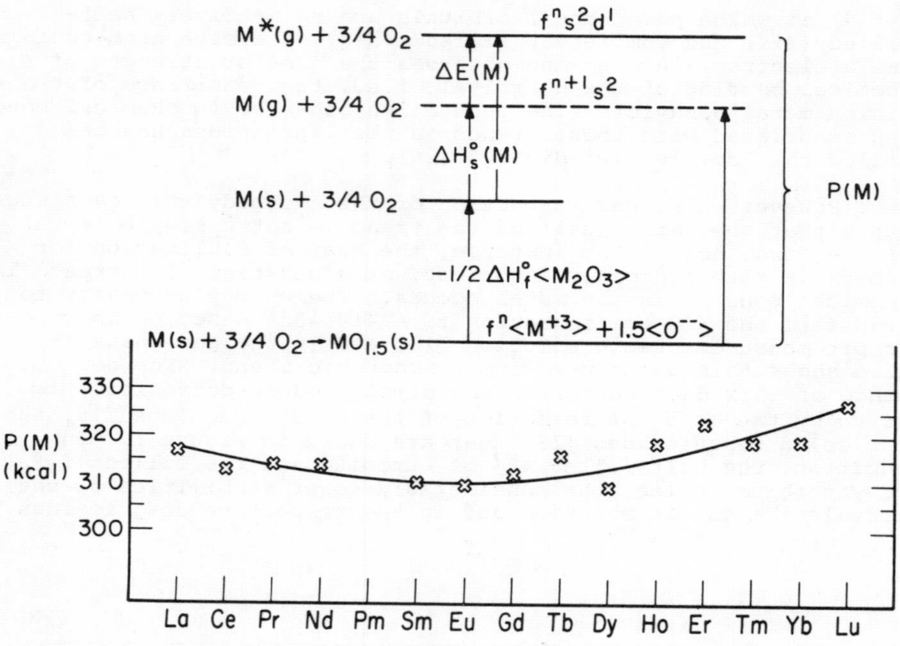

FIG.2. Plot of P(M) for the solid lanthanide sesquioxides.

A valuable correlation of the thermodynamic properties of lanthanide and actinide metals referred to common valence states or equivalent electronic configurations has been developed recently by Nugent, Burnett, and Morss [18]. The enthalpy of sublimation of the metal, the enthalpy of formation of the metal aquo ion, and the difference in energy in the gaseous atom between the lowest-lying energy level of the trivalent $f^n s^2 d^1$ electronic configuration and the lowest-lying energy level of the divalent $f^{n+1} s^2$ configuration are combined to form a quantity P(M) that varies smoothly across both series of metals. We have adapted this concept to the lanthanide sesquioxides and have utilized the enthalpies of formation of the sesquioxides instead of the trivalent aquo ions. The necessary thermodynamic data are taken from Table I and the results are shown in Figure 2. The resulting smooth correlation suggests that the energy associated with the process in the sesquioxides, $3/4\ O_2(g) + 3e^- \rightarrow 3/2\ \langle O^=\rangle$ is essentially constant, or at least smoothly varying throughout the series. A similar correlation for the gaseous monoxides for the divalent state yields a form of the double periodicity of P(M) across the series.

It is with the more subtle and complicated appreciation of chemical bonding of the lanthanides that we begin to examine the chemical and thermodynamic behavior of the actinide sequence. The elements and oxides of this series will be examined first in order to ascertain the status and reliability of thermodynamic properties, after which an attempt at systematization will be presented.

1. The Thorium-Oxygen System

The dioxide ThO_2 is the only thermodynamically stable solid oxide. The phase diagram displays a simple miscibility gap between the metal and dioxide [28] and resembles that of the uranium-oxygen system [22]. Ackermann and Rauh [23] reported that the congruently vaporizing composition of the dioxide departs slightly from the stoichiometric composition at 2820 K at which O/Th = 1.994 which value confirms the earlier observations [24]. Aitken et al. [25] demonstrated that the dioxide phase can exist at grossly substoichiometric compositions; a limiting composition O/Th = 1.96 was observed at 3000 K in a "dry" hydrogen atmosphere. This observation is consistent with the more reduced compositions at the lower phase boundary reported by Benz [21].

Solid thorium metal apparently does not dissolve any significant amount of oxide, but the liquid phase shows increasing solubility of oxide at higher temperatures [21]. In many respects the chemistry of thorium is often compared to that of titanium, zirconium, and hafnium. However, the solubility of oxygen as well as the refractory metals W and Ta in the three transition metals is substantially larger in both solid and liquid phases [26,27].

The solid monoxide is metastable, has a cationic radius and molar volume intermediate between those the metal and dioxide, and disproportionates to Th(s) + ThO_2(s) above 1000°C [28].

The vapor pressure measurements of Darnell et al. [29] are inconsistent with the heat content measurements of α, β and liquid thorium, reported by Levinson [30]. Therefore, a redetermination of the vapor pressure of pure thorium was undertaken by Ackermann and Rauh [31] from which was derived an enthalpy of sublimation, $\Delta H^\circ_{298} = 142.6 \pm 1.0$ kcal mol^{-1}, a value consistent with the thermal functions recently assessed by Rand [32]. The vaporization behavior of the dioxide phase has been studied by Wolff and Alcock (2170-2400 K) [33] by Darnell and McCollum (2268-2593°C) [34] and by Ackermann et al. (2180-2860 K) [24]. The measured effusion rates in the latter two studies agree within a few percent over a wide range of temperature. The Langmiur measurements of Wolff and Alcock [33], however, are lower by approximately 60%.

A revision of the earlier values of the thermodynamic properties of ThO(g) has resulted from a study of the vaporization of the univariant system, Th(l) + ThO_2(s) $\rightarrow$ 2ThO(g) and the isomolecular equilibrium, YO(g) + Th(g) = Y(g) + ThO(g) [23,35]. The partial pressures of the ThO_2(g), ThO(g), and Th(g) over the entire system have been more reliably established. The free energies of formation of the gaseous metal, monoxide, dioxide, and the solid dioxide are given in Table II.

2. The Protactinium - Oxygen System

The thermodynamic behavior of the metal and its oxides is not well known. The enthalpy of sublimation of the metal at 298K, 143 kcal mol^{-1}, has been estimated by Nugent et al. [18].

TABLE II. FREE ENERGIES OF FORMATION OF GASEOUS ACTINIDE METALS AND GASEOUS AND SOLID OXIDES

$[\Delta G_f^0 = \Delta H_f^0 - T\Delta S_f^0$ (cal/mol)]

Compound	ΔH_f^0 (cal/mol)	ΔS_f^0 (cal/K mol)	Temperature (K)	References
O(g)	61245	16.07	1500-3000	[36]
Th(g)	136230	27.5_7	2010-2500	[31]
U(g)	116700	26.5	1600-2400	[48]
	118770	26.8	1720-2340	[44]
Np(g)	102370	23.7_8	1540-2140	[70]
Pu(g)	78300	21.0	1210-1620	[70]
Am (g)	51700	18.9	1450-1820	[95]
	60230	21.4	1100-1450	[96]
Cm (g)	82640	21.69	1179-2068	[105]
ThO(g)	-16000	12.6	1900-2300	[23,35]
UO(g)	- 4400	15.6	1700-2150	[44]
	-10870	11.22		[66]
	- 7800	13.8	1800-2200	[53]
	- 8800	10.3	1800-2200	[54]
NpO(g)	-12890	13.7_3	1700-2000	[74]
PuO(g)	-28500	9.7	1600-2150	See text
AmO(g)				
CmO(g)	-29200	10.7_5	1800-2600	This work
ThO_2(g)	-132120	-7.2_3	2000-3000	[23]
UO_2(g)	-121500	-5.4_5	1600-2200	[67]
	-115500	-1.9	2080-2705	[61,68]
	-116300	-0.5	1700-2150	[44]
NpO_2(g)	-114800	-3.4	1800-2500	[72,74]
PuO_2(g)	-112600	-6.6	1600-2150	See text
AmO_2(g)				
UO_3 (g)	-200000	-19.4	1200-1500	[62]
ThO_2(s)	-292600	-43.7	2000-3000	[23]
PaO_2(s)	-259800	-39.7	1560-1715	[39]
UO_2(s)	-258600	-41.3	1800-2400	[12,36,56-58]
NpO_2(s)	-256000	-40.4	1500-2000	[74]
PuO_2(s)	-249000	-42.6	1600-2150	See text
AmO_2(s)	(-233000)	(-42.5)	> 1500	This work
U_2O_3(s)	(-372000)	(-58)	> 1500	This work

TABLE II. (cont.)

Compound	ΔH_f^0 (cal/mol)	ΔS_f^0 (cal/K mol)	Temperature (K)	References
	$[\Delta G_f^0 = \Delta H_f^0 - T\Delta S_f^0$ (cal/mol)]			
Np_2O_3(s)	(-378 000)	(-58)	> 1500	This work
Pu_2O_3(s)	-399600	-63.8	1600-2150	See text
Am_2O_3(s)	-404 000	-64.0	> 1500	This work
Cm_2O_3(s)	(-396 000)	(-58)	1500-2500	This work
UO(s)	-120 500	-22.8	1473-1648	[55]
PuO(s)	-133 500	-22.1	> 1500	[94]
AmO(s)	(-137 000)	(-22)	> 1500	This work

Recent effusion measurements of the vapor pressure in the range 2100-2600K [37] scatter over a power of ten, but the lowest pressures are consistent with the estimated enthalpy. Vaporization of PaO(g) and the creep of liquid metal through the porous tungsten effusion cells may account for the higher pressures. The earlier estimate of the vapor pressure by Cunningham [38] yields a value of ΔH_{298}° near 155 kcal mol^{-1}.

The free energy of formation of the dioxide PaO_2(s) has been derived by Lorenz and Scherff [39] from the carbothermic reduction, $PaO_2(s) + 3C(s) = PaC(s) + 2CO(g)$, from which the equation for $\Delta G_f^\circ \langle PaO_2 \rangle$ is given in Table II. The result is based on an estimate of $\Delta G_f^\circ \langle PaC \rangle$ from those of UC(s) and ThC(s) and the implicit assumption that the monocarbide is not stabilized by dissolved oxygen. Nevertheless, the entropy of formation of the dioxide $\Delta S_f^\circ \langle PaO_2 \rangle = 39.7$ cal mol^{-1} K^{-1} is nearly comparable with those of other actinide dioxides as seen in Table II. The relatively short temperature range 1560-1715 K of the measurements restricts somewhat the reliability of both the enthalpy and entropy terms.

Knock and Schieferdecker [40] studied the reduction of Pa_2O_5 from 700 to 1400°C with H_2O-H_2 mixtures and were able to derive partial molar quantities over a limited range of composition, $2.30 < O/Pa < 2.40$. It was not possible to completely reduce Pa_2O_5 to PaO_2 with hydrogen.

3. The Uranium-Oxygen System

Since earlier reviews of the uranium-oxygen system [1,3,5, 8] have dealt largely with the properties of UO_2, its superstoichiometric homogeneity range UO_{2+x} and the higher valence states, the present discussion will deal primarily with the substoichiometric UO_{2-x} phase and some of the discrepancies that persist in uranium-oxygen system after nearly two decades of extensive research.

The phase diagram corresponding to the U-UO_2 region displays a miscibility gap between the metal and dioxide below 2740°C and monotectic behavior above this temperature as reported by Martin and Edwards [22]. Latta and Fryxell [41] have determined the solidus-liquidus curves for $UO_{2\pm x}$, and Edwards, et al. [42] have measured the congruently vaporizing composition of $UO_{2-x}(s)$. The lower phase boundary of the UO_{2-x} phase has been confirmed by subsequent studies, but the solubility of oxide in liquid uranium reported by Guinet et al. [43] and Pattoret et al. [44] is considerably greater.

The vapor pressure of liquid uranium from the absolute effusion and mass spectrometric measurements of Ackermann and Rauh [45] together with solubility corrections for tantalum [46] yield an enthalpy of sublimation, ΔH°_{298K} = 127 kcal mol^{-1} which compares with the mass spectrometric values of 126 from Drowart et al. [47] and 129 kcal mol^{-1} from Pattoret et al. [44]. The overall uncertainties in these values is very likely 1-2 kcal mol^{-1}, and hence, no substantial disagreement persists. A value near 127 kcal mol^{-1} is presently preferred based on a recent thermodynamic assessment of the uranium-carbon system [48].

The solid monoxide has been reported on many occasions, however, recent studies have demonstrated its metastability in the absence of stabilizing impurities such as carbon [28].

Tetenbaum and Hunt [49] have measured the oxygen potential of the UO_{2-x} phase from 2080 to 2705 K as a function of composition, $1.92 \leq O/U \geq 2.01$ by a carrier gas or transpiration technique. The agreement of these results with the equilibration results of Markin et al. [50] and Wheeler [51] is good but those of Javed [52] yield considerably more negative oxygen potentials, whereas the mass spectrometric values of Pattoret et al. [44] are systematically somewhat more positive. The composition of the lower phase boundary (hereafter referred to as l.p.b.) of the UO_{2-x} phase has been established principally by Martin and Edwards [22] and by Ackermann et al. [53] by different methods. In the present assessment we have used the results of Tetenbaum and Hunt [49] and extended the thermodynamic properties of the UO_{2-x} phase to its lower phase boundary based on the model developed by Blackburn [54] which relates the chemical potential of oxygen with composition. From the plots of $\Delta\bar{G}_{O_2}$ vs T for various compositions in the range, $1.70 \leq O/U \geq 2.00$, values of the partial molar entropy and enthalpy were obtained. It is difficult to assess here meaningful uncertainties in the partial molar quantities primarily because the experimental scatter in the individual values of $\Delta\bar{G}_{O_2}$ and the extrapolation thereof. Fortunately the choice of the oxygen potential at the l.p.b. does not significantly influence the integral free energy, enthalpy, and entropy, of formation of the UO_{2-x} phase which are calculated from the Gibbs-Duhem equation,

$$\Delta Z^\circ_f\langle UO_{2-x}\rangle - \Delta Z^\circ_f\langle UO_{2.00}\rangle = \frac{1}{2}\int_{2-x}^{2.00} \Delta\bar{Z}_{O_2}\, d(O/U) \quad (1)$$

in which Z = G, H, or S. Values of these integral properties are shown in Figure 3. The numerical values of the standard

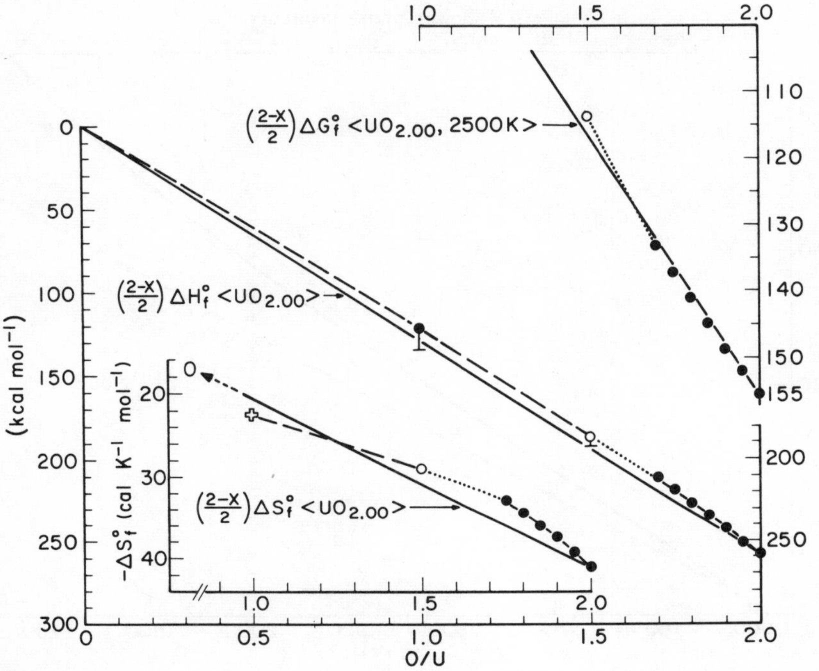

FIG.3. The standard free energy, enthalpy, and entropy of formation versus atomic ratio for the system U-O.

free energy of formation of the UO_{2-x} phase are consistent with its formation at temperatures above approximately 1500 K. Values of the standard free energy of formation of UO_{2-x} at 2500 K shown in Figure 3, lie slightly more negative by several kilocalories than those from the quantity, $[(2-x)/2]\Delta G_f^{\circ}\langle UO_{2.00}\rangle$, which results from the disproportionation,

$$(\tfrac{x}{2})M(\ell) + (\tfrac{2-x}{2})\, MO_2(s) \rightarrow MO_{2-x}(s) . \qquad (2)$$

The values for UO(s) were derived from studies in the U-C-O system reported by Steele et al. [55].

The consistency among the thermodynamic properties of the known vapor species and those of the $UO_{2-x}(s)$ phase at the l.p.b. can be examined by means of the heterogeneous equilibria:

$$UO_{2-x}^{l.p.b.}(s) + (1-x)U(g) = (2-x)UO(g) \qquad (3)$$

$$UO_{2-x}^{l.p.b.}(s) = (1-x)UO_2(g) + xUO(g), \qquad (4)$$

$$UO_{2-x}^{l.p.b.}(s) = (\tfrac{2-x}{2})UO_2(g) + \tfrac{x}{2}U(g) \qquad (5)$$

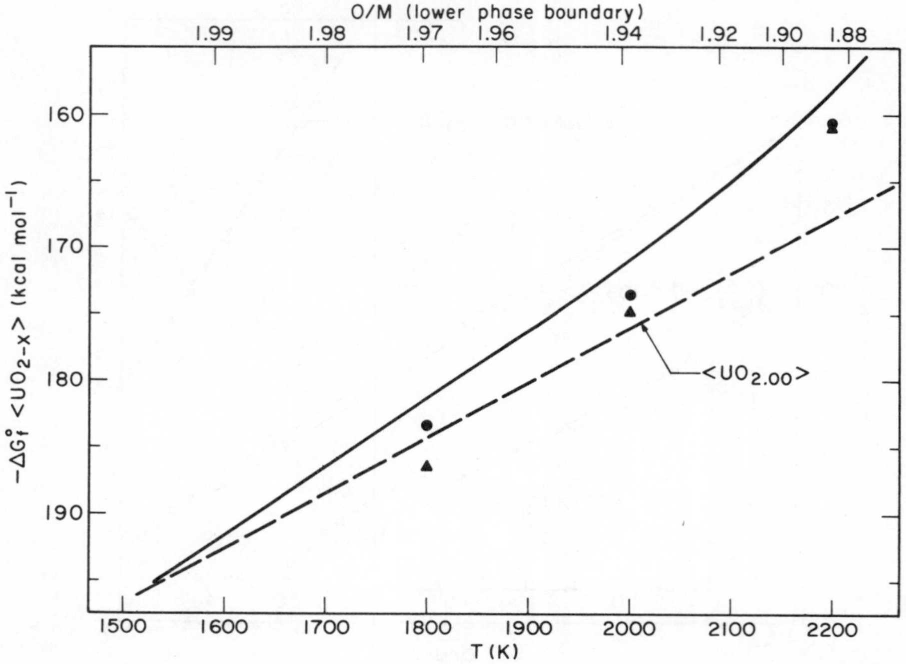

FIG.4. The standard free energy of formation of UO_{2-x} at its lower phase boundary and $UO_{2.00}$ versus temperature (K). The ● points were obtained from vapor phase equilibria of Ackermann et al. [53], and the ▲ points from Pattoret et al. [44].

The standard free energy of formation of $UO_{2-x}^{l.p.b.}$ (s) can be calculated from known values of x, the equilibrium constants for reactions (3-5), and the free energies of formation of the gaseous species given in Table II. Figure 4 shows a comparison of the results of this calculation between the vaporization data of Ackermann et al. [53] and Pattoret et al. [44] with the values obtained from the measurements of oxygen potential [49,54]. The standard free energy of $UO_{2.00}$ from Table II is given as a reference curve. The thermodynamic data given by Ackermann et al. [53] for equations (3), (4) and (5) yield a common answer for ΔG_f° UO_{2-x} but a systematic difference of ∿2 kcal from the integration of the oxygen potential (solid line) is seen to exist over the entire range of temperature. The ionization probability measurements of Blackburn [54] do not significantly improve the situation. The thermodynamic data given by Pattoret et al. [44] for equation (3) yield the points that show a somewhat greater difference especially at the lower temperatures, although their data for equations (4) and (5) (points not shown) superimpose on the solid line. By and large the magnitude of these discrepancies is not entirely unexpected in view of the uncertainties in the relative ionization cross sections and thermal functions of the three vapor species resulting from mass spectrometric analyses together with factor of 2 discrepancies in the measured pressures of U, UO and UO_2. A factor of 2 error in a partial pressure generates an error of approximately 3 kcal in the calculated free energy.

Evaluation of the thermodynamic properties of grossly non-stoichiometric condensed phases stable only at high temperatures via measurement of vapor phase equilibria is presently the most straightforward method available to examine the relative stability of the different valence states present in the solid. Direct calorimetric techniques are usually precluded because it is difficult to quench successfully a sample that is representative of the structure and composition at high temperatures; partial or complete disproportionation invariably occurs, especially in systems that are stable by only a few kilocalories such as UO_{2-x}, La_2O_{3-x} [13], and Ce_2O_{3-x} [16], Y_2O_{3-x} [59], and ThO_{2-x} [60].

The total pressure of uranium-bearing species measured by the transpiration method as a function of temperature and composition ($1.0 \leq O/U \geq 2.0$) reported by Tetenbaum and Hunt [61] agrees closely with the variety of effusion measurements previously reported for UO_2(s). However, a spread of values for the heat of sublimation of UO_2(g), 134.1-147.8 kcal mol^{-1}, indicates systematic errors in composition and/or temperature. The pressure data of Tetenbaum and Hunt [61] extrapolate to those (predominately UO) measured by Ackermann et al. [53] at the l.p.b. of UO_{2-x} rather than those of Pattoret et al. [44].

A rather large discrepancy in the uranium-oxygen system concerns the free energy of formation of the gaseous trioxide. The extent of this discrepancy is as large as 9 kcal mol^{-1} at 1500 K as examined by Ackermann and Chang [62]. These authors measured via thermogravimetric equilibration the chemical potential of O_2 as a function of temperature and composition for the U_3O_{8-z} phase. The temperature- and z-dependence of the free energy of formation of U_3O_{8-z} were evaluated via Gibbs-Duhem integration, and hence, a more accurate evaluation of the free energy of formation of UO_3(g) is possible from the reaction, $(1/3)\ U_3O_{8-z}(s) + \{(1 + z)/6\}\ O_2(g) \rightarrow UO_3(g)$. In previous transpiration measurements of this reaction the substoichiometric index z was estimated [63] or assumed to be zero [64]. The recognition of non-zero values for z change the free energy of UO_3(g) by less than -1 kcal, and therefore, do not significantly reduce the 9 kcal discrepancy between the transpiration results and the mass spectrometric measurements of UO_3(g) over $UO_{2\pm x}$(s) [44]. If the latter results were correct, the observed UO_3(g) transport in eq. (6) should have been larger by a factor of 10. A redetermination of this transport by Chandrasekharaiah et al. [65] confirms the earlier measurements [63, 64]. Therefore, a substantial systematic error appears to be present in the mass spectrometric determinations. The thermodynamic properties of the gaseous uranium oxides from several studies are given in Table II. Clearly, in some cases, particularly for UO(g), clarifying studies are needed.

4. The Neptunium-Oxygen System

The vapor pressure and enthalpy of vaporization of the metal have been measured via the effusion method by Eick and Mulford

[69] and by Ackermann and Rauh [70]. The latter study demonstrated via mass spectrometric analysis the presence of significant amounts of NpO(g) in the vapor phase if oxygen contamination of the metal occurs, and consequently, the effusion rates measured by Eick and Mulford [69] apparently correspond to the Np(ℓ) + NpO_2(s) system. The enthalpy of sublimation derived by Ackermann and Rauh [70], ΔH°_s(298 K) = 110 ± 2 kcal mol^{-1} considerably improves the thermodynamic correlations of the actinide metals by Nugent et al. [18].

The condensed oxides have not been adequately characterized. The only known thermodynamic quantity is the enthalpy of formation of the dioxide, $\Delta H^\circ_f < NpO_2, 298\ K > = -256\ 700$ cal mol^{-1} [71]. Earlier studies indicate that the dioxide phase does not form superstoichiometric compositions [3], which seems remarkable in view of the existence of higher valence states present in the oxide Np_3O_8, and other behavior that closely parallels the uranium-oxygen system.

The vaporization behavior, indicates that a substoichiometric dioxide is stable at high temperatures (∿2000 K), and that a hexagonal, rare earth type A Np_2O_3(s) is precipitated upon cooling to room temperature [72]. The lattice constants of this phase a = 4.23_4 Å and c = 6.10 Å, are several percent larger than those of Pu_2O_3(s). The solid monoxide has been identified by Zachariasen [73] with a = 4.979 Å, which value correlates well with those for other metastable actinide monoxides [28]. The thermodynamic properties of the gaseous monoxides NpO(g) and NpO_2(g) have been measured from vaporization studies [72,74]. The results have been shown to be consistent with the enthalpy of formation of NpO_2(s) [71], a variety of measured pressures including the vapor pressure of the metal, and isomolecular equilibria involving the known thermodynamic properties of LaO(g) and YO(g) [35]. The results suggest that NpO(g) contains many low-lying electronic states. The gaseous trioxide, NpO_3(g) has not been observed but correlations of the ionization potentials of gaseous oxides suggest the likelihood of its stability [75]. Linear equations for the standard free energies of formation of the gaseous metal, monoxide, and dioxide are given in Table II.

5. The Plutonium-Oxygen System

The vapor pressure and enthalpy of vaporization of liquid plutonium have been reported by Phipps et al. [76], Mulford [77], Kent [78], and Ackermann and Rauh [70]. The agreement among all the investigations is good, differing at most by 20-30 per cent in the measured pressures. The determinations by Kent and by Ackermann and Rauh are most consistent with a recent assessment of the thermal functions of solid and liquid plutonium [12].

In the anhydrous Pu-O system, the highest oxidation state of metal so far established is +4. The four known oxide structures are the hexagonal Pu_2O_3, b.c.c. $PuO_{1.52}$, cubic $PuO_{1.6}$, and fluorite PuO_2 [2,3]. The dioxide phase exhibits an extensive substoichiometric homogeniety range above 900 K. The cubic $<PuO_{1.52}>$ phase is stable only below 500 K. A tentative phase diagram has been reported by the 1966 IAEA Panel [2].

Above 925 K, the <PuO_2> phase exhibits considerable oxygen deficiency. Because of the similarity in structure among <PuO_{2-x}>, <$PuO_{1.6}$> and <$PuO_{1.52}$>, it is plausible that a continuous transition from the fluorite <PuO_{2-x}> to the cubic <$PuO_{1.6}$> can exist if the structure of <$PuO_{1.6}$> is f.c.c. If,on the other hand, it is b.c.c. a miscibility gap should exist since a continuous transition from f.c.c. to b.c.c. involving a reconstruction type rearrangement is not likely. This point is yet not completely resolved, although recent studies [79] indicate the existence of a miscibility gap.

In common with many other refractory oxides the study of the vaporization behavior of the Pu-O system couples the thermodynamic data for condensed phases with that for the vapor phase. In addition to the component elements, Pu(g) and O_2(g), two oxide species, PuO(g) and PuO_2(g), constitute the equilibrium vapor phase at high temperatures. The congruently vaporizing composition, $1.9 \leq O/Pu \geq 1.8$; lies within the extensive substoichiometric region of the dioxide phase at 1600 and 2400 K, respectively [80,81]. Previous measurements of the vapor pressure of the substoichiometric dioxide have shown approximately factor of two agreement [82]. Ackermann et al. [80] and two assessments [2,83] have combined vapor pressure data with reported calorimetric and Galvanic cell data [84] to derive the free energies of formation of various solid phases and vapor species.

The high specific activity of plutonium (^{239}Pu) causes self-heating and radiation damage at low temperatures, $T < 50$ K, and these effects may have contributed appreciable uncertainty in the thermal functions of the metal and dioxide. The use of lon-longer-lived isotopes (^{242}Pu and ^{244}Pu) will reduce these effects and heat capacity measurements using these isotopes are presently in progress [85]. The 1966 IAEA Panel [2] adopted the value of 19.7 ± 1.0 cal mol^{-1} K^{-1} for the absolute entropy of PuO_2(s) at 298 K. This value is based on the measured value of 16.3 cal mol^{-1} K^{-1} reported by Sandenaw [86] and a contribution due to magnetic ordering not observed,but predicted [87, 88]. However, the absence of magnetic ordering, and hence, the lower value of the entropy is strongly supported by a constant magnetic susceptibility from 4 to 1000 K reported by Raphael and Lallement [89]. The thermal functions reported by Kruger and Savage [90] are consistent with those of Sandenaw [86], and therefore, in the present assessment we have revised the standard free energy of formation based on the thermal functions of Kruger and Savage [90] for PuO_2(s) up to 1500 K and those of Ogard [91] from 1500 to 2700 K. For the metal we have used the thermal functions given by Hultgren et al. [12] and for O_2(g) those of Stull and Sinke [36]. The resultant equations for the free energies of formation are given in Table II.

The most comprehensive thermodynamic data for the solid phases between $PuO_{1.5}$ and PuO_2 above room temperature result from Galvanic cell measurements of Markin and Rand [84]. The temperature dependence of the partial molar free energy of oxygen at fixed compositions yields values for the partial molar entropy and enthalpy. Quite recently, Chereau, Dean and Gerdanian [92] reported a direct microcalorimetric measurement at 1100°C of the enthalpy which differs significantly from that of

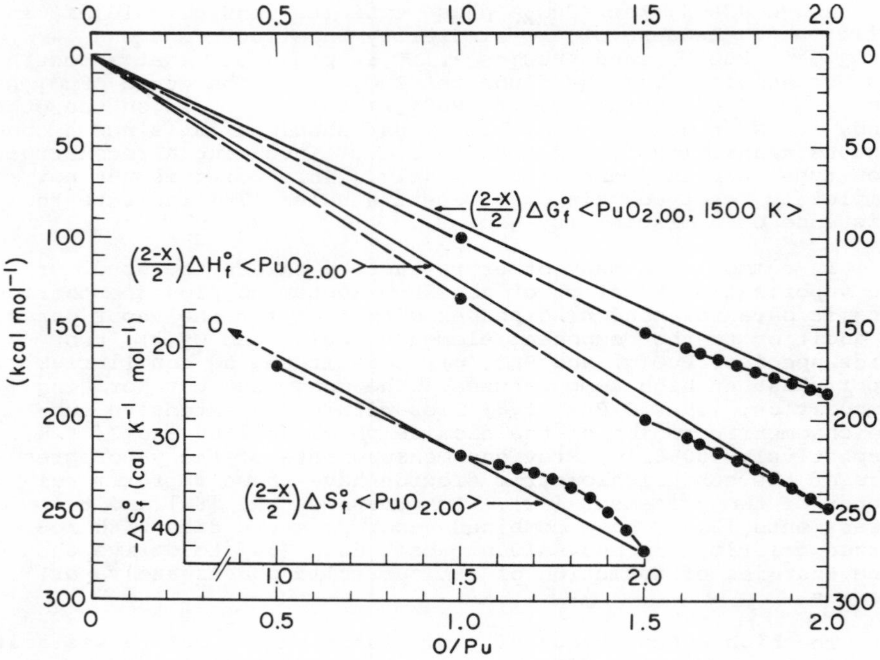

FIG.5. The standard free energy, enthalpy, and entropy of formation versus atomic ratio for the system Pu-O.

Markin and Rand and shows a more complicated dependence on composition. These latter results tend to verify certain aspects of the statistical mechanical model developed by Atlas [93]. Composition-structure detail of this kind should also be present in the partial molar entropy. However, a net cancellation and less sensitive dependence of these more detailed effects occurs in the integral molar properties.

The integral molar properties obtained from the integration of the partial molar quantities and the revision of the thermodynamic data for PuO_2(s) are shown in Figure 5. The values for PuO(s) are taken from the work of Potter [94] which is based on the analysis of the PuO_xC_{1-x} system. The solid monoxide is seen to be just unstable to disproportionation to Pu(ℓ) and Pu_2O_3(s).

6. The Americium-Oxygen System

The vapor pressure and enthalpy of vaporization of liquid americium have been reported by Erway and Simpson [95] and by Carniglia and Cunningham [96]. Ward [97] quite recently has carried out extensive effusion measurements on the solid metal which support better the higher pressures of Erway and Simpson. No measured values of thermodynamic properties exist for any of the solid or gaseous oxides. Therefore, we have estimated the enthalpy of formation of AmO_2(s) via the correlation of the

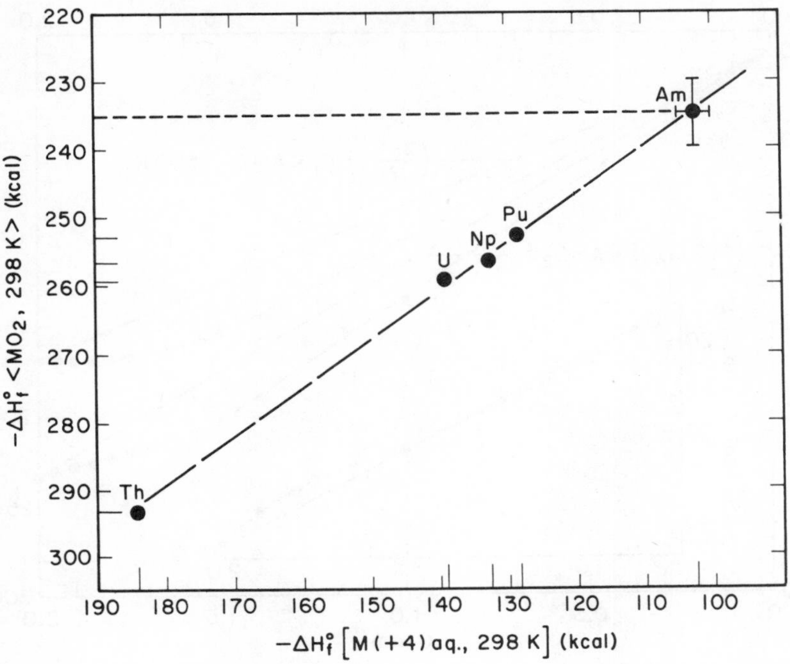

FIG.6. Correlation of the enthalpies of formation of the solid actinide dioxides with those of the +4 aquo ions at 298 K.

enthalpies of formation of other actinide dioxides with the enthalpies of the +4 aquo ions [10] as seen in Figure 6. The value obtained by the apparently linear extrapolation, $\Delta H_f^\circ <AmO_2$, 298 K > = -235 ± 5 kcal mol^{-1}, will be shown later to be consistent with correlations of the +3 valence-states and with vaporization data.

Chikalla and Eyring [98,99] have studied the phase behavior and have measured the partial molar free energy of molecular oxygen for $AmO_{2-x}(s)$ between 1140 and 1445 K. From these data the partial molar enthalpies and entropies have been evaluated and then integrated via equation (1) to give the values of the integral molar quantities shown in Figure 7. The values of $\Delta S_f^\circ < AmO_2 > \simeq 42.5$ cal mol^{-1} K^{-1} is estimated from those of the other dioxides. The proposed values of the integral molar quantities for $Am_2O_3(s)$ are shown in Figure 7.

These values derived from those of the dioxide phases (Figure 6) and the partial molar quantities are closely supported by the measurements of the pressure of americium species vaporizing from dilute concentrations of americium ($x_{Am} = 3.8 \times 10^{-4}$ and 6.5×10^{-4}) contained in the $PuO_{1.6}(s)$ - $Pu_2O_3(s)$ - vapor system and treated in accordance with Raoult's Law [80]. The oxygen potential ($\Delta\bar{G}_O$) for this system insures that the americium in solid solution will be predominantly in the +3 state and the

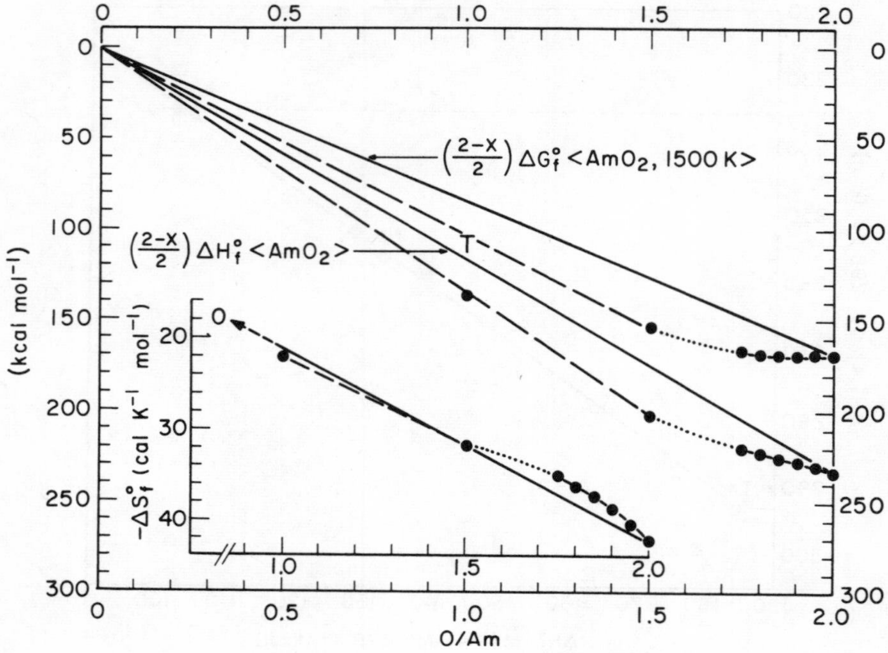

FIG.7. The standard free energy, enthalpy, and entropy of formation versus atomic ratio for the system Am-O.

greater stability of Am(g) relative to Pu(g) (Table II) suggests the principal mode of vaporization to be

$$\frac{1}{2}\,Am_2O_3(s) \rightarrow Am(g) + \frac{3}{2}\,O(g)\,, \tag{6}$$

from which one obtains for the standard free energy of formation the equation,

$$\frac{1}{2}\Delta G^\circ_f < Am_2O_3 > = \Delta G^\circ_f(Am) + \frac{3}{2}\,\Delta G^\circ_f(O) + \frac{3}{2}\Delta\overline{G}_O + RT\ln p_{Am}. \tag{7}$$

The terms in equation (7), evaluated at 1500 K from Table II and reference [80], yield $1/2\Delta G^\circ_f$ < Am_2O_3, 1500 K > = −152 ± 3 kcal mol^{-1} which value agrees remarkably well with that obtained in Figure 7. Linear free energy equations for gases and solids are given in Table II. The calculations of partial pressures of various americium-containing species by Ozhegov et al. [100] are based on assumptions and estimates that are inconsistent with the present treatment.

The solid monoxide has been considered just stable to disproportionation to Am(ℓ) + Am_2O_3(s) at 1500 K based on its direct synthesis by Akimoto [101] and the sequential trend toward stability of MO(s) [28] which, however, are presently not as conclusive as the stability of its lanthanide analogue, EuO(s) [102,103].

7. The Curium-Oxygen System

The vapor pressure, crystal structure, and melting point of the metal are reported [104,105]. Oxygen decomposition pressures and phase relationships of the oxides have been reported by Chikalla and Eyring [106]. The postulated phase diagram is characterized by two phases $CmO_{1.72}$ and $CmO_{1.82}$ between the sesquioxide and dioxide. Pronounced hysteresis in the heating and cooling cycles of the oxygen isobars are typical of those observed in some of the lanthanide oxides [107].

The vaporization behavior of the sesquioxide reported by Smith and Peterson [108] contributes valuable insight to its high temperature stability. From thermodynamic considerations these investigators concluded that the most likely mode of vaporization is $Cm_2O_3(s) \rightarrow 2CmO(g) + O(g)$. Since the thermodynamic properties are unknown for both solid and vapor phases, a knowledge of one is required to evaluate the other. Those for $Cm_2O_3(s)$ and CmO(g) are derived in the following section.

8. Thermodynamic Correlation of the +3 Valence-States

A closer examination of the reliability of the measured and estimated thermodynamic properties of the actinide oxides is instructive. In fact, no calorimetric measurement of the enthalpy of formation of any sesquioxide exists. The question to be considered is whether a systematic formalism, albeit empirical, is capable of demonstrating a consistency that can be used to guide further experimentation and development. Such an approach appears possible as seen in Figure 8 wherein are plotted the enthalpies of formation of the sesquioxide phases derived herein versus the enthalpies of formation of the M(+3) aquo ions taken from references [10] and [109]. The apparent linear correlation of a common valence state (+3) in both the solid sesquioxides and in aqueous solution is seen to extend through the first half of the lanthanide series as well. However, some important differences are indicated. The stability of Am(+3) aquo ion is greater than Cm(+3) whereas one might have expected the opposite from analogous behavior of the lanthanides at Eu and Gd. Apparently there is a decided folding back of the chemical stability in the actinide series with the reappearance of the d-electron in Cm. The details of this reverse trend are obviously not well understood and need to be resolved by further studies. The thermodynamic properties of Cm_2O_3 had been estimated previously by Smith and Peterson [108] to be near those of $Pu_2O_3(s)$. As seen from Figure 8 this appears to be a valid assumption. The estimated free energy of formation of $Cm_2O_3(s)$ is given in Table II and together with the measured vapor pressure of the sesquioxide corresponding to the mode of vaporization, $Cm_2O_3(s) \rightarrow 2CmO(g) + O(g)$, one obtains the standard free energy of formation of CmO(g) given in Table II.

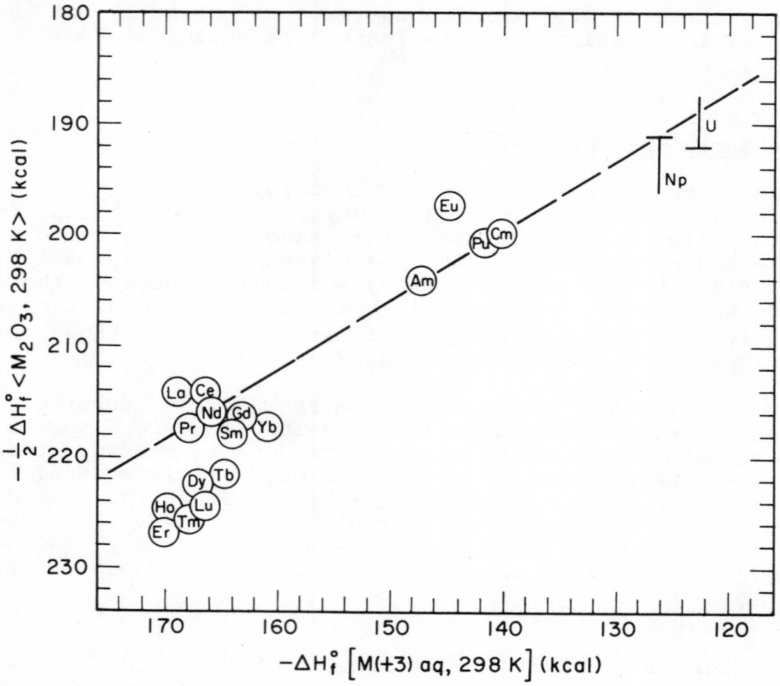

FIG. 8. Correlation of the enthalpies of formation of the solid lanthanide and actinide sesquioxides with those of the + 3 aquo ions at 298 K.

9. Systematic Behavior of Actinide Metal and Oxides at High Temperatures

The thermodynamic properties of the actinide metals (Th-Cm), the gaseous monoxides and dioxides, and the M(+3) and M (+4) aquo ions are compiled in Table III. The enthalpies of sublimation of the metals at 298 K decrease in the sequence Th-Am. The reappearance of d-orbital occupation at Cm produces a significant increase, but the trend is not as pronounced as that in the lanthanide metals (Table I). A similar comparative trend occurs for the dissociation energies of the monoxides [$D_T(MO) = 1/2\ D(O_2) - \Delta H_f^\circ(MO) - \Delta H_v^\circ(M)$] for which the quoted values correspond to the temperatures of measurement (~2000 K) because the enthalpy functions needed for conversion to 298 K are unknown. It may be rather perilous to assume these functions are closely approximated by those of the lanthanide analogues in view of the preferred d-orbital occupation in the actinides.

The dissociation energies of the gaseous dioxides also show a sequential decrease. The comparative trend in the lanthanides has not been sufficiently studied because the +4 valence state is not sufficiently stable to permit the usual mass-spectrometric and thermodynamic analysis of the species with the notable exception of CeO_2(g) [16].

TABLE III. SOME THERMODYNAMIC PROPERTIES OF ACTINIDE METALS AND OXIDES IN VAPOR, SOLID AND AQUEOUS PHASES

	Electronic configuration 5f	6d	7s	$\Delta H^0_{298}(M)$ (kcal)	$D_T(MO)$ (kcal)	$D_T(MO_2)$ (kcal)	$-\Delta H^0$ [M^{3+}, aq.] [10] (kcal)	$-\Delta H^0$ [M^{4+}, aq.] [10] (kcal)	P(M) (kcal)
Th	0	2	2	142.6	213	390	-	184.0	
Pa	2	1	2	(143)	-	-	-	-	
U	3	1	2	127	188	360	122.6	138.8	≤319
Np	4	1	2	110.0	176	340	126.2	132.3	≥301
Pu	6	0	2	83.3	168	313	141.7	128.2	303
Am	7	0	2	(55)	(164)	(290)	147.4 [109]	103	302
Cm	7	1	1	89	173	-	140.2		293

It is the free energies of formation rather than the dissociation energies that reflects the thermodynamic stabilities of the gaseous molecules, and from an inspection of the equations thereof in Table II, a clear perspective of the high temperature behavior of the actinide oxides emerges; the +2 valence state in MO(g) becomes increasingly more stable whereas the +4 state in MO_2(g) becomes less stable. The nature of these trends closely resembles those of the +3 and +4 states in aqueous solution, respectively, as seen in Table III. As in aqueous solution the +6 state in MO_3(g) becomes rapidly less stable, being observed only as UO_3(g), possibly as NpO_3(g) but not as PuO_3(g) [75]. Similar trends in stability occur in the condensed phases as seen from Table II and from Figures 3, 5, 6, 7 and 8. The sequential increase in stability of the +3 valence state in the solid and in aqueous solution is apparently paralleled by a similar trend in the solid monoxides from metastability to real stability at AmO(s) [28]. The stabilization of the solid monoxides by carbon occurs with a significant decrease in the observed lattice parameters (and hence the cationic radii) which is tantamount to oxidation of the +2 valence to the more stable +3 and +4 states.

The correlation scheme of Nugent et al. [18] has been applied to the solid sesquioxides as outlined in Figure 2. Numerical values of P(M) given in the last column of Table III indicate a gradual decreasing trend resembling that of the lanthanides, which demonstrates another measure of thermodynamic consistency.

The thermodynamic behavior displayed by the substoichiometric dioxide phases, UO_{2-x}, PuO_{2-x}, and AmO_{2-x} illustrates the relative stabilities of the +4 with lower valences. The fact that UO_{2-x} is just thermodynamically stable to disproportionation at 2500 K (Figure 3) is a manifestation of the much lesser stabilities of the +2 or +3 valence states. Inspection of Figures 3, 5, and 7, indicates that these comparative stabilities are predominantly influenced by the enthalpy terms; the entropy dependences are quite similar. Furthermore, the nature of the defect fluorite structure of UO_{2-x} may differ considerably from those of PuO_{2-x}, AmO_{2-x}, and CeO_{2-x} especially near the respective lower phase boundaries. The lack of relative stability of lower valence states in UO_{2-x} forces coexistence with liquid metal whereas the others coexist with lower oxides whose defect structures are closely related to oxygen-deficient fluorite structures. Kofstad [110] has pointed out that the compositional dependence of the partial pressure of oxygen over UO_{2-x} is not easily rationalized in terms of oxygen vacancies, but fits rather well a defect model involving uranium interstitials.

At the present time the estimation of the thermodynamic properties of the higher actinides (Bk → Lw) appears quite difficult. Certain features of a "double periodicity" are apparent but the lanthanide analogues do not themselves display a simple, regular behavior that can be effectively borrowed. Additional experimental studies using rather small amounts of the higher actinides will be necessary in order to further characterize their chemistry. Vaporization studies of dilute solutions appear to hold promise as illustrated herein by the Am-O system. Chikalla and Turcotte [111] have pointed out the possible linear correlation of the partial molar free energy of oxygen of the $MO_{1.9}$ phases at 900°C with the M^{+3}-M^{+4} redox potentials from aqueous solutions. At the moment, however, little more of a quantitative nature can be said.

REFERENCES

[1] Thermodynamic and Transport Properties of Uranium Dioxide, Technical Reports Series No.39, IAEA, Vienna (1965).

[2] The Plutonium-Oxygen and Uranium-Plutonium-Oxygen Systems: A Thermochemical Assessment, Technical Reports Series No. 79, IAEA, Vienna (1967).

[3] ROBERTS, L.E.J., Q. Rev., Chem. Soc. 15 (1961) 442.

[4] ASPREY, L.B., PENNEMAN, R.A., Chem. Eng. News 45 (1967) 75.

[5] EYRING, L., Chapter 10, Solid State Chemistry (RAO, C.N.R., Ed.), Marcel Dekker, New York (1974) 565.

[6] FRED, M., in Lanthanide/Actinide Chemistry, Advances in Chemistry Series, No.71, Am. Chem. Soc., Washington, D.C. (1967) 180.

[7] MOELLER, T., J. Chem. Educ. 47 (1970) 417.

[8] EYRING, L., in High Temperature Oxides, Part II (ALPER, A.M., Ed.), Academic Press, New York (1970) 41.

[9] BREWER, L., J. Opt. Soc. Am. 61 (1971) 1101.

[10] FUGER, J., Chapter 5, Lanthanides and Actinides (BAGNALL, K.W., Ed.), Butterworths, London (1972) 157.

[11] MOORE, C.E., Appl. Optics 2 (1963) 665.

[12] HULTGREN, R., DESAI, P.D., HAWKINS, D.T., GLEISER, M., KELLEY, K.K., WAGMAN, D.D., Selected Values of the Thermodynamic Properties of the Elements, Am. Soc. Met. (1973).

[13] ACKERMANN, R.J., RAUH, E.G., J. Chem. Thermodyn. 3 (1971) 445.

[14] ACKERMANN, R.J., KOJIMA, M., RAUH, E.G., WALTERS, R.R., J. Chem. Thermodyn. 1 (1969) 527.

[15] BREWER, L., ROSENBLATT, G.M., in Advances in High Temperature Chemistry (EYRING, L., Ed.) 2, Academic Press, New York (1969).

[16] ACKERMANN, R.J., RAUH, E.G., J. Chem. Thermodyn. 3 (1971) 609.

[17] JORGENSEN, C.K., Mol. Phys. 7 (1964) 417; Mol. Phys. 5 (1962) 271.

[18] NUGENT, L.J., BURNETT, J.L., MORSS, L.R., J. Chem. Thrermodyn. 5 (1973) 665.

[19] GSCHNEIDNER, K.A., Jr., KIPPENHAN, N., McMASTERS, O.D., Thermochemistry of the Rare Earths, Molybdenum Corporation of America, White Plains, N.Y. (1973).

[20] AMES, L.L., WALSH, P.N., WHITE, D., J. Phys. Chem. 71 (1967) 2707.

[21] BENZ, R., J. Nucl. Mater. 29 (1969) 43.

[22] MARTIN, A.E., EDWARDS, R.K., J. Phys. Chem. 69 (1965) 1788.

[23] ACKERMANN, R.J., RAUH, E.G., High Temperature Science 5 (1973) 463.
[24] ACKERMANN, R.J., RAUH, E.G., THORN, R.J., CANNON, M.C., J. Phys. Chem. 67 (1963) 762.
[25] AITKEN, E.A., BRASSFIELD, H.C., FRYXELL, R.E., "Thermodynamic behaviour of hypostoichiometric UO_2", Thermodynamics (Proc. Symp. Vienna, 1965) 2, IAEA, Vienna (1966) 435.
[26] KOMAREK, K.L., SILVER, M., "Thermodynamic properties of zirconium-oxygen, titanium-oxygen and hafnium-oxygen systems", Thermodynamics of Nuclear Materials (Proc. Symp. Vienna, 1962), IAEA, Vienna (1962) 749.
[27] ACKERMANN, R.J., RAUH, E.G., High Temp. Sci. 4 (1972) 272.
[28] ACKERMANN, R.J., RAUH, E.G., J. Inorg. Nucl. Chem. 35 (1973) 3787.
[29] DARNELL, A.J., McCOLLUM, W.A., MILNE, T.A., J. Phys. Chem. 64 (1960) 341.
[30] LEVINSON, L.S., J. Nucl. Mater. 19 (1966) 50.
[31] ACKERMANN, R.J., RAUH, E.G., J. Chem. Thermodyn. 4 (1972) 521.
[32] RAND, M.H., "Part I. Thermochemical properties", Thorium: Physico-Chemical Properties of its Compounds and Alloys, At. Energy Rev., Special Issue No.4 (1975) XX.
[33] WOLFF, E.G., ALCOCK, C.B., Trans. Br. Ceram. Soc. 61 (1962) 667.
[34] DARNELL, A.J., McCOLLUM, W.A., Atomics International Report NAA-SR-6498 (15 Sep. 1961).
[35] ACKERMANN, R.J., RAUH, E.G., J. Chem. Phys. 60 (1974) 2266.
[36] STULL, D.R., SINKE, G.C., Thermodynamic Properties of the Elements, Am. Chem. Soc., Washington, D.C. (1956).
[37] DOD, R.L., PARSONS, T.C., Lawrence Berkeley Laboratory Rep. LBL-666 (1971) 252-3.
[38] CUNNINGHAM, B.B., Tagungsbericht der 3. Internationalen Protaktiniumkonferenz (Schloss Elmau bei Mittenwald, April 1969), H.-J. Born Institut für Radiochemie der Technischen Universität München (1971) paper No. 14.
[39] LORENZ, R., SCHERFF, H.L., Tagungsbericht der 3. Internationalen Protaktiniumkonferenz (Schloss Elmau bei Mittenwald, April 1969), H.-J. Born Institut für Radiochemie der Technischen Universität München (1971) paper No.15.
[40] KNOCH, W., SCHIEFERDECKER, B., Tagungsbericht der 3. Internationalen Protaktiniumkonferenz (Schloss Elmau bei Mittenwald, April 1969), H.-J. Born, Institut für Radiochemie der Technischen Universität München (1971) paper No.10.
[41] LATTA, R.E., FRYXELL, R.E., J. Nucl. Mater. 35 (1970) 195.
[42] EDWARDS, R.K., CHANDRASEKHARAIAH, M.S., DANIELSON, P.M., High Temp. Sci. 1 (1969) 98.
[43] GUINET, P., VAUGOYEAU, H., BLUM, P., Commissariat à l'énergie atomique Rep. CEA-R-3060 (Nov. 1966); see also C. R. 257 (1963) 3401.
[44] PATTORET, A., DROWART, J., SMOES, S., "Etudes thermodynamiques par spectrométrie de masse sur le système uranium-oxygène", Thermodynamics of Nuclear Materials, 1967 (Proc. Symp. Vienna, 1967), IAEA, Vienna (1968) 613; see also PATTORET, A., Doctoral Thesis, Free University of Brussels (1969).
[45] ACKERMANN, R.J., RAUH, E.G., J. Phys. Chem. 73 (1969) 769.
[46] ACKERMANN, R.J., RAUH, E.G., High Temp. Sci. 4 (1972) 496.
[47] DROWART, J., PATTORET, A., SMOES, S., J. Chem. Phys. 42 (1965) 2629.
[48] Assessment of the Thermodynamic Properties of the U-C, Pu-C, and U-Pu-C Systems (Proc. Panel Vienna, 1974), to be published in Technical Reports Series, IAEA, Vienna (1975).
[49] TETENBAUM, M., HUNT, P.D., J. Chem. Phys. 49 (1968) 4739.
[50] MARKIN, T.L., WHEELER, V.J., BONES, R.J., J. Inorg. Nucl. Chem. 30 (1968) 807.
[51] WHEELER, V.J., J. Nucl. Mater. 39 (1971) 315.
[52] JAVED, N.A., J. Nucl. Mater. 43 (1972) 219.
[53] ACKERMANN, R.J., RAUH, E.G., CHANDRASEKHARAIAH, M.S., J. Phys. Chem. 73 (1969) 762.
[54] BLACKBURN, P.E., J. Nucl. Mater. 46 (1973) 244; see also BLACKBURN, P.E., DANIELSON, P.M., J. Chem. Phys. 56 (1972) 6156.
[55] STEELE, B.C.H., JAVED, N.A., ALCOCK, C.B., J. Nucl. Mater. 35 (1970) 1.
[56] HUBER, E.J., Jr., HOLLEY, C.E., Jr., J. Chem. Thermodyn. 1 (1969) 267.
[57] FREDRICKSON, D.R., CHASANOV, M.G., J. Chem. Thermodyn. 2 (1970) 623.
[58] HEIN, R.A., FLAGELLA, P.N., CONWAY, J.B., J. Am. Ceram. Soc. 51 (1968) 291.
[59] ACKERMANN, R.J., RAUH, E.G., J. Chem. Thermodyn. 5 (1973) 331.
[60] RAUH, E.G., ACKERMANN, R.J., The measurement of chemical and thermodynamic properties of refractory oxides above 1500 K, to be published in Can. Metall. Q. (1975).
[61] TETENBAUM, M., HUNT, P.D., J. Nucl. Mater. 34 (1970) 86.
[62] ACKERMANN, R.J., CHANG, A.T., J. Chem. Thermodyn. 5 (1973) 873.
[63] ALEXANDER, C.A., Doctoral Dissertation, Vapor Solid Equilibria in the Uranium Oxide-Oxygen System, The Ohio State University, Columbus, Ohio (1961).

[64] ACKERMANN, R.J., THORN, R.J., TETENBAUM, M., ALEXANDER, C.A., J. Phys. Chem. 64 (1960) 350.
[65] DHARWADKAR, S.R., TRIPATHI, S.N., KARKHANAVALA, M.D., CHANDRASEKHARAIAH, M.S., these Proceedings, Vol.2., paper IAEA-SM-190/32.
[66] RAND, M.H., MARKIN, T.L., "Some thermodynamic aspects of $(U,Pu)O_2$ solid solutions and their use as nuclear fuels", Thermodynamics of Nuclear Materials, 1967 (Proc. Symp. Vienna, 1967), IAEA, Vienna (1968) 637.
[67] ACKERMANN, R.J., GILLES, P.W., THORN, R.J., J. Chem. Phys. 25 (1956) 1089.
[68] REEDY, G.T., CHASANOV, M.G., J. Nucl. Mater. 42 (1972) 341.
[69] EICK, H.A., MULFORD, R.N.R., J. Chem. Phys. 41 (1964) 1475.
[70] ACKERMANN, R.J., RAUH, E.G., to be published in J. Chem. Thermodyn.
[71] HUBER, Jr., E.J., HOLLEY, Jr., C.E., J. Chem. Eng. Data 13 (1968) 545.
[72] ACKERMANN, R.J., FAIRCLOTH, R.L., RAUH, E.G., THORN, R.J., J. Inorg. Nucl. Chem. 28 (1966) 111.
[73] ZACHARIASEN, W.H., Acta Crystallogr. 5 (1952) 664.
[74] ACKERMANN, R.J., RAUH, E.G., to be published in J. Chem. Phys.
[75] RAUH, E.G., ACKERMANN, J. Chem. Phys. 60 (1974) 1396.
[76] PHIPPS, T.E., SEARS, G.W., SEIFERT, R.L., SIMPSON, O.C., Int. Conf. Peaceful Uses At. Energy (Proc. Conf. Geneva, 1955) 7, UN, New York (1956) 382.
[77] MULFORD, R.N.R., "The vapour pressure of plutonium", Thermodynamics (Proc. Conf. Vienna, 1965), IAEA, Vienna (1966) 231.
[78] KENT, R.A., High Temp. Sci. 1 (1969) 169.
[79] DEAN, G., BOIVINEAU, J.C., CHEREAU, P., MARCON, J.P., Plutonium 1970 and Other Actinides (Proc. Conf. Santa Fe, 1970: MINER, W.N., Ed.) Part II, Nucl. Metall. 17 (1970) 753.
[80] ACKERMANN, R.J., FAIRCLOTH, R.L., RAND, M.H., J. Phys. Chem. 70 (1966) 3698.
[81] OHSE, R.W., CIANI, C., "Evaporation behaviour and high-temperature thermal analysis of substoichiometric plutonium oxide for $1.51 < O/Pu < 2.00$", Thermodynamics of Nuclear Materials, 1967 (Proc. Symp. Vienna, 1967), IAEA, Vienna (1968) 545.
[82] MESSIER, D.R., J. Am. Ceram. Soc. 51 (1968) 710.
[83] RAND, M.H., At. Energy Rev. 4 (1966) 7.
[84] MARKIN, T.L., RAND, M.H., "Thermodynamic data for plutonium oxides", Thermodynamics (Proc. Symp. Vienna, 1965) 1, IAEA, Vienna (1966) 145.
[85] FLOTOW, H.E., OSBORNE, D.W., FRIED, S.M., MALM, J.G., these Proceedings, Vol. 2, paper No. IAEA-SM-190/46.
[86] SANDENAW, T.A., J. Nucl. Mater. 10 (1963) 165.
[87] OSBORNE, D.W., WESTRUM, E.F., J. Chem. Phys. 20 (1952) 695.
[88] WESTRUM, Jr., E.F., GRØNVOLD, F., "Chemical thermodynamics of the actinide element chalcogenides", Thermodynamics of Nuclear Materials (Proc. Symp. Vienna, 1962), IAEA, Vienna (1962) 3.
[89] RAPHAEL, G., LALLEMENT, R., Solid State Commun. 6 (1968) 383.
[90] KRUGER, O.L., SAVAGE, H., J. Chem. Phys. 49 (1968) 4540.
[91] OGARD, A.E., Plutonium 1970 and Other Actinides (Proc. Conf. Santa Fe, 1970: MINER, W.N., Ed.) Part I, Nucl. Metall. 17 (1970) 78.
[92] CLEREAU, P., DEAN, G., GERDANIAN, P., C. R. Ser. C. 272 (1971) 515.
[93] ATLAS, L.M., J. Phys. Chem. Solids 29 (1968) 91.
[94] POTTER, P.E., "The ternary system plutonium-carbon-oxygen", Thermodynamics of Nuclear Materials, 1967 (Proc. Symp. Vienna, 1967), IAEA, Vienna (1968) 337.
[95] ERWAY, N.D., SIMPSON, O.C., J. Chem. Phys. 18 (1950) 953.
[96] CARNIGLIA, S.A., CUNNINGHAM, B.B., J. Am. Chem. Soc. 77 (1955) 1502.
[97] WARD, J.W., Los Alamos Scientific Laboratory, private communication.
[98] CHIKALLA, T.D., EYRING, L., J. Inorg. Nucl. Chem. 30 (1968) 133.
[99] CHIKALLA, T.D., EYRING, L., J. Inorg. Nucl. Chem. 29 (1967) 2281.
[100] OZHEGOV, P.I., MYASOEDOV, B.F., ZAKHAROV, E.A., Dokl. Akad. Nauk SSSR 212 (1973) 1122.
[101] AKIMOTO, Y., J. Inorg. Nucl. Chem. 29 (1967) 2650.
[102] HUBER, E.J., Jr., HOLLEY, C.E., Jr., J. Chem. Thermodyn. 1 (1969) 301; J. Chem. Thermodyn. 2 (1970) 896.
[103] HASCHKE, J.M., EICK, H.A., J. Phys. Chem. 73 (1969) 374.
[104] CUNNINGHAM, B.B., WALLMANN, J.C., J. Inorg. Nucl. Chem. 26 (1964) 271.
[105] SMITH, P.K., HALE, W.H., THOMPSON, M.C., J. Chem. Phys. 50 (1969) 5066.
[106] CHIKALLA, T.D., EYRING, L., J. Inorg. Nucl. Chem. 31 (1969) 85.
[107] EYRING, L., J. Solid State Chem. 1 (1970) 376.

[108] SMITH, P.K., PETERSON, D.E., J. Chem. Phys. 52 (1970) 4963.
[109] FUGER, J., SPIRLET, J.C., MUELLER, W., Inorg. Nucl. Chem. Lett. 8 (1972) 709.
[110] KOFSTAD, P., Nonstoichiometry, Diffusion and Electrical Conductivity in Binary Metal Oxides, Wiley-Interscience, New York (1972) 301-04.
[111] CHIKALLA, T.D., TURCOTTE, R.P., Solid State Chemistry (ROTH, R.S., SCHNEIDER, Jr., S.J., Eds), National Bureau of Standards (US) Special Publication 364 (Jul 1972) 319.

DISCUSSION

A. PATTORET (Chairman): May I call attention to what appear to be certain errors in your paper concerning interpretation of our measurements on the uranium-oxygen system. Firstly, we did not determine liquidus compositions; however, our observations led us to conclude that the oxygen in the liquid uranium was slightly soluble. These observations therefore supported the liquidus proposed by Edwards et al. and not that proposed by Blum et al. Secondly, we did determine several points of the UO_{2-x} solidus from the intersection of the curves representing the thermodynamic values in the adjacent fields $UO_{2-x}(s)$ and $U(l) + UO_{2-x}(s)$. If these data alone are used the solidus obtained is in excellent agreement with those determined by Edwards et al. and by yourself and co-workers.

R.J. ACKERMANN: There is no significant disagreement as to the location of the solidus which we have called the lower phase boundary. As we pointed out in our paper (Section 3, para. 2) all subsequent measurements, including your own, confirm the results of Martin and Edwards. The three previous measurements of the liquidus are quite diverse and our interpretation thereof is based on a figure which you give in your doctoral thesis and which we have acknowledged in the last part of Ref.[44].

A. PATTORET: In the case of gaseous uranium trioxide, a species which plays an important part in migrations to the interior of fuel pins, there is a real discrepancy between the results obtained by transpiration on $U_3O_8(s)$ and by mass spectrometry on $U_{2\pm x}$. This question is also linked with the problem of the congruence of dioxide evaporation. An effort should be made to eliminate this rather disturbing inconsistency.

R.J. ACKERMANN: The free energies of formation of UO(g) and of UO_3(g) are not known precisely, as can be seen from our Table II.

J.J. FUGER: I should like to comment on the heat of formation of AmO_2(s) at 298 K. The value which you obtain using your linear relationship, - 235 kcal/mol, is still 11 kcal/mol more negative than that deduced from the heat of solution of AmO_2(s) in an acid solution. I am referring to the early measurements of Eyring, Lohr and Cunningham (J. Am. Chem. Soc. 74 (1952) 1186). My feeling is that now that pure americium metal is available in large amounts, the obvious thing to do is to obtain $\Delta H_f(AmO_2(s))$ by combustion calorimetry. This measurement appears to me to be essential, since AmO_2 will be for many years to come the heaviest actinide dioxide for which we shall be able to obtain the heat of formation by a direct experimental technique.

R.J. ACKERMANN: A measurement of the enthalpy of combustion of americium metal would be extremely valuable in testing the validity of our correlations.

J.J. FUGER: In your correlation of the heat of formation of the solid sesquioxides (Fig.8) with the heat of formation of the aqueous trivalent ions, is it not likely that part of the apparent scatter is due to the fact that not all these oxides are isostructural?

R.J. ACKERMANN: That is a good point. Perhaps I should have made the corrections and referred all the lanthanide sesquioxides to the same crystal structure. The reason I did not do so is that the experimental scatter, perhaps 5 kcal/mol, appeared to be larger than these corrections.

H.E. FLOTOW: We have recently completed a series of measurements on the heat capacity of $^{242}PuO_2$ and have obtained a value of 15.8 cal/K·mol for the entropy of $^{242}PuO_2$ at 298.15 K. This is approximately 0.5 cal/K·mol less than the values reported by Sandenaw and by Kruger and Savage.

R.J. ACKERMANN: In our assessment we used Sandenaw's value. Your value will change the thermodynamic quantities by about one kilocalorie, but will not alter any of the systematic trends.

H.E. FLOTOW: Would the absolute entropy of $^{242}Pu_2O_3$ be of value for a better assessment of the thermodynamic properties of the plutonium metal-oxygen system? It might be possible to convert our $^{242}PuO_2$ into $^{242}Pu_2O_3$.

R.J. ACKERMANN: Yes, it would be most valuable to measure the absolute entropy of Pu_2O_3.

M. KATSURA: In discussing the relationship between electronic configuration and thermodynamic quantities, you use the electronic configuration of pure atomic state. In the solid state, the electronic configuration characteristic of atomic state changes. However, the consistency you obtained is good. Could you comment on this point?

R.J. ACKERMANN: The electronic configurations in the solid metals are indeed different from those of the gaseous atoms. My impression is that they are not well established for the actinide metals. Fortunately, the correlations which I have presented do not specifically require the configurations for the solid metal, as can be seen in Fig.2. The consistency of the correlations which we obtained for the lanthanide sesquioxides with those for the actinide sesquioxides given in Table III to a large extent supports the view that atomic configurations are sufficient.

M. KATSURA: You have discussed your calculations mainly in relation to the enthalpy term. Enthalpy corresponds to the chemical bond. The phase stability must be considered in terms of the Gibbs free energy. How do you take into account the entropy term, since at higher temperatures entropy contributes greatly to the Gibbs free energy?

R.J. ACKERMANN: As I stressed in my oral presentation, the enthalpy term appears to be dominant in the assessment of the relative stability of valence states. As can be seen in Figs 3, 5 and 7 the integral molar entropy curves are virtually the same. Hence the question of relative stability expressed in the Gibbs free energy in these actinide-oxide systems is governed largely by the enthalpy and not by the $T\Delta S$ term.

THEORETICAL STUDIES OF POINT-DEFECT PROPERTIES OF URANIUM DIOXIDE

C.R.A. CATLOW*, A.B. LIDIARD
United Kingdom Atomic Energy Authority,
Atomic Energy Research Establishment,
Harwell, Didcot, Oxon.,
United Kingdom

Abstract

THEORETICAL STUDIES OF POINT-DEFECT PROPERTIES OF URANIUM DIOXIDE.

Calculations are reported firstly of the energies of defects important both for stoichiometric and non-stoichiometric crystals. These are performed using computer simulation techniques using an ionic model for the crystal. The results confirm the simple models for the defect structure suggested by transport and neutron-diffraction data: anion Frenkel disorder predominates over Schottky and cation Frenkel disorder; oxidation leads to the introduction of O^{--} interstitials and U^{5+} cations which cluster to give structures closely related to those found in other anion excess fluoride phases. However, alternative analyses are suggested of certain of the experimental diffusion studies, leading to the prediction of higher defect formation energies — a prediction supported by our calculations. Also the value calculated for the hole-electron disorder reaction of ~ 2 eV is lower than that for any other defect pair. This is important in the second part of the discussion where the partial molar entropies and enthalpies of oxidation of UO_2 are calculated using a simple mass action model. The inclusion of electronic disorder is shown to be essential if the main features of the experimental work are to be explained.

INTRODUCTION

The defect properties of UO_2 are important in determining mass transport and thermodynamic properties both of the stoichiometric and non-stoichiometric material. There has consequently been a wide range of structural, transport and thermodynamic experiments; for a review we refer to a report of an earlier IAEA panel [1]. However, there remain several problems of both a quantitative and qualitative nature concerning the defect structure. As examples we may give the uncertainties about both defect formation and activation energies, and also about the charge states and structures of defects in the oxidized phase. In this paper, therefore, we shall use theoretical methods to investigate defect properties of UO_2 and related non-stoichiometric phases. We shall calculate firstly defect formation energies and then, using simple models for the defect structure of the phase, we shall estimate partial molar enthalpies and entropies of oxidation. Our main aim is to use both sets of calculations, together with other experimental data, to provide a coherent model of the defect structure of the phase.

The discussion is divided into the two main sections — calculations firstly of defect structures and energies, and secondly of thermodynamic properties. Our concern is mainly with the stoichiometric and oxidized

---

* On attachment from: Inorganic Chemistry Laboratory, South Parks Road, Oxford, England.

crystals, as there is comparatively little data on the reduced phase. And at the beginning of each section we shall briefly mention the experimental data that is relevant to our discussion.

A. DEFECT STRUCTURES AND ENERGIES

A-1. Experimental data

The defect structure of UO_{2+x} was investigated by Willis [2]. The neutron diffraction studies showed a structure based on anion interstitials; this confirms that the intrinsic disorder of the crystal is of the anion Frenkel type. Willis' results also suggested that interstitial clustering occurred, of a type similar to that subsequently detected in interstitial excess CaF_2 by Cheetham et al. [3], and which has been treated theoretically by Catlow [4]. The results we report later in this section show that the resemblance between the defect structures of isostructural fluoride and oxide crystals is indeed very close; and that the same mechanisms account for the characteristic interstitial clusters structures in both phases.

Defect formation and activation energies are obtained from transport studies — principally anion and cation diffusion and also creep; a recent review is given by Matthews [5]. The analysis of the experimental data requires basic assumptions about the defect structure of the material. Thus transport data are analysed assuming point defect mechanisms for ion migration; and in all discussions anion Frenkel disorder is taken as predominant. The validity of these assumptions is supported by experimental data, such as that of Willis [2], and is confirmed by our calculations.

From anion diffusion measurements (see, for example, Auskerne and Belle [6]) a value of 3.0 eV was deduced for the Frenkel formation energy, and an activation energy of 1.3 eV was obtained for anion interstitial migration. The cation migration experiments, reviewed by Matzke [7], generally support the value of ~ 3.0 eV for the Frenkel energy; they also suggest values of ~ 6.5 eV for the Schottky energy and ~ 3 eV for the energy of activation for cation vacancy migration. The results for the measurements of creep, for which cation migration is the rate determining step, are in line with those obtained from the direct measurements of cation migration; for a discussion we refer to Matzke [8].

However, we believe that certain of these energies may possibly be in error. For although the models used in analysing the experimental data are fundamentally correct, our calculations suggest that there have been mistakes in the assignments made of the mechanisms responsible for certain transport processes: we will suggest alternative analyses which result in defect energies which are in better agreement with our calculations, and which are more in line with the values obtained for other fluorite crystals.

A-2. Calculations

The calculations are performed using computer simulation methods developed by Norgett. The program, HADES[1], is general for strongly ionic cubic crystals; a brief description is given by Norgett [9]. The methods

[1] Harwell Automatic Defect Examination System.

used by HADES are discussed by Lidiard and Norgett [10] and in more detail by Norgett [11]. The basis of the method originates from the work of Boswarva and Lidiard [12], and consists of an atomistic simulation of an inner region of crystal surrounding the defect, in which the relaxation is treated by explicitly relaxing every ion to zero force, with a continuum description of the remainder outer region. Here the relaxation is calculated using the macroscopic dielectric constant by a procedure closely related to that of Mott and Littleton [13]. An essential feature of the program is the use of fast matrix methods in the Newton-Raphson minimization performed in calculating the relaxation of the inner region; for a discussion we refer to Norgett and Fletcher [14].

The explicit lattice simulation of the inner region requires that a lattice potential be specified. The potential we used is based on the fully ionic model, while the short range repulsions between ions are taken to be of the Born-Mayer form which, for the case of the anion-anion interaction, is supplemented by an attractive r^{-6} term. Ionic polarization is treated by the shell model of Dick and Overhauser [15]. The advantages of this model for defect calculation are discussed by Faux and Lidiard [16] and by Catlow and Norgett [17]. The important points are: firstly the possibility of correctly describing both elastic and dielectric properties of the material (a feature not found with simpler models based on a point dipole treatment of ionic polarization), and secondly the much smaller susceptibility of defect calculations using the shell model to an instability in which two dipoles increase without bound — the polarization catastrophe discussed by Faux [18].

Details of the potentials are reported by Catlow [19]. The parameters are set up by fitting to empirical crystal data — principally elastic and dielectric constants. The empirical method is, however, supplemented by the use of ab-initio Hartree-Fock calculations of the anion-anion interaction. These calculations were performed for the interaction not of two O^{--} ions, but of two O^- ions; we assume that the screening of this latter potential provided by the extra unbound electrons is negligible. We then develop two lattice potentials: potential 1, in which only the repulsive term in the anion-anion interaction is fixed from the Hartree-Fock calculation; the attractive term is obtained from the fit to the empirical data. In potential 2, the entire short range anion-anion interaction is obtained from the molecular orbital calculations. Potential 1 will be more reliable for anion-anion separations close to those occurring in the perfect lattice, while potential 2 is used for calculations where interionic distances occur which are much shorter than the second neighbour separation in the crystal. The calculations reported below were all performed with potential 1, except for those on distorted interstitial structures.

The main factor affecting the reliability of the potentials will unquestionably be the extent to which the ionic model is a valid description of the bonding in the crystal. The fully ionic model will clearly exaggerate the ionic charges in UO_2. But experience gained in work on the alkaline earth fluorides [17] has shown that calculated defect energies are not very sensitive to details of the potential; and provided the ionic model is not a grossly incorrect description of the bonding in UO_2, our calculation should still be reliable. We require, therefore, some evidence for the crystal being at least strongly ionic. This we believe is provided from three separate sources: firstly, the crystal structure, as the 8:4 co-ordination of the fluorite lattice is typical of ionic materials; secondly the cohesive energy, as the value calculated using our lattice models is in good agreement with experimental estimates [20] and thirdly the phonon dispersion curves [21], as these may be fitted with reasonable accuracy to a potential which uses the fully ionic model. This model is therefore suggested by both static and dynamic properties of the crystal.

Finally we should note that calculations were performed of the energy of an electronic defect — the F^+ centre, or singly charged anion vacancy. Quantum mechanical methods were therefore necessary. The methods we used were based on the point-ion treatment of Gourary and Adrian [22]; these were combined with estimates obtained from HADES of the lattice relaxation around the F^+ centre to give a value for the formation energy of this defect.

Thus, using the methods we have described and the potentials for the crystal that we derived, we performed calculations firstly of the intrinsic disorder energies — energies, that is, of defect formation and activation; secondly we investigated the important extrinsic defects, principally oxidized and reduced lattice cations. Finally we examined interstitial clustering in in the oxidized phase. Our results are now discussed separately under these three headings.

A-3. Results and discussion

A-3.1. Intrinsic defects

Formation energies were obtained from the results of calculations performed for isolated cation and anion vacancies and interstitials. Initially, we investigated the doubly charged anionic defects and also took the most highly symmetric structure possible for these defects. This latter condition was then relaxed, and calculations were performed for the anion vacancy and interstitial in which the defects were allowed to adopt a lower symmetry structure. These calculations showed that, for the vacancy, the symmetric structure was of lowest energy; for the interstitial a reduction in the energy of $\sim$ 0.3 eV was obtained by a small $\langle 110 \rangle$ distortion of $\sim$ 0.2 Å. This energy was used in obtaining the anion Frenkel energy, which is reported with that for the other defects (cation Frenkel and Schottky) in Table I. Also reported is the energy for the singly charged Frenkel pair. This is large compared with that for the doubly charged model, and such defects will therefore be unimportant in the stoichiometric crystal.

We then calculated activation energies for defect migration. The mechanisms were shown to be identical to those demonstrated by Catlow and Norgett [17] for the isostructural alkaline earth fluorides: vacancy migration occurs by direct jumps of a lattice ion to the vacancy, with the structure of the saddle points for both cation and anion comprising a migrating ion equidistant between two vacancies. Anion interstitial migration occurs by a concerted or 'interstitialcy' mechanism in which the migrating interstitial displaces a lattice ion into a neighbouring interstitial site, itself finally occupying the site of the displaced lattice ion. Calculations showed that the activation energy for this mechanism was less than that for a direct $\langle 110 \rangle$ jump between interstitial sites, and an extensive examination of the potential surface demonstrated a symmetric structure for the saddle point in which the two ions are equally displaced from the lattice sites involved in the process. The results for these calculations are collected in Table II.

TABLE I. DEFECT FORMATION ENERGIES

Defect	Energy (eV)
Anion Frenkel pair[a]	5.0
Cation Frenkel pair	18.5
Schottky trio	10.3

[a] Obtained using doubly charged model; formation energy for singly charged pair 8.9 eV.

TABLE II. DEFECT ACTIVATION ENERGIES

Mechanism	Activation energy (eV)
Anion vacancy	0.25
Anion interstitial	0.6
Cation vacancy	5.6

Our results in this section demonstrate the validity of analyses of transport data for UO_2 based on point defect models; and more particularly, they confirm the anion Frenkel model. For, the energy of formation for the anion Frenkel defect pair reported in Table I is less than half that of the Schottky trio; and low energy processes exist for the migration of the defects. The quantitative features of our calculations appear, however, to be less satisfactory. Both Frenkel and Schottky energies are considerably higher than experimental estimates. But we believe that these discrepancies may result from incorrect interpretation of the experimental data. The anion Frenkel energy was deduced from Arrhenius energies obtained from the diffusion measurements, and the assumption was made that the same diffusion mechanism was operative in both stoichiometric and oxidized crystals. But our calculated activation energies reported in Table II show a considerably lower value for vacancy than for interstitial migration — a feature found in both experimental and theoretical work on the alkaline earth fluorides [19,23]. This suggests that interstitial migration will be replaced by the vacancy mechanism on passing from the oxidized to the stoichiometric region. And we may re-analyse the experimental data using our calculated vacancy activation energy to give a revised Frenkel energy of 5.1 eV, a value in good agreement with our calculated energy.

Experimental support for our vacancy activation energy comes from recent work of Chereau and Wadier [24] on non-stoichiometric PuO_2. Activation energies of between 0.2 and 0.3 eV were found for anion transport in the reduced crystal. In a simple model which omits defect association the value may be equated with the vacancy activation energy. From the similarity between the two oxides we would expect a similar value for this energy in UO_2.

We should also note that association was neglected by Auskerne and Belle in their suggestion of an activation energy for the anion interstitial of 1.3 eV. Our lower value reported in Table II is not therefore, necessarily incorrect.

The results for cation migration, however, are almost certainly in error; the value obtained for the Arrhenius energy for the oxidized crystal is $\sim$ 5.3 eV, compared with an experimental value $\sim$ 3.0 eV. We may attribute this error to our high cation vacancy activation energy; while this in turn is probably due to inadequacies in our model — principally the neglect of covalence. Further difficulties here are, firstly, our prediction of high Arrhenius energies for the reduced oxide; we found no mechanism for which a value of < 10 eV could be obtained, compared with the experimental value of ~ 4 eV. Two possible explanations may be suggested: either we have seriously overestimated the cation Frenkel formation energy, due to a neglect of covalence which we might expect to be particularly important for the cation interstitial; or, in the reduced crystal, vacancy clustering allows a low energy mechanism for cation mobility. A second problem concerns the change in Arrhenius energy on passing from the hyper- to the hypostoichiometric material. This is measured as ~ 5 eV [7]. But if the cation vacancy mechanism were operative throughout this transition this value would correspond to twice the formation energy of the anion Frenkel pair. The measurement is therefore considered to favour the lower value for the Frenkel energy. However, we believe that it is probable that the cation vacancy mechanism is not operative in the reduced phase, even in the near stoichiometric material. Also it is possible that the measured change of Arrhenius energy corresponds to the transition from the hyperstoichiometric to the stoichiometric material, for there may be uncertainties about the position of the stoichiometric point. If this interpretation were correct, the value of 5 eV would correspond to the anion Frenkel energy, in agreement with our calculated value.

Thus to summarize our results on the intrinsic disorder energies, we believe that re-examination of the experimental data provides some support for our higher calculated defect formation energies. The higher melting point and cohesive energy of UO_2 compared with other fluorite crystals also suggests our higher values. Our proposals must clearly be tentative, owing to the uncertainties in our model. But we believe that the points we have raised are sufficient to justify further careful experimental investigations.

A-3.2. Extrinsic defects

We determined the charge states of the isolated defects formed on oxidation and reduction of UO_2; the interaction between defects is considered for the case of the oxidized crystal in the following section. Calculations were performed of the energies of formation of di-, tri-, penta- and hexavalent uranium at lattice cation sites. These values were then combined with ionization potentials for the metal atom estimated by Childs [25], and with electron affinities and dissociation energy for oxygen used by Dickens et al. [26] in similar calculations. The reactions examined and the energies calculated are collected in Table III. The energies calculated for the first two disproportionation reactions strongly support a defect structure for oxidized UO_2 based on doubly charged interstitials and pentavalent cations, a structure also indirectly indicated by experiments of Anderson et al. [27]. For the reduced crystal we suggest the formation of trivalent cations; and the energy of reaction [4] suggests that these will be compensated by singly charged vacancies. This latter result will, however, be critically dependent on our value for the fourth ionization potential of uranium; in contrast the other energies are probably more reliable, as errors in ionization potentials may be self-cancelling. Determination of the charge state of the vacancy will need more reliable ionization potentials, and also more accurate calculations of the F^+ centre formation energy.

Finally we draw attention to the energy calculated for the disproportionation of lattice cations. For, since both the conduction and the highest valence band in UO_2 will be constructed largely from metal orbitals, this

TABLE III. REDOX REACTION ENERGIES[a]

A. OXIDIZED PHASE

1. $2U^{5+} \rightarrow U^4 + U^{6+}$: E = +0.97 eV

2. $2O^-_{int} \rightarrow \frac{1}{2} O_2 + O^{2-}_{int}$: E = -2.73 eV

B. REDUCED PHASE

1. $2U^{3+} \rightarrow U^4 + U^{2+}$: E = 1.01 eV

2. $U^{3+} + O^{2+}_V \rightarrow U^{4+} + O^+_V$: E = -2.22 eV

C. HOLE-ELECTRON DISORDER REACTION

$2U^4 \rightarrow U^{3+} + U^{5+}$: E = 1.99 eV

[a] O_{int} refers to oxygen interstitial; O_V to oxygen vacancy. All uranium ions are at lattice sites.

energy should give an estimate of the band gap in UO_2. And the value obtained is indeed close to that measured by conductivity studies of Bates et al. [28]. This energy one should note is considerably lower than that for the formation of any other defect pair; and in Section B we shall show how this has important consequences for the thermodynamics of oxidation of the crystal.

A-3.3. Interstitial clusters

We performed calculations firstly of a complex of two U^{5+} cations and one interstitial, and then of a dimer and trimer of this defect. The simple interstitial monomer cluster is bound, but only provided coupled ⟨110⟩ interstitial and ⟨111⟩ lattice relaxations were allowed to give the structure shown in Fig.1, the 2:1:2 cluster originally proposed by Willis [2]. This relaxation mode, discussed by Catlow [4], occurs generally in anion excess fluorite compounds.

In Table IV we give the binding energy of the 2:1:2 cluster with respect to isolated component defects. Also reported are the binding energies with respect to component 2:1:2 clusters of the interstitial dimer and trimer. For the dimer, the minimum energy configuration is the '2:2:2' structure shown in Fig.2, a complex which is again stabilized by the coupled lattice-interstitial relaxation mode. The defect is bound with respect to the monomer. However, we find a very much stronger binding energy for the trimer shown in Fig.3. This defect can be described as a tetrahedron of interstitials around an anion vacancy, whose formation is again stabilized by the interstitial-lattice relaxation, giving the structure shown in the figure. The formation of this defect was suggested by Cheetham et al.[3] to account for their results on the more heavily doped CaF_2 solutions. We propose, however, that in UO_{2+x} this defect will be more important than the 2:2:2 cluster; and, in particular, we suggest that the ordered U_4O_9 phase may be based on these or larger interstitial clusters.

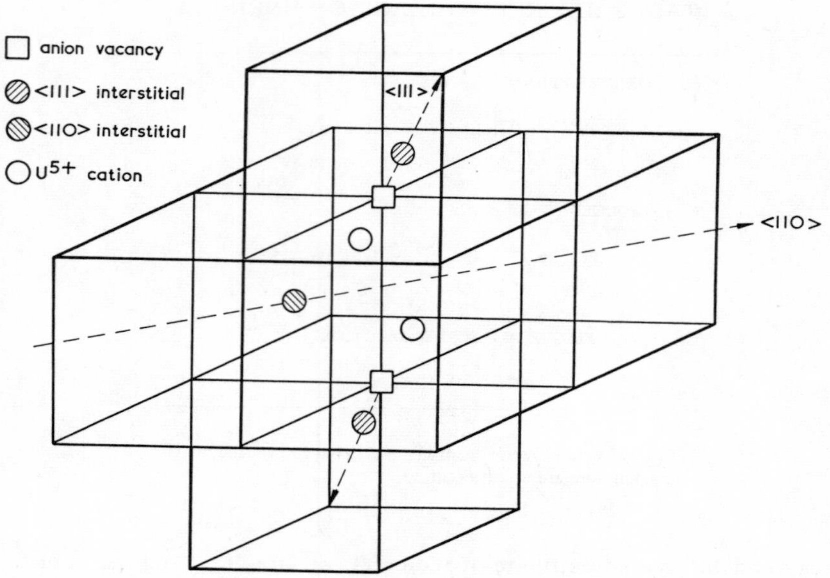

FIG. 1. The 2:1:2 cluster.

TABLE IV. DEFECT CLUSTER ENERGIES

Defect	Binding energy (eV)
Complex of one interstitial and two holes: the 2:1:2 cluster	-1.22 (footnote a)
Complex of two interstitials and four holes: the 2:2:2 cluster	-1.29 (footnote b)
Complex of three interstitials and six holes: the 4:3:2 cluster	-3.11 (footnote b)

[a] Binding energy with respect to isolated component defects.
[b] Binding energy with respect to component 2:1:2 clusters.

To summarize, therefore, our calculations on UO_2 and UO_{2+x} show that the phase has the defect structure which appears common to all fluorite crystals and related anion excess phases. The basic disorder is of the Frenkel type; but the formation energy is larger, in line with the higher cohesive energy and melting point. Anion mobility in the stoichiometric crystal occurs by a vacancy mechanism, while cation mobility is also vacancy assisted for the oxidized and stoichiometric crystals – but uncertain for the reduced phase. Finally in the interstitial excess phase, the basic defects are U^{5+} cations and O^{--} interstitials, which cluster to produce interstitial trimers or possibly larger interstitial aggregates.

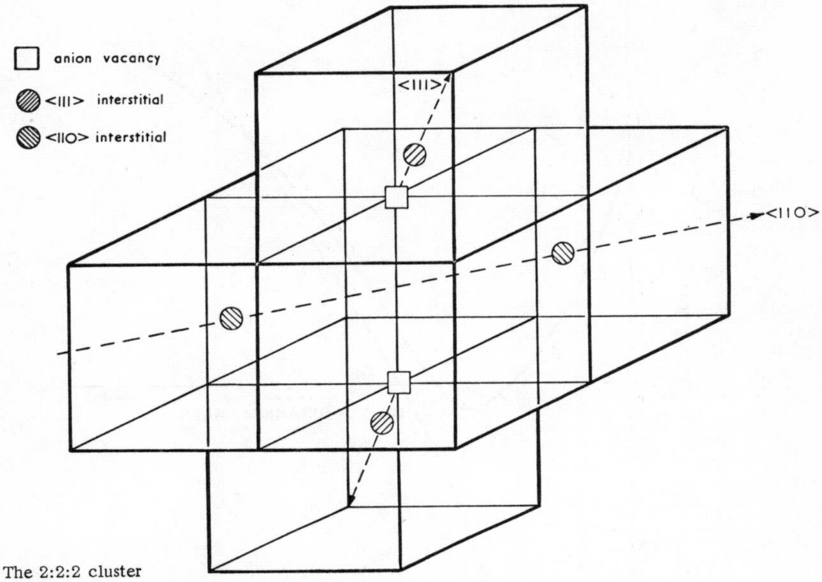

FIG. 2. The 2:2:2 cluster (U^{5+} cations not shown).

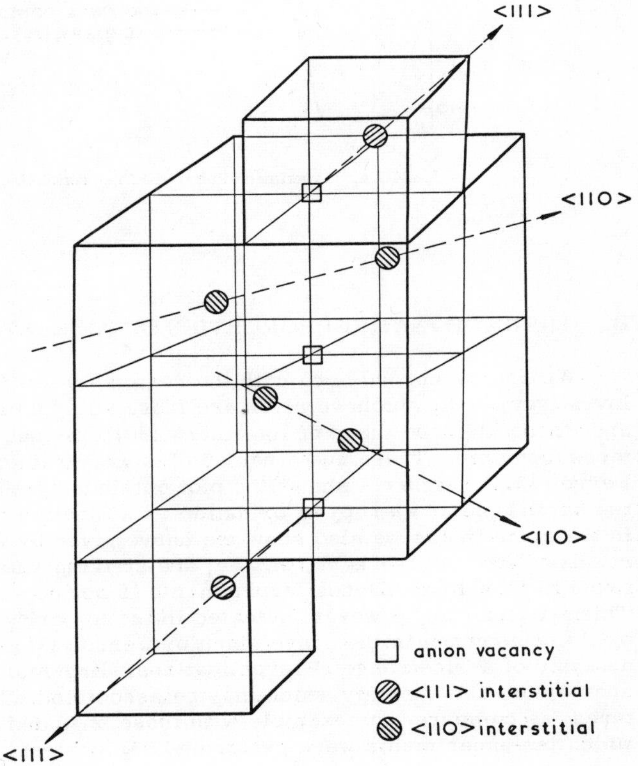

FIG. 3. The 4:3:2 cluster (U^{5+} cations not shown).

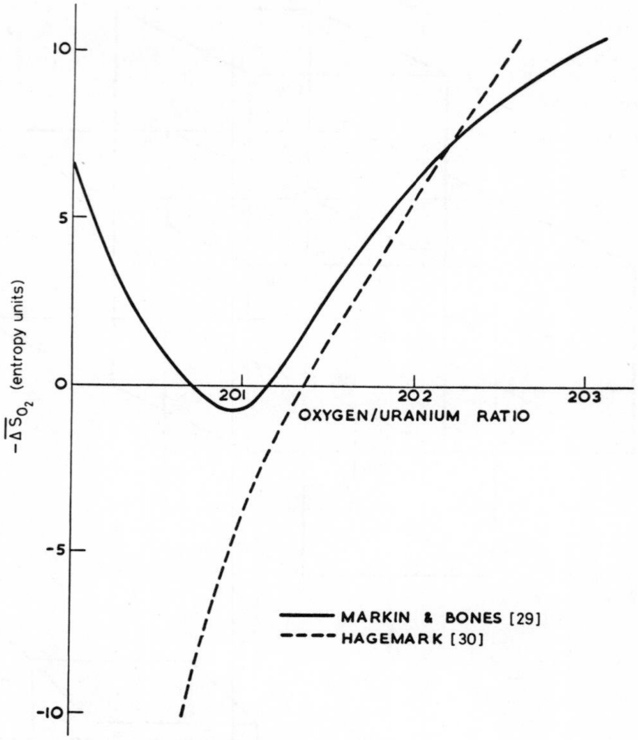

FIG.4. Experimental plots of $-\overline{\Delta S}_{O_2}$ versus O/U ratio.

B. THERMODYNAMIC PROPERTIES OF NEAR STOICHIOMETRIC UO_{2+x}

We are concerned here with the region $0.0 < x < 0.03$. The experimental investigations to which we refer are those using emf measurements of Markin and Bones [29] and those of Hagemark [30] obtained from thermogravimetric measurements. These and other studies were reviewed and analysed by Perron [31]. Figure 4 shows the plot obtained by Markin and Bones [29] of the partial molar entropy of oxidation as a function of non-stoichiometry; in the same figure we also show the curve given by Hagemark [30]. The results of the two workers, we see, are strikingly dissimilar; the minimum found in Markin and Bones' experiments is not reproduced by Hagemark. This minimum is, however, detected in the majority of studies, as is shown by its occurrence in the curve given by Perron [31] obtained from a statistical analysis of a wide range of investigations. Hagemark's results are therefore anomalous — an anomaly which may be associated with the higher temperatures, as compared for example with those of Markin and Bones [29], at which his experiments were performed. In contrast, the enthalpy curves

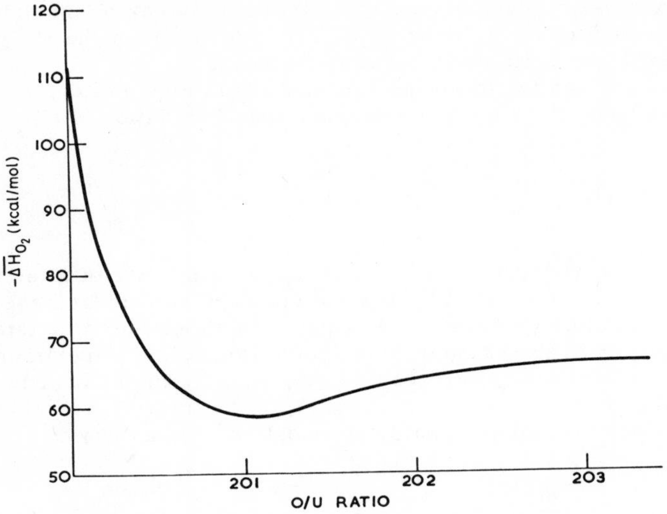

FIG. 5. Experimental enthalpy, $-\overline{\Delta H}_{O_2}$ versus O/U ratio (from [29]).

obtained by different workers are generally in agreement; in Fig.5 we show the plot obtained by Markin and Bones [29]. Thus in the theoretical studies reported in this section our aim will be firstly to explain the minimum observed by Markin and Bones and the majority of other studies, then to comment on the apparent contradiction provided by Hagemark's work, and finally to assess the extent to which the models we propose are compatible with the results of our previous section.

The methods used in calculating thermodynamic properties of the phase differ from those used in previous studies of, for example, Thorn and Winslow [32] and Atlas [33]. These latter methods were directed towards giving a reliable treatment of defect interactions, based in the case of Thorn and Winslow on earlier methods of Anderson [34], and with Atlas on a statistical mechanical model which allows distributions over various local defect concentrations. However, in both models, severely restrictive assumptions were made as to the basic defect structure of the phase: in particular, hole-electron disorder was ignored, and hole interstitial clustering equilibria were neglected. In our treatment we allow greater flexibility in our models by using a general mass action formalism to calculate defect concentrations, with activity coefficients given by Debye-Hückel theory — the procedure discussed by Lidiard [35]. The method is obviously only applicable to the near-stoichiometric region, as discussed here. But, given a more sophisticated treatment of activity coefficients, our method could be extended to more concentrated defect solutions; and it provides a better basis for theories of the phase than do the more restrictive models.

The mass action equations included in our model are those for electronic disorder, anion Frenkel and Schottky disorder and finally for two association reactions — the first to form the neutral interstitial monomer cluster, and the second that for association of three of these to give the 4:3:2 complex.

The equations were solved computationally using iterative numerical methods based on steepest descent techniques. The calculation of the Debye-Hückel activity coefficient again used iterative methods.

We then derived the following expressions in terms of defect concentrations and derivatives for the entropies and of oxidation of UO_2:

$$\overline{\Delta S}(O_2) = -2R \sum_i \frac{dn_i}{dx} \log_e \frac{n_i}{f_i} - S(O_2) \qquad (1)$$

where n_i is the activity of the ith defect, f_i is the number of sites for this defect per formula unit, and $S(O_2)$ is the standard entropy at the temperature of the experiment of gaseous oxygen (the value chosen for this temperature was midway in the range of Markin and Bones [29] experiments, i.e. 1100°K). This expression, we should note, neglects the standard entropies of defect formation.

For the partial molar enthalpy of oxidation we may write:

$$\overline{\Delta H}(O_2) = 2 \sum_i \frac{dn_i}{dx} h_i + H_0 \qquad (2)$$

where h_i is the heat of formation of the ith defect and H_0 is the energy to produce an O^{--} ion at infinity from molecular oxygen.

On differentiating the mass action equations, we obtain a set of linear equations in the derivatives; these are solved and the values substituted in Eqs (1) and (2). Entropies and enthalpies of oxidation are then calculated mainly for simple models which omit association and the Debye-Hückel refinement; such calculations, however, demonstrate the main qualitative features.

Our calculations in Section A suggested that hole-electron disorder would predominate in the stoichiometric crystal. In Fig. 6, therefore, we show the calculated entropy curve for a model in which all disorder constants are below 10^{-10}, except that for electronic disorder; this was allowed to vary from 10^{-3} to 10^{-7}. Also, the relative values of anion Frenkel and Schottky disorder constants were chosen so as to ensure the anion Frenkel mode of charge compensation.

We see that the model generates a minimum, as detected by Markin and Bones [29]; the best resemblance between theory and experiment is for there to be a disorder constant, K_e, of $\sim 10^{-5}$. Furthermore, this minimum could not be reproduced by any other intrinsic disorder model. In particular, a model which omitted hole-electron disorder — thus leaving anion Frenkel pair formation as the only important intrinsic disorder reaction — could produce only monotonic curves. An extrinsic vacancy population with a Frenkel disorder constant of $\sim 10^{-5}$ was required by this model if a minimum was to be obtained. Neither of these features, however, appears reasonable. For the Frenkel disorder constant is far higher than would be predicted from our calculated energy of formation of this defect, while the vacancy population could only be produced by a rather high level — 500 ppm — either of high-valent impurities or of metastable cation vacancies. There is no evidence either from spectrographic analyses [29] or from density measurements [1] to support either of these alternatives.

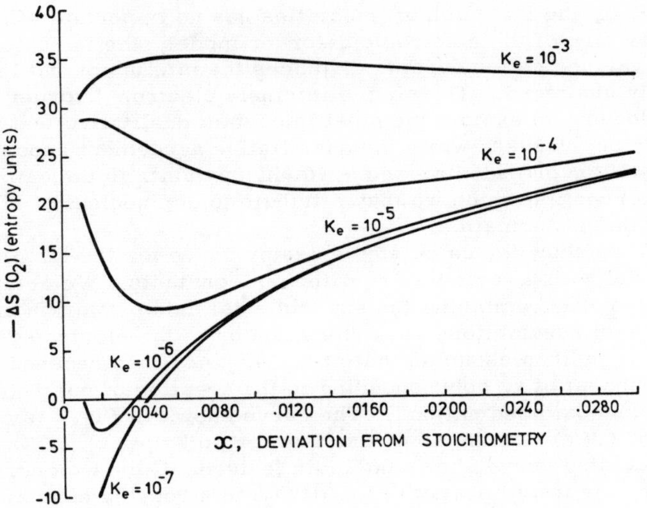

FIG. 6. Calculated entropy versus x, the deviation from stoichiometry (O/U-2) (K_e is the electronic disorder equilibrium constant).

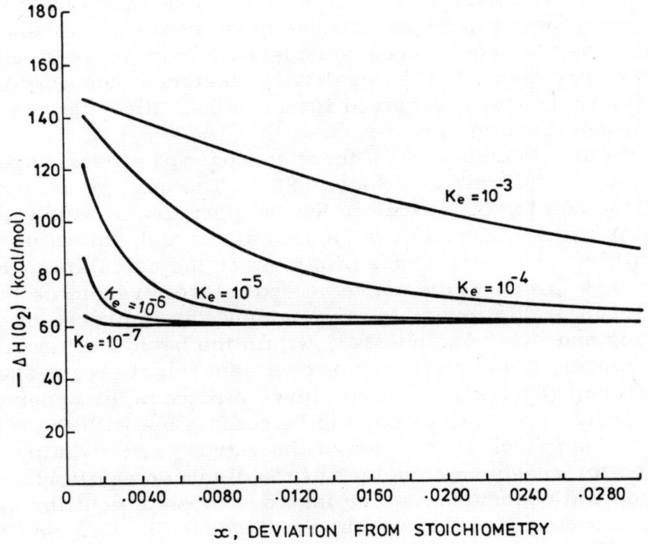

FIG. 7. Calculated enthalpy versus x, the deviation from stoichiometry (O/U-2) (K_e is the electronic disorder equilibrium constant).

In contrast, the inclusion of impurities has no important effect on the qualitative feature of the electronic disorder model; the inclusion of either low or high valence cation dopants displaces the minimum, but leaves the shape largely unaltered. The simple intrinsic electron disorder model is therefore adequate to explain the most important qualitative features of the observed entropy curves, while the quantitative agreement becomes acceptable for $K_e \sim 10^{-5}$, provided we add ~ 10 entropy units to the calculated curves — a correction which we may attribute to our neglect of standard entropies of defect formation.

In Fig. 7 we show the calculated enthalpy curve for the simple electron disorder model with a variety of equilibrium constants. We obtained the defect heats, h_i, by combining the experimental data of Markin and Bones [29] for $x = 0.03$ with assumptions as to the enthalpy of the electronic disorder reaction. Our method assumes that at $x = 0.03$, $\overline{\Delta H}$ has reached a limiting value, independent of x, corresponding to the creation of non-interacting interstitials and oxidized cations. The curve shown in Fig. 7 is calculated for a value of 2.0 eV for the electron disorder enthalpy.

The calculated curve shows the main features of the experimental data. In particular, the initial decrease in $\overline{\Delta H}(O_2)$ to a roughly constant value is reproduced, with reasonable quantitative agreement being obtained for a value for K_e of 10^{-5}. The small minimum at $x \simeq 0.01$ is not obtained, but this is due to our omission of defect clustering from our simple model. Finally we should note the value of 2.0 eV for the enthalpy of the electronic disorder reaction; calculations showed that this gave the best agreement between theoretical and experimental enthalpy curves. It also agrees with the values obtained from the experimental data, and from our calculated energies reported in Section A.

Calculations were performed in which both association and the Debye-Hückel refinement were included. These more realistic models gave quantitative results which differed considerably from those discussed above for the simple treatment, but the qualitative features remained unaltered. Electronic disorder is still required if an intrinsic disorder model is to reproduce the observed minimum.

Finally we should comment on the discrepancy between Hagemark's [30] results and those of Markin and Bones [29]. The only explanation we can propose is based on the difference in the temperature at which the two sets of experiments were made. For we note that, at high values of the electronic disorder equilibrium constant, the minimum in the calculated $\overline{\Delta S}(O_2)$ versus x plot disappears (although the curves do not closely resemble that found by Hagemark [30]). However, the resemblance is improved by the inclusion of association, and it is possible that, within the basic framework of the mass action model, a more elaborate treatment might give reasonable agreement with Hagemark's curve for large electronic disorder constants. Our simple model appears therefore to be compatible with the main experimental data. The principal features of the entropy and enthalpy curves can be reproduced for reasonable choices of the disorder equilibrium constants. Thus all dependent observations, for instance those of equilibrium pressure of oxygen versus concentration which are discussed by Kröger [36], may be accounted for.

To summarize, therefore, our aim in this section has not been to provide quantitative agreement with experiment, but to obtain guidance as to the implications of the experimental thermodynamic measurements for the defect structures of the material. And we have shown how the main feature of the experimental work can be accounted for by a model in which electronic disorder predominates in the region very close to stoichiometry, with oxidation resulting in the formation of holes and anion interstitials. This was the simple model which was suggested by our calculation discussed in the first section.

CONCLUSION

The main points in this study that we wish to emphasize are, firstly our confirmation of the analogy between the defect properties of UO_2 and those of the isostructural alkaline earth fluoride crystals. We have shown that this applies to both transport and structural properties. Quantitative differences are apparent; but these may be attributed largely to the differences in cohesive energies for the oxide and fluoride crystals. The disagreements between calculated and certain experimental data might suggest deficiencies in our lattice models. But we have seen how certain of the discrepancies may be attributed to errors in the analysis of the experimental data.

Our second main result has been the explanation of the occurrence of the minimum in the experimental $\overline{\Delta S}(O_2)$ versus $\times$ plot. For we have shown how this is generated by a model in which the predominant intrinsic disorder is electronic — a feature omitted by most previous models — but which is suggested by our calculations of defect energies.

Our calculations have therefore more firmly established the nature of the defect properties of UO_2, and have suggested areas in which future experimental work would be of use.

REFERENCES

[1] INTERNATIONAL ATOMIC ENERGY AGENCY, Thermodynamic and Transport Properties of Uranium Dioxide and Related Phases, Technical Reports Series No.39, IAEA, Vienna (1965).
[2] WILLIS, B.T.M., Proc. Br. Ceram. Soc. 1 (1964) 9.
[3] CHEETHAM, A.K., FENDER, B.E.F., COOPER, M.J., J. Phys., C 4 (1971) 3107.
[4] CATLOW, C.R.A., J. Phys., C 6 (1973) L64.
[5] MATTHEWS, J., UKAEA, AERE Report (to be published).
[6] AUSKERNE, A.B., BELLE, J., J. Nucl. Mater. 3 (1961) 267.
[7] MATZKE, H., J. Phys. (Paris) 34 (1973) C9-317.
[8] MATZKE, H., AECL Rep. AECL-2585 (1966).
[9] NORGETT, M.J., UKAEA Rep. AERE-R.7015 (1972).
[10] LIDIARD, A.B., NORGETT, M.J., Computational Solid State Physics (HERMAN, F., DALTON, N.W., KOEHLER, T.R., Eds), Plenum Press, New York (1972) 385.
[11] NORGETT, M.J., UKAEA Rep. AERE-R.7650 (1974).
[12] BOSWARVA, I.M., LIDIARD, A.B., Philos. Mag. 16 (1967) 805.
[13] MOTT, N.F., LITTLETON, M.J., Trans. Farad. Soc. 34 (1938) 385.
[14] NORGETT, M.J., FLETCHER, R., J. Phys., C 3 (1970) L190.
[15] DICK, B.G., OVERHAUSER, A.W., Phys. Rev. 112 (1958) 90.
[16] FAUX, I.D., LIDIARD, A.B., Z. Naturforsch., A 26 (1971) 62.
[17] CATLOW, C.R.A., NORGETT, M.J., J. Phys., C 6 (1973) 1325.
[18] FAUX, I.D., J. Phys., C 4 (1971) L211.
[19] CATLOW, C.R.A., D. Phil. Thesis, University of Oxford, 1974.
[20] BENSON, G.C., FREEMAN, P.I., DEMPSEY, E., J. Am. Ceram. Soc. 46 (1963) 43.
[21] DOLLING, G., COWLEY, R.A., WOODS, A.P.B., Can. J. Phys. 43 (1965) 1397.

[22] GOURARY, B.S., ADRIAN, F.J., Phys. Rev. 105 (1957) 1180.
[23] BOLLMAN, W., GÖRLICH, P., HAUK, W., MÖTHES, H., Phys. Status Solidi 2 (1970) 157.
[24] CHEREAU, P., WADIER, J.P., J. Nucl. Mater. 46 (1973) 1.
[25] CHILDS, B.G., AECL Rep. AECL-680 (1958) CRMet-788.
[26] DICKENS, P.G., HECKINGBOTTOM, R., LINNETT, J.W., Trans. Farad. Soc. 64 (1968) 1489.
[27] ANDERSON, J.S., EDINGTON, D.N., ROBERTS, L.E.J., WAIT, E., J. Chem. Soc. (1954) 3324.
[28] BATES, J.L., HINMAN, C.A., KAWADA, T., J. Am. Ceram. Soc. 50 (1967) 652.
[29] MARKIN, T.L., BONES, R.J., UKAEA Rep. AERE-R.4178 (1962).
[30] HAGEMARK, K., Kjeller Rep. KR-67 (1964).
[31] PERRON, P.O., AECL Rep. AECL-3072 (1967).
[32] THORN, R.J., WINSLOW, G.H., J. Chem. Phys. 44 (1966) 2632.
[33] ATLAS, L.M., J. Phys. Chem. Solids 29 (1968) 1349.
[34] ANDERSON, J.S., Proc. R. Soc. A185 (1946) 69.
[35] LIDIARD, A.B., Handbuch der Physik 20 (1957) 246.
[36] KRÖGER, F.A., Z. Phys. Chem. (Neue Folge) (Frankfurt) 49 (1966) 178.

DISCUSSION

F. SCHMITZ: You mentioned that your model gives an excellent prediction for the lattice parameters of UO_2. Does this refer to the stoichiometric composition or are you able to predict the variation as a function of the O/M ratio?

C.R.A. CATLOW: The lattice parameter we predicted was calculated for the perfect lattice. If we wished to predict the variation of the lattice parameter with stoichiometry we would have to calculate the volumes of formation of the defects in the non-stoichiometric phases. Attempts have been made to calculate defect formation volumes, but such calculations have been far less successful than calculations of energies. I doubt whether they could be successfully applied to such complex defect structures as exist in the non-stoichiometric UO_2 phases.

F. SCHMITZ: Is it possible to apply your model to (U, Pu) mixed oxides? If so, it would be extremely interesting to calculate in addition to the lattice energies the function $a = f(O/M)$, for which considerable experimental data already exist. The result might be regarded as a valuable test of your model.

C.R.A. CATLOW: Your suggestion that we extend our calculations to examine the mixed U/Pu oxides obviously follows from the importance attached to these materials, which has been emphasized at this Symposium. I think, however, that there are better tests of the model than the one you propose and that an examination of the mixed-oxide phases should concentrate on the more fundamental problem of defect structures and migration mechanisms.

Hj. MATZKE: The energy of formation of Frenkel defects is a crucial value in all defect calculations. I agree that the previous interpretation of the data of Auskern and Belle needs to be reviewed, and to do so, you use your calculated migration energy for oxygen vacancies. Could you comment on the fact that this value seems small compared with that obtained for alkali halides (~0.6 eV) or for ZrO_2 (~1.3 eV), all of which have lower melting temperatures? A higher value of the migration energy would reduce the formation energy.

C.R.A. CATLOW: While there is obviously some empirical justification for assuming that the melting points and the defect formation energies are proportional to each other, I think it is very doubtful whether we can assume a similar general relationship between melting points and activation energies; in particular I think that no useful comparison of this type can be made between activation energies in compounds of different crystal structure such as UO_2 and the alkali halides. The value you quote for ZrO_2 does appear to make it difficult to apply my analysis. However, I cannot accept that this is the energy for isolated vacancy activation. There is considerable evidence from other crystals, e.g. CeO_2 and PuO_2, that vacancy activation energies are much lower, nearer the value obtained in my calculation. I suspect that the value of 1.3 eV is obtained by partially or wholly omitting from the Arrhenius energies deduced from transport data the term due to defect association. This could easily contribute 1 eV to the measured energy.

Hj. MATZKE: The values calculated for the formation of metal defects seem high in comparison with the values deduced from experiments. Would it be possible to obtain better agreement if other potentials were used and greater allowance were made for some covalency?

C.R.A. CATLOW: Let us consider the Schottky energy first. This must actually be a lower limit if the revised Frenkel energy of ~5 eV is correct, as we have the requirement that for the anion interstitial rather than cation vacancy compensation on oxidation:

$$g_{F^-} < \tfrac{1}{2} g_s$$

where g_{F^-} and g_s are the free energy of Frenkel and Schottky defect formation. Thus if we disregard defect entropies, we require a Schottky enthalpy of at least 10 eV.

As to the cation Frenkel energy, I think you are probably right. The value we calculated is almost certainly an overestimate as a result of disregarding a covalence contribution to the cation interstitial.

INTERPRETATION OF PROPERTIES OF AND LATTICE DEFECTS IN UO_2 AND PuO_2 ON THE BASIS OF A COVALENT MODEL FOR OXIDES

H. BLANK
Commission of the European Communities,
European Institute for Transuranium Elements,
Karlsruhe,
Federal Republic of Germany

Abstract

INTERPRETATION OF PROPERTIES OF AND LATTICE DEFECTS IN UO_2 AND PuO_2 ON THE BASIS OF A COVALENT MODEL FOR OXIDES.

Arguments are presented, based on the spectroscopic theory of the ionicity, that metal oxides should in most cases be treated as covalent compounds with regard to their crystal radii. It is shown that most 5f fluorite-type oxides should not be stable in the ionic picture. A covalent lattice model for these oxides is proposed and the tetrahedral covalent radius of oxygen is deduced. A discussion of the covalent lattice geometry and of some geometrical aspects of point defects and of dislocations in UO_2 and PuO_2 is given.

1. INTRODUCTION

The first successful treatment of the cohesive force in a solid was made on the basis of the ionic model — many years before the development of quantum mechanics — and the important role of this model has been firmly established in chemistry and solid-state physics ever since. Right from the beginning it was accepted that not only alkali halides, i.e. compounds of type A^1B^{8-1}, but also compounds of type A^2B^{8-2}, etc., including the chalcogenides, should be treated as ionic solids. In this way it was taken for granted that all metal oxides are ionic compounds. Since the bond lengths and crystallography of their structures can be explained rather well on the basis of ionic radii, the ionic model for metal oxides has become firmly accepted even if there are certain difficulties in understanding the properties of the double negatively charged oxygen ion. For instance, it is difficult to assign to this oxygen ion a definite value of its polarizability [1]. Recently Levin, Syrikin and Dyatkina criticized the concept of the O^{--} ion in solids [2]; however, in spite of the clear arguments they presented in their paper, it is difficult to draw practical conclusions for choosing a new structural model for the metal oxides.

The analysis of the stability of the 5f fluorite-type oxides [3] (see Section 2 of this paper), showing their apparent instability in the ionic model, drops another hint to indicate that something is probably wrong in applying this model to metal oxides. However, one could argue that, with an ionicity f_i = 0.67 on the Pauling scale, there seems sufficient justification for treating UO_2 as an ionic compound, with no convincing justification for requiring a drastic change in this model. The difficulty with the stability of the 4f and 5f fluorite-type of dioxides could, perhaps, have been caused by a wrong choice or by an unrealistic definition of the ionic radii employed. A set of atomic radii established by Slater on empirical grounds [4] appeared not to be a possible alternative; besides it is too coarse for a quantitative application.

Eventually, it was found that the spectroscopic theory of ionicity of Phillips [5] and Van Vechten [6, 7], with its application to A_nB_m compounds by Levine [8, 9], could lead to the answer to why a rather drastic change in the ionic model for metal oxides is justified [10]. The reason lies in the fact that the onset of true ionic behaviour in crystalline solids occurs only at a rather high ionicity, i.e. at a Phillips ionicity $f_i > 0.911$, and that, in contradistinction, true covalent behaviour already exists at $f_i = 0.786$. But it is just between these two limits, i.e. for Phillips ionicities of $0.786 \leq f_i \leq 0.911$, that the f_i values of many metal oxides can be found. In this range the ionicity of the chemical bond is still high enough to give the atoms a good approximation to a spherical shape, and their size relations are important enough to determine the stability of the various structures in the compounds. However, the ionicity is low enough for the directional covalent forces to be the main determinants of the bonding.

In purely covalent compounds, i.e. for $f_i < 0.786$, the atoms cannot any longer be regarded as having an approximately spherical shape, and unfavourable size relationships can be overshadowed to a large extent by the valence bridges extending from the outer electron shells.

In purely ionic compounds, with $f_i > 0.911$, directional short-range forces are weak, since electrostatic forces have spherical symmetry. From these arguments, the conclusion should be drawn that metal oxides are sufficiently covalent to describe them by covalent crystal radii as long as their ionicity, f_i, is less than 0.911, which certainly applies to the 4f and 5f fluorite-type oxides (for UO_2, $f_i = 0.88$). Of course, this qualitative statement has to be substantiated by a quantitative model for the bonding in oxides, a model which specifies the ionic and covalent contributions and their roles in the various solid-state properties in greater detail. Such a model does not seem to be available at present and will certainly require some time in development. For these reasons, this paper only presents in a rather qualitative fashion some simple geometrical consequences of having a covalent lattice model for UO_2 and PuO_2.

In Section 2, the tetrahedral covalent crystal radius for oxygen in the 5f fluorite-type oxides is deduced after the deficiency of the ionic model with the seemingly best available ionic radii for oxides has been demonstrated. In Section 3, some relations between the geometry of the covalent model and point defects are established, while in Section 4 some geometrical aspects of dislocations in UO_2 are treated. Section 5 summarizes the similarities and dissimilarities between the ionic and the covalent models as far as these can be discussed now.

Since only the geometrical aspect of the problems can be discussed at present, proponents of the ionic model for these oxides may find many points to which they can raise objections. Since it took more than 50 years to develop the ionic model to its present stage, the covalent model for these oxides, only a few months old, must be regarded as still being in its infancy!

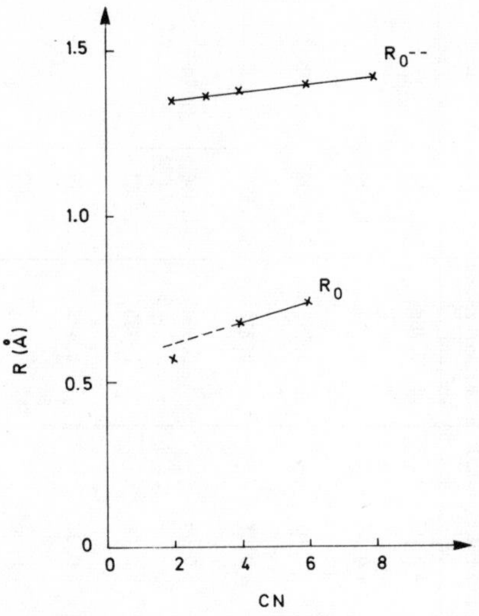

FIG. 1. Covalent and ionic radii of oxygen as a function of the coordination number (CN). The cross indicates an oxygen radius for double bonding.

2. IONIC VERSUS COVALENT LATTICE GEOMETRY IN FLUORITE-TYPE OXIDES

As a basis for this analysis one needs the radii of the atomic species constituting the structure. Therefore, in Fig.1, the ionic and covalent radii of oxygen are plotted against the coordination number (CN) and, in Fig.2, ionic and covalent crystal radii of the actinides are plotted against their atomic numbers. The ionic radii have been taken from the table of Shannon and Prewitt [11] for ions with a charge of 4+ and CN8, and those for CN6 are due to Ahrens [12]. Radii for ions with charge 3+ were estimated for CN7 from Ref. [11]. It should be recalled that the choice of a certain set of ionic radii for a given purpose is still a matter of debate. I have used here those of Ref. [11] since these radii were especially derived in order to fit the lattice parameters of oxides and fluorides. With the ionic radii for M^{4+} and CN8 and the ionic oxygen radius for CN4 $R_{O^{--}}$ = 1.38 Å the experimental bond lengths d_{MO} of the dioxides of the actinide series can be reproduced with little error (see Table I, columns 5, 6, 7). The usual assumption:

$$d_{MO} = R_{M^{+}} + R_{O^{--}} \qquad (1)$$

has been used, with $R_{O^{--}}$ kept constant along the series, and these ionic radii, indeed, appear to give a good description of the lattice geometry of the 5f fluorite-type oxides.

TABLE I. ANALYSIS OF THE IONIC MODEL OF THE ACTINIDE FLUORITE-TYPE OXIDES

	1	2	3	4	5	6	7	8	9	10
	$R_{M^{4+}}$ (8)	$R_{M^{4+}}$ (6)	$R_{M^{3+}}$ (7)	$R_{M^{5+}}$	d_{MO} (th)	d_{MO} (exp)	d_{MO} (th) - d_{MO} (exp)	$R_{M^{4+}}/R_{O^{--}}$	R_M^{IV}	
Th	1.06	1.02			2.44	2.42	+ 0.02	0.768	1.748	
Pa	1.01	0.98			2.39	2.39	+ 0.00	0.732	1.709	
U	1.00	0.97	1.075	0.96 (7)	2.38	2.37	+ 0.01	0.725	1.692	(1.670)[a]
Np	0.98	0.95	1.055		2.36	2.35	+ 0.01	0.710	1.673	
Pu	0.96	0.93	1.035		2.34	2.34	+ 0.00	0.696	1.660	(1.746)[b]
Am	0.95	0.92	1.025		2.33	2.33	+ 0.00	0.689	1.652	
Cm	0.95	0.92			2.33	2.32	+ 0.01	0.689	1.644	
Bk	0.93	0.90			2.31	2.31	+ 0.00	0.674	1.633	

[a] Estimated covalent crystal radius of U^{V} } 'effective radii'.
[b] Estimated covalent crystal radius of Pu^{III} }

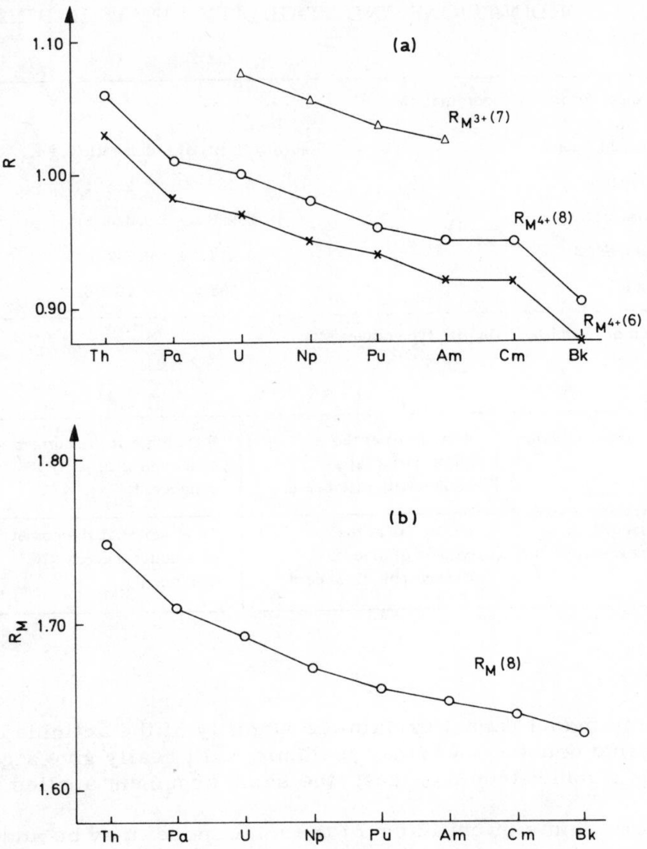

FIG. 2. Ionic [11] and covalent radii of the actinides. The covalent radii are taken from Table I.

Ionic crystals should obey geometrical stability criteria [13, 14], and these are collected, together with a description of the coordination of metal and non-metal atoms in the fluorite lattice, in Table II. The stability limit relevant to the ionic model of the fluorite lattice is:

$$R_{M^{4+}}/R_{O^{--}} = 0.732 \tag{1a}$$

because here the structural unit is the cube, with eight oxygen ions at its corners and a cation at its centre. The condition (1a) has special importance since, for ratios of the ionic radii < 0.732, the cation would no longer make contact with its eight anion next-neighbours and, hence, the structure would lose its shear resistance.

Comparing the value (1a) with column 8 of Table I shows that, according to this stability criterion, all actinide dioxides beyond PaO_2 should be unstable, whereas, in fact, the last stable dioxide in the series is BkO_2.

TABLE II. COORDINATIONS AND STABILITY LIMITS IN BINARY COMPOUNDS AB_n

A) Stability ranges for several coordinations

Coordination	Stability limits of radius ratio $\lambda = R_A/R_B$
triangular	$0.155 \leq \lambda \leq 0.225$ (a)
tetrahedral	$0.225 \leq \lambda \leq 0.414$ (b)
octahedral	$0.414 \leq \lambda \leq 0.732$ (c)
cubic	$0.732 \leq \lambda \sim 1.00$ (d)

B) Coordination of O and M in fluorite type oxides MO_2

	O	M
next-neighbours	4 M atoms at the corners of a tetrahedron with distance d_{MO}	8 O atoms at the corners of a cube with distance d_{MO}
second next-neighbours	6 O atoms at the corners of an octahedron with distance d_{OO}	12 M atoms at the corners of a dodecahedron with distance d_{MM}

Thus the ionic model cannot explain the stability of the actinide dioxides and this casts some doubts on whether the ionic radii really give a good description of the 5f fluorite-type dioxides; the same argument applies to the 4f series.

Two more points of criticism of the ionic model may be added.

(a) It is difficult to understand that refractory compounds like the actinide dioxides (or corundum), having high values for the elastic constants as compared with true ionic crystals like CaF_2, should have a structural frame (see Table II) which is based on the large soft O^{--} ions and which has no shear resistance in itself.

Manes [15] has recently analysed the ion-ion force constant R_{12} and the shell-core force constants K_1, K_2 in the shell model for oxides (UO_2, PuO_2). It is well known that the ratios R_{12}/K_1, R_{12}/K_2 can be related to the polarizabilities of the ions and, for purely ionic solids, $R_{12}/K_{1,2} \ll 1$. For oxides, however, Manes finds $R_{12}/K_{oxygen} \approx 1$, with a drastic reduction in ionic charge, and the ionic model is no longer able to describe the lattice dynamics correctly.

Looking for an alternative description of these oxides, one may start from the metal sublattice. Then the structural unit becomes the tetrahedron of four metal atoms with an oxygen atom in the centre; this should be better suited as a basis for refractory compounds. The stability range of this geometry is given by relation (b) in Table II. However, the problem here is that the atomic radii are not known.

Using the same argument as before, i.e. that it is unlikely that the oxygen radius will change along the series of the actinite dioxides, it is possible to calculate the oxygen radius by applying the upper limit of relation (b) of Table II to the experimental stability limit of the fluorite structure in the actinide series — that is, to the bond length of BkO_2. That BkO_2 is indeed the last stable dioxide in the series is shown by the work of Turcotte and Chikalla [16] and of Baybarz, Haire and Fahey [17]. BkO_2 is stable at normal oxygen pressures up to about 1000°C [16], whereas CfO_2 can be prepared only with difficulty at 300°C under 100 atm oxygen pressure for 100 hours [17].

Thus writing, together with Eq.(1),

$$\lambda = 0.414 = \frac{R_O}{d_{MO} - R_O} \tag{2}$$

yields, with the d_{MO} value of BkO_2 of Table I,

$$R_O = 0.676 \text{ Å} \tag{2a}$$

for the covalent tetrahedral radius of oxygen.

A covalent tetrahedral value of R_O had been established recently by Van Vechten and Phillips on the basis of completely different arguments as 0.678 Å [18]. Thus the agreement between the two values is very good, and this may be taken as confirmation that our covalent description of the fluorite oxides is physically meaningful.

The covalent metal radii along the actinide series up to berkelium for CN8 can now be obtained from the experimental bond lengths using Eq.(1); they are shown in Fig.2b and in column 9 of Table I. A comparison of these covalent metal radii with the metallic radii of the actinides and the covalent radii in the carbides and nitrides will be given elsewhere [19].

3. COVALENT LATTICE GEOMETRY OF UO_2 AND PuO_2 AND POINT DEFECTS[1]

The radius ratios R_O/R_M in the structures of UO_2 and PuO_2 are 0.40 and 0.407, respectively; hence, to a good approximation, the two structures have the same geometry. Several aspects of the geometry of the covalent model for UO_2 are shown in Fig.3 as an illustration for the following considerations.

3.1. The various views of the structural model indicate that the oxygen diameter is larger than the tetrahedral interstitial site of the fcc sublattice of the metal atoms. In a close packed fcc lattice of spheres with radius R_M, the radius r_t of the tetrahedral hole is given by:

$$r_t = R_M\left(\sqrt{\frac{3}{2}} - 1\right) = 0.223\ R_M \tag{3}$$

[1] The geometrical relations of this section and of Section 4 are most conveniently visualized with the aid of a structural model as shown in Fig.3, i.e. a model with the ratio R_O/R_M = 0.40 for the atomic radii.

(a)

(b)

FIG. 3. Covalent structure model of UO_2 with radius ratio $R_O/R_U = 0.40$, i.e. the large spheres represent uranium atoms.

(a) View of the {100} and the {111} slip surfaces; each is constituted by an oxygen and a uranium lattice plane. Note the 'perforated' geometry of the {100} slip surface and the very smooth aspect of the {111} slip surface. Each 'opening' in the {100} surface leads to an octahedral hole of the structure. (For the meaning of the arrows and 'rings', see text in Section 4.3.)

(b) View of the {110} slip surface which is constituted alone by a {110} lattice plane. A corrugated topology of this slip surface is observed, with the valleys extending along the ⟨110⟩ direction of the lattice plane. Each rectangle defined by the four oxygen atoms at the bottom of the valleys represents a diagonal cut through an octahedral hole of the structure.

whereas we have in the UO_2 lattice $R_O = 0.40\ R_M$. However, as seen from Fig.3, the structure can still approximately be regarded as an interstitial compound similar to UC or UN, where R_X/R_M is 0.488 and 0.45, respectively.

3.2. As a consequence of the size of the oxygen atoms, the atoms on the metal sublattice cannot be in close contact if one assumes them to have a spherical shape (which, however, may not be necessarily exactly true). In any event the metal radius defined by Eqs (1) and (2a) and the experimental metal-metal distance d_{MM} can be used to define an overlap parameter Ω_{MM} for the atomic wave functions of adjacent metal atoms. In UO_2 we have:

$$\Omega_{MM} = \frac{2R_M}{d_{MM}} = 0.875 \qquad (4)$$

and in PuO_2 the value is 0.869. These values suggest that some overlap of the wave functions of neighbouring metal atoms should still exist, giving rise to a weak metal-metal bonding, the overwhelming part of the bonding residing, of course, in the metal-oxygen bonds. The corresponding overlap parameter Ω_{OO} for O-O interaction is:

$$\Omega_{OO} = \frac{2R_O}{d_{OO}} = 0.495 \qquad (4a)$$

and indicates, there should be no O-O bonding in the structure.

3.3. The radius of the octahedral interstitial hole is given in the covalent model by:

$$r_0 = \frac{1}{2}(a_0 - 2R_M) \qquad (5)$$

This results, for UO_2, in a value of r_0 of:

$$r_0 = 1.043\ \text{Å} \qquad (5a)$$

The corresponding value in the ionic model is, with:

$$r^+ = \frac{1}{2}\left(\frac{\sqrt{3}}{2}\,a_0 - 2R_O\right) \qquad (6)$$

$$r_0^+ = 0.99\ \text{Å} \qquad (6a)$$

It is obvious from Fig. 3 that an interstitial covalent oxygen atom in such a hole most occupy a strongly eccentric position in order to make bonds with uranium atoms. The symmetry of the octahedron defined by the six uranium atoms adjacent to the hole requires that the oxygen interstitial can make contacts only with any two neighbouring metal atoms of the octahedron, resulting in a displacement from the centre of the hole in a $\langle 110 \rangle$ direction. One can compare the possible amount δ' of this displacement with the experimental value of Willis [20], $\delta' = 0.85$ Å. Since no exact covalent radius of oxygen for CN2 is known, one might take its radius for the double bond $R_O'' = 0.57$ Å [21] and, furthermore, since the uranium radius should change somewhat due to the additional U-O bond, no

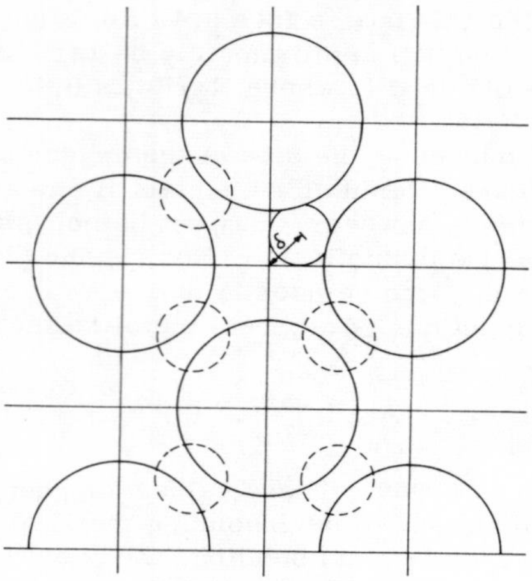

{100} - plane

FIG. 4. Atomic arrangement in a {100} uranium lattice plane with indication of an interstitial oxygen in O' position. (Oxygen atoms on normal lattice sites above the uranium plane are indicated as broken circles.)

exact agreement with the experimental value can be expected.[2] However, when placing the double bonded oxygen atom at $\delta' = 0.85$ Å, the fit is quite good (see Fig. 4). The oxygen radius is probably somewhat too small and the uranium radius somewhat too large in this representation of the O' type of interstitial.

The formation of Willis complexes [20] associated with the excess oxygen on octahedral sites must be explained by the covalent model on the basis of the disturbance that the symmetry of the bonding hybrids suffers on the metal atoms when the extra bond to the interstitial oxygen is established. However, there are, in addition, geometrical restrictions enforced by the symmetry of the two uranium tetrahedra to which the two displaced lattice oxygen atoms belong. The only way for them to escape from their lattice sites into the respective adjacent octahedral holes leads them in a ⟨111⟩ direction through the middle of the corresponding faces of their uranium tetrahedra. Since the displaced oxygen should remain bonded to the three uranium atoms of their former tetrahedra, but now from 'outside', their new position corresponds to changing sides of a {111} uranium atomic plane, say from above to a roughly equivalent position below the plane. Locally this change of oxygen position in ⟨111⟩ direction corresponds to a stacking fault for the sequence of ⟨111⟩ planes of the lattice (see Table IV). In an undisturbed lattice the distance of an oxygen {111} plane above a uranium plane to an oxygen plane directly below, according to Table IV, is $d_{AC} = a/2\sqrt{3}$ and amounts in UO_2 to $d_{AC} = 1.58$ Å. The experimental value given by Willis for the ⟨111⟩ displacement of the lattice oxygen into the O'' type position is $\delta'' = 1.05$ Å [20], and it appears reasonable that the displacement in the defect complex is smaller than the

[2] See subsection 3.6 and column 10 of Table I.

distance d_{AC} since the oxygen atom should be bonded more strongly in the direction of its old site and have a smaller radius due to the lower coordination. Thus the geometry of the covalent lattice model accounts semi-quantitatively reasonably well for the existence of the Willis complexes of type 2:1:2. Without a detailed model for the lattice forces it seems of little use to speculate on more complex defect clusters.

3.4. The formation of oxygen Frenkel pairs in the perfect lattice with a formation energy ΔG_F should formally occur in two steps:

(a) Breaking the U-O bonds to the four next uranium neighbours within the tetrahedron;

(b) Pushing aside three of the four uranium atoms for the jump from the tetrahedral site through the corresponding face of the uranium tetrahedron in the $\langle 111 \rangle$ direction into an adjacent octahedral hole, as already described in subsection 3.3.

In the perfect lattice there are four equivalent directions for the oxygen atom to separate from its lattice site in this process. Once it has gained the octahedral hole, there are nine equivalent sites where it could make bonds to two uranium atoms as for a normal interstitial O' position (see 3.3 above); three sites are blocked because they are directly adjacent to the hole through which the oxygen atom has just jumped from its lattice site, and where the oxygen atom is roughly in a O'' position. Here, of course, there is a high probability for recombination with the vacancy which it has just left at the other side of the uranium {111} plane.

3.5. Oxygen self-diffusion can, in principle, occur either by the migration of vacancies or interstitials. For a review of the present experimental situation with regard to oxygen and metal self-diffusion in UO_2 and $(U,Pu)O_2$ see Ref.[22].

An oxygen vacancy has six (second next) oxygen neighbours in a perfect lattice. The direct jump from one of these six lattice sites into the vacancy requires the movement of the oxygen atom along a $\langle 100 \rangle$ direction and squeezing two uranium atoms apart. Such a direct jump, however, is rather unlikely for geometrical reasons, and therefore the movement should work via the creation of a Frenkel defect with energy $\Delta G_F^{\prime} < \Delta G_F$, the lowering of the formation energy coming from the relaxation of the uranium atoms (especially at high temperatures) around the oxygen vacancy. The lattice oxygen moves through the centre of the face of its own tetrahedron into the octahedral hole, and then leaves it through the face of the tetrahedron where the oxygen vacancy is to be filled. Oxygen self-diffusion via interstitials may occur either by the movement of O' type oxygen atoms or O'' type atoms, depending upon the geometry of the covalent lattice. The details should sensibly depend upon the bond strength of the corresponding U-O bonds, O' type oxygens having two bonds, and O'' type oxygens having three U-O bonds. In any case, there will be no direct jumps from one fixed position to a next one, but again two-stage processes are necessary:

(a) Movement of the interstitial within the octahedral hole, changing between positions O' and O''; and

(b) Jumps along $\langle 111 \rangle$ directions through the faces of uranium tetrahedra from one octahedron into an adjacent one.

It seems likely that the directions of the migration of excess oxygen on the basis of these geometrical conditions can be influenced by the presence of mechanical stress fields, such as are found with dislocations.

3.6. The creation of a metal Frenkel defect in the metal sublattice at first requires that the eight U-O bonds of the uranium atom must be broken, and then the adjacent octahedral site must be enlarged in order to accommodate the interstitial. From Eq. (4) one obtains, for the ratio of the radius of the octahedral hole r_0 and of the metal radius R_M in UO_2:

$$r_0 / R_M = 0.619$$

Thus the diameter of the octahedral hole must be enlarged by 38% to fit the interstitial. In the ionic model the corresponding radius ratio from Eq. (6a) is:

$$r_0^+ / R_{M^{4+}} \approx 1.00$$

Thus, from the geometrical point of view, the probability for creating metal Frenkel pairs is sensibly higher in the ionic model than in the covalent one; there are, however, other factors which contribute to the defect energies which could change this picture. Taking into account substoichiometry gives two contributions to the geometry which act against each other. Adjacent to an oxygen vacancy the number of bonds to be broken by the metal atom is reduced by one and a certain relaxation of the neighbouring lattice positions is possible (especially at high temperatures), which might weaken the other bonds. This should favour the formation of metal interstitials. On the other hand, the removal of one oxygen atom results in an increase of the covalent radii of at least two neighbouring metal atoms (see 3.7). This increases the strain energy necessary to place the interstitial into the octahedral hole, provided the interstitial itself does not suffer a drastic change of radius due to electronic effects. Thus the issue is not clear and needs a detailed quantitative treatment.

3.7. Effective metal radii in UO_{2+x} and in PuO_{2-x} are the final consideration. In the ionic model the change of the lattice parameters with x, da/dx, for UO_{2+x} and PuO_{2-x} is closely related to the change of the ionic radii from charge states 4+ to 5+ and 4+ to 3+, respectively. Depending on the set of ionic radii used, the change of $R_{Pu^{4+}}$ to $R_{Pu^{3+}}$ in particular would attain values up to 14% [13] which however are very unlikely to be realistic.

The exact relation between the change of the metal radius in the defects and the lattice parameter, a, in non-stoichiometric structures involves:
(a) The radius change ΔR_M;
(b) The relaxation of the lattice positions around the interstitial or vacancy;
(c) Elastic effects.
Thus it can only be determined by detailed defect calculations, which necessitate a complete quantitative model of the structure. Since these results are not available for the covalent model, a first rough estimate of ΔR_{Pu}^{IV-III} and ΔR_U^{IV-V} is made in Table III without taking into account effects (b) and (c). The covalent model is better suited to such a procedure than the ionic model, since in the former relaxation effects should be smaller

TABLE III. ESTIMATION OF EFFECTIVE RADIUS DIFFERENCES

$\Delta R_U = R_U^{IV} - R_U^V$ AND $\Delta R_{Pu} = R_{Pu}^{IV} - R_{Pu}^{III}$

ΔR_M	da/dx	Remarks
+0.061	-	Pu_2O_3 treated as pseudo fluorite structure
+0.0865	+0.40	for PuO_{2-x} near x=0
+0.022	-0.1	variation between UO_2 and U_4O_9

and the relative change of R_M also will be smaller than in the ionic model. Because of the errors involved we shall call R_U^V and R_{Pu}^{III} 'effective' covalent crystal radii, they are upper and lower limits, respectively, with regard to the true radii.

In the case of Pu^{III} a rough check of the method can be made by regarding C-Pu_2O_3 as a pseudo-fluorite structure with all plutonium atoms in the valence state III and assuming unrelaxed metal positions, whereas the oxygen coordination is practically unchanged. Under these assumptions the effective covalent metal radius of Pu^{III} in cubic Pu_2O_3 must be clearly smaller than the value obtained from the change of the lattice parameter of PuO_{2-x} near x=0. The results for ΔR_M are given in Table III and the effective values for R_U^V and R_{Pu}^{III} in column 10 of Table I. The two values for ΔR_{Pu}^{IV-III} show qualitatively the expected behaviour, indicating that the relaxation of the Pu^{III} atoms in PuO_{2-x} is much smaller than in cubic Pu_2O_3.

4. DISLOCATIONS

4.1. Dislocation behaviour in UO_2

The technological aspect of the plasticity of UO_2 and (U, Pu)O_2 has received wide attention during the past five years and, consequently, interest in the investigation of the dislocation behaviour in UO_2 was renewed. Up to now the interpretation of the experiments was always based on the ionic model of UO_2 [23-25], even if sometimes "a considerable covalent contribution to the bonding" was admitted [23]. The following brief and non-exhaustive collection of experimental results is intended to give an illustration of the experimental background on dislocation behaviour with which the lattice geometry of UO_2 has to be confronted:

(a) UO_2 has three slip systems, which are in the order of importance: (i) $\{100\}\langle 110\rangle$; (ii) $\{110\}\langle 110\rangle$; (iii) $\{111\}\langle 110\rangle$ [26, 27]; In addition it can deform under certain conditions also by composite or non-crystallographic slip [28, 29].

(b) The slip system $\{100\}\langle 110\rangle$ has the lowest critical resolved shear stress (CRSS). At low deformations the CRSS and the resolved flow stress depend little on temperature and on excess oxygen [27]. In the

{100} slip planes, the edge components of the dislocations are slightly more mobile than the screw components, leading to dislocation loops elongated in the direction of the Burgers vector [30]. Stress-free dislocations on a (100) glide plane tend to take a straight line geometry along [010] and [001] directions [24] (see also figure 3 of Ref. [30]). When deforming stoichiometric UO_2, the glide system {100}⟨110⟩ shows a yield point effect after reloading the crystal or after a change in strain rate [31, 32]. This effect seems not to have been observed in other glide systems, and vanishes in hyperstoichiometric UO_2.

(c) The CRSS of the {110}⟨110⟩ slip system is the double or triple of that of the {100}⟨110⟩ system [27, 28], and may be greatly lowered by (i) increasing temperature, and (ii) increasing amounts of excess oxygen [27].

(d) UO_{2+x} oriented for {100} slip at low temperatures (800-1000°C) may deform non-crystallographically near {111} slip planes, but at 1200°C slip returns to the {100}-plane [29].

(e) On deformation of UO_{2+x} dislocation networks are produced more easily than in stoichiometric UO_2, hence cross slip in general is favoured by excess oxygen. In particular, cross slip of screw dislocations from {100} planes into {111} planes has been proposed at T ≤ 1100°C [29].

(f) The activation energy, Q, for steady-state creep in UO_{2+x} single crystals (i.e. creep after deformation of at least 15%) depends strongly on x, i.e. the amount of excess oxygen. The value of Q drops from Q = 130 kcal/mol at $x \approx 0$ to Q = 62 kcal/mol at x = 0.01 in the temperature range 1100 to 1400°C [33].

(g) The activation energy for metal self-diffusion is closely related to that of high-temperature steady-state creep (see point f). Additional evidence on this point has been provided by Routbort [34] and Matzke [22] on steady-state creep and metal self-diffusion in sub-stoichiometric mixed oxides $(U, Pu)O_{2-x}$. In both cases $\dot{\epsilon}$ and D fall to a minimum near O/M = 1.98 and then increase again for lower O/M values.

In the following sections some general and some very special remarks on dislocation behaviour in UO_2 are given based on the covalent model, without trying to explain all experimental details given above. For obvious reasons the purpose of this section must be rather to contribute to the discussion on plastic properties than to present a complete picture.

4.2. Slip planes and geometry of slip surfaces in UO_2

In an oxide lattice the frictional force acting on a dislocation (Peierls force) should have the following contributions:

(i) Topology of the atomic arrangement in those lattice planes which constitute the surface on which the core of the dislocation must move (slip surface);

(ii) The number and strength of covalent bonds which must be broken when the dislocation moves by a distance b, where b is the Burgers vector;

(iii) The change of the electrostatic lattice energy during the movement of the dislocation core over the distance b;

(iv) The change of the elastic energy during the movement of the dislocation core over the distance b.

In a purely ionic model contribution (ii) does not exist, while in a purely covalent model contribution (iii) is absent. In oxides both contributions should probably be taken into account; however, contribution (ii) may be more important. Here, where only geometrical aspects are discussed, only point (i) and the geometrical aspect of point (ii) will be considered, since a more detailed discussion has to wait for the development of a reliable model for the lattice potential. It is felt, however, that even this very restricted reasoning is useful, since geometrical features are more important in a covalent model than in an ionic one.

For a comparison of the slip planes {100}, {110} and {111}, their geometrical parameters have been listed in Table IV, which was taken from Ref. [24], and in Table V. The information contained in Table IV is independent of the structural model since no atomic radii are involved. Table V, however, was established with the covalent model in view. The fact that the slip surfaces of type {100} and {111} are composed of two lattice planes, of which one contains only uranium atoms and the other only oxygen atoms, comes out very clearly in the covalent model but is less obvious in the ionic one. The slip surface of type {110} is identical with a {110} lattice plane for both models. The bond geometry described in Table V clearly is more important for the covalent model, since the electrostatic forces in the ionic model have no directional character.

Comparing Table V with the experimental results summarized in subsection 4.1, one finds a clear correlation, for stoichiometric UO_2, between the number of bonds per atom in the core of an edge dislocation across a given slip plane and the relative value of the CRSS in the corresponding slip system. The {100}⟨110⟩ slip system having the highest number of bonds per atom in the core possesses the lowest CRSS and the {111}⟨110⟩ system with only one bond per atom in the core is the most difficult slip system, the {110}⟨110⟩ system lying between the two others with regard to bonds and CRSS. The reason for this behaviour appears obvious if one remebers that, in the {100}⟨110⟩ system with the highest number of bonds per atom and the lowest CRSS, the dislocation core has the smallest number of atoms per unit length of dislocation line. Due to stress concentration it seems easier to break four bonds on one atom than 4 bonds on three or four atoms. From this typical covalent argument one may conclude that, in fact, covalency should play an important role in determining the crystal plasticity of UO_2.

Next one may consider the influence of the different radius ratios, $R_{M^{4+}}/R_{O^{--}} = 0.725$ in the ionic case and $R_O/R_M = 0.40$ in the covalent one, on the geometry of the slip surfaces. In the ionic structure model, the geometrical differences between the three types of slip planes are less pronounced than in the covalent one, (i) because the ratio of cation radius to anion radius is relatively close to 1.00 and (ii) oxygen ions with the larger ionic radius are in close contact and are twice as numerous as the smaller cations. These two features, and especially the first one, make the slip surfaces in all three slip systems relatively 'rough' in the ionic model, and the pronounced experimental differences especially between the {100} and {110} slip planes should be understood on the basis of electrostatic effects, and hardly on the basis of geometrical effects.

In the covalent model, see Fig. 3, the character of UO_2 as an 'interstitial compound' makes the geometrical differences between the three slip planes more evident, in a way, however, which, at first sight, is just

TABLE IV. RELATIONS BETWEEN LATTICE PLANES IN THE FLUORITE STRUCTURE

Lattice plane	{100}	{110}	{111}
1. Stacking sequence	... $A\gamma A\beta A\gamma A\beta$ ...	... a b a b a b ...	... $\gamma AC\alpha BA\beta CB\gamma AC\alpha$ ...
2. Atomic density, B^+, of lattice plane in fcc lattice	$B^+_{\beta,\gamma} = \frac{2}{a^2}$	$B^+_{110} = \frac{\sqrt{2}}{a^2}$	$B^+_{\alpha,\beta,\gamma} = \frac{4}{\sqrt{3}}\frac{1}{a^2}$
3. Atomic density, B^+, of lattice planes in fluorite lattice	$B^+_A = \frac{4}{a^2}$; $B^+_\beta = \frac{2}{a^2}$	$B^+_{a,b} = \frac{3\sqrt{2}}{a^2}$	$B^+_{A,B,C} = B^+_{\alpha,\beta,\gamma} = \frac{4}{\sqrt{3}}\frac{1}{a^2}$
4. Distance, d, between lattice planes in fcc lattice	$d_{100} = \frac{a}{2}$	$d_{110} = \frac{a}{4}\sqrt{2}$	$d_{111} = \frac{a}{\sqrt{3}}$
5. Distance, d, between (non-equivalent) lattice planes in fluorite structure	$d_{A\gamma} = d_{A\beta} = \frac{a}{4}$	$d_{ab} = d_{110} = \frac{a}{4}\sqrt{2}$	$d_{\gamma A} = d_C = \frac{a}{4\sqrt{3}}$ $d_{AC} = d_{BA} = d_{CB} = \frac{a}{2\sqrt{3}}$
6. Shear vector, R, producing stacking sequence	$\gamma + R = \beta$ $R = \frac{a}{2}\langle 100\rangle$	$a + R = b$ $R = \frac{a}{4}\langle 112\rangle$	$\alpha + R = \beta,\ \beta + R = \gamma$ $R = \frac{a}{12}\langle 112\rangle$

TABLE V. SLIP PLANE PARAMETERS FOR {100}, {110} AND {111} SLIP PLANES IN FLUORITE STRUCTURE WITH REGARD TO EDGE DISLOCATIONS

	{100}	{110}	{111}
Number of lattice planes constituting the slip surface	2	1	2
Density of atoms in slip surface	$\frac{6}{a^2}$	$\frac{3\sqrt{2}}{a^2} = \frac{4.24}{a^2}$	$\frac{8}{\sqrt{3}a^2} = \frac{4.62}{a^2}$
Number and type of atoms in the dislocation core per unit length	$\frac{2\,U}{\sqrt{2}\,a}$	$\frac{1\,U + 2\,O}{a}$	$\frac{2\,U + 2\,O}{\sqrt{3}\,a}$
Bonds across slip plane	2 × 4	1 × 2 + 2 × 1	2 × 1 + 2 × 1
Number of bonds per core atom across slip plane	4	1.33	1
Bond angle with respect to the normal of the slip plane	55°	35°	0
Number of bonds per unit length of dislocation core across slip plane	5.65	4.0	2.31

contrary to what one would expect from the experimental results in subsection 4.1. As one can clearly see from Fig. 3, the slip surface of the most difficult slip system, {111}⟨110⟩, appears extremely smooth, because the small oxygen atoms fit relatively well into the tetrahedral interstitial sites of the {111} planes of the fcc metal sublattice. Also the {110} type slip surface is rather smooth along its ⟨110⟩ direction, and, from the geometrical point of view, an edge dislocation should be able to move rather freely in the direction of its Burgers vector. In the covalent model, only the experimentally 'easy' glide system {100}⟨110⟩ shows relatively large potential troughs similar to those in the ionic model. However, these troughs are welcome in both models in order to explain the preferred alignment of stress-free dislocations along ⟨010⟩ directions in the (001) plane [24].

Thus we are led to the conclusion (which is not too surprising for a structural model with covalent atomic radii) that the <u>bond geometry</u> defined in Table V is more important in determining the differences in behaviour of the three slip systems than the differences in the geometry of the slip surfaces. An analogous observation can be made for the slip surfaces and bond geometry in the case of the covalent structure model for MgO.

4.3. Dislocations on {100} planes

The behaviour of dislocations in the {100}⟨110⟩ slip system is of special interest because: (i) this is the 'easy' slip system, i.e. the Peierls force should be low and, in spite of this, as special effects are known to obtain; (i) there is an alignment of stress-free single dislocations along crystallographic ⟨100⟩ directions; (iii) the yield point effect exists; and (iv) possible charge effects are under discussion.

Within the frame of the ionic model, it has been repeatedly proposed that dislocations on {001} slip planes should carry an electrical charge. From the alignment of stress-free dislocations along [100] and [010] directions in a (001) slip plane, it was concluded in Ref. [24] on the basis of the slip surface geometry in the ionic model that the dislocation core should only contain uranium atoms. This interpretation was questioned later by Ashbee and Frank [35], who pointed out that such highly charged dislocation cores appeared unlikely; they in turn proposed a model in which the oxygen ions make a correlated movement at right angles to that of the uranium ions in the moving dislocation core. In this way the dislocation core should, in effect, have sufficient oxygen ions to make it charge-free.

This question may now be reconsidered in the light of the covalent model. To this end, one considers the geometry of the proposed slip surface of the {100} plane in the covalent model. From Fig. 3 it can be seen that this slip surface consists of a chess-board arrangement of two types of squares, one with a rather flat, 'smooth' surface (constituted by a uranium atom and four oxygen atoms), and the other represented by the openings which lead to the octahedral holes. The next plane above this is composed only of uranium atoms sitting on these openings. If such a lattice plane contains the core of a dislocation, this core will only consist of a string of uranium atoms, i.e. it will contain no oxygen atoms. For the movement of the dislocation on the {100} plane shown in Fig. 3, this string of uranium atoms must be lifted out of the octahedral troughs and then be placed further over into a line of adjacent troughs. At first sight this movement is strongly impeded by oxygen atoms — if it occurs in one of the ⟨110⟩ directions of the (001) plane — but it should be relatively easy in the [100] and [010] directions. For the [100] and [010] directions there are 45° dislocations. As a first point it should be remarked that, in the covalent model, charge effects at the dislocation core should be less important than in the ionic model because the charges on the atoms are certainly lower, perhaps by a factor of about one half.

Since in the fluorite lattice only dislocations of type $\frac{a}{2}\langle 110\rangle$ are known, the movement of the core atoms can effectively occur only in the two ⟨110⟩ directions of the (001) plane, as indicated by arrows in Fig. 3. This movement is strongly impeded by the oxygen atoms marked with 'rings'. In principle two possibilities arise:

(a) In order to circumvent the 'oxygen' obstacles, the dislocation dissociates into:

$$\tfrac{1}{2}[110] = \tfrac{1}{2}[100] + \tfrac{1}{2}[010]$$

Yet a real dissociation that is observable in the electron microscope on stationary dislocations cannot exist, because a position which is $\frac{1}{2}$[100] away from the starting point of the Burgers vector corresponds to the saddle point of the movement and not to a stable equilibrium position in the lattice.

(b) The impeding oxygen atoms are pushed by the uranium atoms of the moving dislocation core roughly in a {111} direction down into the octahedral holes and they regain their old lattice position once the dislocation core has passed.

The second mechanism might be the cause of the yield point effect found for the {100}⟨110⟩ slip system in stoichiometric UO_2. Certainly more energy is needed to produce the relaxation of the oxygen atoms along a stationary dislocation line than on a moving one. In the latter case the elastic waves emitted from the core of the moving dislocation should aid the relaxation processes. Such a mechanism could possibly also explain why the yield point effect becomes weaker or is absent in hyperstoichiometric UO_2. The regular square pattern of the {100} oxygen lattice planes, constituting the important requirement for such a behaviour of the slip surface, is partly destroyed if Willis complexes in statistical orientation are introduced by the excess oxygen. Thus the number of oxygen atoms to be relaxed, their direction of relaxation and also the energy needed for the relaxation process are changed.

Summarizing the results for dislocations on {100} planes one can state:

(i) That charge effects should be less important in the covalent model then in the ionic one;

(ii) The geometrical interpretation given in Ref. [24] about the preferred alignment of stress-free dislocations along [100] and [010] directions can be retained in the covalent model;

(iii) The 'ionic' model of Ashbee and Frank [35] concerning the role of oxygen in the movement of dislocations on {100} slip surfaces has to be modified, but then may explain the yield point effect and its variation with excess oxygen.

4.4. Dislocation behaviour in UO_{2+x}

Discussion of the dislocation behaviour in non-stoichiometric UO_2 and in the mixed oxides $(U, Pu)O_{2\pm x}$ covers a wide field; there is room here for only one general remark concerning the differences between the {100}⟨110⟩ and the {110}⟨110⟩ slip systems.

In subsection 4.1 it was stated that the CRSS of the {110}⟨110⟩ slip system is strongly reduced by excess oxygen and by increasing temperatures, whereas the {100}⟨110⟩ system depends less drastically both on x and T. Without going into details of proposing models for the interaction of Willis complexes with stationary and moving dislocations at temperatures $T < 1150°C$, it can be stated that:

(i) {100}-type slip surfaces more or less 'touch' the octahedral holes of the fluorite structure (see Fig.3), whereas

(ii) {110}-type slip surfaces cut diagonally through these octahedral holes.

Thus the probability of interaction between the excess oxygen atoms sitting somewhere in the octahedral holes on O' or O'' positions (see subsection 3.3) with the uranium atoms of a dislocation core is higher for the {110} slip system than for the {100} system; that is, the {110}⟨110⟩ slip system is most sensible to the presence of excess oxygen, while the {100}⟨110⟩ is less so.

Finally it should be shown that the CRSS of the {110}⟨110⟩ system is in fact more strongly reduced by the excess oxygen than the CRSS of the

{100}⟨110⟩ system. This effect may be understood by consulting Table V. In the discussion in subsection 4.2, it was concluded that the differences between the various slip systems are based:

(a) On the number of bonds per atom in the core of an edge dislocation which act across the slip plane, and

(b) On the number of atoms in the core per unit length of dislocation line; and, in the discussion on oxygen self-diffusion via interstitials in sub-section 3.5, it was mentioned that

(c) By introduction of excess oxygen into the lattice, U-O bonds are, in effect, eliminated from the basic UO_2-lattice.

In a {100}⟨110⟩ dislocation, the bonds across the slip plane are concentrated on relatively few uranium atoms – each carrying four bonds. In the {110}⟨110⟩ dislocation on the other hand, less bonds per unit length of dislocation line are distributed over more atoms. Hence, eliminating, due to the presence of excess oxygen, one bond in the {110}⟨110⟩ dislocation core has more severe geometrical consequences and should, therefore, weaken the forces across the slip plane more than in the case of the {100}⟨110⟩ dislocation core.

Thus it appears that the covalent model could, in principle, explain many features of the crystal plasticity of UO_2. The final answers probably can only be given when a quantitative model for the lattice forces is available.

5. CONCLUSIONS

The geometrical relations developed in the previous sections for the covalent lattice model, and their reasonable qualitative agreement with experimental results, appear encouraging enough for the model to be further developed to give a quantitative interpretation of lattice and defect properties. At this stage it is too early to attempt a true comparison between the ionic and the covalent model for the fluorite type oxides. However, it seems possible, briefly to define possible areas of agreement and of disagreement between the two models.

5.1. There should be differences in the area of the lattice potential. The ionic model has to assume direct interaction between next neighbour (nn) oxygen ions but no interaction between nn uranium ions. The covalent model on the contrary should work with attractive forces between nn uranium atoms but without interaction between nn oxygen atoms. It has still to be seen whether the covalent model will or will not encounter difficulties in reproducing the experimental values of the elastic constants.

5.2. Agreement between the two models is expected in the field of simple defect thermodynamics, as developed by Matzke [36] and by Lidiard [37]. In addition, the covalent model should require charge compensation and an electrical neutrality condition, even if all charges should be lower.

5.3. The geometry of the point defects should differ in some details, the main difference coming from the fact that, in the ionic model, a missing oxygen ion leaves much space into which, in principle, surrounding ions may relax; the geometrical constraints for the covalent model are more severe since, to a first approximation, the atoms may be regarded as rigid spheres. These geometrical differences may lead, for certain types of defects, to differences in the defect entropies. However, no general conclusion can be drawn: each defect species has to be compared in greater detail between the two models.

5.4. It appears that details of the dislocation geometry and the influence of non-stoichiometry are easier to understand in the covalent model than in the ionic model because of the clearer geometrical limitations of the possible movement of the various atoms in the covalent model.

REFERENCES

[1] TESSMAN, J.R., KAHN, A.H., SHOCKLEY, W., Phys. Rev. 92 (1953) 890.
[2] LEVIN, A.A., SYRIKIN, Y.K., DYATKINA, M.E., Russ. Chem. Rev. 38 (1969) 95.
[3] BLANK, H., 3me Journée d'actinides, Karlsruhe, Dec. 1973, unpublished.
[4] SLATER, J.C., J. Chem. Phys. 41 (1964) 3199.
[5] PHILLIPS, J.C., Rev. Mod. Phys. 42 (1970) 317.
[6] VAN VECHTEN, J.A., Phys. Rev. 182 (1969) 981.
[7] VAN VECHTEN, J.A., Phys. Rev. 187 (1969) 1007.
[8] LEVINE, B.F., J. Chem. Phys. 59 (1973) 1463.
[9] LEVINE, B.F., Phys. Rev. B7 (1973) 2591.
[10] BLANK, H., Solid State Commun. (in press, 1974).
[11] SHANNON, R.D., PREWITT, C.T., Acta Crystallogr., Sect. B 25 (1969) 925.
[12] AHRENS, L.H., Geochim. Cosmochim. Acta 2 (1952) 155.
[13] WELLS, A.F., Structural Inorganic Chemistry, Oxford University Press (1962).
[14] PAULING, L., The Nature of the Chemical Bond, 3rd ed. (1960).
[15] MANES, L., private communication.
[16] TURCOTTE, R.P., CHIKALLA, T.D., J. Inorg. Nucl. Chem. 33 (1971) 3749.
[17] BAYBARZ, R.D., HAIRE, R.G., FAHEY, J.A., J. Inorg. Nucl. Chem. 34 (1972) 557.
[18] VAN VECHTEN, J.A., PHILLIPS, J.C., Phys. Rev., B 2 (1970) 2160.
[19] BLANK, H., 4me Journée d'actinides, Harwell, June 1974, to be published.
[20] WILLIS, B.T.M., J. Phys. (Paris) 25 (1964) 431.
[21] COULSON, C.A., Valence, Oxford University Press (1961).
[22] MATZKE, Hj., J. Phys. (Paris) 34 (1973) C9-213.
[23] ASHBEE, K.H.G., Proc. R. Soc. (London), Ser. A 280 (1964) 37.
[24] BLANK, H., RONCHI, C., J. Nucl. Mater. 31 (1969) 1.
[25] SAWBRIDGE, P.T., SYKES, E.C., J. Nucl. Mater. 35 (1970) 122.
[26] RAPPERPORT, E.H., HUNTRESS, A.M., NMI-1242 (1960).
[27] NADEAU, J.S., J. Am. Ceram. Soc. 52 (1969) 1.
[28] SAWBRIDGE, P.T., SYKES, E.C., Philos. Mag. 24 (1971) 33.
[29] YUST, C.S., McHARGUE, C.J., J. Am. Ceram. Soc. 56 (1973) 161.
[30] YUST, C.S., McHARGUE, C.J., J. Nucl. Mater. 31 (1969) 121.
[31] DE NOVION, Ch., private communication, 1973.
[32] YUST, C.S., McHARGUE, C.J., J. Am. Ceram. Soc. 54 (1971) 628.
[33] SELTZER, M.S., CLAUER, A.H., WILCOX, B.A., J. Nucl. Mater. 44 (1972) 43.
[34] ROUTBORT, J., JAVED, N.A., VOGLEWEDE, J.C., J. Nucl. Mater. 44 (1972) 247.
[35] ASHBEE, K.H.G., FRANK, F.C., Philos. Mag. 21 (1970) 211.
[36] MATZKE, Hj., Atomic Energy of Canada Ltd. Rep. AECL-2585 (1966).
[37] LIDIARD, A.B., J. Nucl. Mater. 19 (1966) 106.

DISCUSSION

C.R.A. CATLOW: While welcoming an attempt to treat the problem of covalency in oxides, I cannot accept some of the arguments you have presented in support of a predominantly covalent model. In particular, I would like to raise two points. First, the failure of the structure of oxide crystals to conform with radius ratio rules is, I think, not surprising, nor is it evidence for the large deviation from ionic bonding. Radius ratio rules are based on the concept of hard sphere models of the anion-anion interactions. Oxide-oxide interactions are, however, soft; indeed ab initio molecular orbital calculations (Catlow, D. Phil. Thesis, University of Oxford, 1974) have suggested that the interaction is attractive up to rather short internuclear separations. Secondly, I do not believe that the Phillips prescription for calculating ionicities can be applied to a crystal in which both valence and conduction bonds are predominantly metallic in nature, as is almost certainly the case in UO_2.

H. BLANK: Let me first correct a misunderstanding. I still regard the actinide dioxides as rather strongly ionic compounds (UO_2 has $f_i = 0.67$ on the Pauling scale of ionicity, but roughly $f_i \simeq 0.88$ on the Phillips scale). The point I wished to stress was that one should use covalent crystal radii for the description of the structure in spite of the rather high ionicity.

Turning now to the matters you specifically raise: You state that geometrical stability rules are based on the concept of hard spheres. I do not entirely agree with this statement, since these rules were established for and applied to the classical ionic compounds such as the alkali halides, alkaline-earth fluorides etc., and anions per se are not hard spheres. This can be recognized in the values of the elastic constants for these compounds and their melting points in comparison with those of the oxides. Thus the

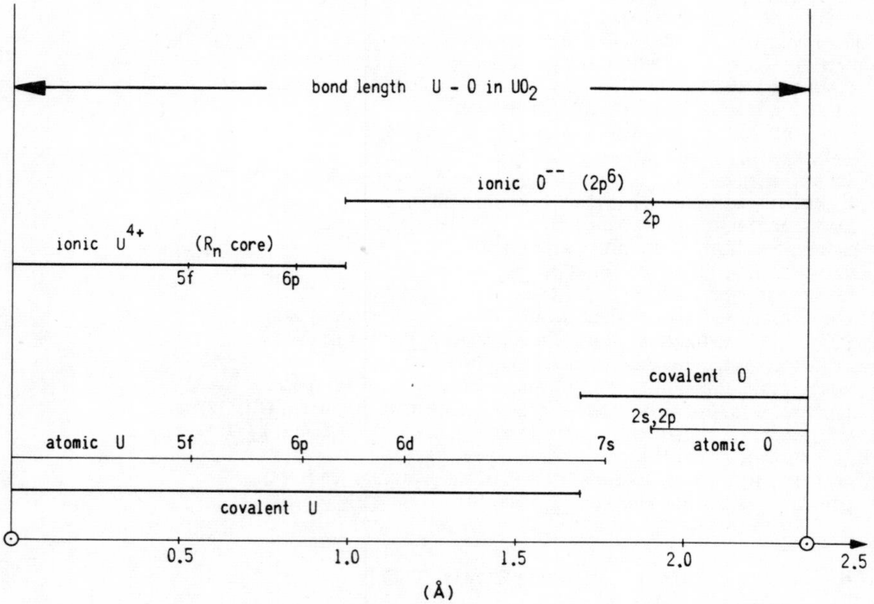

FIG. A. Comparison of the bond length d_{UO} in UO_2 with ionic and covalent crystal radii and with the radii of the free atoms.

softness of the anion-anion interaction in a model using ionic crystal radii for UO_2 appears less to contradict the geometrical stability rules than the refractory properties of this compound. On the other hand it seems to me that one introduces quite a strong covalent element into the ionic model if one calculates the attractive forces for the O^-–O^- interaction in the lattice from the bonding in a free O_2^- molecule and uses the results to modify the lattice potential of the ionic model (the two 'unbound' electrons of the O^{--}–O^{--} pair are put separately into the lattice), as shown in Mr. Catlow's thesis. The interaction between neighbouring oxygen ions in the lattice may in fact, be somewhat different from this since these have six equivalent oxygen neighbours, apart from the four uranium next neighbours. Since, at first sight, there seems to be such a frightening difference between ionic and covalent crystal radii, one should consider carefully the basic meaning of these two types of radii and see the consequences of, for instance, reducing the full charges of the pure ionic model. In order to illustrate this, the bond length d_{UO} in UO_2 is compared with the ionic radii, with the covalent crystal radii deduced in my paper and with the radii of the free atoms of uranium and oxygen in Fig.A. The 'radius' of the oxygen 2s and 2p states has been calculated by the 'Slater rules' and the 'radii' of the 5f, 6p, 6d and 7s states of uranium are taken from Waber and Cromer (J. Chem. Phys. 42 (1965) 4116). Each 'radius' of an electronic state corresponds to the outermost maximum of its radial charge density.

The figure shows that (a) the covalent crystal radii of uranium and oxygen deduced in my paper are rather close to the 'radii' of the free atoms (see also Ref.[4] of my paper), and (b) one can easily imagine that the ionic radius of the O^{--} anion should reduce directly to the covalent radius if a fraction of the ionic charge on the oxygen ion is returned to the cation.

With regard to your second point, I agree that at present there is no straightforward method of applying the Phillips theory to the actinide dioxides (a formal use of its relations leads to $f_i = 0.88$ for UO_2). However, this theory has been applied successfully to the 3d oxides (see Ref.[9] of my paper). This fact and the general behaviour of the simple metal oxides across the periodic table (see Ref.[10]) support the view that the ionicity of the U-O bond in UO_2 and of the other actinide dioxides lies within the range defined in the introduction of my paper.

C.R.A. CATLOW: I also doubt whether the metal-metal interaction can be as large as suggested by your model. This does not agree with the observed Curie-Weiss paramagnetism of the crystal. Clearly, further experiments to determine the 'size' of the uranium 'ion' would be of value; in particular, I think that neutron diffraction studies to determine the extent of spin delocalization from uranium might end the controversy over the bonding in this crystal.

H. BLANK: I cannot yet comment on the magnetic behaviour of the actinide dioxides; clearly, however, any model using a charge distribution other than the full ionic charges must also provide an explanation of the magnetic properties of UO_2. On the point regarding the size of the uranium ion in UO_2, besides neutron diffraction studies, there are certainly other methods which can give information on the U-O bond type. As many methods as possible should be used to improve our knowledge.

A.S. PANOV: What can you say about the correlation of the ionic and covalent components in non-stoichiometric uranium dioxide? Which bonding is predominant?

H. BLANK: The ionicity of the bonding in UO_2 certainly changes if one goes, for instance, from UO_2 to U_4O_9, but this does not shift the ionicity of U_4O_9 out of the range $0.786 < f_i < 0.911$, which I defined at the beginning of my paper. Thus the use of covalent crystal radii as proposed in the paper is not invalidated by a change in stoichiometry.

With regard to the second part of your question, I should emphasize once more that a high numerical value of the ionic contribution to the bonding ($f_i \simeq 0.88$ for UO_2 or $f_i = 0.839$ for MgO) does not mean that one has to use ionic radii to describe the structure. Even compounds with clearly covalent structures (ZnS, CdSe etc.) may have a high ionic component in the bonding. According to the Phillips-Van Vechten spectroscopy theory of ionicity this means at $f_i = 0.786$ an ionic contribution of 65.6%, since $f_i = C^2/(C^2+E_h^2)$, where C and E_h are the ionic and covalent components, respectively, to the mean gap between valence band and conduction band.

Thus ionic bonding is predominant in the actinide dioxides, but this does not necessarily mean that one has to use ionic radii.

A.S. PANOV: I think that your arguments in favour of considering the covalent model rather than the ionic model are not very well founded. Data on the physical and chemical properties of oxides show the predominance of ionic bonding. I believe that Mr. Catlow's ionic model gives a better description of the properties of uranium dioxide.

H. BLANK: Permit me to list some of the reasons which led me to look for an alternative method of describing oxides with regard to the ionic crystal radii:

(a) In the normal ionic model the frame of the structure in all oxides is invariably constituted by the lattice of the large and soft O^{--} ions, while the small cations are found in interstitial positions. It is difficult to understand why such structures should be able to account for the refractory nature, high elastic constants and hardness of compounds such as ThO_2, UO_2, corundum, etc.

(b) It is also difficult to understand why in the normal ionic model the fluorite-type dioxides of the rare earth and actinide series can be reduced to the C-type sesquioxides, i.e. one quarter of the oxygen ions constituting the structural frame in the usual ionic model is removed, virtually without changing the positions of the cations in the lattice.

(c) The difficulties involved in describing the properties of the doubly charged negative oxygen ion in the crystal lattice have been known for a long time (see Ref.[1] of my paper). Mr. Catlow must also have encountered these difficulties in his calculations. They are further discussed by Levin, Syrikin and Dyatkina (see Ref.[2] of the paper).

(d) In bond calculations it has been found by several authors that it is better to start with the wave functions of the ions only in cases of very high ionicity ($f_i \gtrsim 0.911$); in other cases it is easier to start with the atomic wave functions.

(e) The discussion on covalent atomic and ionic crystal radii contained in the paper by Slater which I quote as Ref.[4] also gave me food for thought.

J.S. de la TIJERA: According to your covalent model, the interstitial sites for oxygen would have to be occupied by uncharged oxygen, while several experiments based on Kröger and Vinck's formalism show doubly ionized oxygen, which doesn't fit in that site. How do you explain this?

H. BLANK: Firstly, I do not deny that there is a large ionic contribution to the bonding in my structural model, and I suggested that the charges on the ions in UO_2 may perhaps be roughly one half of the charges used in the full ionic picture. This should also apply to interstitial oxygen.

You refer in your question to 'experiment', but I do not know exactly which experiments you have in mind. Certainly there is a mass of experimental data which has so far been interpreted on the basis of the ionic model with full charges. It has still to be worked out what other interpretations are possible for these data.

J. de la TIJERA: It is noteworthy that the diffusion mechanisms proposed in your paper are completely at variance with those I suggest, which are based on thermal diffusion data. What is your opinion on this?

H. BLANK: The details of the mechanism of oxygen diffusion in UO_{2+x} and $(U, Pu)O_{2\pm x}$ should be closely related to the interstitial positions which oxygen can occupy in the Willis complexes (see Ref.[19] of the paper). These positions are reproduced both by the calculations Mr. Catlow presented in paper IAEA-SM-190/13 and in a semi-quantitative way by my geometrical considerations. Thus any model for oxygen diffusion should take them into account, even if there are some differences in the details of the jump geometry between the ionic lattice model and the covalent lattice geometry. Since I cannot make any predictions on the energies of migration for oxygen vacancies and interstitials on purely geometrical grounds, and since there is as yet no quantitative model for the lattice potential of my geometrical model, it is not possible to comment on your model on the basis of my paper.

INTERSTITIAL AND SUBSTITUTIONAL METALLIC IMPURITY DIFFUSION IN THE GROUP IV ELEMENTS ZIRCONIUM AND TITANIUM

Ruth TENDLER, E. SANTOS*
Comisión Nacional de Energía Atómica,
Buenos Aires,

J. ABRIATA, C.F. VAROTTO
CNEA — Centro Atómico Bariloche,
San Carlos de Bariloche, Río Negro,
Argentina

Abstract

INTERSTITIAL AND SUBSTITUTIONAL METALLIC IMPURITY DIFFUSION IN THE GROUP IV ELEMENTS ZIRCONIUM AND TITANIUM.

Impurity diffusion measurements of ^{51}Cr in β-Zr, ^{65}Zn in both α-Zr and β-Zr and ^{59}Fe, ^{60}Co and ^{115}Cd in α-Zr are reported, together with self-diffusion measurements of ^{44}Ti in diluted Ti(Ni) alloys. An analysis is made of available impurity diffusion data in zirconium. It is concluded that the interstitial-like or substitutional behaviour of the impurities may be understood by a judicious use of Anthony and Turnbull conditions, atomic-size parameters, features of the corresponding phase diagram, and the Hume-Rothery rules controlling substitutional dissolution. It is shown that the interstitial-like mechanism does not necessarily mean a simple interstitial mechanism.

1. INTRODUCTION

In the last decade the results of many diffusion experiments in which fast diffusion of metals has been found were explained on the basis of the interstitial diffusion mechanism. Anthony and Turnbull [1-4] proposed three basic conditions that need to be satisfied by solute and solvent for this mechanism to be operative: there has to be (i) a polyvalent or very electropositive solvent, (ii) a sufficiently small ion size and (iii) a low solute valence.

The Anthony and Turnbull rules should not necessarily be independent of the classical Hume-Rothery [5] rules for substitutional dissolution. In the authors' opinion, to be advanced in the present article, the first and third of the Anthony and Turnbull rules just given can be interpreted in terms of the Hume-Rothery scheme, and can therefore be considered as independent of the actual model for the non-substitutional fast-diffusing dissolution state of the impurity.

The diffusion behaviour of a series of impurity elements with two Group IV elements, zirconium and titanium, as solvents has been reported, and, when fast diffusion was apparent, their behaviour was explained by the Anthony and Turnbull scheme [6,7].

---

* Present address: Centro Atómico Bariloche, CNEA, San Carlos de Bariloche, Río Negro, Argentina.

In this paper, new experimental diffusion data in titanium and zirconium are presented which, together with previous data, allow of a more general analysis of impurity dissolution and diffusion in zirconium. It would be possible to apply this analysis to other solvents where impurity fast-diffusion was also found.

Two different sets of diffusion experiments were carried out. One consisted of the measurement of impurity diffusion of chromium in β-Zr, zinc in α-Zr and β-Zr, and cadmium, cobalt and iron in α-Zr. These experiments were made to obtain additional information on impurity diffusion in Zr. The second set of experiments were made in order to study the self-diffusion behaviour of an impurity-doped Group IV solvent in which the impurity was a fast diffusing solute.

Though the experimental data analysed in this paper refer mainly to zirconium as solvent, because of availability of materials, the Ti-Ni system was chosen for the second set of experiments. Since the diffusion behaviour of the Ti-Ni and Zr-Ni systems is very similar [8,9], we can extend the conclusions derived from the enhanced diffusion experiments in the Ti-Ni system to the case of zirconium as solvent. The previously reported data indicate that nickel behaves as a fast diffusing solute in both zirconium and titanium solvents and, besides, both systems do satisfy the Anthony and Turnbull conditions for the interstitial diffusion mechanism.

It is known that one way of studying the actual mechanism for impurity diffusion is to measure the solvent self-diffusion modification in dilute alloys for low solute concentrations (less than 2 at.%). The self-diffusion dependence with the solute concentration, c, for dilute alloys may be expressed as:

$$D_{self} = D_{self}(0)\,(1+bc) \tag{1}$$

where, for bcc structures and a vacancy diffusion mechanism, the coefficient b is related to the impurity diffusion correlation factor f_{imp} by [10]:

$$f_{imp} = \frac{D_{imp}}{D_{self}(0)} \; \frac{6\ f_{self}(0)}{b+15+\dfrac{1.53b+9.18}{b+9.06}} \tag{2}$$

$D_{self}(0)$ being the self-diffusion coefficient for the pure solvent, and D_{imp} the impurity diffusion coefficient in the pure solvent. If a different mechanism is assumed, for example the vacancy-interstitial impurity pair originally proposed by Miller [11] for the fcc structure and extended by Dyment et al. [12] to bcc structures, an expression similar to Eq.(1) can be derived. For both structures the coefficient b is given by:

$$b \cong f_{self}(0)\,\frac{D_{imp}}{D_{self}(0)} \tag{3}$$

From the above considerations it can be seen that the study of the concentration dependence of D_{self} may suggest different models for the impurity diffusion mechanism.

2. EXPERIMENTAL PROCEDURE

Impurity diffusion experiments were made using zirconium polycrystals of 99.99 wt% purity of grain size greater than 2 mm diameter. (In a few cases 99.92 wt% zirconium was employed and attention will be drawn to these experiments.) The lapping and measuring techniques as well as the tracer deposition and annealing methods for the experiments on the diffusion of chromium, iron, cobalt and titanium were the same as reported elsewhere [6]. For zinc and cadmium, the diffusion experiments were made starting with these elements in the vapour phase. Two different methods were employed, one suitable for the annealings in the α-Zr phase and the other for the β-Zr phase. The reported conditions [13,14] for gaseous sources in diffusion experiments were taken into account both for instant and infinite source situations. For the measurements in the α-Zr phase, the experimental set-up employed is similar to the one reported in Ref.[14], but the zirconium wafer was enclosed in tantalum foil. For the temperature range and the mass of zinc employed (less than 2×10^{-3} g) all the zinc was in vapour phase. So small a quantity of zinc was necessary in order to minimize the formation of layers of Zn-Zr compounds, as is apparent from the zinc-rich zone of the Zr-Zn phase diagram [15]. In a few cases a previously prepared Zn-Zr alloy containing the active zinc, in face-to-face contact with the zirconium wafer, was employed as tracer source. No significant differences were found between the two methods. For the diffusion experiments in the β-phase, a different approach was employed. Figure 1 is a schematic view of the apparatus with which it was possible to pre-anneal the sample in the high-temperature phase with no tracer transport to it. To start the diffusion experiments, it was only necessary to remove the quartz holder from the cooling system in order to allow tracer evaporation to take place. A careful furnace calibration ensured control of the pre-annealing and diffusion annealing temperatures to ±1°C. In this class of experiments the mass of solute was less than 10^{-3} g. Tests were also made to ensure that no tracer transport to the

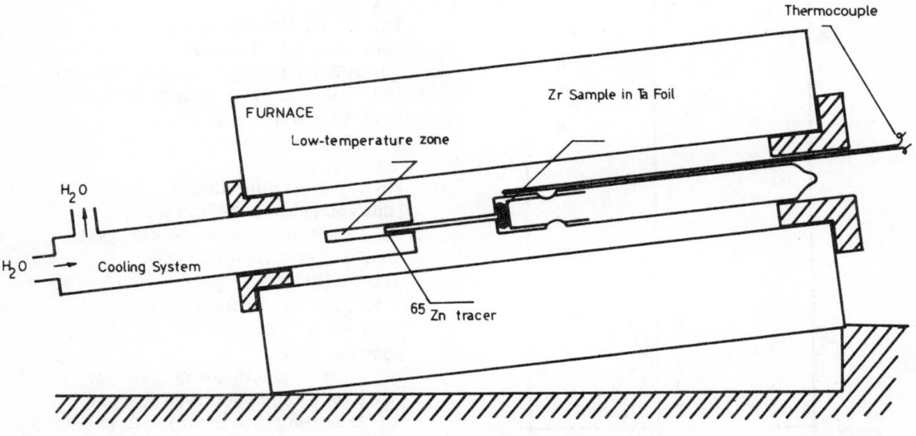

FIG. 1. Schematic view of the experimental arrangement for the pre-annealing and diffusion annealings in β-Zr, showing gaseous tracer source.

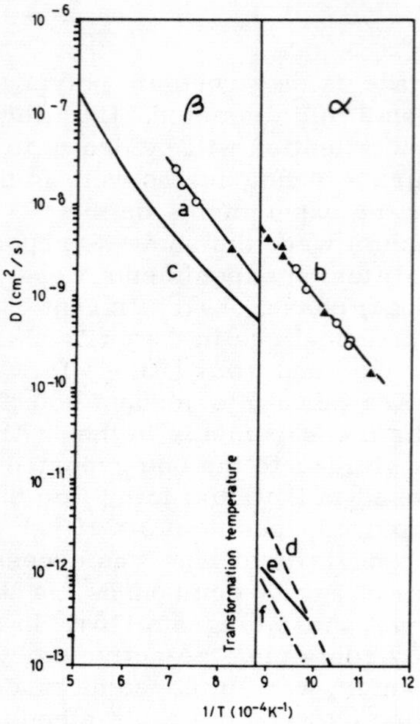

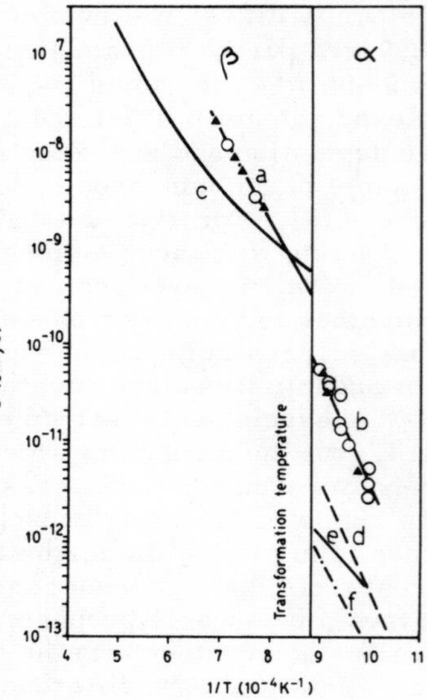

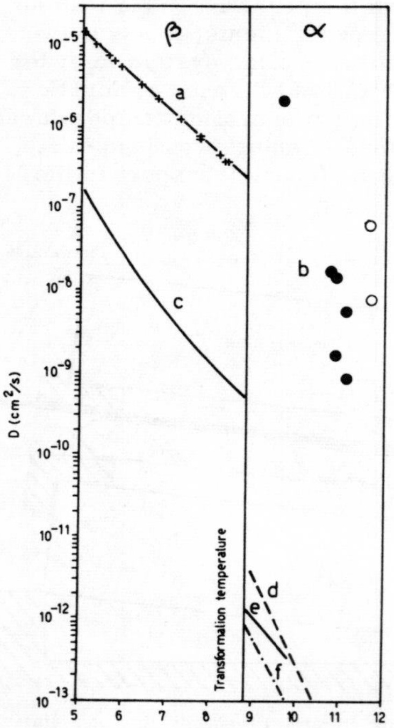

TOP LEFT

FIG. 2. (i) ^{51}Cr diffusion in: (a) β-Zr [this work] and (b) in α-Zr [6]. (ii) Zr self-diffusion in: (c) β-Zr [34]; (d) α-Zr [22]; (e) α-Zr [20]; (f) α-Zr [21].
○ 99.92 wt% Zr purity
▲ 99.999 wt% Zr purity

TOP RIGHT

FIG. 3. (i) ^{65}Zn diffusion in: (a) β-Zr [this work]; (b) α-Zr [this work].
(ii) Zr self-diffusion in: (c) β-Zr [34]; (d) α-Zr [22]; (e) α-Zr [20]; (f) α-Zr [21].
○ 99.92 wt% Zr purity
▲ 99.999 wt% Zr purity

BOTTOM LEFT

FIG. 4. (i) ^{60}Co diffusion in: (a) β-Zr [23]; (b) α-Zr: ● [this work], o [8].
(ii) Zr self-diffusion in: (c) β-Zr [34]; (d) α-Zr [22]; (e) α-Zr [20], (f) α-Zr [21].

zirconium wafer could take place during the pre-annealing treatment. For all the experiments (in both the α and the β-phases) a careful autoradiographic analysis of the sample showed that there was a good homogeneous deposit on the surfaces.

For the titanium self-diffusion experiments the Ti-Ni alloys were prepared (starting with titanium of 99.998 wt% and nickel of 99.99 wt% purity) by the electron beam technique in a vacuum better than 5×10^{-7} torr. The alloys were then homogenized at 1200°C for a week in a sealed quartz tube under a high-purity argon atmosphere. The samples were wrapped in tantalum foil to avoid any contact with the silica walls. Zirconium turnings were used as a getter. The sample homogeneity was controlled by an electron microprobe. Six different alloys were prepared, with nickel contents of 0.41, 0.65, 0.92, 1.18, 1.64 and 3.29 at.%. Two diffusion annealings (in the β-Ti phase) were made, at 911°C and 1060°C. In order to avoid errors, all the samples, including a pure titanium wafer, were annealed at one time. The temperature differences between the samples were always less than 2°C and the temperature regulation was better than 0.5°C. The experimental error in the titanium self-diffusion values was estimated to be less than 5%. It is important to note that, while nickel in titanium forms solid solutions up to 12.3 at.% of nickel in the bcc β-Ti phase, it has a very low solubility in the Ti hcp phase (less than 0.3 at.%). The grain sizes of the Ni-Ti alloys ranged from 3 to 4 mm diameter.

3. EXPERIMENTAL RESULTS

3.1. ^{51}Cr diffusion in β-Zr

Chromium-51 diffusion in β-Zr was measured over the temperature range from 936 to 1136°C. No grain boundary contribution was apparent from the penetration profiles. The temperature dependence of ^{51}Cr diffusion in β-Zr may be expressed as:

$$D_{\beta\text{-Zr}}^{\text{Cr-51}}\,(\text{cm}^2/\text{s}) = 2.07 \times 10^{-3}\ \exp\left(\frac{-31\,900}{RT}\right)$$

and is plotted in Fig.2. These results, while differing by 10 kcal from those reported in Ref.[16], are almost equal to the data reported in Ref.[17]. It is important to notice that, as with ^{54}Mn diffusion in zirconium [6], the diffusion parameters are almost insensible to the phase transition, indicating that the same diffusion mechanism operates in both zirconium phases, over the temperature range studied.

3.2. ^{65}Zn diffusion in α-Zr and β-Zr

The diffusion measurements in α-Zr were made over the temperature range from 729 to 841°C. After an initial zone (less than 30 μm deep) disturbed by possible oxide and compound formation [18], good penetration profiles were obtained (up to 300 μm) from where the bulk diffusion coefficients could be computed. For those cases in which the Zn-Zr compound was used as tracer source, no anomalous initial zone appeared.

Many careful and different experiments were made using different stabilization procedures, grain sizes and zirconium purities; no meaningful differences were found in the results [9]. No grain boundary contribution to the diffusion was apparent from the penetration profiles.

The temperature dependence of the ^{65}Zn diffusion coefficient in the α-Zr phase may be expressed as:

$$D^{Zn\text{-}65}_{\alpha\text{-}Zr}(cm^2/s) = 1.65 \exp\left(-\frac{53\,500}{RT}\right)$$

and it is plotted in Fig.3.

^{65}Zn diffusion measurements in β-Zr were made over the temperature range from 1000 to 1180°C. The penetration profiles showed no initial anomalous zone for the full penetration distances, which ranged from 300 to 400 μm. No grain boundary contribution to the diffusion was apparent here either. The experimental results were neither affected by zirconium purity in the range studied (zirconium wafers of 99.99 wt% and 99.92 wt% purity were employed), nor by an increase in the pre-annealing time and temperature. The temperature dependence of the ^{65}Zn diffusion coefficient in β-Zr may be expressed as:

$$D^{Zn\text{-}65}_{\beta\text{-}Zr}(cm^2/s) = 8.2 \times 10^{-2} \exp\left(-\frac{43\,900}{RT}\right)$$

and it is plotted in Fig.3, also.

There are some aspects that differentiate zinc diffusion in zirconium from both fast and slow diffusing impurities. For example:

(i) The ratio of the zinc diffusion coefficient in the β-phase to the one in the α-phase at the phase transition temperature is about 3.5, while the ratio for slowly diffusing silver is 24.5 [19].

(ii) The diffusion activation energy for zinc in α-Zr is 26.5 kcal larger than the activation energy for zirconium self-diffusion as reported by Dyment et al. [20], and it is almost coincident with the values reported by Fulbacher [21] and by Lyashenko et al. [22]. This activation energy behaviour compares with the behaviour of silver [19].

(iii) The phase transition does affect the zinc diffusion activation energy, it being 9.6 kcal larger in α-Zr than in β-Zr.

3.3. ^{59}Fe diffusion in α-Zr

Two experiments were made to study ^{59}Fe diffusion in α-Zr. The values obtained were: 3.59×10^{-9} cm^2/s and 5.40×10^{-9} cm^2/s at 618°C and 2.9×10^{-7} cm^2/s at 823°C. The value at the higher temperature is coincident with a previously reported value by Hood et al. [8]. Since some peculiar segregation effects appeared, additional measurements are under way.

3.4. ^{60}Co diffusion in α-Zr

^{60}Co diffusion was measured over the temperature range of 624 to 803°C. The first experiments immediately showed that this element was

TABLE I. COBALT DIFFUSION COEFFICIENTS IN α-Zr

Sample	Annealing temperature (°C)	$D^{Co-60}_{\alpha-Zr}$ (cm^2/s)
Co-5	655.5	1.6×10^{-8}
Co-4	644	1.6×10^{-9} 1.4×10^{-8}
Co-3	624	8.4×10^{-10} 5.5×10^{-9}
Co-8	761	2.1×10^{-6}

a very fast diffusing one. In effect, two experiments, at 803°C for 30 hours and at 653.5°C for 50 hours, resulted in homogeneous distribution of the tracer in the zirconium wafers. Many attempts to measure ^{60}Co diffusion at temperatures above 770°C failed, since the times that would have been required were very short, which in turn would have made it very difficult to evaluate the diffusion coefficients. Four experiments were made and the results are given in Table I. The corresponding penetration profiles [9] in the cases of experiments named Co-5 and Co-8 (after an anomalous initial zone of about 30 μm in depth, which may be due to some oxide surface effect) showed a good Gaussian distribution behaviour. In the Co-3 and Co-4 experiments, after the initial short anomalous zone, a curved zone follows. This second zone may well be resolved into two Gaussian zones of different slopes [9]. For these two experiments, after lapping, autoradiographic and metallographic analyses of the lapped surfaces were made. The micrographs were obtained, using polarized light, before and after a slight chemical polishing. An inhomogeneous blackening of the surface resulted in the autoradiographs, indicating that the diffusing impurity concentrates in certain zones of the wafer. A comparison of the autoradiographs with the micrographs showed a one-to-one correspondence of the blackened and non-blackened zones with crystal grains that reflected polarized light differently, indicating that different grains had different crystal orientations. This seems to indicate an anisotropy effect; hence, it may be concluded that, at large distances from the initial surface, the sample activity arises almost only from those grains with the more favourable orientation for diffusion. The measured values for ^{60}Co diffusion in α-Zr as a function of temperature are plotted in Fig.4, together with the reported values at 583°C by Hood et al. [8] measured in zirconium single crystals, and the values reported by Kidson et al. [23] for Co diffusion in β-Zr. Assuming that diffusion along different crystal grains does not correlate, it is possible to separate, using a simple mathematical analysis [9] on the penetration profiles, the two diffusion coefficients. In Table II, the ^{60}Co diffusion coefficients obtained in this way for the Co-4 and Co-5 experiments are presented (D_1 refers to the less favourable orientation and D_2 to the more favourable). In the same table are included the values reported by Hood et al. [8]. It is interesting to observe that the D_2/D_1 ratio obtained

TABLE II. FAST AND SLOW DIFFUSION COMPONENTS OF ^{60}Co DIFFUSION IN α-Zr

Sample	Temperature (°C)	D_2 (cm^2/s)	D_1 (cm^2/s)	D_2/D_1	Ref.
Co-4	644	1.4×10^{-8}	1.6×10^{-9}	8.7	this work
Co-3	624	5.5×10^{-9}	8.4×10^{-10}	6.5	this work
Single crystal	583	7.4×10^{-8}	5.8×10^{-9}	7.8	[8]

TABLE III. ^{44}Ti SELF-DIFFUSION COEFFICIENTS IN Ti(Ni) ALLOYS

Alloy composition (at.% Ni)	D for annealing at 911°C (cm^2/s)	D for annealing at 1060°C (cm^2/s)	Ref.
0.00	3.84×10^{-10}	$2.16 - 10^{-9}$	[24]
0.00	6.90×10^{-10}	3.16×10^{-9}	[25]
0.00	9.59×10^{-10}	3.08×10^{-9}	this work
0.41	1.01×10^{-9}	3.61×10^{-9}	↓
0.65	9.98×10^{-10}	3.14×10^{-9}	
0.92	7.89×10^{-10}	2.09×10^{-9}	
1.18	1.15×10^{-9}	3.52×10^{-9}	
1.64	9.88×10^{-10}	3.88×10^{-9}	
3.29	2.29×10^{-9}	8.66×10^{-9}	

in the present work is similar to the ratio found by Hood. The dispersion of the experimental points in Fig.4 should be attributed to the anisotropy. This may be considered additional support for the hypothesis that cobalt diffuses, at least partially, as an interstitial.

3.5. ^{115}Cd diffusion in α-Zr

Two diffusion experiments were made with ^{115}Cd in α-Zr. The values obtained were 2.9×10^{-13} cm^2/s at 813.5°C and 5.4×10^{-12} cm^2/s at 839°C.

Unfortunately, many difficulties arose in the course of the experiments with this impurity, making the work very time consuming. In spite of the scatter observed for the two measured values, it can be stated that cadmium appears as a slow diffusing element in α-Zr, and may be compared with silver [19].

3.6. ^{44}Ti self-diffusion in nickel-doped titanium alloys

The measured values for ^{44}Ti self-diffusion in nickel-doped titanium alloys at 911 and 1060°C (β-Ti phase) are indicated in Table III and plotted

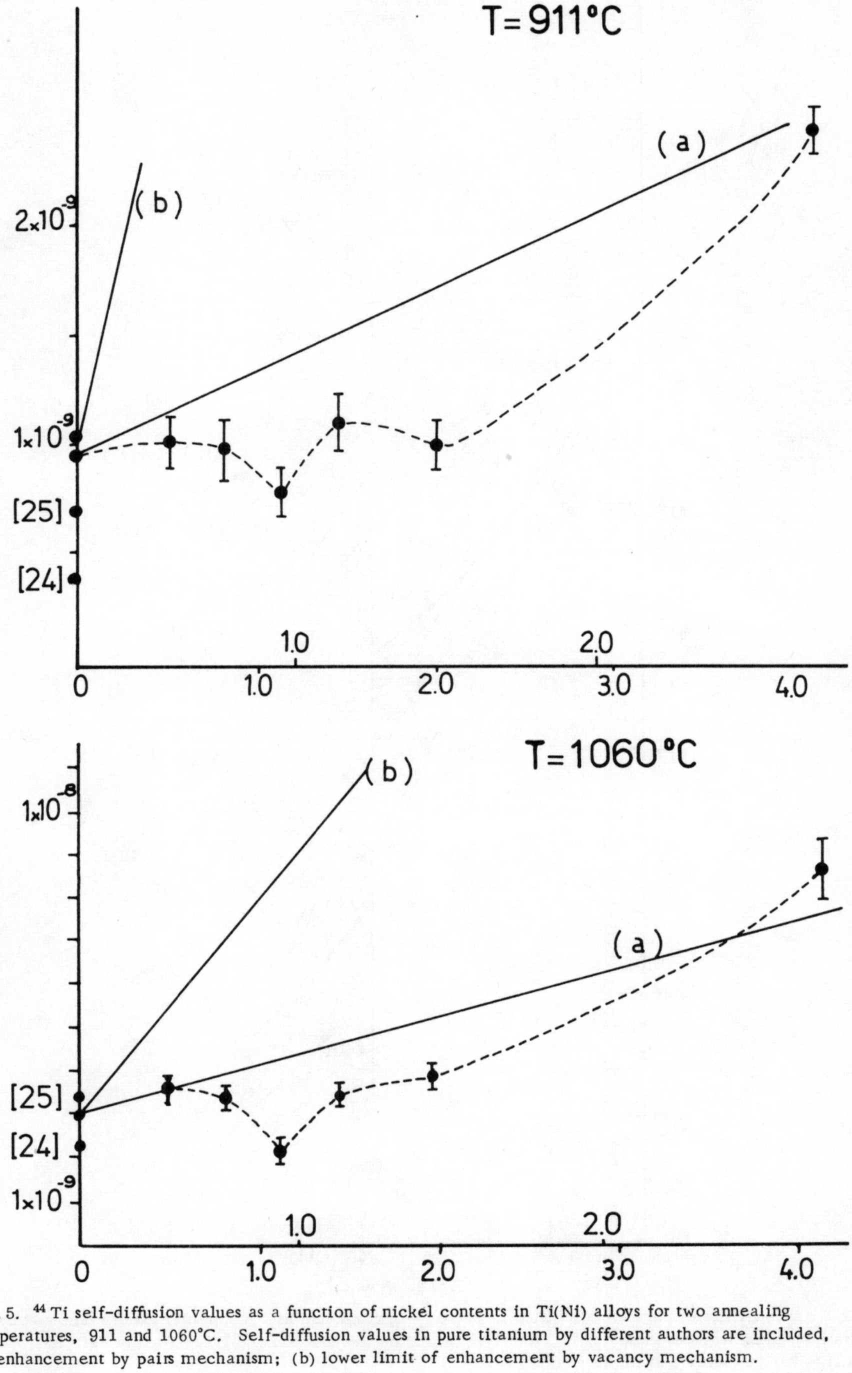

FIG. 5. [44]Ti self-diffusion values as a function of nickel contents in Ti(Ni) alloys for two annealing temperatures, 911 and 1060°C. Self-diffusion values in pure titanium by different authors are included, (a) enhancement by pairs mechanism; (b) lower limit of enhancement by vacancy mechanism.

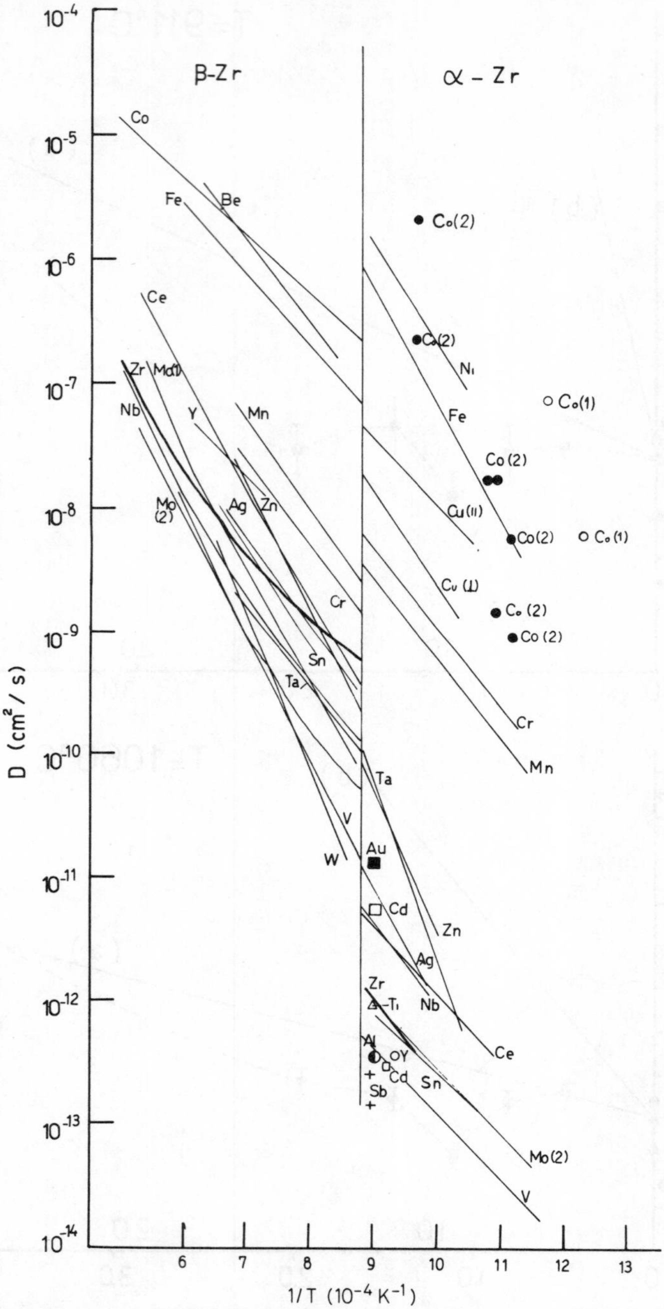

FIG. 6. Diffusivities of the elements listed in Table IV in zirconium as a function of inverse temperature. Data from the references indicated in Table IV. (Mo(1) corresponds to Ref. [17]; Mo(2) to Ref. [35]; Co(1) to Ref. [8] and Co(2) to this work).

in Fig.5 as a function of nickel contents. No indication of grain boundary diffusion was found in the penetration profiles, in agreement with the existence of large grains in the samples. The resulting dependence of the titanium self-diffusion coefficient on nickel concentration is not a simple one.

4. DISCUSSION

In this section, the experimental results presented above are discussed after assembling them together with existing data for other metallic solutes. The aim is to look for trends in the diffusion behaviour in zirconium of these solutes as a function of their intrinsic physical properties.

Experiments indicate that, in primary substitutional metallic alloys, the diffusion coefficients associated with the substitutional impurities are akin to the self-diffusion value of the matrix [26]. For the case of zirconium, the existence of various metallic solutes with D_{imp} values which are several orders of magnitude higher than D_{self} (see Fig.6) has suggested [6-9] an essentially different bulk mechanism for impurity diffusion. The basic idea is that at least one additional non-substitutional state should be available in zirconium for these 'faster' impurities. A similar situation exists in other metallic systems, where the non-substitutional fast diffusing state has been conjectured to be closely related to the interstitial configuration [1-4]. Describing the diffusion behaviour of the metallic solutes listed in Table IV by assuming that two types of state are available for the impurities in zirconium, namely, substitutional and interstitial-like states, one has [4]:

$$D_{imp} = X_s D_s + X_i D_i \tag{4}$$

where X_sD_s and X_iD_i are the fractions of solute atoms and the respective diffusion coefficients for the substitutional and interstitial states, respectively. X_s and X_i are considered to be mutual equilibrium values, as determined simultaneously by the properties of the given solute and of the zirconium matrix. As a rule, X_i is less than X_s and D_i is some orders of magnitude greater than D_s.

4.1. Anthony and Turnbull conditions for interstitial dissolution in zirconium

In one way or another, most authors have used the Anthony and Turnbull conditions to discuss fast diffusing metallic impurities in metals. We are going to discuss now how each of these conditions applies to the case of zirconium doped with the metallic impurities listed in Table IV.

Condition 1. A polyvalent or very electropositive solvent

Various values have been suggested for the valence of zirconium in the metallic state. They range from 0 to 4, depending on the particular property discussed and on how the electrons are thought to divide between tightly bound d-type and s-type electron states [27]. Recently a calculation of the bulk moduli (a quantity rather sensitive to valence) for a number of metals has been reported [28], and zirconium and titanium do fit the

TABLE IV. IMPURITY DIFFUSION IN ZIRCONIUM: DATA FOR METALLIC IMPURITIES

Solute	Atomic radius (Å)	Ionic radius (Å)	Electro-negativity (Pauling)	$\rho^{\alpha} = \frac{D_{imp}^{\alpha\text{-}Zr}}{D_{self}^{Zr\text{-}95}}$ (Temp. range) Ref.	$\rho^{\beta} = \frac{D_{imp}^{\beta\text{-}Zr}}{D_{self}^{Zr\text{-}95}}$ (Temp. range) Ref.	Maximum solubility in α-Zr at.% (Temp. range) Ref.	Maximum solubility in β-Zr at.% (Temp. range) Ref.
Ag	1.44	1.25 (+1)	1.9	5 - 9 (764-847°C) [19]	0.8 - 1.3 (947 - 1200°C) [19]	1.1 (821°C) [36]	20 (1191°C) [36]
Al	1.43	0.50 (+3)	1.5	0.4 (835°C) [7]	- - -	11 (940°C) [32]	26 (1350°C) [32]
Au	1.44	1.37 (+1)	2.4	16.7 (830°C) [29]	- - -		
Be	1.12	0.31 (+2)	1.5	-	270 - 300 (915 - 1300°C) [42]	<1 (800°C) [36]	<3 (965°C) [36]
Ce	1.81	1.01 (+4)	1.1	4 (763 - 830°C) [44]	0.9 - 6.5 (910 - 1600°C) [44]	6 (850°C) [39]	- - -
Cd	1.54	0.97 (+2)	1.7	0.4 - 6.9 (814 - 839°C) [this work]	- - -	17.5 (905°C) [36]	21 (1050°C) [36]
Co	1.25	0.78 (+2)	1.8	5.3×10^6 - 2.6×10^6 (761 - 830°C) [this work, 8]	96 - 407 (950 - 1704°C) [23]	0.3 (833°C) [38]	3.4 (981°C) [38]

Cr	1.27	0.69 (+3)	1.6	4200 - 5400 (763 - 832°C) [41]	3 - 6 (936 - 1136°C) [this work]	≪1 (830°C) [38]	7.5 (1270°C) [38]
Cu	1.28	0.96 (+1)	1.9	1.5×10^4 (I) $- 4.4 \times 10^4$ (II) (835°C) [29]	- - - 	0.3 (822°C) [32]	5.3 (995°C) [32]
Fe	1.26	0.76 (+2)	1.8	$3.7 \times 10^5 - 8.9 \times 10^5$ (823 - 840°C) [this work]	72 - 150 (890 - 1100°C) [40, 17]	0.03 (795°C) [36]	6.8 (947°C) [36]
Mo	1.39	0.68 (+4)	1.8	0.8 - 1.1 (750 - 840°C) [35]	0.08 - 2.4 (900 - 1600°C) [17 - 35]	0.2 (780°C) [32]	21 (1520°C) [32]
Mn	1.26	0.80 (+2)	1.5	2100 - 3350 (750 - 838°C) [6]	6 - 10 (942 - 1146°C) [6]	≪1 (795°C) [32]	10.2 (1135°C) [32]
Nb	1.46	0.70 (+5)	1.6	2.9 - 4.4 (795 - 857°C) [20]	0.3 - 1.1 (911 - 1758°C) [34]	2 (590°C) [37]	complete miscibility [32]
Ni	1.24	0.78 (+2)	1.8	10^6 (800 - 830°C) [29]	- - -	≪1 (808°C) [32]	2.9 (961°C) [32]
Sb	1.59	0.62 (+5)	1.9	0.1 - 0.2 (847°C) [7]	- - -	0.6 (875°C) [36]	4.4 (1300°C) [36]

Table IV. (cont.)

Solute	Atomic radius (Å)	Ionic radius (Å)	Electro-negativity (Pauling)	$\rho^{\alpha} = \frac{D_{imp}^{\alpha-Zr}}{D_{self}^{Zr-95}}$ (Temp. range) Ref.	$\rho^{\beta} = \frac{D_{imp}^{\beta-Zr}}{D_{self}^{Zr-95}}$ (Temp. range) Ref.	Maximum solubility in α-Zr at.% (Temp. range) Ref.	Maximum solubility in β-Zr at.% (Temp. range) Ref.
Sn	1.62	0.71 (+4)	1.8	1 - 1.1 (763 - 820°C) [46]	0.5 - 1.1 (950 - 1200°C) [47]	7 (980°C) [32]	17 (1590°C) [32]
Ta	1.46	0.73 (+5)	1.5	7 - 24 (750 - 830°C) [45]	0.2 - 0.3 (900 - 1200°C) [45]	1 (800°C) [36]	complete miscibility [36]
Ti	1.47	0.90 (+2)	1.5	0.9 (843°C) [7]	- - -	complete miscibility [32]	complete miscibility [32]
V	1.34	0.74 (+3)	1.6	0.4 (750 - 850°C) [44]	0.03 - 0.5 (900 - 1400°C) [44]	<0.5 (777°C) [32]	16.5 (1230°C) [32]
W	1.39	0.64 (+4)	1.7	- - -	0.02 - 0.5 (900 - 1250°C) [17]	<0.2 (860°C) [32]	4 (1660°C) [32]
Y	1.80	0.93 (+3)	1.3	0.5 (800°C) [43]	4.6 - 4 (1100 - 1335°C) [43]	0.6 (886°C) [36]	3.9 (1363°C) [36]
Zn	1.38	0.74 (+2)	1.6	21 - 54 (763 - 830°C) [this work]	0.5 - 4.5 (963 - 1180°C) [this work]	1 (750°C) [36]	20.1 (1015°C) [36]
Zr	1.60	0.80 (+4)	1.4	1 [20]	1 [34]	-	-

resulting general scheme when a valence close to four is assigned to them. On the other hand, zirconium is a rather electropositive element, and it can be seen that the solutes listed in Table IV (with the exception of yttrium and cerium, which behave substitutionally) are more electronegative than zirconium. Therefore, we consider that, with the possible exception of yttrium and cerium, zirconium satisfies Condition 1 for interstitial dissolution and diffusion for all the solutes listed in Table IV. Since several of the solutes clearly behave as substitutional impurities, the remaining conditions are then of fundamental importance in our discussion of zirconium.

Condition 2. A sufficiently small ion size

The ionic radius for Zr^{4+} is 0.80 Å. Hence, tetrahedral and octahedral interstitial sites in α-Zr can be occupied with little or no overlap between ion cores by solutes whose effective ionic radii are less than 1.16 and 1.46 Å. For β-Zr the corresponding numbers are 1.27 and 1.05 Å for the same type of sites. In Table IV we have listed the ionic radius and the corresponding ionization state for the solutes under consideration. For Ag^+ and Au^+, the ionic radii are 1.26 and 1.37 Å, respectively. Thus, quite reasonably, silver and gold are predicted by Condition 2 to diffuse substitutionally in zirconium, since strong ion core repulsion with this solvent will drastically reduce the ratio X_i/X_s. Recent detailed experimental results have proved that the diffusion behaviour of silver in α-Zr and β-Zr is, in fact, very similar to the self-diffusion behaviour of zirconium itself [19], indicating the substitutional nature of this solute. Also, the existing experimental data for gold confirm this solute to be substitutional in zirconium [29]. These results support the proposition that if Condition 2 is not satisfied then the solute is placed at substitutional sites in the solvent lattice [4]. Therefore, with the exception of silver and gold, all the other metallic solutes of Table IV do not satisfy Condition 2, and in the Anthony and Turnbull scheme Condition 3 should finally decide the type of diffusion behaviour.

Condition 3. A low solute valence

Following an analysis similar to Hood and Schultz [7], Figs 7 and 8 are plots of the ratio $\rho = (D_{imp}/D_{self}^{Zn-95})$ (at the same temperature) versus valence for the following metallic solutes: copper, zinc, cadmium, beryllium, aluminium, tin and antimony; the transition elements iron, cobalt and nickel, all of them with the effective valences as proposed by Barrett and Massalski [30]. For titanium we choose a valence of four [28]. The values of ρ^α and ρ^β are given in Table IV, together with the temperature range for which they have been calculated. We used for this calculation the zirconium self-diffusion values in α-Zr as measured in our laboratory by Dyment and Libanati [20].

α-Zr presents a defined overall trend: ρ^α values tend to decrease with increasing solute valence. The general trend is also indicated in Fig.7, with a broken line. It can be seen that, for valence over three, the solute exhibits substitutional behaviour. In this region ρ^α has a value of around unity. For valence equal to or less than unity, ρ^α is over 10^3, and interstitial behaviour is expected to become the dominant. For the solutes situated in this region, typical interstitial characteristics are observed, i.e. faster solute diffusion and a very slight effect of the α-to-β phase transition on

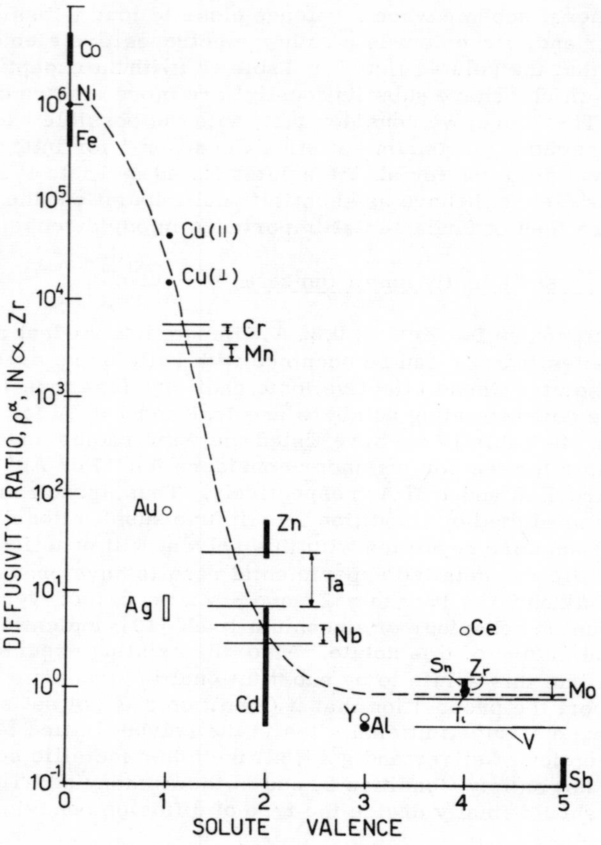

FIG. 7. Diffusivity ratio $D_{imp}^{\alpha-Zr}/D_{self}^{Zr-95}$ of various elements in α-Zr as a function of solute valence. Data from Table IV.

the impurity diffusion as compared with both self-diffusion and with the solutes situated in the substitutional region. Anisotropic diffusion behaviour is also observed, as in the case of cobalt. For valences ranging from one to three, an intermediate situation can exist. Divalent zinc is an example of such intermediate behaviour, i.e. between the substitutional and interstitial. In effect, zinc verifies the first and second of the Anthony and Turnbull conditions, and D_{Zn} is also rather insensitive (as compared with zirconium) to the α-to-β phase transformation. However, ρ^α for zinc is only around 21 to 54, which would suggest substitutional type of diffusion behaviour. Since no significant curvature is observed in the Arrhenius plot, as would be expected for a combined diffusion mechanism, it is, therefore, possible that, for zinc, a more complex diffusion mechanism operates (such as vacancy-interstitial pairs, suggested by Miller [11] for the diffusion of cadmium in lead).

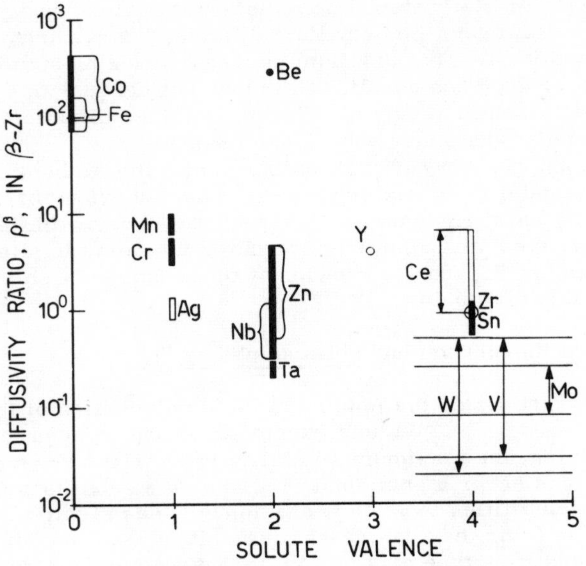

FIG. 8. Diffusivity ratio $D_{imp}^{\beta-Zr}/D_{self}^{Zr-95}$ of various elements in β-Zr as a function of assigned solvent valence. Data from Table IV.

Since the valences for the transition elements manganese, chromium, vanadium, niobium, tantalum and molybdenum are not well defined, we have used the trend of ρ^{α} versus valence as shown in Fig.7 to ascribe to them a valence in zirconium. In this way, we obtained reasonable valences: near unity for manganese and chromium, near two for tantalum and niobium, and over 3 for molybdenum and vanadium. We see that if D_{imp} is measured for another solute of unknown valence and of a small enough ionic radius, the same procedure can be used in order to assign to it a valence in zirconium, which is a quantity of general physical interest.

For comparison, we have also plotted in Fig.7 ρ^{α} values for gold, silver, yttrium and cerium. As expected, gold and silver depart significantly from the behaviour associated with the Anthony and Turnbull Condition 3.

In β-Zr, the variation in diffusion behaviour with valence is less defined than in α-Zr. However, it can be seen in Fig.8 that a general trend similar to that in α-Zr can be suggested:

$$\rho^{\beta} \approxeq 10^2 \text{ for Fe and Co} > \rho^{\beta} \approxeq 10 \text{ for Mn and Cr} > \rho^{\beta} \approxeq 1 \text{ for Zn, Nb and Ta} > \rho^{\beta} \approxeq 10^{-1}/10^{-2} \text{ for Mo, V and W}$$

The reason for such low values of ρ^{β} is not clear, but they can probably be associated with the high values in β-Zr and with the smaller interstitial holes of the bcc structure. A more significant item of data in Fig.8 is the unexpectedly high value of ρ^{β} for beryllium. This element satisfies Conditions 1 and 2 of Anthony and Turnbull, but since it has a normal chemical valence of two it should be an intermediate case, such as, for example, zinc. However, the beryllium diffusion behaviour in β-Zr should lead to the classification of beryllium as a fast diffusing solute in zirconium, as is the case for iron and cobalt. This does not agree with the diffusion description just given in terms of valence. Further diffusion experiments to clarify this point are under way in our laboratory.

In conclusion, we can say that, on the whole, the Anthony and Turnbull conditions are useful in describing the diffusion behaviour of highly diluted metallic impurities in zirconium. However, they are of limited application when the effective solute valence is not known a priori; an alternative way of assessing X_i/X_s from a knowledge of the impurity solid-solubility behaviour is discussed below.

4.2. Interstitial dissolution and solid solubility

In Table IV are listed the maximum solid solubilities of the metallic solutes in α-Zr and β-Zr. Without exception, it can be seen that, for the interstitial solutes, the maximum solubility in α-Zr is very small (below 1%), while in β-Zr it is never higher than 11%. For the substitutional solutes the maximum solubilities in α-Zr are in many cases significantly higher than 1%, while in β-Zr they can reach 100%. It follows that, if there is substantial interstitial-type dissolution, this always correlates with a significant tendency to having a relatively low solid solubility. This fact suggests a relationship between the Anthony and Turnbull conditions and the well-known Hume-Rothery's rules on metal-metal solid solubility. (Alternatively, although the same correlation has been also noted in other systems with solvents other than zirconium, it has been recently claimed that no direct link exists between interstitial dissolution and low solubility [31].) Let us recall that at very high dilution [4]

$$\frac{X_i}{X_s} \alpha \exp\left(-\frac{h_i - h_s}{kT}\right) \exp\left(\frac{s_i - s_s}{K}\right) \tag{5}$$

where $h_i - h_s$ is the difference in partial enthalpies between an interstitial and substitutional solute atom in the zirconium matrix, $s_i - s_s$ is the corresponding difference in non-configurational partial entropies, and k is the Boltzmann's constant.

The radio X_i/X_s compared with D_s/D_i decides the type of diffusion behaviour to be observed. For the metallic solutes considered here it is quite reasonable to assume that $h_i > h_s$. But according to Eq.(5), the higher h_s, the more favoured is the interstitial state of the impurity: in turn, the higher h_s, the lower the possibility of extended solid solubility. Therefore, when the low solid solubility of an impurity is due to a high h_s, there is also a greater tendency toward interstitial dissolution, and vice-versa. In the general case, however, low solubility does not necessarily imply interstitial behaviour. For example, this is the case when primary solid solubility is limited by Hume-Rothery's 'electronegativity difference effect'. This occurs when the electronegativity difference between the elementary components results in a very stable intermediate compound which severely limits primary solid solubility simply through the usual free-energy curve-tangent principle.

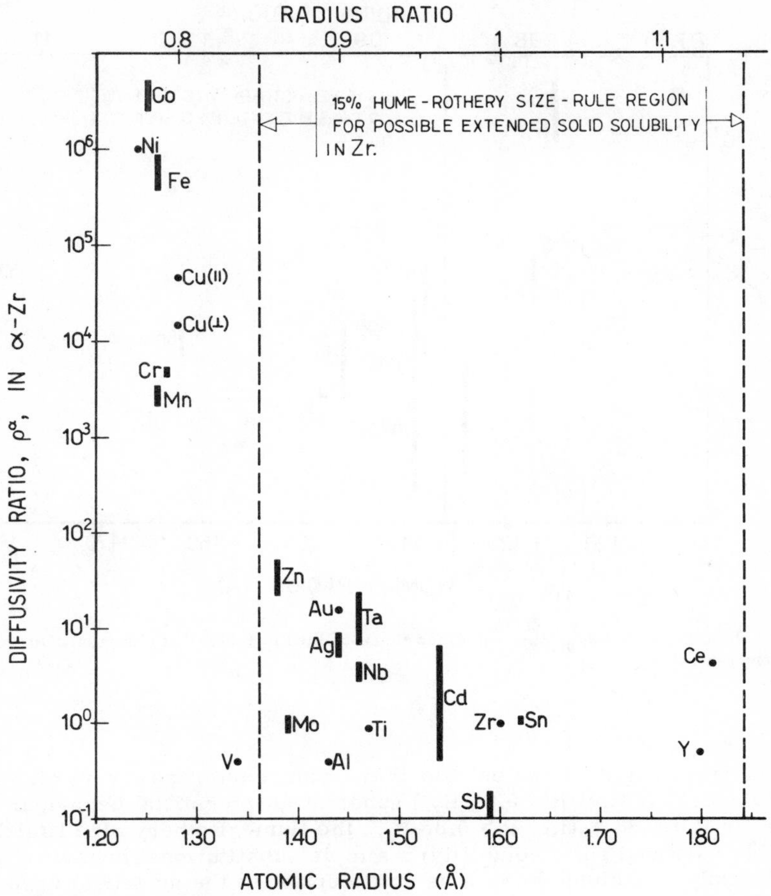

FIG. 9. Diffusivity ratio D_{imp}/D_{self}^{Zr-95} of all the elements listed in Table IV as a function of atomic radius. Data from Table IV.

From the above, we recognize that the Hume-Rothery '15% atomic-size rules' [30] must be particularly useful when assessing the ratio X_i/X_s. When the atomic radius difference between solute and solvent exceeds 15%, this rule implies a large h_s — and a very restricted solid solubility results. Thus, in this context a significant dependence of the diffusion behaviour (interstitial-like or substitutional) of the solute on its atomic radius is to be expected. Figures 9 and 10 are plots of ρ versus the atomic radius, R, for the different impurities given in Table IV in α-Zr and β-Zr, respectively. For α-Zr, a defined separation between substitutional and interstitial behaviour can be established. This scheme has the advantage over the valence-dependence one, discussed previously, that the atomic radii are known a priori from diffusion measurements.

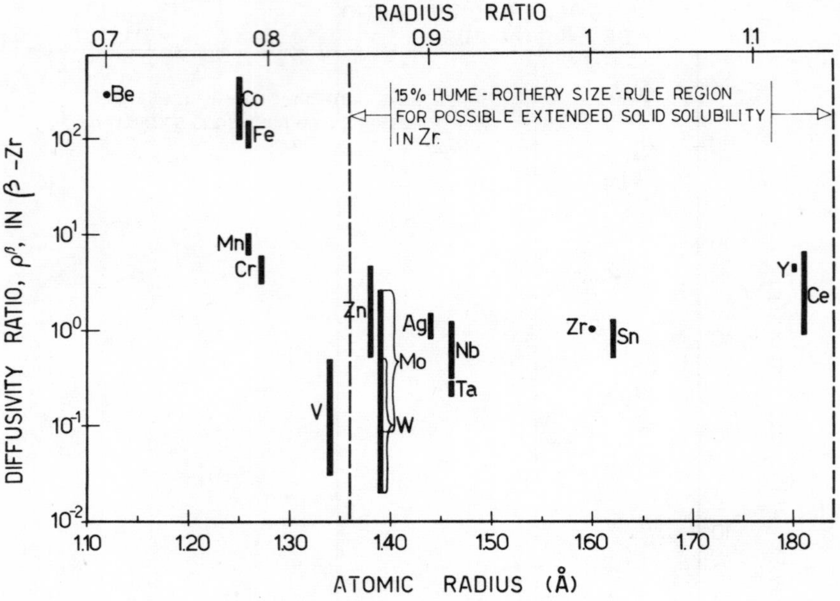

FIG. 10. Diffusivity ratio D_{imp}/D_{self}^{Zr-95} of all the elements listed in Table IV as a function of atomic radius. Data from Table IV.

Judging from Fig.9, it can be said that, in zirconium and at very high dilutions, when $R < 1.30$ Å (i.e. $R < 0.81\ R_{Zr}$) substantial interstitial behaviour occurs; while for $R > 1.36$ Å (i.e. $R > 0.85\ R_{Zr}$, the Hume-Rothery size limit for possible extended solid solubility) a simple substitutional behaviour is found. Only vanadium appears as an exception to the argument based on atomic size: the atomic radius of vanadium is small enough to give a low mutual solid solubility with zirconium and, hence, a tendency to show interstitial behaviour is to be expected. However, looking at the phase diagram of V-Zr [32], the relatively high solubility of zirconium in vanadium (>5% at 1300°C) indicates that the V-Zr system can be considered as an exception to the 15% atomic-size rule and, therefore, to our arguments relating atomic radius to interstitial dissolution.

Generalizing, it can be suggested that in very dilute metal-metal alloys a significant interstitial dissolution is expected when the Hume-Rothery's 15% size rule prohibits extended solid solubility and when, at the same time, the atomic radius of the impurity is smaller than that of the solvent. (When the solute atomic radius is greater than that of the solvent, one can reasonably expect that low solubility does not necessarily imply interstitial behaviour.) It is interesting to note that Anthony [4] had already pointed out that, on the whole, the new metal-metal interstitial alloys have an atomic-size range which appears to start somewhere above 0.59 of the radius of the solvent, and ends with a slight overlap into the size range allowing extensive substitutional alloy formation. Accordingly, Hagg's rule for extensive interstitial dissolution ($R_{solute} \leq 0.59\ R_{solvent}$) cannot be extended to the case of extremely dilute solutions.

Obviously, within these size criteria, the greater the atomic radius of the solvent, the greater the number of metallic elements which could dissolve interstitially.

It must also be pointed out that the level of independence or otherwise of the various criteria based on size or valence of the metallic solute has not been clarified. For example, an empirical correlation has been proposed which relates valence and atomic radius [33].

As was the case using the valence scheme, the variation in solute behaviour with R is less defined for β-Zr (Fig.10) than for α-Zr, although a general trend similar to that in the α-phase can be ascertained. The main difference between this and the valence scheme is found for the case of beryllium, which now appears logically placed as an interstitial solute. It is most probable that beryllium already lies on the plateau that one would reasonably expect to reach as R_{solute} continues to decrease.

Another of the Hume-Rothery rules controlling primary substitutional solid solubility is of interest in relation to interstitial dissolution. This is the 'effect of relative valence' [30], which says that, for two metals A and B, if the valence of B is higher than the valence of A, the solid solubility of A in B is much lower (i.e. the corresponding h_s is much higher) than that of B in A. This rule is well substantiated for alloys of the noble metals with metals of higher valence (e.g. with lead, indium and tin), systems already discussed by Anthony [4] to give experimental support to the 'low solute valence' and 'polyvalent solvent' Anthony and Turnbull conditions. Recalling that when restricted solid solubility is associated with a high h_s, a relatively high ratio X_i/X_s is expected, these two conditions appear, in our view, to be closely related to the relevant Hume-Rothery rule.

Concerning the Anthony and Turnbull Condition 1, that the solvent should be very electropositive, we note that Hume-Rothery [5] had already pointed out that primary substitutional solid solubility is favoured by electronegative rather than by electropositive solvents. Thus, since a higher h_s is likely to exist when the solvent is electropositive, appropriately higher X_i/X_s ratios should also be observed for these kinds of solvents.

It is of particular interest to note that all the considerations that have been discussed are fully independent of the non-substitutional state assumed to be competing with the simple substitutional one. In fact, for simplicity, we have kept using the term 'interstitial', but the reasoning remains unaltered if a much more complex, fast-diffusing defect is the real entity to which the substitutional impurities in fact give rise. Therefore, the Anthony and Turnbull conditions and the atomic-size scheme still remain valid in the event that much more complex fast-diffusing defects exist.

This is supported by the ^{44}Ti self-diffusion measurements in nickel-doped titanium alloys. From previously reported data [8], nickel in titanium behaves as a fast-diffusing solute and satisfies the Anthony and Turnbull conditions. In effect, nickel should behave as a low-valence solute in the titanium lattice, since the atomic and ionic radii are small enough for no ion core overlapping to occur, and titanium may be considered a high-valence element. The electronegativity condition is also fulfilled, the Pauling electronegativities being 1.5 for titanium and 1.8 for nickel. The ratio of atomic radii is $R_{Ni}/R_{Ti} = 0.84$. Moreover, as previously mentioned, the fact that phase transition does not affect the diffusion parameters indicates that the same diffusion mechanism is operating in both phases. This allows one to make self-diffusion-enhancement measurements as a function of impurity within a temperature range in which the solid solubility is high enough to allow for alloy preparation and homogenization without having to resort to extremely elaborate techniques.

In Fig.5 the previously reported self-diffusion values for ^{44}Ti in pure titanium and those measured in this work are given, together with their dependence on nickel concentration. The theoretical limiting values of titanium self-diffusion as a function of nickel concentration, for both the vacancy and pair mechanisms, are also indicated. An interstitial mechanism corresponds to a line having a lower slope than the one for pairs [12]. It can be seen that, even using the lowest self-diffusion values for pure titanium (obtained by Reca and Libanati [24]), the vacancy mechanism can still fit the experimental results. It must, therefore, be concluded that a different fast-diffusion mechanism is operating.

From the above considerations, we confirm that the fulfilment of either the Anthony and Turnbull conditions or the advanced size correlation does not imply a simple interstitial mechanism.

5. CONCLUSIONS

It is concluded that the 'interstitial' or substitutional behaviour of very dilute metallic impurities in zirconium (as well as in other metallic solvents) can be understood by a judicious use of Anthony and Turnbull conditions, atomic-size parameters, and the general features of the corresponding phase diagrams.

It has been shown that the extent of the solid solubility, and consequently the impurity-host radius ratio, are important and well-founded hints towards assessing the existence of 'interstitial' dissolution.

These conclusions are not affected by the true structure of the fast-diffusing 'interstitial' defect.

ACKNOWLEDGEMENTS

Thanks are given to Dr. F. Dyment for experimental help and stimulating discussions and to Ms. Susana Bermudez for assistance in the metallographic work.

REFERENCES

[1] DYSON, B.F., ANTHONY, R.T., TURNBULL, D., J. Appl. Phys. 37 (1966) 2370.
[2] ANTHONY, T.R., TURNBULL, D., Phys. Rev. 151 (1966) 495.
[3] ANTHONY, T.R., MILLER, J.W., TURNBULL, D., Scr. Metall. 3 (1969).
[4] ANTHONY, T.R., Vacancies and Interstitials in Metals, North Holland, Amsterdam (1969) 935.
[5] HUME-ROTHERY, A., RAYNOR, G.V., The Structure of Metals and Alloys, The Institute of Metals, ed. Richard Clay and Co., London (1954).
[6] TENDLER, R., VAROTTO, C.F., J. Nucl. Mater. 46 (1973) 107; ibid. 44 (1972) 99.
[7] HOOD, G.M., SCHULTZ, R.J., Acta Metall. 22 (1974) 459.
[8] HOOD, G.M., SCHULTZ, R.J., Philos. Mag. 26 (1972) 239.

[9] TENDLER, R., Thesis, Fac. de Ciencias Exactas y Naturales, UNBA, Argentina (1974).
[10] LE CLAIRE, A.D., Philos. Mag. 21 (1970) 819.
[11] MILLER, J.W., Phys. Rev. 188 (1970) 1074.
[12] DYMENT, F., SANTOS, E., submitted for publication to Philos. Mag.
[13] REYNOLDS, R.A., Ph. D. Thesis, Dept. Mater. Sci., Stanford University (1965).
[14] BORSENBERGER, P.M., Ph. D. Thesis, Dept. Mater. Sci., Stanford University (1967).
[15] Metallurgy of Zirconium (LUSTAN, B., KERZE, F., Eds), McGraw Hill, New York (1955).
[16] AGARWALA, R.P., MURAROKA, S.P., ANAND, H.S., Trans. Am. Inst. Mech. Eng. (1965) 986.
[17] PAVLINOV, L.V., Fiz. Met. Metalloved. 24 (1967) 272.
[18] CHIOTTI, P., KILP, G.R., Trans. Am. Inst. Met. Eng. 215 (1959) 892; ibid. 218 (1960) 41.
[19] TENDLER, R., VAROTTO, C.F., J. Nucl. Mater., to be published.
[20] DYMENT, F., LIBANATI, C.M., J. Mater. Sci. 3 (1968) 349.
[21] FULBACHER, P., European Company for the Chemical Processing of Irradiated Fuels Rep. ETR-Bericht N° 49 (1963).
[22] LYASHENKO, V.S., BIKOV, B.N., PAVLINOV, L.V., Fiz. Met. Metalloved. 8 (1959) 362.
[23] KIDSON, G.V., KIRKALDY, J.S., Philos. Mag. 20 (1969) 1057.
[24] RECA, E.W., LIBANATI, C.M., Acta Metall. 16 (1968) 1297.
[25] STANSBURY, et al., Diffusion in bcc Metals, ASM, Metals Park, Ohio 3 (1965).
[26] ADDA, V., PHILIBERT, J., La diffusion dans solides, Presses Universitaires de France (1966).
[27] DOUGLASS, D.L., The Metallurgy of Zirconium, Atomic Energy Review, Suppl. 1971, IAEA, Vienna (1971).
[28] MEYER, A., UMAR, I.H., YOUNG, W.H., Phys. Rev., B 4 (1971) 3287.
[29] HOOD, G.M., Diffusion Processes (SHERWOOD, J.N. et al., Eds) 1, Gordon and Breach (1971).
[30] BARRETT, C.S., MASSALSKI, T.B., Structure of Metals, McGraw-Hill (1966).
[31] MUNDY, J.N., McFALL, W.D., Phys. Rev. 88 (1973) 5477.
[32] HANSEN, M., ANDERKO, K., Constitution of Binary Alloys, McGraw-Hill (1958).
[33] SHAW, R.W., Phys. Rev. 58 (1972) 4856.
[34] FEDERER, J.I., LUNDY, T.S., Trans. Am. Inst. Mech. Eng. 227 (1963) 592.
[35] AGARWALA, R.P., PAUL, A.R., Nuclear and Radiation Chemistry (Proc. Symp. Poona), (1967).
[36] ELLIOTT, R.P., Constitution of Binary Alloys, 1st Suppl., McGraw-Hill (1965).
[37] RICHTER, H., WINCIERZ, P., ANDERKO, K., ZWICKER, U., J. Less-Common Met. 4 (1962) 252.
[38] SHUNK, F.A., Constitution of Binary Alloys, 2nd Suppl., McGraw-Hill (1966).
[39] HARRIS, I.R., RAYNOR, G.V., J. Less-Common. Met. 6 (1964) 70.
[40] BLINKIN, A.M., VOROBIOV, B.B., Ukr. Fiz. Zh. 9 (1964) 191.
[41] TENDLER, R., VAROTTO, C.F., J. Nucl. Mater. 44 (1972) 99.
[42] PAVLINOV, L.V., GRIGORIEV, G.V., GROMYKO, G.O., Izv. Akad. Nauk SSSR, Met. 3 (1969) 207.
[43] FEDOROV, G.B., ZHOMOV, F.I., SMIRNOV, E.A., Metall. Metalloved. Chist. Met., Sb. Nauchn. Rab. 5 (1966) 22.
[44] AGARWALA, R.P., MURARKA, S.P., ANAD, M.S., Acta Metall. 16 (1968) 61.
[45] BORISOV, E.V., GODIN, Y.U.G., GRUZIN, P.L., YEVSTUYUKHIN, A.I., EMELAYANOV, V.S., Trans. All-Union Conf. on Radioactive and Stable Isotopes (1957) 292.
[46] FEDOROV, G.B., ZHOMOV, F.I., Metall. Metalloved. Chist. Met., Sb. Nauchn. Rab. 1 (1959) 162.
[47] FEDOROV, G.B., GULAYAKIN, V.D., Metall. Metalloved. Chist. Met., Sb. Nauchn. Rab. 1 (1959) 170.

DISCUSSION

D. CALAIS: Regarding the heterodiffusion of the rapid solutes in the two zirconium phases, I think you accept, as I do, an interstitial type of diffusion mechanism. The ratio of the heterodiffusion coefficient of the rapid solute to the self-diffusion coefficient in the lattice decreases regularly with the phase change $\alpha\text{-Zr}_{hex} \rightarrow \beta\text{-Zr}_{bcc}$. Self-diffusion occurs by single vacancies in α-zirconium. Do you think that from this we can deduce a contribution of the auto-interstitials in the self-diffusion process of β-zirconium, which is moreover abnormally fast? I personally believe that your results support a contribution of the auto-interstitials in the self-diffusion of body-centred cubic zirconium.

E. SANTOS: In our paper we state that the diffusion mechanism of the impurity is not necessarily a simple interstitial mechanism. We have no evidence on the basis of which we can define the self-diffusion mechanism in the bcc phase. This is in any case not a part of the present work.

SOLUTION AND REACTION THERMODYNAMICS OF HYDROGEN WITH SELECTED FUSION REACTOR MATERIALS*

V.A. MARONI, E. VELECKIS, F.A. CAFASSO
Argonne National Laboratory,
Argonne, Ill.,

C.C. HERRICK, R.C. FEBER
Los Alamos Scientific Laboratory,
Los Alamos, New Mex.,
United States of America

Abstract

SOLUTION AND REACTION THERMODYNAMICS OF HYDROGEN WITH SELECTED FUSION REACTOR MATERIALS.

Experimental data on the thermodynamic properties of solutions of protium, deuterium and tritium in liquid lithium, and calculated stabilities of four insulator materials toward atomic and molecular hydrogen are reported. The significance of the results is discussed with regard to problems that are pertinent to the development of controlled thermonuclear fusion reactors. Values are given for equilibrium constants, free energies of formation, and Sieverts' Law parameters in the binary systems Li-H, Li-D and Li-T. The observed isotope effects in these systems appear to originate predominantly from the gas phase. Calculations of the stability of alumina, beryllia and boron nitride in a molecular hydrogen environment, typical of a theta-pinch confined fusion plasma, indicate that all three materials would: (i) perform satisfactorily at 1100 K; (ii) show modest reaction at 1500 K; and (iii) be unsuitable at 2000 K. In a hydrogen-atom environment, all three materials are found to be unsatisfactory over the temperature range studied (1100 to 2000 K).

1. INTRODUCTION

1.1. Lithium-hydrogen studies

In a number of recently published conceptual designs of deuterium-tritium (DT)-fueled fusion reactors [1,2], liquid lithium has been selected as the blanket medium and primary coolant. This selection stems mainly from (i) the utility of the $^{6}Li(n,\alpha)T$ and $^{7}Li(n,n'\alpha)T$ nuclear reactions in breeding tritium and (ii) the fact that lithium is an excellent heat-transfer medium [3]. It is likely [4] that a large liquid-lithium system, operated as the blanket of a DT-fueled fusion power plant, will contain significant quantities of protium and deuterium in addition to the continuously bred tritium. The successful development of efficient recovery methods and sensitive monitoring equipment for the hydrogen isotopes in the blanket region requires exact knowledge of their solution behavior in liquid lithium. Results of thermodynamic studies of the lithium-protium (Li-H) and lithium-deuterium (Li-D) systems, carried out at the Argonne National Laboratory (ANL), are reported in section 2.1. of this paper.

*Work performed under the auspices of the USAEC.

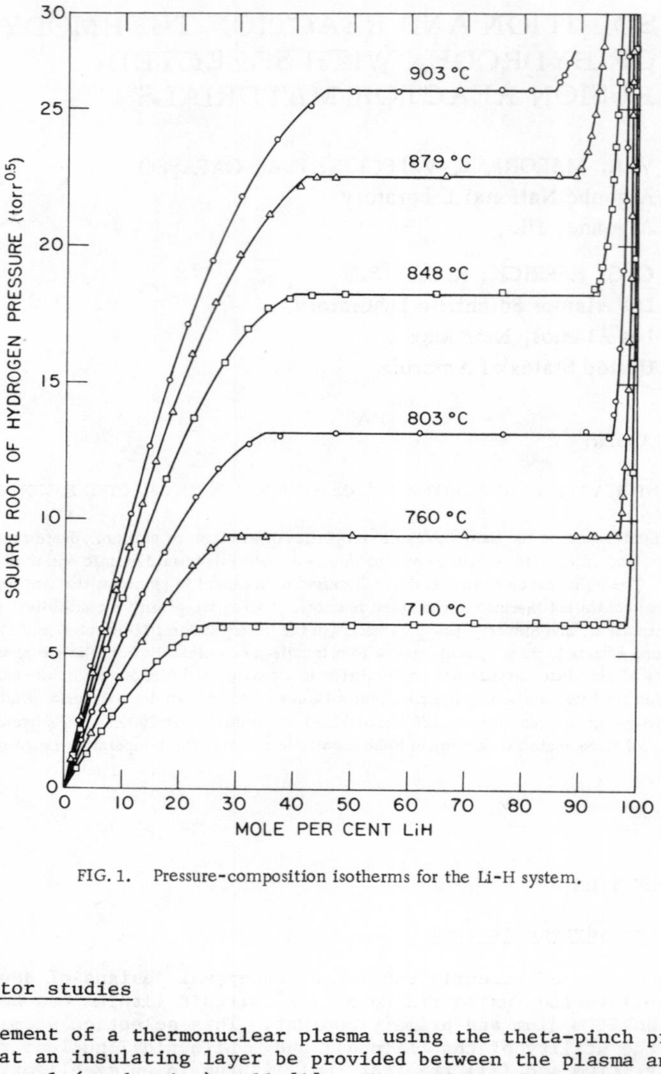

FIG. 1. Pressure-composition isotherms for the Li-H system.

1.2. Insulator studies

Confinement of a thermonuclear plasma using the theta-pinch principle requires that an insulating layer be provided between the plasma and the first structural (conductive) wall [1]. The surface of the insulator material facing the plasma is in contact with molecular hydrogen (D_2, DT, and T_2) for an appreciable fraction of each theta-pinch duty cycle and is subjected to bombardment by atomic hydrogen (mostly D and T) throughout the burn stage of the cycle. In addition, during the duty cycle, the temperature of the insulator-layer is expected to rise above 1000 K in power-producing reactor systems. A detailed knowledge of the interactions between candidate insulating materials and the environment created by a thermonuclear plasma is essential to the development of a suitable insulator. The results of calculational studies carried out at the Los Alamos Scientific Laboratory (LASL) to evaluate the stabilities of selected insulator materials toward atomic and molecular hydrogen are presented in section 2.2. of this paper.

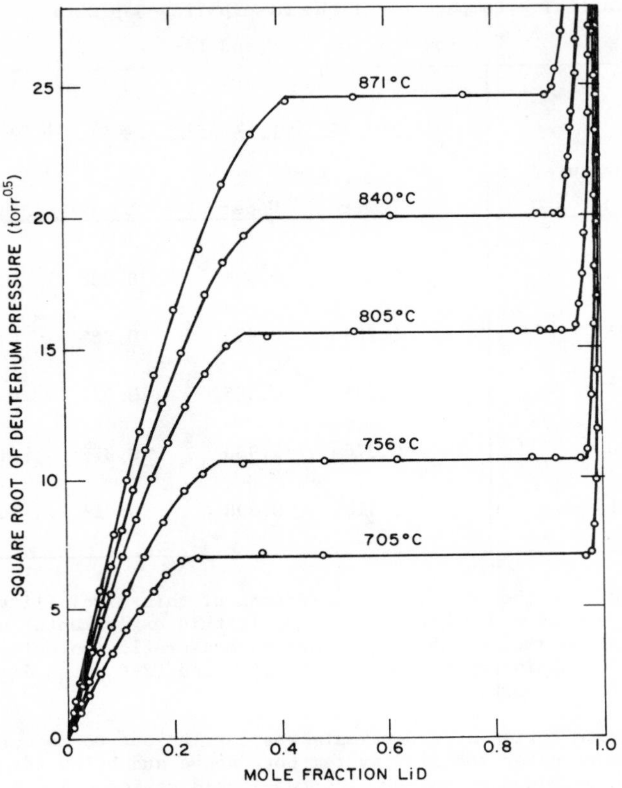

FIG. 2. Pressure-composition isotherms for the Li-D system.

2. RESULTS AND DISCUSSION

2.1. Lithium-hydrogen studies

An experimental study of the Li-H, Li-D, and Li-T systems has been undertaken at ANL to obtain reliable pressure-composition-temperature data that are amenable to the derivation of useful thermodynamic quantities for these systems. The experimental data were obtained by a tensimetric method using a modified Sieverts' apparatus. The equilibrium hydrogen pressure over lithium-lithium hydride mixtures, sealed in iron capsules, was measured as a function of temperature and composition. The apparatus and experimental procedures are described in detail elsewhere [5,6].

Pressure-composition isotherms obtained for the Li-H and Li-D systems are shown in Figs. 1 and 2, and selected data from these isotherms are listed in Table I. The general features of the two families of isothermal curves in Figs. 1 and 2 are analogous. There are two terminal solutions separated by an extensive two-liquid region (miscibility gap). The data in

TABLE I. Comparison of the Miscibility-Gap Data for the Systems Li-H and Li-D

Temperature, (°C)	Miscibility Gap Limits (mol% LiH or LiD)			
	Li-H		Li-D	
	Lower	Upper	Lower	Upper
705	0.240	0.986	0.238	0.979
756	0.289	0.973	0.285	0.972
805	0.339	0.955	0.337	0.951
840	0.376	0.930	0.372	0.930
871	0.410	0.898	0.415	0.908

Table I show that the boundary compositions of this miscibility gap are essentially the same for the two systems (within experimental uncertainty). This observation implies that there are no measurable isotopic differences (isotope effects) in the regions of the Li-H and Li-D phase diagrams investigated in our work.

A separate series of experiments was carried out to measure the plateau pressures for the Li-D system both above and below the monotectic (i.e. the phase boundary between the two-liquid region and the liquid-solid region [3]). These results are shown in Fig. 3, where the logarithm of the plateau pressure, P_{pl}, is plotted against the reciprocal of the Kelvin temperature. The monotectic temperature, identified by the intersection of the two straight-line portions of the curve for the Li-D system in Fig. 3, occurs at 690°C. This value is only slightly higher than that reported [7] for the Li-H system [5,6]. The ratio of the plateau pressures for the two systems, $P_{pl,D}/P_{pl,H}$, ranges from 1.40 to 1.34 over the temperature range from 700 to 900°C.

The data in Figs. 1 and 2 have been used to calculate activity coefficients for Li and LiH (or LiD) in each of the homogeneous terminal solutions as follows. The reaction between hydrogen gas and liquid lithium is assumed to result in a solution of hydride (H^-, D^-, or T^-) in lithium according to eq. 1.

$$Li_{(soln)} + \frac{1}{2} H_{2(g)} \rightleftharpoons LiH_{(soln)} \tag{1}$$

The equilibrium constant, K_{eq}, for this reaction is given by

$$K_{eq} = \frac{N_2\gamma_2}{N_1\gamma_1 \, P_{H_2}^{1/2}} \tag{2}$$

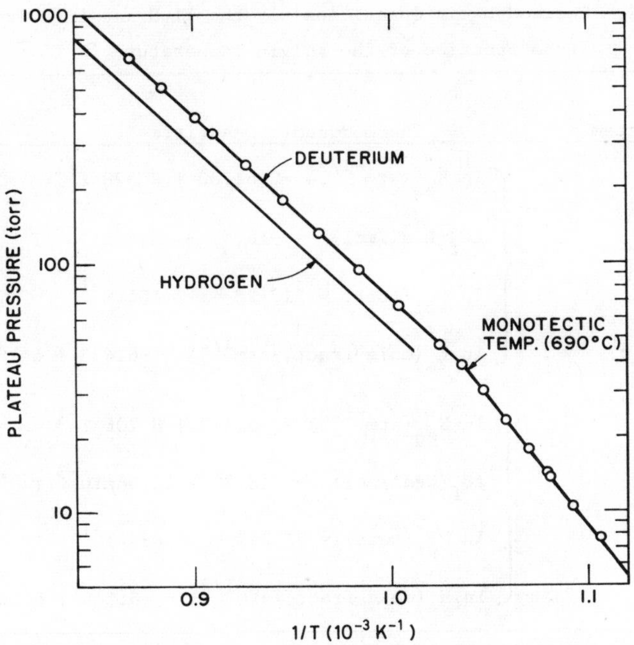

FIG. 3. Curves for the plateau dissociation pressures in the Li-H and Li-D systems as a function of the reciprocal Kelvin temperature.

where N_1 and N_2 are the mole fractions of Li and LiH, respectively, γ_1 and γ_2 are the corresponding activity coefficients, and p_{H_2} is the equilibrium hydrogen pressure. Equation 2 can be written in the form

$$\ln (p_{H_2}^{1/2} N_1/N_2) = \ln K_{eq} + \ln \frac{\gamma_1}{\gamma_2} \tag{3}$$

In accord with the Margules-type treatment of binary solutions [8], the activity coefficients may be represented by power series in N_1 and N_2. When these series are truncated at their cubic terms, the activity coefficients are consistently given by

$$\ln \gamma_1 = \alpha N_2^2 + \beta N_2^3 \tag{4}$$

and

$$\ln \gamma_2 = (\alpha + \frac{3}{2} \beta) N_1^2 - \beta N_1^3 \tag{5}$$

Substitution of eqs. 4 and 5 into eq. 3 gives

$$\ln (p_{H_2}^{1/2} N_1/N_2) = -\ln K_{eq} + \alpha(1-2N_2) + \frac{1}{2}\beta(1-3N_2)^2 \tag{6}$$

The data in Figs. 1 and 2 have been refined [6] to give values for the parameters α and β which allow calculation of the activity coefficients.

TABLE II. Thermodynamic Quantities for the Li-H and Li-D Systems As a Function of the Kelvin Temperature, T

System	Thermodynamic Quantities[a]
Li-H	$\ln K_{eq}(\text{atm}^{-1/2}) = -6.780 + 8\ 328\ T^{-1}$
	$\Delta G_f^\circ(\text{kcal/mol}) = -16.55 + 13.47\times10^{-3}\ T$
	$\ln P_{pl}(\text{torr}) = 21.128 - 17\ 186\ T^{-1}$
	$\ln K_s(\text{mole fract./atm}^{1/2}) = -6.498 + 6\ 182\ T^{-1}$
Li-D	$\ln K_{eq}(\text{atm}^{-1/2}) = -6.822 + 8\ 206\ T^{-1}$
	$\Delta G_f^\circ(\text{kcal/mol}) = -16.30 + 13.56\times10^{-3}\ T$
	$\ln P_{pl}(\text{torr}) = 21.211 - 16\ 694\ T^{-1}$
	$\ln K_s(\text{mole fract./atm}^{1/2}) = -6.540 + 6\ 060\ T^{-2}$

[a] The quantities K_{eq}, ΔG_f°, P_{pl}, and K_s are defined in the text.

Equations derived from these activity coefficients for (i) the equilibrium constant, K_{eq}, (ii) the standard free energy of formation, ΔG_f°, of LiH (and LiD), (iii) the plateau pressure, P_{pl}, and (iv) the Sieverts' Law parameter, K_s, ($K_s = N_2/p^{1/2}$ for $N_2 << N_1$) are given in Table II for both the Li-H and Li-D systems.

Accurate knowledge of the thermodynamic properties of the Li-H and Li-D systems has also permitted reliable estimates of properties of the Li-T system to be made. An approximate relationship for the Sieverts' Law parameter for tritium solutions in lithium was obtained from the Li-H and Li-D data using the procedure described in a previous ANL report [3]. This relationship, given by eq. 7, is probably reliable to within ± 20%.

$$\ln K_{s,T_2}\ (\text{mole fraction } T^-/\text{atm}^{1/2}) = -6.558 + 6\ 005\ T^{-1} \tag{7}$$

Precise knowledge of K_{eq} and ΔG_f° are important in determining the stabilities of the hydrides of lithium relative to the stabilities of hydrides of other structural and impurity elements in the blanket of a fusion reactor. The Sieverts' Law parameter provides the important relationship between hydrogen (H_2, D_2, or T_2) pressure and hydride (H^-, D^-, or T^-) concentration for dilute solutions of H, D, or T in lithium. The latter relationship is likely to be useful in the conception and development of recovery systems and monitoring equipment for hydrogen isotopes in a liquid-lithium blanket.

2.2. Insulator studies

An ideal surface on an insulated first-wall of a controlled thermonuclear reactor must withstand degeneration from bremsstrahlung and neutron fluxes, possess good dielectric and thermal conductivity properties, and not deteriorate chemically by reaction with a deuterium-tritium gas mixture. With regard to the latter requirement, in the absence of either experimmental or theoretical rate data, a limit to potential chemical corrosion of first-wall insulators in a reducing atmosphere is sought by treating the system as a complex equilibrium problem in which many gaseous constituents are in equilibrium with a pure solid. The principles used to calculate these complex chemical equilibria, which were established by Gibbs, have been applied many times before [9-11]. All calculations assume equilibrium is reached, i.e. the ratio of computed to actual extent of reaction is unity. The reasonableness of this approximation can be assessed only when the requisite experimental rate data are available.

Previous calculations [12] on a niobium-aluminum oxide composite first-wall showed the surface temperature to be substantially higher than that of the liquid lithium heat sink. In this case the sink temperature was limited to 1100 K by the properties of the niobium. However, a more refractory metal-metal oxide first-wall would permit higher operating temperatures and consequently a higher Carnot efficiency. Therefore, these stability calculations were made at 1100, 1500, and 2000 K. Because actual fuel pressures are uncertain, calculations were usually made at 1, 5 and 10 torr, but some were extended to $1x10^{-5}$ torr to further assess pressure effects. Since the geometry of theta-pinch type fusion reactors are not yet fixed, the initial fuel gas/solid ratios, expressed as equivalent moles of hydrogen atoms per mole of solid, were varied from 0.1 to 10.

2.2.1. Thermodynamic relations

Brinkley [9c, 9d] first systematized the calculation of complex gaseous equilibria. Briefly, the equilibrium problem consists of finding a set of N concentrations which minimizes the total Gibbs free energy, G,

$$G = \sum_{i=1}^{N} \mu_i n_i, \tag{8}$$

subject to the constraints

$$\sum_{i=1}^{N} a_{ie} n_i = Q_e \tag{9}$$

and

$$n_i \geq 0, \tag{10}$$

where G = Gibbs free energy, $\mu_i = (\partial G/\partial n_i)_{T,P}$, n_i = moles of substance i, a_{ie} = number of atoms of element e in gas i, and Q_e = total moles of element e.

Recently, the number and variety of methods for solving this problem have proliferated greatly, but because there is only one equilibrium state, all methods must be equivalent, regardless of the numerical method used.

The methods may be divided into two general types: a minimization of the total free energy as given by eq. 8, or the solution of a set of N simultaneous equations containing non-linear equilibrium expressions and linear mass balance equations. The latter method was chosen for present purposes.

In a system of m elements distributed among s gases, the number of components, c, has been shown by Brinkley [9c, 9d] to be equal to the rank $R \leq m$ of the matrix of the stoichiometric coefficients a_{ie} ($i=1,---, s$; $e=1,----,m$). A row of the matrix:

$$\alpha_i = (a_{i1}, ---, a_{ie}, ---, a_{im}) \tag{11}$$

has been called the formula vector of the i-th substance. Of these vectors, c are linearly independent (the component vectors). The remaining (s−c) substances, hereafter subscripted j and designated as species, have linearly dependent formula vectors α_j which may be expressed as linear combinations of the independent vectors:

$$\alpha_j = \sum_{i=1}^{c} \nu_{ji}\alpha_i \qquad j = c+1, c+2, ---, s \tag{12}$$

The choice of components is arbitrary in principle, but, because of numerical problems, not in practice. Equation 12 corresponds to a reaction between components A_i to form the species A_j.

$$A_j = \sum_{i=1}^{c} \nu_{ji}A_i, \tag{13}$$

where the ν_{ji}'s are the stoichiometric coefficients of the j-th reaction which produces only A_j. A ν_{ji} will be zero if component A_i does not appear in the reaction or negative if it is a product.

Any particular reaction j will proceed to an extent ξ_j corresponding to its final equilibrium state. In the final equilibrium state the system will contain $\sum_{i=1}^{c} \nu_{ji}(1-\xi_j)$ total moles of components and ξ_j moles of A_j.

At a given temperature and pressure the Gibbs free energy for the j-th reaction ($G_j = \sum_i \mu_i n_i$, $i=1, ---, c, j$) becomes

$$G_j = \sum_{i=1}^{c} \nu_{ji}(1-\xi_j)\mu_i + \xi_j\mu_j \quad . \tag{14}$$

The equilibrium state is by definition the state of minimum G_j, i.e.

$$\left(\frac{\partial G_j}{\partial \xi_j}\right)_{P,T} = 0. \tag{15}$$

With this condition, eq. 14 becomes

$$\sum_{i=1}^{c} \nu_{ji}\mu_i = \mu_j \quad . \tag{16}$$

For an ideal gas

$$\mu_k = \mu_k^\circ + RT \ln p_k, \tag{17}$$

so that eq. 16 becomes

$$\sum_{i=1}^{c} \nu_{ji}(\mu_i^\circ + RT \ln p_i) = \mu_j^\circ + RT \ln p_j,$$

or

$$RT \ln \frac{p_j}{\prod_{i=1}^{c} p_i^{\nu_{ji}}} = -\left(\mu_j^\circ - \sum_{i=1}^{c} \nu_{ji}\mu_i^\circ\right) . \tag{18}$$

The right hand side of this equation will be recognized as the standard free energy change, $(\Delta G_j^\circ)_T$, for the j-th reaction. Equation 18 is then rewritten as

$$\ln \frac{p_j}{\prod_{i=1}^{c} p_i^{\nu_{ji}}} = -(\Delta G_j^\circ)_T = \ln (K_p^{(j)})_T . \tag{19}$$

Tabulated values of Gibbs free energies, such as those in the JANAF Thermochemical Tables [13], were used to compute $(\Delta G_j^\circ)_T$ and hence $(K_p^{(j)})_T$. Because all gases are assumed to be ideal

$$p_k = x_k P = \frac{n_k}{\sum_{\ell=1}^{s} n_\ell} P,$$

where P is the total pressure. Therefore for the j-th species, eq. 19 reduces to

$$n_j = (K_p^{(j)})_T \left[\frac{P}{\sum_{\ell=1}^{s} n_\ell}\right]^{\nu'} \prod_{i=1}^{c} n_i^{\nu_{ji}}, \tag{20}$$

where $\nu' = \sum_{i=1}^{c} \nu_{ji} - 1.$

It follows from the above derivation that there are as many mass action equations of this type as there are gaseous species, i.e. (s-c). It is only necessary then to obtain $c \leq m$ more equations to completely define the system. An obvious constraint on a closed system is that of constant mass. The set of s concentrations which form the solution must satisfy c mass balance equations of the form

$$\sum_{\ell=1}^{s} \beta_{\ell m} n_\ell = Q_m . \tag{21}$$

If a solid is present, these mass balance equations are not all independent. Here $\beta_{\ell m}$ is the number of atoms (or molecules) of m present in substance ℓ and Q_m is the number of gram-atoms (or moles) of m in the system.

2.2.2. Numerical method

To calculate a solution to the set of (s-c) non-linear and c linear equations, we devised a method which, although not necessarily the most computationally efficient possible, was, to our knowledge, novel and therefore interesting in its own right. By using a method of steepest descents [14], the equations were converted to a set of pseudo-rate equations. The latter equations were in turn solved using Gear's algorithm [15] for a set of stiff ordinary differential equations.

If in an iterative solution to the set of equations the current values for the moles of the components are substituted in the mass action expression for the j-th species, we may define an error as

$$\varepsilon_j = n_j^* - f(n_1^*, ---, n_c^*), \tag{22}$$

where the asterisks indicate the current values of the numbers of moles and f is the mass action expression. For a mass balance equation (no solid present)

$$\varepsilon_m = Q_m - \sum_{\ell=1}^{s} \beta_{\ell m} n_\ell^* \ . \tag{23}$$

A solution to the set is reached when the ε's are all zero or less than some acceptable maximum.

In the method of steepest descents used [14], a local minimum is sought to the sum of squares of the errors

$$S = \sum_{i=1}^{s} \varepsilon_i^2 \ . \tag{24}$$

This minimum will be reached by iteration if $dS/d\lambda \leq 0$ for all λ. At the minimum the derivative is zero. The search parameter λ can be regarded as a pseudo-time, and the derivative of S with respect to λ is

$$\frac{dS}{d\lambda} = 2 \sum_{i=1}^{s} \varepsilon_i \frac{d\varepsilon_i}{d\lambda} \ . \tag{25}$$

But

$$\frac{d\varepsilon_i}{d\lambda} = \sum_{j=1}^{s} \frac{\partial \varepsilon_i}{\partial n_j} \frac{dn_j}{d\lambda} \tag{26}$$

and therefore

$$\frac{dS}{d\lambda} = 2 \sum_{i=1}^{s} \varepsilon_i \sum_{j=1}^{s} \frac{\partial \varepsilon_i}{\partial n_j} \frac{dn_j}{d\lambda} = 2 \sum_{j=1}^{s} \frac{dn_j}{d\lambda} \sum_{i=1}^{s} \varepsilon_i \frac{\partial \varepsilon_i}{\partial n_j} \ . \tag{27}$$

TABLE III. Formula Matrix for the BN-H_2 System

Species	Components	BN(s)	B 1	H 0	N 1
	1	BH_3	1	3	0
	2	NH_3	0	3	1
3		BN	1	0	1
4		H	0	1	0
5		NH	0	1	1
6		NH_2	0	2	1
7		N_2H_2	0	2	2
8		N_2	0	0	2
9		N_2H_4	0	4	2
10		B	1	0	0
11		BH	1	1	0
12		BH_2	1	2	0
13		B_2	2	0	0
14		B_2H_6	2	6	0
15		B_5H_9	5	9	0
16		$B_{10}H_{14}$	10	14	0
17		H_2	0	2	0
18		$B_3N_3H_6$	3	6	3

A method of steepest descents is introduced by defining

$$\frac{dn_i}{d\lambda} = -\sum_{i=1}^{s} \varepsilon_i \frac{\partial \varepsilon_i}{\partial n_j} , \qquad (28)$$

and therefore eq. 27 becomes

$$\frac{dS}{d\lambda} = -2 \sum_{j=1}^{s} \left(\frac{dn_i}{d\lambda}\right)^2 , \quad j=1, ---, s. \qquad (29)$$

Because the total derivative of S with respect to the search parameter is always ≤ 0, regardless of the sign of the derivatives of the n_i, S will continue to decrease as the iterations proceed until a local minimum is reached.

The set of non-linear, coupled, ordinary differential equations to be solved are of the form of eq. 28. At present the necessary partial derivatives of the ε_i are obtained analytically. The iterative procedure is terminated when S has reached an acceptably low value.

TABLE IV. Chemical Reactions for $BN\text{-}H_2$ System

	$(\Delta G°)$ 1100, kcal
$BH_3 + NH_3 = 6H + BN(s)$	142.781
$2NH_3 + BN(s) = 3NH + BH_3$	288.319
$5NH_3 + BN(s) = 6NH_2 + BH_3$	279.055
$4NH_3 + 2BN(s) = 3N_2H_2 + 2BH_3$	307.934
$NH_3 + BN(s) = N_2 + BH_3$	51.231
$5NH_3 + BN(s) = 3N_2H_4 + BH_3$	226.888
$BH_3 + BN(s) = 2B(g) + NH_3$	209.427
$2BH_3 + BN(s) = 3BH + NH_3$	229.375
$5BH_3 + BN(s) = 6BH_2 + NH_3$	119.665
$BH_3 + BN(s) = B_2(g) + NH_3$	167.694
$2BH_3 = B_2H_6$	-4.056
$4BH_3 + BN(s) = B_5H_9 + NH_3$	43.074
$22BH_3 + 8BN(s) = 3B_{10}H_{14} + 8NH_3$	292.433
$BN(s) = BN(g)$	121.878
$BH_3 + NH_3 = 3H_2 + BN(s)$	-86.419
$2BN(s) + NH_3 + BH_3 = B_3N_3H_6$	15.274

Gear's algorithm was used to solve the system of differential equations of the form of eq. 28. This algorithm was devised for the numerical solution of a set of ordinary differential equations using multistep predictor-corrector methods whose order and step-size are automatically varied, consistent with specified tolerance on the estimated error. The user has the option of using either a form of the Adams-Bashforth method or a method suitable for a system of "stiff" equations, i.e. a system in which the solutions have widely different time constants. It is the latter option which makes Gear's algorithm appropriate, when coupled to a method of steepest descents, to the solution of a set of complex chemical equilibrium equations. Thus, a system which required 244 iterations and 20 seconds (CDC 7600) to reach convergence ($S < 10^{-20}$) with the stiff option failed to converge in 5 minutes and 60 000 iterations with the Adams-Bashforth method.

2.2.3. Outline of the method

1. Identify all chemical substances that may be important to the system under investigation.

2. Determine the rank of the formula matrix, if the number of components is not intuitively obvious. Table III illustrates the formula matrix for the BN-H_2 system, with rank 3.

3. Select the components and write a set of chemical reactions, each of which contains one species on the right-hand side and whichever components are necessary (see Table IV). To optimize the calculations, the components should be selected from among those gases expected to be present at higher concentrations at equilibrium. If, as is true here, a solid phase is also present at equilibrium, many of the reactions will include a solid. The latter will not appear in the equilibrium constant, because its activity is taken to be unity. However, the number of components is effectively reduced because of the additional constraint on mass balance equations imposed by the fixed composition of the solid.

4. Compute the Gibbs free energies of the reactions and the corresponding equilibrium constants. The results in Table IV are calculated using the JANAF Thermochemical Tables [13].

5. Write as many mass balance equations as there are components, i.e. for hydrogen, boron and nitrogen. As noted, the presence of a solid will modify one or more of the mass balance equations. The modified equations will be obtained by algebraically eliminating the unknown amount of solid remaining from the mass balance equations of the elements making up that solid. For each pair of mass balance equations so combined, the number of components will decrease by one.

6. Rewrite the mass action and mass balance equations as error expressions.

7. Differentiate each error expression with respect to each substance in the gas phase and assemble into the set of differential equations.

8. Iterate this set of equations until the sum of errors is sufficiently small.

2.2.4. Results of stability calculations

A summary of the results of the calculations is given in Tables V and VI. The chemical reactivity of hydrogen has been substituted for that of a deuterium-tritium mixture. The last columns of the Tables contain estimates of total amount of insulator eroded per year, assuming equilibrium during each duty cycle (0.01 sec burn time and 0.09 sec rest time) and a renewable insulator surface.

2.2.4.1. Beryllia

One suggested surface material, with a large free energy of formation, good thermal conductivity (for a ceramic), and low electrical conductivity, is beryllia. Results of calculations on the beryllium-oxygen-hydrogen system (Table V) show that approximately 1.9×10^{-10} mole of an initial 1 mole of BeO(s) is eroded by 0.5 mole of H_2 at a pressure of 1 torr and a temperature of 1100 K. At 1100 K the extent of erosion is not significantly affected by changes in H_2 pressure and gas/solid ratio. The extents of

TABLE V. Summary of Results

Insulator + H_2
Pressure = 1 torr Reactant Ratio: 0.5 mole H_2/1 mole Insulator

Material	Temperature (K)	Insulator Eroded (Mole Fraction)	Insulator Loss (g/a)
BeO	1100	1.91×10^{-10}	1.59
BeO	1500	3.84×10^{-7}	3.20×10^{3}
BeO	2000	3.00×10^{-4}	2.50×10^{6}
Al_2O_3	1100	1.27×10^{-10}	4.03
Al_2O_3	1500	5.42×10^{-7}	1.72×10^{4}
Al_2O_3	2000	2.46×10^{-4}	7.80×10^{6}
BN	1100	1.11×10^{-9}	8.57
BN	1500	4.71×10^{-6}	3.63×10^{4}
BN	2000	2.01×10^{-3}	1.55×10^{7}

TABLE VI. Summary of Results

Insulator + H
Pressure = 1 torr Reactant Ratio: 1 mole H/1 mole Insulator

Material	Temperature (K)	Insulator Eroded (Mole Fraction)	Insulator Loss (g/a)
BeO	1100	2.27×10^{-1}	1.89×10^{9}
BeO	1500	1.57×10^{-1}	1.31×10^{9}
BeO	2000	1.56×10^{-2}	1.30×10^{8}
Al_2O_3	1100	8.90×10^{-2}	2.82×10^{9}
Al_2O_3	1500	3.51×10^{-2}	1.11×10^{9}
Al_2O_3	2000	2.15×10^{-2}	6.8×10^{8}
BN	1100	1.67×10^{-1}	1.29×10^{9}
BN	1500	1.59×10^{-1}	1.23×10^{9}
BN	2000	1.43×10^{-1}	1.10×10^{9}

erosion at 1500 and 2000 K (under the same conditions as stated above for 1100 K) are $\sim 4 \times 10^{-7}$ mole and $\sim 3 \times 10^{-4}$ mole respectively. The major products of the reaction at 2000 K are hydrogen atoms, water, beryllium, and BeOH, all of which are in the vapor phase.

Thus, at 1100 K beryllia is a suitable insulator in H_2 in thermal equilibrium with atomic hydrogen. At 1500 K there are potential problems, while at 2000°K substantial erosion losses could occur unless the reaction probability is of the order of 10^{-3} to 10^{-4}. Reaction probabilities of this magnitude have been found experimentally by Rosner [16] for the fluorination of refractory metals.

We also computed the stability of BeO(s) in an environment consisting entirely of hydrogen atoms. A more appropriate calculation might use various other H/H_2 ratios. But because the ratio to be expected in an actual device was not available, the limiting and worst case - pure hydrogen atoms - was calculated. The total amounts of solid eroded by a 1 torr hydrogen atom pressure at 1100, 1500, and 2000 K are summarized in Table VI. Additional calculations show that as the hydrogen-atom pressure is reduced to 10^{-5} torr the extent of erosion drops to 4×10^{-3} mole per mole of BeO(s) originally present. These results pose a potential corrosion problem if the hydrogen atom - beryllia system approaches equilibrium at any of the proposed temperatures or pressures. It is unlikely that the probabilities for this reaction would be low enough to reduce it to insignificance.

We did not include ions in the calculations, as it has been assumed that ions in the fuel gas cannot pass the magnetic barrier. Thermal ionization is probably negligible. There is little doubt, however, that if these assumptions are not valid, ion-insulator reactions could add appreciably to the products formed by hydrogen-insulator reactions and would have higher reaction probabilities than the latter.

2.2.4.2. Alumina

Alumina is a second prominently mentioned material for a first-wall insulator. An attractive property of this oxide is illustrated by the fact that it can be bonded to niobium metal through the intermediate alloy $NbAl_3$.

The molecules included in the equilibrium calculations for the aluminum-oxygen-hydrogen system were: Al_2O_3(s), H, H_2, H_2O, AlH, Al, Al_2O, AlOH, AlO_2H, AlO, AlO_2, HAlO, Al_2O_2, OH, O_2, O, HO_2, H_2O_2. Including these possible species, our calculations, summarized in Table V, predict that at 1100 K alumina will be stable in molecular hydrogen in thermal equilibrium with hydrogen atoms, less so at 1500 K and require reaction probabilities of 10^{-3} to be useful at 2000 K.

Again, as was the case for beryllia, if hydrogen atoms are allowed to react with alumina, unacceptable consequences are predicted (Table VI) unless an extremely favorable reaction probability is involved.

2.2.4.3. Boron nitride

Boron nitride is also a good conductor and has low electrical conductivity. It has long been a favorite material of high-temperature chemists because it is easily fabricated and is inexpensive.

Calculated stabilities in molecular hydrogen, summarized in Table V, are similar to those for the oxides under the same conditions. Species used in the calculations included BN(s), H, H_2, BH_3, N_2, NH, NH_2, N_2H_2, NH_3, N_2H_4, B, BH, BH_2, B_2, B_2H_6, B_5H_9, $B_{10}H_{14}$, BN, $B_3N_3H_6$.

Again, as was the case for the oxides, erosion in the presence of a hydrogen atom environment is predicted to be catastrophic at all temperatures and pressures (Table VI) unless the reaction probability is extremely low. In this system the major reaction products are BH_3, BH_2, B_2H_6, NH_3 and some $B_3N_3H_6$.

The actual calculations in this system proved to be more difficult than in the oxide systems. In the beryllia-hydrogen system, for example, it was necessary to redefine components only once over the range of pressures and temperatures of interest. However, the temperature dependences for ΔG_f° of such species as BH_3, BH_2, NH_3 and NH_2 were such that several different combinations of components were required both as a function of temperature and reacting gas.

2.2.5. Conclusions

The conditions imposed on the first-wall are sufficiently demanding that a choice of a particular material for either the wall or its insulating surface requires experimental data. At temperatures up to 1500 K any of the materials considered here are candidates in an environment of molecular deuterium and tritium in thermal equilibrium with the atoms. An environment of the atoms, however, poses a problem at all temperatures unless the reaction probabilities are unexpectedly small. Increasing the insulator thickness is of limited usefulness, as the surface temperatures would thereby increase. Obviously, experiments or more definitive calculations are required to determine

1) the actual surface temperature,

2) the H/H_2 ratio impinging on the wall,

3) the probabilities of hydrogen-insulator reactions,

4) the possibility that ions will reach the surface from the fuel gas.

On the basis of available information, beryllia, with its better thermal conductivity and more negative free energy of formation, would seem to be the prudent insulator choice.

3. ACKNOWLEDGEMENTS

The authors wish to thank Prof. Kermit L. Holman of New Mexico State University for suggesting the applicability of Gear's algorithm and John W. Starner, Jr. for programming assistance.

References

[1] KRAKOWSKI, R. A., RIBE, F. L., COULTAS, T. A., HATCH, A. J., An Engineering Design Study of a Reference Theta-Pinch Reactor (RTPR), USAEC Rep. LA-5336/ANL-8019 (1974).

[2] BADGER, B. et al., A Wisconsin Toroidal Fusion Reactor Design, University of Wisconsin (USA) Rep. UWFDM-68 (1973).

[3] MARONI, V. A., CAIRNS, E. J., CAFASSO, F. A., A Review of the Chemical, Physical and Thermal Properties of Lithium That Are Related to Its Use in Fusion Reactors, USAEC Rep. ANL-8001 (1973).

[4] MARONI, V. A., An Analysis of tritium distribution and leakage characteristics for two fusion reaction reference designs, Proceedings of the Fifth Symposium on Engineering Problems of Fusion Research, IEEE Pub. No. 73CHO843-3-NPS (1974) 206.

[5] VELECKIS, E. et al., Physical Inorganic Chemistry Semiannual Report (July-December 1972), USAEC Rep. ANL-7978 (1973) 4-9.

[6] VELECKIS, E., VANDEVENTER, E. H., BLANDER, M., The lithium-lithium hydride system, J. Phys. Chem. 78 (1974) 1933.

[7] MESSER, C. E., et al., J. Phys. Chem. 62 (1958) 220.

[8] HILDEBRAND, J. H., SCOTT, R. L., The Solubility of Nonelectrolytes, 3rd Ed., Reinhold Publishing Corp., New York (1950) 34.

[9] a. ZELEZNIK, F. J., GORDON, S., An Analytical Investigation of Three General Methods of Calculating Chemical Equilibrium Compositions, NASA Report NASA-TN-D473 (1960).
b. VAN ZEGGEREN, F., STOREY, S. H., The Computation of Chemical Equilibrium, Cambridge University Press, London (1970).
c. BRINKLEY, S. R., Calculation of the equilibrium composition of systems of many constituents, J. Chem. Phys. 15 (1947) 107.
d. KANDINER, H. J., BRINKLEY, S. R., Calculation of complex equilibrium relations, Ind. Eng. Chem. 42 (1950) 850.

[10] ZELEZNIK, F. J., GORDON, S., A General IBM 704 or 7090 Computer Program for Computation of Chemical Equilibrium Compositions, Rocket Performance, and Chapman-Jouget Detonations, NASA Report NASA-TN-D1454 (1962).

[11] VILLARS, D. S., Computation of complicated combustion equilibria on a high-speed digital computer, Kinetic, Equilibria and Performance of High Temperature Systems, Proc. 1st Conf. (BAHM, G. S., ZUKOSKI, E. E., Eds.), Butterworths, London (1960) 141.

[12] TESTER, J. W., FEBER, R. C., HERRICK, C. C., Heat Transfer and Chemical Stability Calculations for Controlled Thermonuclear Reactors (CTR), Los Alamos Sci. Lab. Report LASL-5328-MS (1973).

[13] JANAF THERMOCHEMICAL TABLES, Dow Chemical Co., Midland, Michigan (June, 1974).

[14] STEINMETZ, H. L., Using the method of steepest descent, Ind. Eng. Chem. 58 (1966) 33.

[15] GEAR, C. W., The automatic integration of ordinary differential equations, Common. ACM 14 (1971) 176.

[16] ROSNER, D. E., ALLENDORF, H. D., Kinetics of the attack of refractory solids by atomic and molecular fluorine, J. Phys. Chem. 75 (1971) 308.

DISCUSSION

A. PATTORET (Chairman): Were the curves shown in Figs 1 and 2 drawn by hand, or smoothed out by adjusting the α and β-parameters of the solution model?

V.A. MARONI: The curves shown in Figs 1 and 2 were drawn 'by hand' through the data. The curves obtained using α and β give an excellent fit to the isotherms below 850°C for Li-H. There is some deviation from experimental data (less than 3%) for the 879°C and 903°C isotherms for Li-H. In the case of Li-D (Fig.2) the refined values of α and β also correspond very well to the isotherms.

A. PATTORET: Does your programme for calculating complex equilibria enable you to deal with systems in which the activities of the constituents in the condensed phase are lower than unity and are variable?

V.A. MARONI: In the studies reported in our paper, the only condensed phase considered to be present was the base material (BeO, BN, or Al_2O_3) at unit activity. My impression of the work by Feber and Herrick is that only gas phase species may be introduced into the calculation unless the activity of any added condensed phases is fixed at unity throughout the iteration procedure.

HIGH-TEMPERATURE THERMOPHYSICAL PROPERTIES OF TUNGSTEN, MOLYBDENUM, NIOBIUM AND TANTALUM

M. HOCH
University of Cincinnati,
Cincinnati, Ohio
United States of America

Abstract

HIGH-TEMPERATURE THERMOPHYSICAL PROPERTIES OF TUNGSTEN, MOLYBDENUM, NIOBIUM AND TANTALUM.

Specific heat data for tungsten, molybdenum, niobium and tantalum, obtained in various laboratories using calorimetry, extremely rapid transient techniques, and the modulation method, are compared and evaluated. For these elements, the data agree to within 1%, 3%, 5% and 3%, respectively, and confirm the author's previously developed equation for the specific heat of solids between room temperature and the melting point. Similarly, a comparison of the values of the total emittance measured by various methods, and also computed from electrical resistivity measurements, reveals close agreement in the case of molybdenum and tungsten, though not for niobium and tantalum. Finally, the thermal conductivity of the four metals was evaluated using values derived from diffusivity and direct conductivity measurements as well as some computed from lattice theory and utilizing available data on electrical resistivity. With the exception of niobium, the results are in good agreement.

1. Introduction

The specific heat, total emittance, thermal conductivity, and thermal diffusivity are important thermophysical quantities because they describe the heat storing ability as well as the heat and energy transmitting ability of the material. Large amounts of data on these properties have been collected for tungsten, molybdenum, niobium, and tantalum during recent years. This store of data suggested that evaluation of the best current values for Cp, ε, and λ would be appropriate.

2. Experimental Data and Discussion

2.1. Specific Heat

In an earlier paper[1] Hoch showed that the high temperature specific heat of refractory materials can be expressed between $T = \theta_D$ and the melting point by an equation of the form

$$Cp = 3RF(\theta/T) + bT + dT^3 \qquad (1)$$

where $F(\theta/T)$ is the Debye function, b reflects the electronic contribution, and d is the anharmonic term. Values of the specific heat can be determined experimentally by drop calorimetry, by extremely rapid transient measurements, or by the modulation method.

It was shown earlier[2] that for tungsten the values obtained by calorimetric and transient techniques agree quite well; in fact, within 1%. In Fig. 1 we replot the data on

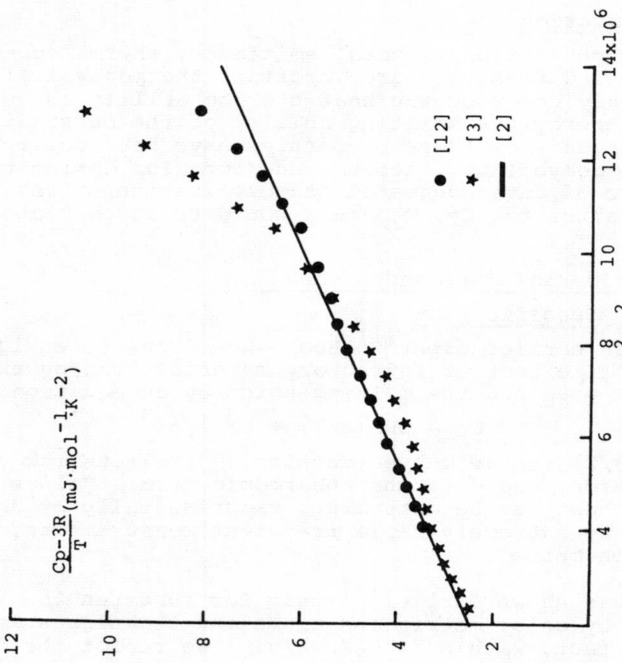

FIG.1. Specific heat of tungsten.

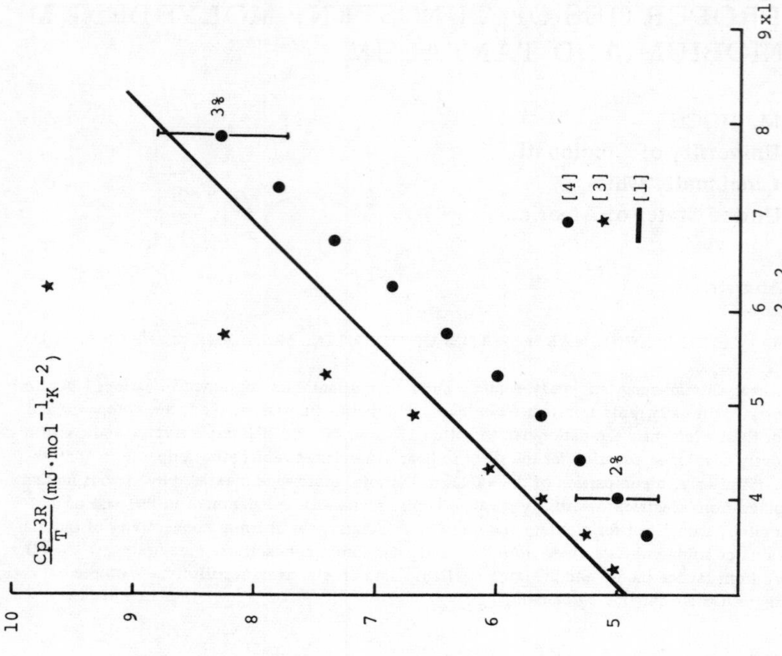

FIG.2. Specific heat of molybdenum.

$$Y = \frac{1}{T^2}\left[H_T - H_{273} - 3R\left(T + \frac{\theta_D^2}{20T}\right) + 3R\left(273 + \frac{\theta_D^2}{20\times 273}\right)\right] = \frac{b}{2} + \frac{d}{4}T^2$$

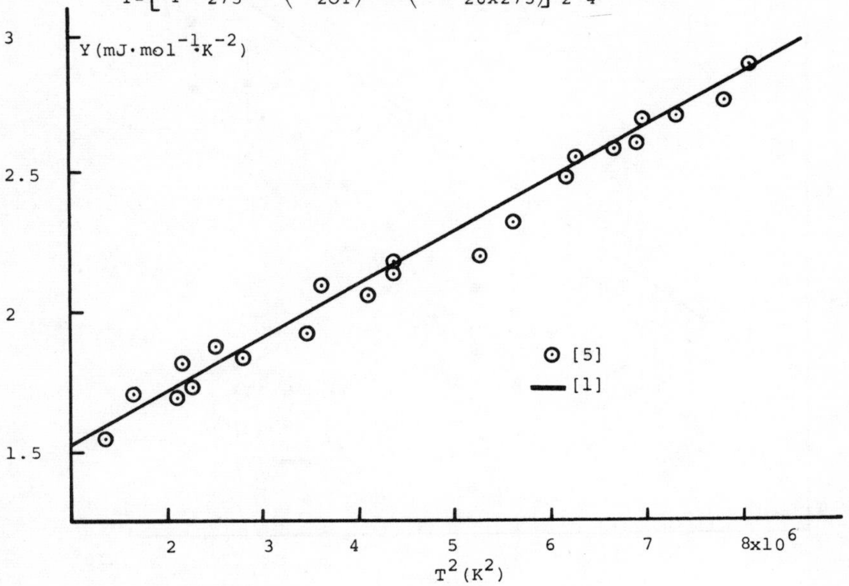

FIG.3. Heat content of molybdenum.

tungsten obtained by calorimetric and transient techniques[2] together with that from the modulation method of Kraftmakher.[3] At low temperatures the agreement between the modulation method and the previously published results is extremely good. At elevated temperatures, however, the data obtained by the modulation technique shows increasing deviation toward higher values.

For molybdenum, in Fig. 2 the high temperature specific heat data of Cezairliyan[4] and Kraftmakher[3] is plotted according to Eq. (1). For comparison the line is reproduced from our earlier results.[1] Agreement between our results and those of Cezairliyan[4] is within the experimental errors given by Cezairliyan.[4] Kraftmakher's[3] data shows, again, deviation at higher temperatures. For the same metal, Fig. 3 presents the high temperature heat content data of Kirrillin et. al.[5] along with the line developed previously in our laboratory.[1] Here the agreement is extremely good. Figure 4 shows data on the high temperature specific heat of niobium from Cezairliyan[11] and Kraftmakher,[3] again with the line being from our earlier result.[1] The agreement here is within 5%. Figure 5 compares the heat content data of Kirillin[6] on niobium with our earlier line.[1] Here the agreement is extremely good.

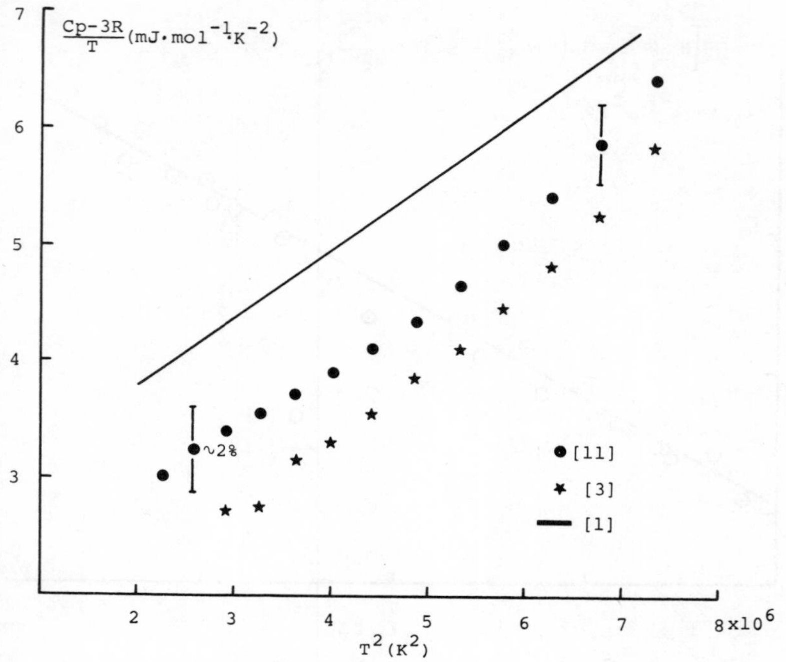

FIG. 4. Specific heat of niobium.

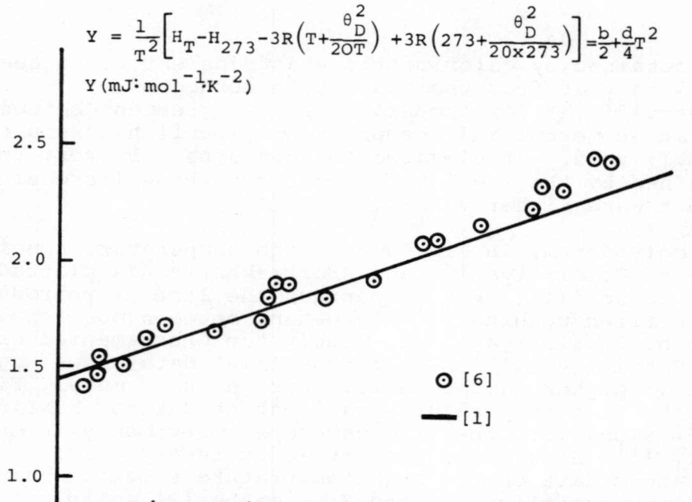

FIG. 5. Heat content of niobium.

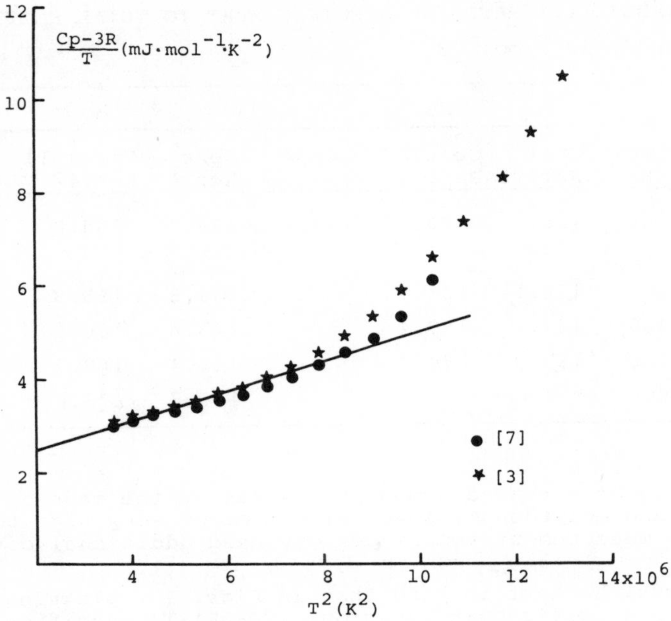

FIG. 6. Specific heat of tantalum.

TABLE I. SPECIFIC HEAT OF REFRACTORY METALS

$Cp = 3RF(\theta_D/T) + bT + dT^3$

Material	θ_D (K)	b (J·mol^{-1}·K^{-2})	d (J·mol^{-1}·K^{-4})	Accuracy of Cp in %	$\gamma_{e\ell}$ (J·mol^{-1}·K^{-2})[8]
W	310	1.68×10^{-3}	4.25×10^{-10}	1	1.3×10^{-3}
Mo	380	2.67×10^{-3}	7.65×10^{-10}	3	2.0×10^{-3}
Nb	300	2.61×10^{-3}	5.93×10^{-10}	5	7.8×10^{-3}
Ta	245	1.85×10^{-3}	3.15×10^{-10}	3	5.9×10^{-3}

The high temperature data on tantalum from the previous report[1] showed that the excess specific heat function, (Cp-3R)/T, unexplainably decreases with increasing temperature. Figure 6 presents the high temperature specific heat data of Cezairliyan[7] and of Kraftmakher[3] on tantalum. The line through the data represents the points of Cezairliyan quite well and agrees with the low temperature data of Kraftmakher. At high temperatures Cezairliyan's data deviate from the straight line, possibly due to vacancy formation. On the other hand, a 3% error in the specific heat at 3100 K would bring that point onto the line. Kraftmakher's data, however, deviate

TABLE II. RATIO OF SPECIFIC HEAT TO TOTAL EMITTANCE, Cp/ε, in $J \cdot mol^{-1} \cdot K^{-1}$

	Nb			Mo		
Temp. (K)	Cp/ε Ref.11	Cp/ε(ρT) Ref.11	Cp/ε Ref.10	Cp/ε Ref.4	Cp/ε(ρT) Ref.4	Cp/ε Ref.10
2000	128.9	140.0	163.1±5	136.5	168.8	154.6±7
2200	126.1	138.4		138.8	165.3	
2400	127.4	139.4		143.6	166.0	
2600	132.8	144.2		149.4	168.7	
2800				156.7	173.6	

drastically at elevated temperatures (as in the case of tungsten and molybdenum) suggesting that at very high temperatures his measurement techniques may need additional correction.

It must be kept in mind that in Figs. 1-6 plotting only the excess specific heat function, $(Cp-3R)/T$, magnifies deviations and scatter. Table I contains the coefficients, b and d, of the specific heat equations for tungsten, niobium, molybdenum, and tantalum, as well as the electronic specific heat, γ (last column), determined from low temperature specific heat data.[8] In the case of tungsten and molybdenum, the agreement between b and γ is very good. In the case of tantalum and niobium, γ is larger than b. This is because in niobium and tantalum the Fermi level is probably within a peak of density of states and, therefore, as the temperature is raised the electronic specific heat decreases.[9] It is, however, noteworthy that the value b for niobium and tantalum at high temperature is close to the value of molybdenum and tungsten, respectively.

2.2 Total Emittance

It has been shown previously[10] that the ratio of specific heat to total emittance, Cp/ε, is constant over a wide temperature range. Since specific heat data are much more accurate and easily obtained than total emissivity data, the use of Cp/ε to obtain ε is recommended.

Total emittance for tungsten, molybdenum, niobium, and tantalum was measured by Cezairliyan et. al.[4,7,11,12] The total emittance, $\varepsilon(\rho T)$, can also be calculated independently from the electrical resistivity measurements of Cezairliyan using an equation derived by Abott and Parker:[2,13]

$$\varepsilon(\rho T)=0.766(\rho T)^{1/2}-[0.309-0.091g\ \rho T]\rho T-0.0175(\rho T)^{3/2} \quad (2)$$

where ρ is the resistivity in ohm-cm.

Combining this with his specific heat data, Cp/ε can be calculated from Cezairliyan (their resistivity and total

W				Ta			
Temp. (K)	Cp/ε Ref.12	Cp/ε(ρT) Ref.12	Cp/ε Ref.14	Temp. (K)	Cp/ε Ref.7	Cp/ε(ρT) Ref.7	Cp/ε Ref.15
2336	105.2	129.4		2400	107.2	121.2	153.3±6
2673	110.5	126.5		2600	108.9	120.1	
3013	120.4	126.6	128±3	2800	112.0	121.0	
3323	131.3	130.9		3000	116.3	124.0	
3418	133.0	133.0					

emittance data were from the sample on which they measured emittance). Hoch et. al.[10,14,15] also directly measured Cp/ε, by cooling techniques on large solid samples. All those data are given in Table II. The agreement is best in the case of tungsten and least satisfactory in the case of tantalum.

2.3. Thermal Conductivity

The thermal conductivity of tungsten was evaluated both from direct measurements and from thermal diffusivity data by Hoch and Minges[16] in their correlation of the data from the large international program to measure thermal properties of tungsten (and other materials) at high temperatures. (This program was supported by the United States Air Force Materials

TABLE III. THERMAL CONDUCTIVITY
λ, W $cm^{-1}K^{-1}$)

Material	2000 K	2500 K	Ref.	Method
W	1.00 ±.03	1.00 ±.03	14	direct
	.991±.13	.914±.13	16	direct
	1.06 ±.07	1.07 ±.07	16	diffusivity
	.985	.913		Eq. 3
Mo	.950±.05		17	direct
	.984±.05	.925±.05	18	diffusivity
	1.03	.971		Eq. 3
Nb	.921±.05	.943±.05	18	diffusivity
	.737	.767		Eq. 3
Ta	.631±.05	.609±.05	15	direct
	.631±.03	.619±.03	18	diffusivity
	.68	.70		Eq. 3

Laboratory and AGARD.) The results are given in Table III, together with the data of Jun and Hoch.[14] The molybdenum and tantalum data shown are those of Hoch and Jun[15,17] obtained by direct measurements and of Kraev[18] and coworkers utilizing diffusivity measurements. The niobium data is that of Kraev[18] and coworkers.

The thermal conductivity of metals at elevated temperature can be expressed as $\lambda=\lambda_e+\lambda_p$ where λ_e and λ_p are the thermal conductivity components due to electron transport and phonon transport, respectively. For λ_e we can use the classical Wiedemann-Franz-Lorenz relationship, while at high temperatures λ_p is inversely proportional to the temperature. Thus (at high temperature),

$$\lambda = \frac{A}{T} + \frac{2.45 \times 10^{-8} T}{\rho} \qquad (3)$$

where ρ is the resistivity.

The constant, A, has been evaluated from a simple anharmonic model by Leibfried and Schlomann[19] and its value is 225 for tungsten, 215 for molybdenum, 113 for tantalum, and 99 for niobium when the thermal conductivity is expressed in $W\ cm^{-1}K^{-1}$. Using those values and the electrical resistivity data of Cezairliyan for the various metals, a theoretical thermal conductivity was computed and is also given in Table III. Considering the difficulty in measuring thermal conductivity, either by direct methods or by thermal diffusivity measurements, the theoretical and experimental results for tungsten, tantalum, and molybdenum are in excellent agreement.

REFERENCES

[1] HOCH, M., "The high temperature specific heat of body-centred-cubic refractory metals", High Temperatures-High Pressures 1 (1969) 531.

[2] HOCH, M., "High temperature thermophysical properties of tungsten", High Temperatures-High Pressures 4 (1972) 659.

[3] KRAFTMAKHER, Ya A., "The modulation method for measuring specific heat", High Temperatures-High Pressures 5 (1973) 433.

[4] CEZAIRLIYAN, A., MORSE, M.S., BERMAN, H.A., BECKETT, C.W., "Hich-speed (subsecond) measurement of heat capacity, electrical resistivity, and thermal radiation properties of molybdenum in the range 1900 to 2800°K", J. of Research NBS 74A 1 (1970) 65.

[5] KIRILLIN, V.A., SHEINDLIN, A.E., CHEKHOVSKOI, V. Ya., PETROV, V.A., "Enthalpy and heat capacity of molybdenum at extremely high temperatures", Proceedings of Fourth Symposium on Thermophysical Properties, The American Society of Mechanical Engineers, New York (1968) 54.

[6] KIRRILLIN, V.A., SHEINDLIN, A.E., CHEKHOVSKOI, V. Ya., ZHUKOVA, I.A., "Experimental determination of the enthalpy of niobium in the temperature range 600-2600°K, Taplofizika Vysokikh Temperatur 3 3 (1965) 395.

[7] CEZAIRLIYAN, A., McCLURE, J.L., BECKETT, C.W., "High-speed (subsecond) measurement of heat capacity, electrical resistivity, and thermal radiation properties of tantalum in the range 1900 to 3200°K", J. of Research NBS 75A 1 (1971) 1.

[8] KITTEL, C., Introduction to Solid State Physics, 4th Ed., John Wiley, New York (1971).

[9] HOCH, M., "Equilibrium measurements in high melting materials", Vacancies and Interstitials in Metals, A. Seeger, D. Schumacher, W. Schilling, and J. Diehl, Eds., North Holland, Amsterdam (1969) 81.

[10] NARASIMHAMURTY, H.V.L., IYER, A.S., HOCH, M., "Relation between specific heat and total emittance in tantalum, niobium, tungsten, and molybdenum", J. Phys. Chem. 69 (1965) 1420.

[11] CEZAIRLIYAN, A., "High-speed (subsecond) measurement of heat capacity, electrical resistivity, and thermal radiation properties of niobium in the range 1500 to 2700°K", J. of Research NBS 75A 6 (1971) 565.

[12] CEZAIRLIYAN, A., McCLURE, J.L., "High-speed (subsecond) measurement of heat capacity, electrical resistivity, and thermal radiation properties of tungsten in the range 2000-3600°K", J. of Research NBS 75A 4 (1971) 283.

[13] ABOTT, G.L., PARKER, W.J., "Theoretical and experimental studies of the total emittance of metals", NASA-SP-55 (1965).

[14] HOCH, M., JUN, C.K., EBRAHIM, S., "The thermal conductivity and total emittance of tungsten from 1800-2800°K, High Temperatures-High Pressures 2 (1970) 43.

[15] JUN, C.K., HOCH, M., "Thermal conductivity and total emittance of tantalum, tungsten, rhenium, Ta-10W, T_{111}, T_{222}, and W-25Re in the temperature range 1500-2800°K", High Temperature Technology, Butterworths, London (1969) 535.

[16] HOCH, M., MINGES, M.L., "Accuracy and precision in high temperature specific heat and thermal conductivity measurements", III International Conference on Chemical Thermodynamics, Congress Center, Baden (1973) to be published.

[17] HOCH, M., JUN, C.K., "The thermal conductivity of molybdenum at elevated temperatures: influence of grain size, Advances in Thermophysical Properties at Extreme Temperatures and Pressures, ASME, New York (1965) 296.

[18] KRAEV, O.A., STEL'MAKH, A.A., "Thermal diffusivity of tantalum, molybdenum, and niobium at temperatures above 1800°K, Teplofizika Vysokikh Temperatur 2 2 (1964) 302.

[19] LEIBFRIED, G., SCHLOMANN, E., Nachr. Akad. Wiss, Gottingen IIa, 4 (1954) 71.

DISCUSSION

R.J. ACKERMANN: What experimental method are you at present using or do you intend to use for measuring the heat capacities of liquid metals?

M. HOCH: Drop calorimetry is the only method to use, either with levitation heating or using a crucible to contain the liquid. In the first case the problem is emissivity, while in the second it is metal-container interaction. Furthermore, in order to obtain meaningful data a temperature range of 300-500°C is required. For liquid refractory metals we shall model the alkali metals and predict the values. For oxides, Al_2O_3 can be heated in tungsten containers to 2500°C without difficulty. It must be borne in mind that careful work and analysis is needed to obtain C_p from heat content data.

M. KATSURA: When you apply your theoretical equation to the high-temperature compounds, for example the oxides, how can you determine the Debye temperature of the compounds?

M. HOCH: We use a Debye temperature determined from low-temperature measurements.

M. KATSURA: As far as I am aware, the electronic contribution to the specific heat is predominant only at low temperatures. Is it necessary to take into account the electronic contribution in the higher temperature range between the Debye temperature and the melting point?

M. HOCH: The electronic specific heat is γT. This is important at low temperatures, where the lattice contribution (Debye term) decreases to zero, and at very high temperatures, where the Debye term is 3R. γT increases with temperature and can amount to 15-25% of 3R.

A.S. PANOV: I would first like to note the excellent agreement (particularly at low temperatures) of your model calculations with the experimental results. With an increase in temperature, the concentration of structural vacancy defects in metals increases and at high temperatures (2000-2500°C) may reach very substantial values. How did you, in your calculations, take into account the change in the specific heat with the rise in the concentration? Is this not a reason for the deviation of the theory from practice at high temperatures?

M. HOCH: As is pointed out in the paper, the excess specific heat function $(C_p - 3R)/T$ is plotted, which greatly magnifies deviations and scatter. May I in this connection quote from an earlier work of ours (Vacancies and Interstitials in Metals, North Holland Publishing Company, Amsterdam (1969)): "Specific heat data with an accuracy of 0.1% or better from the Debye temperature to the melting point are required to be able to obtain concentration and energies of formation of vacancies from specific heat data."

STABILITY AND SOLUBILITY OF METAL OXIDES IN HIGH-TEMPERATURE WATER

D.D. MACDONALD, T.E. RUMMERY, M. TOMLINSON
Atomic Energy of Canada Limited,
Whiteshell Nuclear Research Establishment,
Pinawa, Manitoba,
Canada

Abstract

STABILITY AND SOLUBILITY OF METAL OXIDES IN HIGH-TEMPERATURE WATER.

Some recent compilations of elementary thermodynamic properties of metal-water systems and derived quantities which are needed for optimum design and operation of nuclear steam generation equipment are summarized. Free energies of iron, cobalt, nickel, copper and aluminium metal, oxides and ions in dilute aqueous solution at temperatures up to 300°C are tabulated. Derived data on potential-pH relations, solubilities, H_2 and O_2 equilibrium pressures for reduction and oxidation processes and compound oxide formation are outlined. The practical significance is indicated. Needs for improved data are noted.

INTRODUCTION

The nuclear power industry has greatly stimulated the evaluation of the thermodynamic properties of reactor core materials, such as nuclear fuel and sheathing materials, as is evident from this conference. Advances in knowledge of the thermodynamics of the heat transport and steam raising parts of the plant are much less in evidence although there is in fact great scope for practical improvements and cost savings here. For example the margin of incentive for improvements in fuel and core components in CANDU reactors [1, 2] is slight because fuelling costs are less than one seventh of the total power generation costs, existing fuel designs are highly reliable and efficient, and if fuel defects should occur, they could be immediately removed by the on-power refuelling equipment. The cost contribution and the scope for saving is significant for the heat transport and the steam raising components.

Almost all power plants, nuclear heated or otherwise, are steam engines and therefore must have enormous areas of metal-oxide-water interfaces at elevated temperature. The most widely used nuclear power plant materials are iron-nickel-chromium and nickel-copper alloys which form the bulk of the heat transport system that transfers the nuclear heat to the steam turbines. The thermodynamic properties of the first row transition metals, their oxides and their ions in solution are thus of prime concern for nuclear power plant design. There are ∿20 000 m^2 of wetted metal surface (Fe and Ni based alloys) outside the core of a CANDU unit compared with ∿3 000 m^2 of wetted surface (Zr alloys) in-core. The 4 unit Pickering Station [2] has 2000 km of boiler tubing. As in most steam generating plants whatever the heat source, this includes about 30% excess boiler tubes to allow for possible tube failures. Here is one large potential cost saving through improved performance of the material. In addition, all water cooled reactors accumulate radioactivity in out-core components [3, 4]. This creates radiation hazards for the operating staff and the preventive and protective measures add to construction costs. Furthermore, in some plants, deposits of corrosion products have caused costly shut-downs.

This paper summarises computed values [5-11] of the free energies and important derived quantities for metal, oxides and ions in dilute aqueous solution for Fe, Co, Ni, Cu and Al at temperatures up to 300°C. These are: free energy changes for reactions in metal-water systems, potential - pH relations, solubilities, H_2 and O_2 equilibrium pressures for reduction and oxidation processes, and stability criteria for compound oxides. The practical significance of such data is briefly indicated. Instances are noted where data are lacking and where improvements of existing information are desirable.

Notable features of the calculations are: firstly, the primary step is the computation of free energies of individual species at elevated temperature rather than integration of the free energy changes for particular reactions as is often done. The present method facilitates application of the primary high temperature free energy tabulations to any reactions of the tabulated species. Secondly, whereas normally the free energies of the elements are defined as zero at the temperature of interest, here the free energies of formation for all temperatures are based on the elements in their standard states at 298 K as the reference state. Thus the $G°(T)$ values tabulated here are for formation of the species at temperature T from the elements at 298 K. Having the same reference state for all temperatures greatly facilitates computations (and, of course, leads to values of the free energy change for reactions which are the same as calculated by other routes from the same basic data.) Thirdly, the Criss-Cobble relationship is used to estimate entropies of ions at elevated temperatures in lieu of specific heat data.

BASIC RELATIONS

The treatment follows that of Pourbaix [12] and Criss and Cobble [13-15]. Free energies at elevated temperature were obtained [5-8] from known free energies at 25°C and heat capacity functions by evaluation of equation (1)

$$G°(T_2) = G°(T_1) - S°(T_1)\,[T_2-T_1] - T_2 \int_{T_1}^{T_2} \frac{C_P^\circ}{T}\,dT + \int_{T_1}^{T_2} C_P^\circ dT \qquad (1)$$

Thermodynamic functions for solids and dissolved species were computed for 0.1 MPa pressure without correction for the $\int VdP$ contribution to the free energy due to the change in vapour pressure of water at elevated temperature. The correction is negligible for the systems of interest at temperatures up to 573 K [11, 13]. The standard free energies of solution species were computed at unit activity. Note that standard free energies of formation, $G°(T)$ have the basis that the free energies of the elements are zero in their normal state at the temperature 298 K and 0.1 MPa pressure (more precisely 0.101325 MPa = 1 atmosphere). Equation 1 was used for pure substances for which accurate experimentally based heat capacity functions of the form $C_P^\circ = A+BT+CT^{-2}$ . . . are available. Such data are not available for ions in solution. For these, the approximate relation 2 was used, in which entropies are substituted for specific heat data.

$$G°(T_2) = G°(T_1) - [T_2S°(T_2) - T_1S°(T_1)] + \frac{(T_2-T_1)}{\ln(T_2/T_1)}\,[S°(T_2)-S°(T_1)] \qquad (2)$$

The entropies of the ions at elevated temperature were estimated by means of the correspondence principle, equation 3, of Criss & Cobble [13-15]

$$S^\circ(T_2) = a(T_2) + b(T_2)\, S^\circ(T_1) \tag{3}$$

The parameters $a(T_2)$, $b(T_2)$ are invariant for a given class of ion and a fixed reference temperature T_1. Here T_1 = 298 K throughout. Note that the entropies used are on the absolute scale [13, 14]: $S^\circ(H^+, 298) \equiv 20.93$ J/K mol

Electrode potentials were derived in order to construct potential-pH (Pourbaix) diagrams [12] and determine potential-pH relationships at elevated temperature. The electrode potentials were obtained from the free energy data using equation 5 for the reduction potential of the half cell reaction 4 for the reduction of A to B.

$$\alpha A + \beta H^+ + \gamma e^- = \delta B + \varepsilon H_2O \tag{4}$$

$$E = -\frac{\Delta G^\circ}{\gamma F} - \frac{2.303\ RT}{\gamma F}\left\{\delta \log a_B - \alpha \log a_A\right\} - \frac{2.303\ RT\beta}{\gamma F} \times pH_T \tag{5}$$

The activity of water was assigned the value 1 (dilute solutions). Reduction potentials are referred to the standard hydrogen electrode at the same temperature so that ΔG° in equation 5 is the standard free energy change, equation 7, for the whole-cell reaction 6.

$$\alpha A + (\beta-\gamma)H^+ + \frac{1}{2}\gamma H_2 = \delta B + \varepsilon H_2O \tag{6}$$

$$\Delta G^\circ = \left\{\delta G^\circ_B + \varepsilon G^\circ_{H_2O}\right\} - \left\{\alpha G^\circ_A + (\beta+\gamma) G^\circ_{H^+} + \frac{1}{2}\gamma G^\circ_{H_2}\right\} \tag{7}$$

Solubilities of solids in aqueous solution were obtained by summation (equation 8) of equilibrium concentrations in solution of all species B which might be formed in solution from solid A by reactions of the general form of reaction 4, the activities a_B of individual species B being obtained from equation 9.

$$C_s = \sum_B a_B/\gamma_B \tag{8}$$

$$\log a_B = \frac{-\Delta G^\circ}{2.303\ RT\delta} + \left(\frac{\beta-\gamma}{\delta}\right) pH_T + \left(\frac{\gamma}{2\delta}\right) \log p_{H_2} \tag{9}$$

The approximation $\gamma_B \simeq 1.0$ was used since we were interested primarily in materials of very low solubility and aqueous solutions (such as 10^{-4} mol/kg LiOH) of low ionic strength (<1 mmol/kg).

Note that because of the water dissociation equilibrium 10, the complete specification of the system requires a hydrogen activity (or alternatively an oxygen activity as determined by equation 11).

$$H_2O \underset{}{\overset{K_{H_2O}}{\rightleftharpoons}} H_2 + \frac{1}{2} O_2 \tag{10}$$

$$K_{H_2O} = a_{H_2} \times a_{O_2}^{\frac{1}{2}} \tag{11}$$

Oxidation - Reduction relationships giving equilibrium H_2 pressures or O_2 pressures were obtained [9, 10] in order to determine the thermodynamic stability of metals and metal oxides in aqueous systems in the presence of hydrogen or in the presence of oxygen.

The reduction of a metal oxide by hydrogen to give a lower oxide or metal plus water is represented by reaction 12.

$$xA + zH_2 = yB + zH_2O \tag{12}$$

The equilibrium partial pressures of hydrogen for such reactions were obtained from equation 13.

$$\log_{10} P(H_2, eq) \simeq \Delta G° / 2.303zRT \tag{13}$$

Here $\Delta G°$ is the standard free energy change for reaction 12. The approximation is in equating partial pressure to fugacity, which is acceptable for conditions of practical interest.

The equilibrium partial pressure data can be accurately described in the form of equation 14 and the appropriate constants have been evaluated [16] for the reactions of interest by least squares fitting.

$$\log_{10} P(H_2, eq) = A + BT\log_{10}T + CT \tag{14}$$

The reaction of oxygen with metals or lower oxides to form oxides is described by reaction 15, where A now represents the reduced component and B the oxidised component.

$$xA + z'O_2 = yB \tag{15}$$

The equilibrium partial pressures of oxygen for such reactions were calculated using equation 16, which has also been evaluated as a series expression in T, analogous to equation 14.

$$\log_{10} P(O_2, eq) \simeq \Delta G° / 2.303z'RT \tag{16}$$

Input data: Specific heats were obtained from the US Bureau of Mines compilations [17, 18]. Free energy and entropy data were obtained principally from NBS compilations of "Selected Values of Chemical Thermodynamic Properties" [19-21]. The free energy of formation of the hydrogen ion at temperature T was based [5] on data for OH^- and experimental values [22] for $K_W^°$, the ionic dissociation constant of water. $G_{OH^-}^°$ was computed by equation 2 then $G_{H^+}^°$ was obtained from equation 17.

$$G^\circ_{H^+} = G^\circ_{H_2O} - G^\circ_{OH^-} - RT\ln K^\circ_w \qquad (17)$$

The occasional use of data from sources other than the above is discussed in the detailed compilations for particular metal systems [5-8, 11]. The extremes of expediency are represented by the use of the entropies of Co_3O_4, Fe_2O_3 and MnO_2 as substitutes for the entropies of Ni_3O_4, Ni_2O_3 and NiO_2, respectively.

Errors and uncertainties depend on the particular system of interest. For the better known substances, free energy data at ordinary temperatures from different sources frequently agree to within 5 kJ/mol. The NBS input data are stated to be internally consistent such that computed free energy changes are independent of the path chosen. The basic heat capacity data are known to better than ± 2%, often better than ± 0.1%. The integration of G° over the temperature range of interest is expected to be accurate to within a few percent. The principal uncertainty in the free energy at elevated temperature probably is the error in the 298 K value.

For ions, the accuracy of the free energy values at elevated temperature depends on the validity of the Criss and Cobble correspondence principle which has not been extensively tested. Agreement to within 5% has been reported [11] for experimental data up to 300°C. Calculated solubilities are thought to be accurate to within 1 order of magnitude in general. Near the pH of minimum solubility, the values may be low due to omission from the summations of neutral and polymeric species in solution, for which there are no data. The change of solubility with pH and temperature is more precisely determined, probably within a few percent.

For the reduction and oxidation processes considered, the uncertainty in ΔG° at elevated temperature is unlikely to exceed 15 kJ/mol. For the equilibrium H_2 and O_2 pressures, the corresponding maximum uncertainty in log P is ∿2.

Specific details are discussed in the compilations for individual metal systems.

RESULTS AND DISCUSSION

Free energies have been computed for temperatures from 298 K to 573 K in 50 K intervals for the systems H_2O, Fe, Co, Ni, Cu and Al [5-8]. The values are given in Table I.

To derive potential, solubility and equilibrium H_2 and O_2 pressure data, the free energy changes for a large number of reactions which are involved in the processes of dissolution, hydrolysis, oxidation, reduction, hydration, corrosion, etc., in the above systems have also been computed. Examples of such data are given in Tables II, III and Figure 1. Table II lists the reactions for which free energy changes have been computed up to 573 K for the copper-water system. Complete tabulations of ΔG°(T) values for these reactions and for analogous reactions of the other metal-water systems are given in the appropriate reports. The corresponding potential-pH relationships are also tabulated there.

The potential-pH relationships for Cu at 573 K are reproduced in Table III. The same data are shown in the form of a Pourbaix diagram in Figure 1. Equilibrium lines involving dissolved ions are plotted for ionic activity = 10^{-6}, this being in the range of practical interest, rather than for the standard activity = 1. The numbers on the lines in the diagram correspond with the reaction numbers in Table III. For example, the vertical line 12 corresponds to 10^{-6} mol/kg Cu^{2+} in

TABLE I. FREE ENERGIES AT ELEVATED TEMPERATURE
$G^0(T)$ (kJ/mol)

Temp (K)	298	333	373	423	473	523	573
H^+	0.0	0.54219	0.56229	-0.31401	-2.5958	-6.0457	-10.135
H_2	0.0	-4.6302	-10.052	-17.001	-24.122	-31.397	-38.812
O_2	0.0	-7.2392	-15.647	-26.338	-37.212	-48.251	-59.441
H_2O	-237.34	-239.94	-243.24	-247.83	-252.86	-258.29	-264.08
OH^-	-157.40	-157.37	-156.24	-153.14	-148.23	-141.50	-132.78
Fe	0.0	-1.0053	-2.2691	-4.0073	-5.9087	-7.9628	-10.1616
Fe_3O_4	-1016.1	-1021.6	-1028.5	-1038.2	-1048.9	-1060.6	-1073.2
Fe_2O_3	-742.74	-746.01	-750.25	-756.25	-762.98	-770.384	-778.43
Fe^{2+}	-78.921	-73.070	-67.697	-63.002	-60.408	-59.972	-61.659
$Fe(OH)^+$	-262.18	-260.75	-260.11	-260.80	-263.07	-266.96	-272.46
$HFeO_2^-$	-378.07	-379.68	-380.11	-378.39	-374.20	-367.62	-363.80
FeO_2^{--}	-295.59	-292.35	-285.52	-272.92	-255.78	-233.56	-206.41
Fe^{3+}	-4.6055	+8.0613	+20.726	+33.723	+43.798	+50.908	+55.084
$Fe(OH)^{++}$	-224.58	-218.31	-212.50	-207.30	-204.25	-203.40	-204.73
$Fe(OH)_2^+$	-438.36	-437.07	-436.59	-437.46	-439.88	-443.91	-449.53
$Fe_2(OH)_2^{4+}$	-467.00	-450.45	-433.63	-415.91	-401.57	-390.66	-383.15
$Fe_3(OH)_4^{5+}$	-930.77	-907.77	-884.04	-858.45	-837.03	-819.79	-806.73
FeO_4^{--}	-467.60	-468.60	-467.65	-463.69	-456.67	-446.29	-432.61
Co	0.0	-1.1017	-2.4750	-4.3483	-6.3810	-8.5606	-10.877
CoO	-214.36	-216.32	-218.80	-222.21	-225.93	-229.94	-234.20
Co_3O_4	-774.56	-778.40	-783.38	-790.43	-798.35	-807.08	-816.56
Co^{++}	-54.428	-49.422	-44.953	-41.286	-39.620	-40.012	-42.426
$Co(OH)^+$	-235.72	-231.33	-227.52	-224.60	-223.61	-224.60	-227.55
$HCoO_2^-$	-347.38	-348.99	-349.42	-347.70	-343.51	-336.93	-333.11
Co^{+++}	133.98	146.29	158.57	171.14	180.81	187.57	191.44
Ni	0.0	-1.0980	-2.4734	-4.3608	-6.4226	-8.6482	-11.029
NiO	-211.85	-213.27	-215.10	-217.69	-220.60	-223.83	-227.38
Ni_3O_4	-712.38	-717.86	-724.71	-734.10	-744.35	-755.42	-767.23
Ni_2O_3	-470.05	-473.33	-477.56	-483.56	-490.30	-497.70	-505.74
NiO_2	-215.29	-217.25	-219.76	-223.27	-227.17	-231.41	-235.99
Ni^{++}	-45.636	-40.085	-35.034	-30.705	-28.441	-28.300	-30.246
$Ni(OH)^+$	-227.76	-224.90	-222.73	-221.68	-222.37	-224.87	-229.13
$HNiO_2^-$	-349.45	-351.07	-351.50	-349.78	-345.58	-339.00	-335.18

TABLE I (cont.)

Temp (K)	298	333	373	423	473	523	573
Cu	0.0	-1.2093	-2.7024	-4.7188	-6.8850	-9.1868	-11.613
Cu_2O	-146.12	-149.52	-153.72	-159.40	-165.51	-172.01	-178.87
CuO	-129.79	-131.37	-133.39	-136.18	-139.25	-142.58	-146.15
Cu^+	50.032	49.068	47.149	43.543	38.656	32.417	24.867
Cu^{++}	65.565	70.114	74.093	77.201	78.364	77.524	74.716
$Cu_2(OH)_2^{++}$	-281.56	-279.02	-277.18	-276.52	-277.55	-280.36	-284.89
$HCuO_2^-$	-258.74	-260.36	-260.79	-259.07	-254.87	-248.30	-244.48
CuO_2^{--}	-183.80	-180.57	-173.73	-161.13	-144.00	-121.77	-94.625
Al	0.0	-1.043	-2.340	-4.120	-6.050	-8.122	-10.32
α-Al_2O_3	-1583.	-1586.	-1588.	-1592.	-1596.	-1601.	-1606.
γ-$Al_2O_3H_2O$ [a]	-1827.	-1830.	-1835.	-1842.	-1849.	-1858.	-1867.
$Al_2O_3H_2O$ [b]	-1842.	-1845.	-1848.	-1853.	-1859.	-1866.	-1872.
$Al_2O_33H_2O$ [c]	-2322.	-2327.	-2333.	-2342.	-2350.	-2360.	-2369.
Al^{3+}	-485.7	-472.7	-460.1	-446.7	-436.3	-429.1	-425.0
$Al(OH)^{2+}$	-694.6	-688.7	-682.9	-678.3	-675.3	-675.0	-676.2
AlO_2^-	-840.3	-844.1	-846.6	-847.4	-746.6	-843.2	-837.4

[a] Boemite [b] Diaspore [c] Gibbsite

equilibrium with solid CuO. This reaction does not involve electrons and is independent of potential. The horizontal line 19 corresponds to equilibrium at the Cu/Cu^{2+} electrode at a potential 0.007 V higher than the standard hydrogen electrode. This reaction does not involve hydrogen ions at the Cu/Cu^{2+} interface and is independent of pH. The sloping lines correspond to reactions involving both electrons and hydrogen ions. Lines 28 and 29 correspond to the equilibrium potentials for H_2 and O_2 formation from H_2O respectively.

Comparison with the potential-pH diagram at 298 K in Figure 2 shows the reduced range of stability of H_2O at elevated temperature and the shift to lower pH of the regions of stability of the anionic copper species.

Such diagrams show the regions of stability of various solid phases and ionic species. They are useful in connection with laboratory studies of the metal-water interface at elevated temperature and in interpreting electrochemical aspects of corrosion under power plant conditions.

The derivation of solubility data is illustrated in Figure 3 which shows the contributions of various copper ions to the total solubility of CuO in oxygenated water at 573 K. The contributions of $Cu(OH)^+$ and $Cu(OH)_2$ (dissolved) are lacking, although these may not be entirely negligible at elevated temperature. There are no basic data for these species. Minimum solubility occurs where the predominant positive and negative ions - Cu^{2+} and $HCuO_2^-$ in this case - are present in nearly

TABLE II. REACTIONS IN THE COPPER-WATER SYSTEM

1	$2Cu^{++} + 2H_2O$	---------->	$2H^+ + Cu_2(OH)_2^{++}$
2	$Cu^{++} + 2H_2O$	---------->	$3H^+ + HCuO_2^-$
3	$Cu^{++} + 2H_2O$	---------->	$4H^+ + CuO_2^{--}$
4	$Cu_2(OH)_2^{++} + 2H_2O$	---------->	$4H^+ + 2HCuO_2^-$
5	$HCuO_2^-$	---------->	$H^+ + CuO_2^{--}$
6	$Cu_2O + 2H^+$	---------->	$2Cu^+ + H_2O$
7	$2Cu^{++} + H_2 + H_2O$	---------->	$4H^+ + Cu_2O$
8	$Cu_2(OH)_2^{++} + H_2$	---------->	$2H^+ + Cu_2O + H_2O$
9	$2HCuO_2^- + 2H^+ + H_2$	---------->	$Cu_2O + 3H_2O$
10	$2CuO_2^{--} + 4H^+ + H_2$	---------->	$Cu_2O + 3H_2O$
11	$2CuO + 2H^+ + H_2$	---------->	$2Cu^+ + 2H_2O$
12	$CuO + 2H^+$	---------->	$Cu^{++} + H_2O$
13	$2CuO + 2H^+$	---------->	$Cu_2(OH)_2^{++}$
14	$CuO + H_2O$	---------->	$H^+ + HCuO_2^-$
15	$CuO + H_2O$	---------->	$2H^+ + CuO_2^{--}$
16	$Cu_2O + H_2$	---------->	$2Cu + H_2O$
17	$CuO + H_2$	---------->	$Cu + H_2O$
18	$2CuO + H_2$	---------->	$Cu_2O + H_2O$
19	$Cu^{++} + H_2$	---------->	$2H^+ + Cu$
20	$Cu_2(OH)_2^{++} + 2H_2$	---------->	$2H^+ + 2Cu + 2H_2O$
21	$HCuO_2^- + H^+ + H_2$	---------->	$Cu + 2H_2O$
22	$CuO_2^{--} + 2H^+ + H_2$	---------->	$Cu + 2H_2O$
23	$2Cu^+ + H_2$	---------->	$2H^+ + 2Cu$
24	$2Cu^{++} + H_2$	---------->	$2H^+ + 2Cu^+$
25	$Cu_2(OH)_2^{++} + H_2$	---------->	$2Cu^+ + 2H_2O$
26	$2HCuO_2^- + 4H^+ + H_2$	---------->	$2Cu^+ + 4H_2O$
27	$2CuO_2^{--} + 6H^+ + H_2$	---------->	$2Cu^+ + 4H_2O$
28	$O_2 + 2H_2$	---------->	$2H_2O$
29	$2H^+ + H_2$	---------->	$2H^+ + H_2$

TABLE III. POTENTIAL-pH EQUATIONS FOR REACTIONS OF COPPER AT 573 K

REACTION	
1	$-2pH = -6.702 + 2 \log [Cu^{++}] - \log [Cu_2(OH)_2^{++}]$
2	$-3pH = -16.27 + \log [Cu^{++}] - \log [HCuO_2^{-}]$
3	$-4pH = -28.99 + \log [Cu^{++}] - \log [CuO_2^{--}]$
4	$-4pH = -25.83 + \log [Cu_2(OH)_2^{++}] - 2 \log [HCuO_2^{-}]$
5	$-pH = -12.73 + \log [HCuO_2^{-}] - \log [CuO_2^{--}]$
6	$2pH = 1.386 + \log [Cu_2O] - 2 \log [Cu^{+}]$
7	$E_T = 0.34148 - 0.05685 \log [Cu_2O] + 0.11370 \log [Cu^{++}] + 0.11370\ pH$
8	$E_T = 0.72249 - 0.05685 \log [Cu_2O] + 0.05685 \log [Cu_2(OH)_2^{++}]$
9	$E_T = 2.1909 - 0.05685 \log [Cu_2O] + 0.11370 \log [HCuO_2^{-}] - 0.22739\ pH$
10	$E_T = 3.6380 - 0.05685 \log [Cu_2O] + 0.11370 \log [CuO_2^{--}] - 0.34109\ pH$
11	$E_T = 0.65794 - 0.11370 \log [Cu^{+}] + 0.11370 \log [CuO] - 0.22739\ pH$
12	$2pH = 2.090 + \log [CuO] - \log [Cu^{++}]$
13	$2pH = 2.521 + 2 \log [CuO] - \log [Cu_2(OH)_2^{++}]$
14	$-pH = 14.18 + \log [CuO] - \log [HCuO_2^{-}]$
15	$-2pH = 26.90 + \log [CuO] - \log [CuO_2^{--}]$
16	$E_T = 0.36058 - 0.11370 \log [Cu] + 0.05685 \log [Cu_2O] - 0.11370\ pH$
17	$E_T = 0.46987 - 0.05685 \log [Cu] + 0.05685 \log [CuO] - 0.11370\ pH$
18	$E_T = 0.57916 - 0.05685 \log [Cu_2O] + 0.11370 \log [CuO] - 0.11370\ pH$
19	$E_T = 0.35103 - 0.05685 \log [Cu] + 0.05685 \log [Cu^{++}]$
20	$E_T = 0.54154 - 0.05685 \log [Cu] + -0.02842 \log [Cu_2(OH)_2^{++}] - 0.05685\ pH$
21	$E_T = 1.2758 - 0.05685 \log [Cu] + -0.05685 \log [HCuO_2^{-}] - 0.17055\ pH$
22	$E_T = 1.9993 - 0.05685 \log [Cu] + -0.05685 \log [CuO_2^{--}] - 0.22739\ pH$
23	$E_T = 0.28180 - 0.11370 \log [Cu] + 0.11370 \log [Cu^{+}]$
24	$E_T = 0.42027 - 0.11370 \log [Cu^{+}] + 0.11370 \log [Cu^{++}]$
25	$E_T = 0.80128 - 0.11370 \log [Cu^{+}] + -0.05685 \log [Cu_2(OH)_2^{++}] - 0.11370\ pH$
26	$E_T = 2.2697 - 0.11370 \log [Cu^{+}] + 0.11370 \log [HCuO_2^{-}] - 0.34109\ pH$
27	$E_T = 3.7168 - 0.11370 \log [Cu^{+}] + 0.11370 \log [CuO_2^{--}] - 0.45479\ pH$
28	$E_T = 1.0127 - 0.05685 \log [H_2O] + -0.02842 \log [O_2] - 0.11370\ pH$
29	$E_T = -0.05685 \log [H_2] + \log [H^{+}] - 0.11370\ pH$

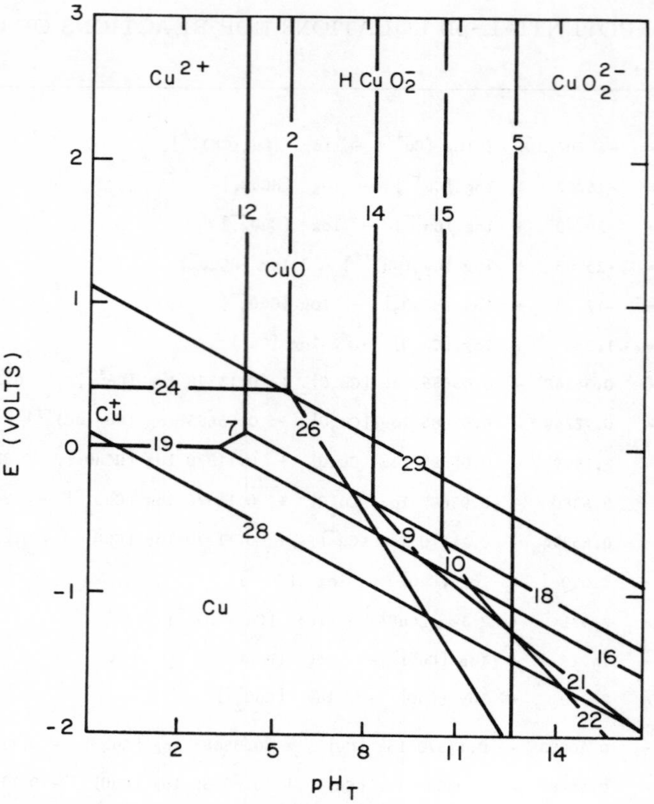

FIG. 1. Potential-pH diagram for Cu-H_2O at 300°C (ionic activity = 10^{-6}; H_2 or O_2 = 0.1 MPa; line numbers refer to reactions in Table II).

equal amounts. The OH^- ion concentrations of minimum solubility for various metals and oxides are given in Table IV. Corrosion rates are generally low near the solubility minimum. For the commoner structural metals, this occurs near 0.1 mmol/kg OH^- concentration. For this reasons, CANDU's and many other power plants operate with ∿0.1 mmol/kg alkali (LiOH or LiOD) in the primary heat transport system water.

Calculated solubilities for various metals and oxides in 10^{-4} mol/kg OH^- solution at temperatures from 298 K to 573 K are plotted in Figure 4. These are of interest in connection with material transport in power plants. Large water flows and temperature changes are needed to transport heat in the plants. Changes in solubility cause transport of metals and corrosion products. For example, iron oxide particles deposited on fuel elements dissolve in 0.1 mmol/kg OH^- solution as a result of the rising water temperature in the reactor core [23]. In less alkaline solutions the solubility gradient is reversed and thick deposits are formed.

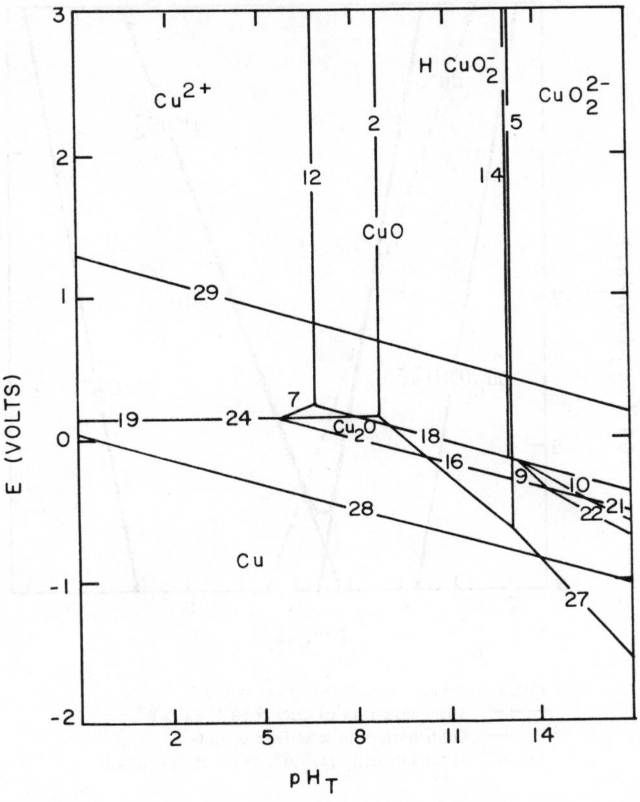

FIG. 2. Potential-pH diagram for $Cu-H_2O$ at 25°C (ionic activity = 10^{-6}; H_2 or O_2 = 0.1 MPa; line numbers refer to reactions in Table II).

Free energy changes and equilibrium O_2 and H_2 pressures have been calculated for the reduction and oxidation reactions listed in Table V. In many cases the equilibrium pressures lie far outside the practical range and the data simply serve to show which are the stable phases. In some cases transitions occur within or near the practical range and the precise numerical values are important in understanding the behaviour of power plants and the effects of operating variables. Figures 5 and 6 show equilibrium pressures for some important reduction and oxidation processes, respectively, of plant interface materials.

The simplest case is that of nickel which is stable as the metal in the presence of moderate H_2 concentrations, for example in the region above the equilibrium line in Figure 5. Under these conditions, not even a monomolecular layer of oxide can be detected, even after exposure to 0.1 mmol/kg LiOH at 250°C [24]. At low H_2 concentrations or in the presence of measurable amounts of oxygen, nickel oxide is the stable form. This transition has a profound influence on the behaviour of the boiler tube alloy Monel (65Ni 33Cu 2Fe). Under reducing conditions, corrosion

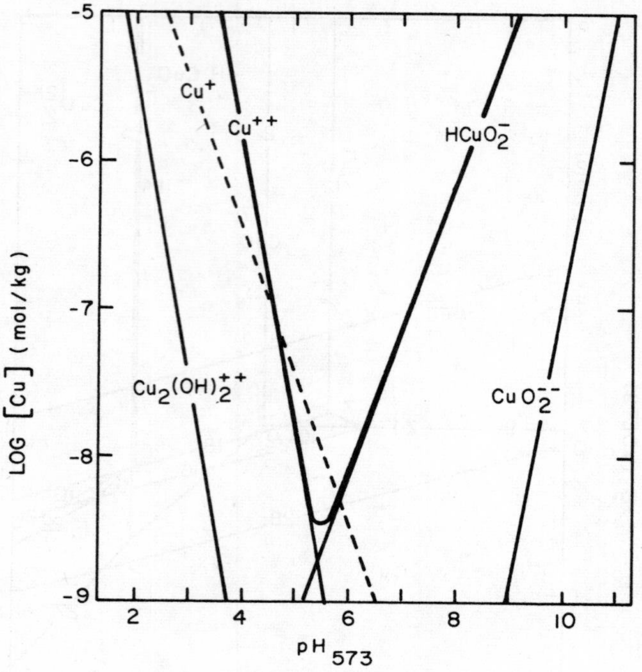

FIG.3. Solubility of CuO and Cu at 573 K.
▬▬ Total solubility of CuO (> 10^{-10} Pa O_2)
—— Contributions to solubility of CuO
------ Total solubility (as Cu^+) of Cu at 0.1 MPa H_2

TABLE IV. HYDROXYL CONCENTRATION FOR MINIMUM SOLUBILITY

$-\log [a_{OH^-} \text{ (mol/kg)}]$

Temp. K	298	373	473	573
Fe_3O_4	3.3	3.5	3.7	4.0
Co,CoO	3.0	3.5	4.0	4.6
Ni,NiO	3.8	4.0	4.3	4.8
CuO	5.1	5.0	5.1	5.5
Cu	0.1	0.2	0.5	1.0
pKw	14.0	12.25	11.26	11.04

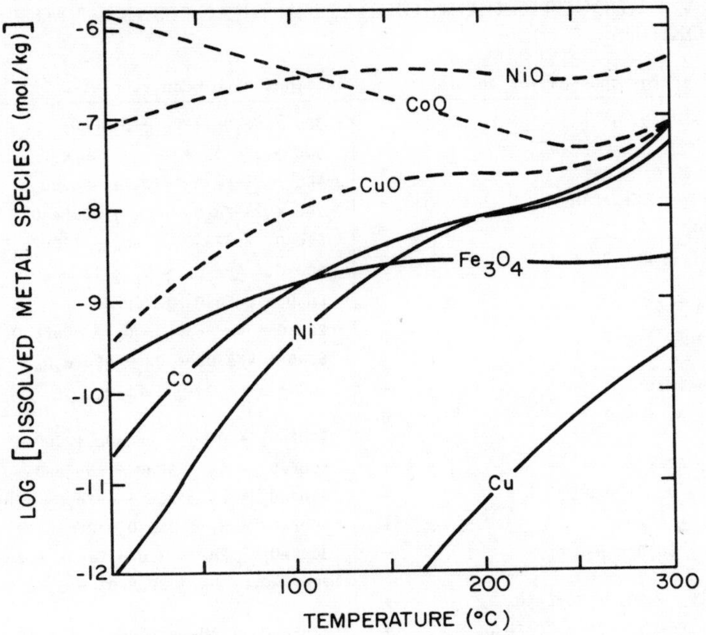

FIG. 4. Solubility of metals and oxides in 10^{-4} mol/kg OH^- solution at elevated temperatures.
——— Hydrogenated solution (0.1 MPa H_2)
------ Oxygenated solution (10^{-10} Pa O_2)

is low, the surface remains free of corrosion products and retains little radioactivity. Under oxidising conditions, corrosion increases, an oxide layer is formed and radioactivity is incorporated at the interface [25, 26]. The behaviour of Cu is similar to that of Ni except that it is over 0.5V more noble. Thus Cu metal is the stable form in the presence of minute amounts of H_2 (O_2 absent) while CuO is stable if O_2 is present. Transitions from reducing to oxidising conditions, or the reverse, cause changes in the interface layer and major movements of corrosion products. Uncontrolled transitions can lead to increased deposits on fuel and excessive radiation fields from boilers [27]. Controlled reduction and oxidation can alleviate such conditions [28].

The other transitions shown in Figures 5 and 6 can also have important practical consequences [9, 10]. Whereas the data for Ni/NiO are fairly satisfactory for practical purposes, the data for some of the spinels are not adequate. We know the Ni/NiO data up to 300°C are accurate because the computed stabilities are consistent with laboratory and plant observations. For some of the nickel ferrite transformations crude approximations were made to obtain input data and the computed values are imprecise. However, they at least serve to show which transitions are important.

TABLE V. REDUCTION AND OXIDATION REACTIONS OF METALS AND THEIR OXIDES

Free energy changes for the following reactions have been tabulated[9,10] to 573K

Reaction	
$2H_2 + O_2 \rightleftarrows 2H_2O$	+
$3Fe + 2O_2 \rightleftarrows Fe_3O_4$	+
$Fe_3O_4 + 4H_2 \rightleftarrows 3Fe + 4H_2O$	-
$4Fe_3O_4 + O_2 \rightleftarrows 6Fe_2O_3$	+
$3Fe_2O_3 + H_2 \rightleftarrows 2Fe_3O_4 + H_2O$	0
$Co + \frac{1}{2}O_2 \rightleftarrows CoO$	+
$CoO + H_2 \rightleftarrows Co + H_2O$	0
$2Ni + O_2 \rightleftarrows 2NiO$	+
$NiO + H_2 \rightleftarrows Ni + H_2O$	0
$4Cu + O_2 \rightleftarrows 2Cu_2O$	+
$Cu_2O + H_2 \rightleftarrows 2Cu + H_2O$	+
$2Cu_2O + O_2 \rightleftarrows 4CuO$	+
$2CuO + H_2 \rightleftarrows Cu_2O + H_2O$	+
$3Co + 2Fe_3O_4 + 2O_2 \rightleftarrows 3CoFe_2O_4$	+
$2Co + 2\alpha - Fe_2O_3 + O_2 \rightleftarrows 2CoFe_2O_4$	+
$6CoO + 4Fe_3O_4 + O_2 \rightleftarrows 6CoFe_2O_4$	+
$CoO + \alpha\text{-}Fe_2O_3 \rightleftarrows CoFe_2O_4$	+
$3CoFe_2O_4 + H_2 \rightleftarrows 3CoO + 2Fe_3O_4 + H_2O$	-
$3CoFe_2O_4 + 4H_2 \rightleftarrows 3Co + 2Fe_3O_4 + 4H_2O$	-
$6CoFe_2O_4 + O_2 \rightleftarrows 2Co_3O_4 + 6\alpha\text{-}Fe_2O_3$	-
$3Ni + 2Fe_3O_4 + 2O_2 \rightleftarrows 3NiFe_2O_4$	+
$2Ni + 2\alpha\text{-}Fe_2O_3 + O_2 \rightleftarrows 2NiFe_2O_4$	+
$6NiO + 4Fe_3O_4 + O_2 \rightleftarrows 6NiFe_2O_4$	+
$NiO + \alpha\text{-}Fe_2O_3 \rightleftarrows NiFe_2O_4$	+
$3NiFe_2O_4 + H_2 \rightleftarrows 3NiO + 2Fe_3O_4 + H_2O$	-
$3NiFe_2O_4 + 4H_2 \rightleftarrows 3Ni + 2Fe_3O_4 + 4H_2O$	0
$6NiFe_2O_4 + O_2 \rightleftarrows 2Ni_3O_4 + 6\alpha\text{-}Fe_2O_3$	0
$4NiFe_2O_4 + O_2 \rightleftarrows 2Ni_2O_3 + 4\alpha\text{-}Fe_2O_3$	0
$3Cu + Fe_3O_4 + O_2 \rightleftarrows 3CuFeO_2$	+
$3Cu + 2Fe_3O_4 + 2O_2 \rightleftarrows 3CuFe_2O_4$	+
$4Cu + 2\alpha\text{-}Fe_2O_3 + O_2 \rightleftarrows 4CuFeO_2$	+
$2Cu + 2\alpha\text{-}Fe_2O_3 + O_2 \rightleftarrows 2CuFe_2O_4$	+
$6Cu_2O + 8Fe_3O_4 + 5O_2 \rightleftarrows 12CuFe_2O_4$	+
$6Cu_2O + 4Fe_3O_4 + O_2 \rightleftarrows 12CuFeO_2$	+
$Cu_2O + \alpha\text{-}Fe_2O_3 \rightleftarrows 2CuFeO_2$	+
$2Cu_2O + 4\alpha\text{-}Fe_2O_3 + O_2 \rightleftarrows 4CuFe_2O_4$	+
$6CuO + 4Fe_3O_4 + O_2 \rightleftarrows 6CuFe_2O_4$	+
$CuO + \alpha\text{-}Fe_2O_3 \rightleftarrows CuFe_2O_4$	-
$4CuFeO_2 + O_2 \rightleftarrows 2CuFe_2O_4 + CuO$	0
$4CuFeO_2 + O_2 \rightleftarrows 4CuO + 2\alpha\text{-}Fe_2O_3$	0
$2CuFeO_2 + H_2 \rightleftarrows 2Cu + \alpha\text{-}Fe_2O_3 + H_2O$	0
$6CuFeO_2 + H_2 \rightleftarrows 3Cu_2O + 2Fe_3O_4 + H_2O$	-
$3CuFeO_2 + 2H_2 \rightleftarrows 3Cu + Fe_3O_4 + 2H_2O$	0
$2CuFe_2O_4 + H_2 \rightleftarrows 2CuFeO_2 + \alpha\text{-}Fe_2O_3 + H_2O$	+
$3CuFe_2O_4 + 2H_2 \rightleftarrows 3CuFeO_2 + Fe_3O_4 + 2H_2O$	+
$2CuFe_2O_4 + H_2 \rightleftarrows Cu_2O + \alpha\text{-}Fe_2O_3 + H_2O$	+
$CuFe_2O_4 + H_2 \rightleftarrows Cu + \alpha\text{-}Fe_2O_3 + H_2O$	+
$3CuFe_2O_4 + H_2 \rightleftarrows 3CuO + 2Fe_3O_4 + H_2O$	+
$6CuFe_2O_4 + 5H_2 \rightleftarrows 3Cu_2O + 4Fe_3O_4 + 5H_2O$	+
$3CuFe_2O_4 + 4H_2 \rightleftarrows 3Cu + 2Fe_3O_4 + 4H_2O$	+
$CuFe_2O_4 + Cu \rightleftarrows 2CuFeO_2$	+

\+ Reaction goes completely to right under practical conditions

\- Equilibrium remains completely to left under practical conditions

0 Equilibrium may be to left or right depending on H_2 or O_2 concentrations

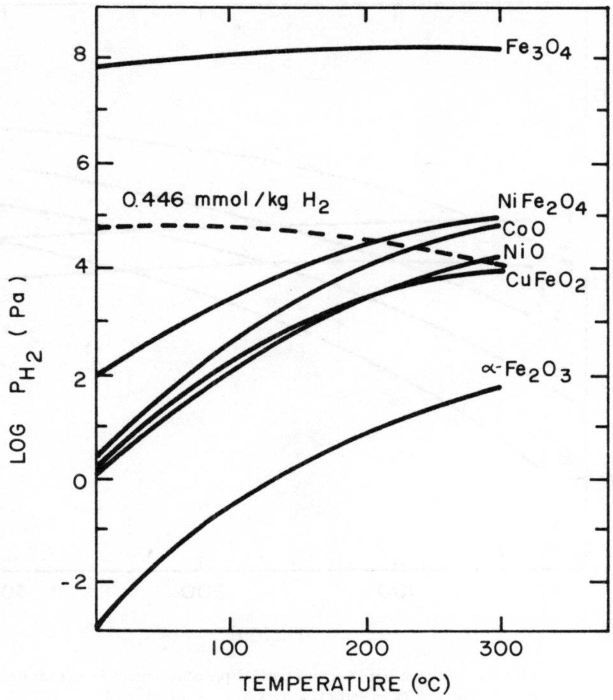

FIG. 5. Hydrogen equilibrium partial pressure versus temperature for the reduction of metal oxides and for 0.446 mmol/kg dissolved H_2.

REFINEMENTS

Data such as those summarised here are proving valuable in understanding the behaviour of metal-oxide-high temperature water interfaces in power plants and improving the design, performance and economics. It is also apparent that there is much scope for improving the quality of these data.

Thermodynamic input data for predicting the properties of the ions of interest in high temperature aqueous solutions are particularly deficient. There do not appear to be any data at all for neutral species such as dissolved $M(OH)_2$ yet these are expected to be the dominant species in solution near the pH of minimum solubility, which is the area of greatest practical interest. Polymeric species presumably become increasingly dissociated and therefore less important with increasing temperature. Confirmation of this is desirable. Conversely, the neglect of the monomeric species $Cu(OH)^+$ in favour of the known dimer $Cu_2(OH)_2^{2+}$, as was done in these calculations, is less admissible at elevated temperature. The need for information on $Cu(OH)^+$ is thus a specific instance of the general need.

Use of the Criss and Cobble equivalence principle, for want of other data, needs validation for hydrolysed transition metal ions and elevated temperatures.

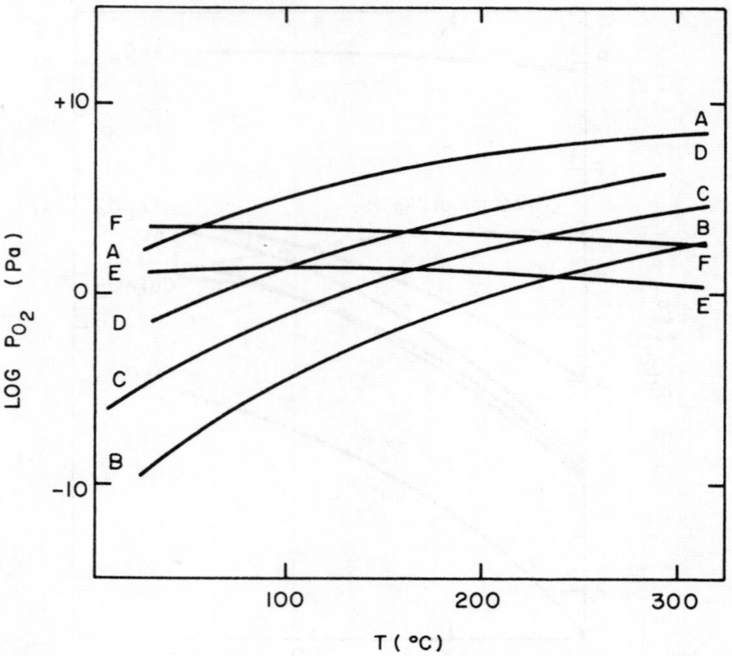

FIG. 6. Oxygen equilibrium partial pressure versus temperature.
A $4NiFe_2O_4 + O_2 = 2Ni_2O_3 + 4\alpha Fe_2O_3$
B $6NiFe_2O_4 + O_2 = 2Ni_3O_4 + 6\alpha Fe_2O_3$
C $4CuFeO_2 + O_2 = 2CuFe_2O_4 + 2CuO$
D $4CuFeO_2 + O_2 = 4CuO + 2\alpha Fe_2O_3$
E 0.125 mol/kg of O_2
F 25.0 mol/kg of O_2

Thermodynamic input data also are lacking, or in need of improvement, for a number of lesser known, or less well defined solids. Conspicuous and important omissions from these tabulations are the data for stability and solubility of metal hydroxides. For example nickel hydroxide is obtainable as a well characterised crystalline solid which is at least kinetically stable [29] and is thought to be thermodynamically stable (with respect to dehydration to the oxide) up to 560 K. Basic thermodynamic data for this and other solid hydroxides are lacking.

The lack of data for higher oxides of nickel has been noted already. The spinels are important since they are the principal corrosion product of many of the corrosion-resistant structural alloys. The computed spinel data summarised here are for spinels of integral stoichiometry e.g. $CoFe_2O_4$ which are obtained only by special preparation. Substitution of one metal for another is possible over a very wide range of composition in the spinel lattice and mixed metal spinels of non-integral stoichiometry such as $(FeNiCo)^{II}(FeCr)_2{}^{III}O_4$ are found in plants and play a major role. Computations for these spinels are being attempted. Basic free energy data are needed for such spinels, as is a knowledge of the thermodynamics of substitution. Data on substitution by small amounts of cobalt in the spinel lattice are particularly important in connection with the preferential uptake of Co^{60} in corrosion products and the dispersal of radioactivity.

The existing data summarised here are of considerable value in building and operating power plants now. Providing improved data such as those outlined above and resolving the theoretical and experimental questions which are implied are challenging and important tasks for thermochemists.

REFERENCES

[1] GRAY, J.L., "Why CANDU? Its Achievements and Prospects", Atomic Energy of Canada Ltd. Rep. AECL-4709 (1974).

[2] MOON, C.L., "Pickering Generating Station", Nuclear Engineering International 15 (1970) 501.

[3] COHEN, P., "Water Coolant Technology of Power Reactors", Gordon and Breach Inc. N.Y. (1969).

[4] TOMLINSON, M., "Transport of Corrosion Products", Presented at The International Conference on High Temperature High Pressure Electrochemistry in Aqueous Solutions, Surrey, England, Jan. 1973. Proceedings to be published by the National Association of Corrosion Engineers (U.S.).

[5] MACDONALD, D.D., SHIERMAN, G.R. and BUTLER, P., "The Thermodynamics of Metal-Water Systems at Elevated Temperatures. Part 1: The Water and Copper-Water Systems", Atomic Energy of Canada Ltd. Rep. AECL-4136 (1972).

[6] MACDONALD, D.D., SHIERMAN, G.R. and BUTLER, P. "The Thermodynamics of Metal-Water Systems at Elevated Temperatures. Part 2: The Iron-Water System", Atomic Energy of Canada Ltd. Rep. AECL-4137 (1972).

[7] MACDONALD, D.D., SHIERMAN, G.R. and BUTLER, P., "The Thermodynamics of Metal-Water Systems at Elevated Temperatures. Part 3: The Cobalt-Water System", Atomic Energy of Canada Ltd. Rep. AECL-4138 (1972).

[8] MACDONALD, D.D., SHIERMAN, G.R. and BUTLER, P., "The Thermodynamics of Metal-Water Systems at Elevated Temperatures. Part 4: Nickel Water System", Atomic Energy of Canada Ltd. Rep. AECL-4139 (1972).

[9] MACDONALD, D.D. and RUMMERY, T.E., "The Thermodynamics of Metal Oxides in Water-Cooled Nuclear Reactors", Atomic Energy of Canada Ltd. Rep. AECL-4140 (1973).

[10] RUMMERY, T.E. and MACDONALD, D.D., "The Thermodynamics of Selected Transition Metal Ferrites in High Temperature Aqueous Systems", Atomic Energy of Canada Ltd. Rep. AECL-4577 (1973).

[11] MACDONALD, D.D. and BUTLER, P., Corrosion Science, 13 (1973) 259.

[12] POURBAIX, M., Atlas of Electrochemical Equilibria in Aqueous Solutions, Pergamon, London (1966).

[13] CRISS, C.M. and COBBLE, J.M., "The Thermodynamic Properties of High Temperature Aqueous Solutions. IV. Entropies of the Ions up to 200° and the Correspondence Principle", J. Am. Chem. Soc. 86 (1964) 5385-90.

[14] CRISS, C.M. and COBBLE, J.M., "The The Thermodynamic Properties of High Temperature Aqueous Solutions. V. The Calculation of Ionic Heat Capacities up to 200°. Entropies and Heat Capacities Above 200°", J. Am. Chem. Soc. 86 (1964) 5390-93.

[15] COBBLE, J.M., "The Thermodynamic Properties of High Temperature Aqueous Solutions. VI. Applications of Entropy Correspondence to Thermodynamics and Kinetics", J. Am. Chem. Soc. 86 (1964) 5394-5401.

[16] RUMMERY, T.E. and MACDONALD, D.D., Unpublished.

[17] KELLEY, K.K., "Contributions to the Data on Theoretical Metallurgy. XIII. High Temperature Heat-Content, Heat-Capacity, and Entropy Data for the Elements and Inorganic Compounds", U.S. Bureau of Mines, Bulletin 584, U.S.G.P.O. Washington (1960).

[18] WICKS, C.E. and BLOCK, F.E., "Thermodynamic Properties of 65 Elements: Their Oxides, Halides, Carbides, and Nitrides", U.S. Bureau of Mines Bulletin 605, U.S. Govt. Printing Office, Washington (1963).

[19] ROSSINI, F.D., "Selected Values of Chemical Thermodynamic Properties", NBS Circular 500, U.S. Dept. of Commerce, U.S.G.P.O. Washington (1961).

[20] WAGMAN, D.D., "Selected Values of Chemical Thermodynamic Properties", NBS Technical Note 270-3, U.S. Department of Commerce, U.S.G.P.O. Washington (1968).

[21] WAGMAN, D.D., "Selected Values of Chemical Thermodynamic Properties", NBS Technical Note 270-4, U.S. Department of Commerce, U.S.G.P.O. Washington (1969).

[22] FISHER, J.R. and BARNES, H.L., "The Ion-Product Constant of Water to 350°", Journal of Physical Chemistry, 76 (1972) 90-9.

[23] BURRILL, K., Reported in 4.

[24] MCINTYRE, N.S. and RUMMERY, T.E., Unpublished.

[25] LESURF, J.E. and TAYLOR, G.F., "Material Selection and Corrosion Control Methods for CANDU Nuclear Power Reactors", Atomic Energy of Canada Ltd. Rep. AECL-4057 (1972).

[26] BAILEY, M.G., MONTFORD, B., TOMLINSON, M. and WALLACE, G., Unpublished.

[27] MONTFORD, B., "Cycling Decontamination Techniques at Douglas Point Generating Station - Their Evolution and Recommended Practice", Atomic Energy of Canada Ltd. AECL-4223, (1973).

[28] MONTFORD, B., "Decontamination of the Douglas Point Generating Station by Cycling Techniques", Atomic Energy of Canada Ltd. Rep. AECL-4435 (1973).

[29] AIA, M.A., Journal of Electrochemical Society, 113 (1966) 1045.

DISCUSSION

N.A. JAVED: Do you believe that, at the operating temperature range in the heat exchangers of CANDU reactors, the inside surface corrosion products may also consist of mixed hydroxides rather than pure oxides? What was the precise chemical composition of the observed corrosion products?

M. TOMLINSON: Hydroxides have not normally been detected in power plant corrosion products. They are not stable at operating temperatures. Traces may occasionally be found in samples subsequently exposed to water at low temperatures. The composition of corrosion products from power plants is complex and varies with both time and location.

C.B. ALCOCK: Could you indicate the spinel phases which are found to be of major importance in corrosion processes in CANDU reactor systems?

M. TOMLINSON: Spinel phases of particular interest are those of intermediate composition between stoichiometric compounds. For example, thermodynamic data are desirable for spinels of the form $(Ni_x Fe_{1-x})Fe_2O_4$ intermediate between stoichiometric nickel ferrite, $NiFe_2O_4$, and magnetite. More complex forms with other admixed metals are also of interest.

Session VII
BASIC PHASE EQUILIBRIA OF NUCLEAR FUELS

Chairman: Y. TAKAHASHI
JAPAN

CONSTITUTIONAL STUDIES ON U-Pu-C-FISSION PRODUCT SYSTEMS

With application to the prediction of the chemical state of irradiated carbide nuclear fuels

H.R. HAINES, P.E. POTTER
United Kingdom Atomic Energy Authority,
Atomic Energy Research Establishment,
Harwell, Didcot, Oxon.,
United Kingdom

Abstract

CONSTITUTIONAL STUDIES ON U-Pu-C-FISSION PRODUCT SYSTEMS WITH APPLICATION TO THE PREDICTION OF THE CHEMICAL STATE OF IRRADIATED CARBIDE NUCLEAR FUELS.

For an understanding of the chemical constitution of an irradiated carbide nuclear fuel, a knowledge of the phase relationships of the individual fission product elements with uranium-carbon and plutonium-carbon is required. Experimental data for the phase diagrams of uranium-carbon, and plutonium-carbon with the lanthanides lanthanum, cerium, praseodymium and neodymium are presented. Some data for the uranium-carbon-gadolinium and yttrium systems are also given. Although there are no lanthanide monocarbide phases analogous to uranium and plutonium monocarbides there is limited solubility of the lanthanides in the actinide monocarbides; this solubility increases with increasing atomic number of lanthanide. The same tendency is observed for the limited solubility of the lanthanides in uranium sesquicarbide. There are regions of complete miscibility between plutonium and the lanthanide sesquicarbides. Limited data for the dicarbide region indicate that the high temperature cubic dicarbide phase can be quenched to room temperatures in the uranium-lanthanum, -cerium, -praseodymium, and -neodymium dicarbide systems and in the plutonium-lanthanum dicarbide system. Isothermal sections for the complete phase diagrams are presented. Experimental data are also presented for the plutonium-Group VIII-fission product-carbon systems, the fission product elements being ruthenium, rhodium and palladium. No ternary compounds analogous to U_2RuC_2 were found but the carbon-filled Cu_3Au structured compounds, $PuRu_3C$ and $PuRh_3C$, were found. Using the phase diagram data presented together with existing phase equilibria data for the fission product elements not considered in this paper, the change in overall carbon potential of fuel compositions $U_{0.8}Pu_{0.2}C$ and $U_{0.8}Pu_{0.2}C_{1.09}$ after irradiation to 10% burn-up has been calculated. For the composition $U_{0.8}Pu_{0.2}C$, containing just a trace of sesquicarbide, the carbon potential will fall during burn-up, whilst that containing the larger amount of sesquicarbide will have almost the same carbon potential after burn-up as in the initial condition.

1. INTRODUCTION

The purpose of this paper is to describe some of our recent phase diagram studies on systems of relevance to the prediction of the chemical state of irradiated carbide nuclear fuels and as an aid to the interpretation of post-irradiation observations.

Because of the higher thermal conductivity of uranium-plutonium monocarbides compared with urania-plutonia solid solutions, the temperatures attained in carbide irradiation experiments are much lower than those for oxide irradiations under similar conditions. Thus any restructuring will be much less than in the oxide; grain growth and fission product diffusion will be slow and thus the growth of fission product phases will be slow and consequently observations of the form of fission product phases within carbide fuels have been limited.

TABLE I

The composition of $U_{0.80}Pu_{0.20}$ carbide fuels at 10% burn-up after fast neutron irradiation and 50 days cooling time

Element	$U_{0.8}Pu_{0.2}C$ at.%	$U_{0.8}Pu_{0.2}C_{1.09}$ at.%
U	34.31	32.84
Pu	8.49	8.14
Am	0.02	0.02
Se	0.02	0.02
Br	0.01	0.01
Kr	0.08	0.08
Rb	0.11	0.10
Sr	0.12	0.11
Y	0.10	0.10
Zr	0.97	0.94
Nb	0.02	0.02
Mo	1.02	0.99
Tc	0.29	0.28
Ru	1.01	0.97
Rh	0.27	0.26
Pd	0.65	0.63
Ag	0.07	0.07
Cd	0.04	0.04
In	< 0.01	< 0.01
Sn	0.04	0.04
Sb	0.02	0.02
Te	0.15	0.14
I	0.08	0.07
Xe	1.07	1.03
Cs	0.91	0.88
Ba	0.31	0.30
La	0.28	0.27
Ce	0.59	0.57
Pr	0.23	0.23
Nd	0.69	0.67
Pm	0.09	0.08
Sm	0.16	0.16
Eu	0.03	0.03
Gd	0.02	0.02
C	47.73	49.87

The calculated fission product yields for $(U^{Nat}Pu^{239})C$ irradiated to ca. 10% burn-up after 50 days cooling time for two different carbon concentrations are shown in Table I.

As a basis for the prediction of the chemical state of an irradiated carbide we have endeavoured to obtain an understanding of the phase relationships of the individual fission product elements with uranium and carbon and plutonium and carbon, and to a limited extent the nature of the interactions of the fission products with one another by studying selected fission product element systems, usually in the presence of carbon. With a knowledge of such phase relationships it should be possible to make some predictions about the nature of any new phases formed during burn-up and the influence of the initial composition of the carbide and the magnitude of burn-up on the chemical constitution of the fuel.

Firstly, in this paper some appropriate phase relationships are discussed and, secondly, the possible constitution of a burnt carbide fuel is considered.

2. PHASE RELATIONSHIPS

The phase relationships for the fission product elements with the U and Pu-carbon systems are best considered in a number of groups. The rare gases, krypton and xenon, together with caesium and rubidium remain in the elemental form in the presence of the carbides.

The other fission product elements may be conveniently considered in the following groups.

a. Yttrium and elements from lanthanum to gadolinium.

b. Molybdenum, technetium, ruthenium, rhodium and palladium.

c. Zirconium, niobium.

d. Barium, strontium.

e. Selenium, tellurium.

f. Bromine, iodine.

g. Silver, cadmium, indium, tin, antimony.

2.1 Experimental techniques

The alloys were prepared by arc melting the elements together in gettered argon at 0.3 atm pressure; the uranium and plutonium contained less than 2000 ppm impurities, the graphite was type EY-9 and contained less than 2000 ppm impurities. The rare earth elements and yttrium were of 99.9 wt.% purity, as were the transition elements used. The alloy buttons were melted and turned several times on the furnace hearth to ensure homogeneity. For annealing under vacuum in either alumina tube furnaces, or in an all metal system with a tungsten heater, the alloys were contained in carburised tantalum crucibles. The alloys were examined after annealing in the 'furnace cooled' condition.

Checks were made on the compositions of the alloys using chemical analysis, and the levels of contamination with oxygen and nitrogen were also determined. The alloys were also examined using ceramography and X-ray powder photography; the photographs were taken with an 11 cm Debye-Scherrer Camera. Electron microprobe analysis was employed for the characterisation of some alloys.

2.2 The systems U-C- and Pu-C- with the rare earth elements (La to Gd) and yttrium

The systems containing uranium are discussed first.

The systems of uranium and carbon with lanthanum, cerium, neodymium, praseodymium, gadolinium and yttrium are considered. These studies extend the authors' previously reported data[1] on La, Ce and Nd. Work has appeared from other laboratories on the systems U-La-C[2,3], U-Ce-C[2,3,4,5], U-Nd-C[2] and U-Gd-C[6].

2.2.1 The systems U-(La, Pr, or Nd)-C

These three systems are best considered together. The relevant binary phase diagrams which must be considered and those for U-C, La-C, Pr-C, Nd-C, U-La, U-Pr and U-Nd. The U-C binary system has been the subject of numerous studies[7]; the system is characterised by three compounds. U monocarbide has a cubic NaCl structure and exists over a narrow composition range at low temperatures but which widens appreciably above ~ 1600°C. U sesquicarbide has a body-centred cubic cell with a very narrow homogeneity range. U dicarbide which is stable only above 1500°C is dimorphic. The lower temperature form α-'UC_2' is tetragonal with a CaC_2 structure and exists over a range of composition (with C/U always < 2.0). At temperatures above 1765°C the tetragonal form transforms to β-UC_2 which has a KCN structure[8]. The three lanthanide-carbon systems are each characterised by two compounds, the sesquicarbide and dicarbide which are isomorphous with the corresponding uranium compounds. The sesquicarbides exist over a range of composition, and La_2C_3 and Nd_2C_3 decompose peritectically at 1415°C[9] and ca. 1840°C[10] respectively. Phase diagrams have been given for the La-C[9] and Nd-C[10] systems. The α-β dicarbide transformation temperatures for LaC_2, PrC_2, NdC_2[11] increase with atomic number of the lanthanide, and are 1060, 1135 and 1150°C respectively. There is limited solid solubility of carbon in La, and probably none in Pr and Nd[12]. There is partial solubility of uranium in La, Pr and Nd, and also of these lanthanides in uranium[13]; at the uranium rich end of the phase diagrams there are monotectics; the region of liquid immiscibility probably exists up to temperatures above 1400°C. The three phase diagrams considered here may be represented as isothermal sections by diagrams of the type shown in Figures 1 and 2. Figure 1 shows a section of the U-Ln-C phase diagram (Ln = Lanthanide, here La, Pr or Nd) at a temperature where C and U_2C_3 are in equilibrium, and there is only a limited region where the lanthanide dicarbide with uranium solubility exists. Obviously the most important features required for an isothermal section are the concentrations of the solid solutions; the monocarbide at A and B, the sesquicarbide at C in the uranium rich region of the diagram, and the composition D. No evidence for a lanthanide monocarbide for these three elements has been found, however the limited solid solution range up to the composition at point B can be regarded as a stabilisation of the Ln monocarbide phase.

The equilibria for the 3-phase region $LnC_{1.5}$, Ln and $U_{1-x}Ln_xC$ (composition at point B) being[1]

$$3[LnC]_{UC} \rightleftharpoons \langle Ln \rangle + \langle Ln_2C_3 \rangle$$

at temperatures below the melting points of the lanthanide elements (< 1050°C).

Figure 2 shows a somewhat more complicated section of the systems for temperatures where the U sesquicarbide and dicarbide coexist; this region exists in the temperature range ca. 1500 - 1800°C; this range will be changed by the solubility of the lanthanides in both phases. There is also a region of dicarbide immiscibility which is found in these systems. A knowledge of the variation of the limits A, B, C, D and E with temperature is required.

1 Throughout the paper the following symbols are used: < > solid (unit thermodynamic activity); [ ] solid solution (suffix denotes solvent).

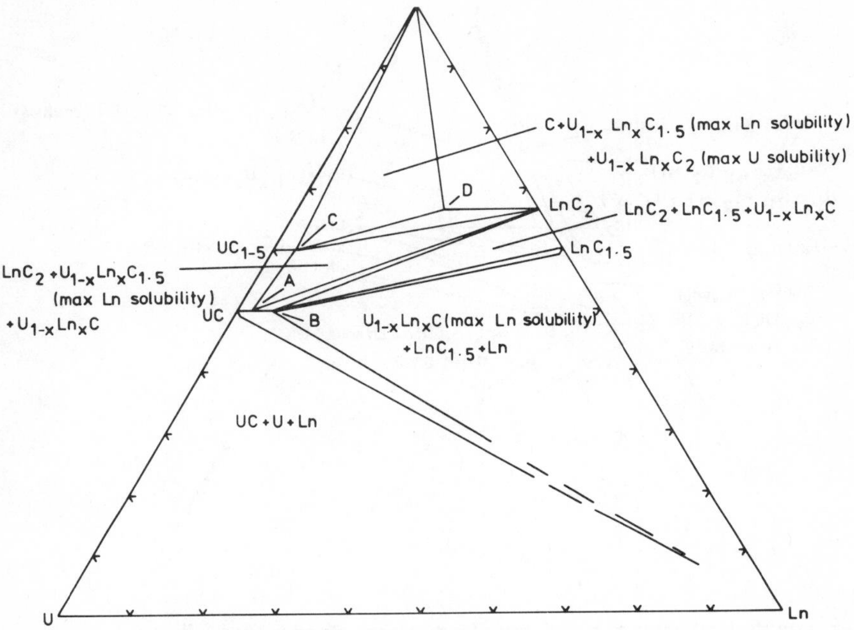

FIG. 1. A section of the U-Ln-C phase diagram for temperatures at which U_2C_3 and C are in equilibrium (Ln = La, Pr or Nd).

2.2.1.1. The U-La-C system. In addition to the authors' preliminary results on this system[1], some additional studies of the phase diagram have appeared[2], and also some studies specific to the dicarbide section of the system[3]. The alloys prepared by Lorenzelli and Marcon[2] were prepared by heating a mixture of uranium and lanthanum hydrides with carbon at 1200°C; the mixtures were then annealed at 1600°C for 10 - 40 hours in a tungsten cell fitted with a lid to prevent loss of material by evaporation. For many arc-melted alloys longer annealing times, 480 hours at 950°C and 1250°C and 240 hours at 1450°C were used. The dicarbide alloys discussed by McColm et al.[3] were also prepared by arc melting. The lattice parameters of the annealed alloys were used to estimate the solubility limits in both these studies and those of Lorenzelli and Marcon. To use the lattice parameter measurements an estimate of the lattice parameter of the hypothetical compound 'LaC' has been made using the relation

$$< \quad a_o\langle UC\rangle - a_o\langle UN\rangle = a_o\langle 'LaC'\rangle - a_o\langle LaN\rangle$$

where a_o is the lattice parameter; $a_o\langle UC\rangle - a_o\langle UN\rangle = 0.071$ Å[14,15], $a_o\langle LaN\rangle = 5.295$ Å[16], thus $a_o\langle 'LaC'\rangle = 5.366$ Å. A Vegard's law relationship between LnC and UC has then been assumed to calculate the limits of solubility in this region of the system.

The results from electron microprobe analysis, at least, for the monocarbides of the U-La-C, U-Nd-C and U-Ce-C systems agree reasonably well with the determinations from the lattice parameter measurements.

The values found for the limits of solubility of La in the U carbides are given in Table II.

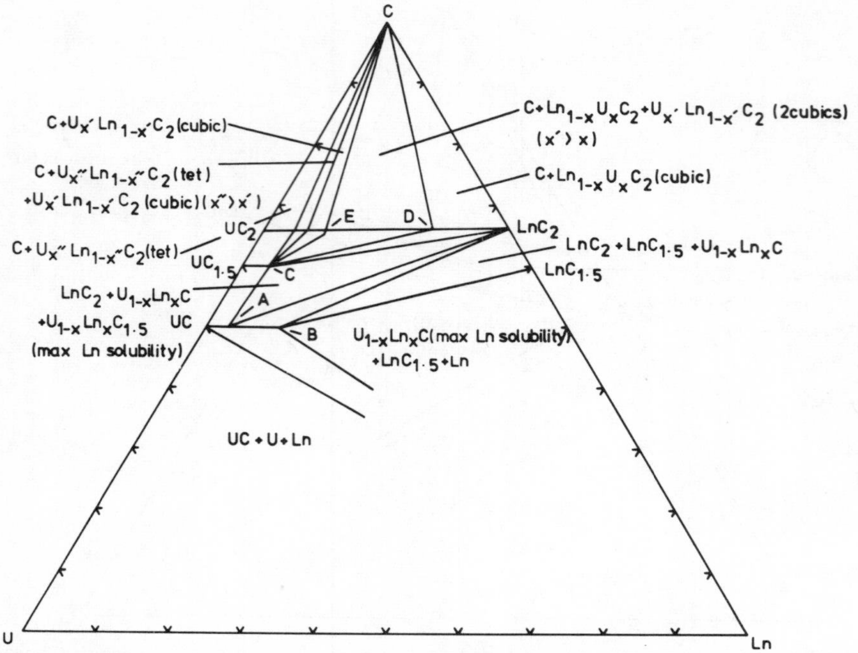

FIG. 2. A section of the U-Ln-C phase diagram for temperatures at which U_2C_3 and UC_2 are in equilibrium (Ln = La, Pr, or Nd).

A comparison of the results of Table II shows that there is agreement that the solubility of La in UC and U_2C_3 is low, but a pronounced decrease of La solubility in UC has been found with increasing annealing temperature; the decrease in the lattice parameters of the monocarbide phase cannot be due to oxygen contamination the amounts of oxygen present after annealing were less than 1000 ppm by weight.

No evidence has been found that the positions of the limits A and B are very different from one another, which is in disagreement with the data of Lorenzelli and Marcon.

Alloys were also examined in the arc melted condition with lanthanum concentrations between 1 and 20 at.% and carbon contents between 50 and 60 at.%. In all these alloys the monocarbide was the primary freezing phase, the lattice parameters of this phase were 4.962–4Å, indicating ca. 1 mol.% 'LaC' solubility.

Alloys with carbon contents of 67 at.% are two phase in the arc melted condition; four alloys of compositions $U_{0.78}La_{0.22}C_2$, $U_{0.7}La_{0.3}C_2$, $U_{0.5}La_{0.5}C_2$, $U_{0.3}La_{0.7}C_2$ were examined and in that containing equi-molar amounts of UC_2 and LaC_2 two f.c. cubic phases were found, whilst the other alloys gave b.c. tetragonal phases. The alloy of composition $U_{0.78}La_{0.22}C_2$ was annealed at 1450°C for 240 h; U sesquicarbide was formed with a_o = 8.090Å together with a tetragonal phase. The maximum solubility of La in U_2C_3 at this temperature is very close to that in the presence of the monocarbide. The f.c. cubic phases observed in the alloys with 67 at.% carbon concentration are the quenched high temperature (or β) forms of the dicarbide stabilised by solid solution formation.

TABLE II

The limits of substitution in the U-La-C system (see Figures 1 and 2)

Temperature (°C)	Limits			Reference
	A (mol.% 'LaC')	B	C (mol.% $LaC_{1.5}$)	
1250 and 1600	0.5 (4.9605Å)	0.9 (4.963Å) 2 ± 0.5*	0.4 (8.091Å)	2
950	As limit B	0.9-1.1 (4.965Å)	ca. 0.25 (8.090Å)	This work 1
1250	"	0.2-0.4 (4.961Å)	"	
1450	"	0 (4.960Å)	"	

*Determination using electron microprobe analysis

N.B. Standard deviation of lattice parameter measurements < 0.001Å

McColm et al. [3] also found that the high temperature dicarbide form could be quenched to room temperature in alloys rapidly cooled in an arc-melting furnace. A cubic dicarbide phase was found in all arc melted alloys with compositions in the range $U_{0.9}La_{0.1}C_2$ - $U_{0.15}La_{0.85}C_2$ although evidence of a miscibility gap was not conclusive.

Further reference will be made to the dicarbide region of the phase diagram in section 2.2.6.

2.2.1.2. The uranium-praseodymium-carbon system. A number of alloys in this system mainly in the uranium rich region of the phase disgram have been examined. Arc melted alloys with compositions in the range 5 to 20 at.% Pr and 50 - 60 at.% C contained U monocarbide as the primary freezing phase; the lattice parameter, a_o = 4.966-4.969Å indicates a solubility of up to 3.3 mol.% of hypothetical 'PrC'.

Alloys were annealed at 1400 - 1450°C for up to 140 h and the form of the isothermal section of the phase diagram is shown in Figure 1. The limits of solubility are shown in Table III, and the lattice parameters of the U monocarbide phase are the same for the phase fields Pr + Pr_2C_3 + UC (Pr saturated) indicating that points A and B are very close in this system. There is very little solubility of uranium in Pr_2C_3.

In arc melted alloys containing 67 at.% carbon, two f.c. cubic phases have been found when there is close to equiatomic amounts of U and Pr.

TABLE III

The Limits of Substitution in the U-Pr-C System (See Fig. 1)

Temperature (°C)	Limits		
	A (mol.% 'PrC')	B	C (mol.% $PrC_{1.5}$)
1400 - 1450	As limit B	3.3 (4.969Å)	2.1 - 2.3 (8.099Å)

TABLE IV

The Limits of Substitution in the U-Nd-C System

Temperature (°C)	Limits			Reference
	A (mol.% NdC)	B	C (mol.% Nd_2C_3)	
1400 and 1600	0.8* (4.962Å)	5.7* (4.975Å)	2.5* (8.099Å)	2
1000 and 1400	As limit B	3.9 (4.970Å)	2.5 (8.099Å)	This work

*These values were obtained using a_o = 5.222Å for the lattice parameter of 'NdC', compared with a_o = 5.15 for 'NdC_x'(x<1) used by Lorenzelli and Marcon.

2.2.1.3 The uranium-neodymium-carbon system. Some preliminary studies on this system have previously been reported by the authors[1].

Arc melted alloys with carbon concentrations in the range 32 - 60 at.%, and Nd concentrations up to 20 at.% contained uranium monocarbide as the primary freezing phase. The lattice parameter of this phase was 4.965-4.970Å, indicating a solubility of hypothetical 'NdC' of 2 - 4 mol.%. Lattice parameters of U_2C_3 (a_o = 8.099Å) and Nd_2C_3 (a_o = 8.515Å) indicate solubilities of 2.4 - 2.6 mol. Nd_2C_3 and as much as 7 mol.% U_2C_3 respectively, assuming a Vegard's law relationship for the sesquicarbide solid solution. The structure of an alloy of composition $U_{0.42}Nd_{0.58}C_{2.04}$ indicates that there is a liquidus valley cutting the dicarbide section slightly on the Nd rich side of a composition containing equal quantities of U and Nd. The alloy had a fine structure of primary graphite

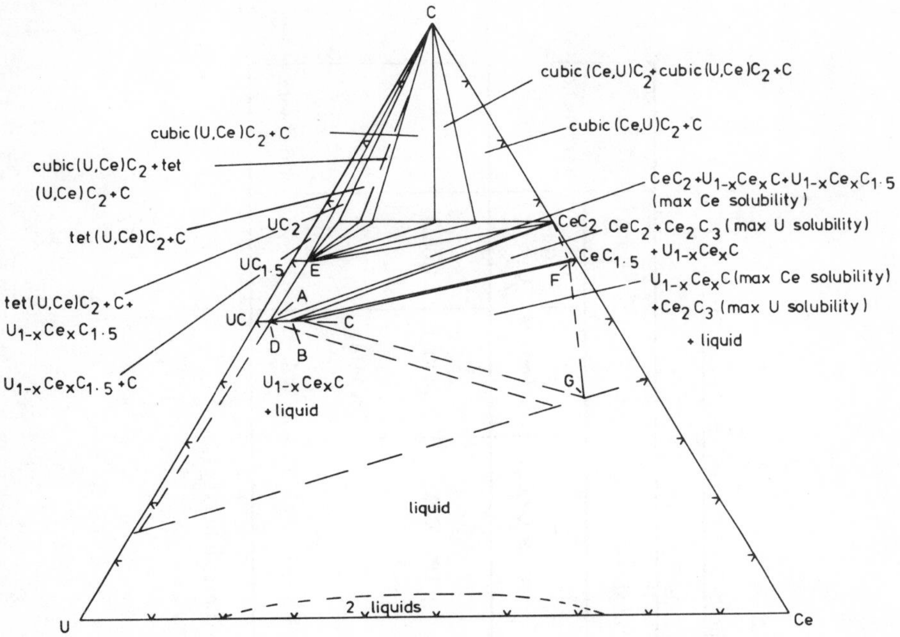

FIG.3. A possible section for the U-Ce-C system for temperatures ≥ 1450°C.

a uranium rich dicarbide phase and a Nd rich eutectic phase; the X-ray powder photograph was very poorly resolved but gave definite evidence of a f.c. cubic phase(a_o = 5.52Å). An alloy of nominal composition $U_{0.92}Nd_{0.08}C_{2.33}$ contained graphite and tetragonal UC_2.

Alloys were annealed at 1000° and 1450°C for 220 - 260 h. An isothermal section of the type shown in Figure 1 is appropriate to describe the phase diagram. There is a tie line between UC (very small Nd solubility) and Nd. The composition at the limits are shown in Table IV and are in quite good agreement with the data of Lorenzelli and Marcon[2] for 1400°C and 1600°C; the only discrepancy is that in our studies the results suggest that the points A and B are practically coincident.

The alloy of composition $U_{0.92}Nd_{0.08}C_{2.33}$ after annealing at 1450°C for 300 h contained U_2C_3 with a_o = 8.099Å, indicating the same solubility of Nd as in the presence of the monocarbide. The uranium solubility in the neodymium carbides has not been examined for annealed alloys.

2.2.2 The uranium-cerium-carbon system

In addition to the studies described here this system has been examined by Lorenzelli and Marcon[2], Stecher et al.[4], Holleck and Wagner[5] and McColm et al.[3].

For temperatures, 1450 and 1600°C the isothermal sections of the phase diagram given by Lorenzelli and Marcon[2], Stecher et al.[4] are compared with

TABLE V

The Limits of Substitution in the U-Ce-C System (See Fig. 3)

Temperature (°C)	Limits							Reference
	A	B (mol.% 'CeC')	C	D	E (mol.% Ce_2C_3)	F	G (mol.% U_2C_3)	
1600	~1 (4.961Å)	~7 (4.969Å)	~14 (4.978Å	2.5	~4 (8.103Å)	~2 (8.440Å)	- (5.08Å)	2
1600	~5	~20	~30 (4.972Å)			~2 (8.441Å)		4
1450 and 1600	~4* (4.965Å)		9 - 10 (4.973Å)					This work

*No evidence of U_2C_3 present - phase field UC_2, CeC_2, UC (Ce solubility).

TABLE VI

Lattice Parameters of the f.c. Cubic Dicarbide Phases of the U-Ce-C System

Alloy Composition	Lattice parameter* (Å)			
	Arc-melted		Annealed (1450°C)	
$U_{0.34}Ce_{0.66}C_2$		5.60		-
$U_{0.50}Ce_{0.50}C_2$	5.54	5.60	5.50	5.65
$U_{0.66}Ce_{0.04}C_2$	5.45		5.48	-

* ±0.01Å

the data of this study in Table V. A section for this system is shown in Figure 3. At these temperatures there is probably considerable solubility of carbon in cerium[4, 12] and the immiscibility in the U-Ce binary system[13] is assumed to disappear above 1450°C.

The agreement between our results and those of Lorenzelli and Marcon is quite good in that the maximum solubility of 'CeC' in UC is much lower than suggested by Stecher et al.; an alloy of nominal composition $U_{0.7}Ce_{0.3}C$ which according to those latter authors should have been single phase was in fact a 3-phase mixture of Ce_2C_3 (U saturated), Ce (C and U saturated), and UC (Ce saturated). In these studies the actual position of limit A is difficult to determine as the lattice parameter of the monocarbide in the phase field UC (Ce solubility), UC_2 (Ce saturated) and CeC_2 varied from 4.961-4.964Å; similar values were obtained for limit B in these studies, suggesting that the positions of limits A and B could be very close to one another. The different values obtained for the limits B and C in our studies and those of Lorenzelli and Marcon could simply be due to the different annealing temperature. The lattice parameters found for γ-Ce (the phase freezing from the liquid) 5.08-5.10Å[2] compared with that of carbon saturated cerium, 5.130Å, suggest some uranium solubility in this phase (point G).

Although the data of Lorenzelli have suggested the existence of a phase field U_2C_3 (Ce saturated), UC (Ce solubility) and CeC_2 at 1600°C, no evidence of this phase field at 1450°C has been found in these studies. However Stecher et al. [4] found these phases in alloys prepared by heating the hydrides and carbon at 1500, 1600 and 1700°C for 20 h. When the alloys were prepared by the carbothermic reduction of the oxides no U_2C_3 was observed. The phase relationships in this region of the phase diagram will be extremely complex; Ce can dissolve in the lattices of U_2C_3 and UC_2, and oxygen in UC and UC_2, and the temperature range of existence of the U carbides will be markedly influenced by these solubilities.

Several arc-melted and annealed alloys with carbon contents between 66.7 and 70 at.% were examined. Alloys with compositions quite close to the binary edges, namely $U_{0.88}Ce_{0.12}C_2$, $U_{0.17}Ce_{0.83}C_2$ and $U_{0.08}Ce_{0.92}C_2$ contained a single b.c. tetragonal dicarbide with a small quantity of graphite in the arc melted

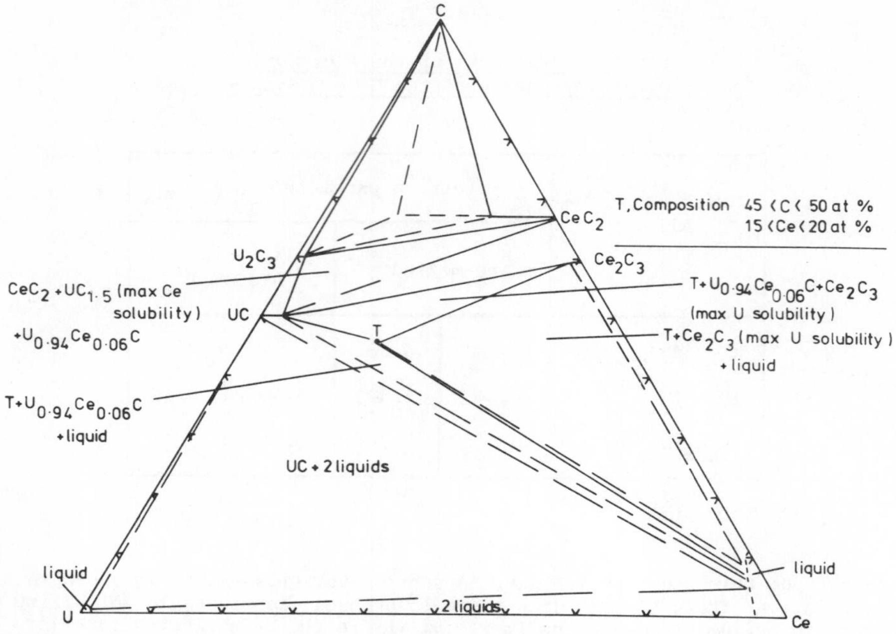

FIG. 4. A possible section of the U-Ce-C phase diagram at 1250°C.

state and after annealing for 480 h at 1450°C. Alloys with compositions $U_{0.34}Ce_{0.66}C_2$, $U_{0.50}Ce_{0.50}C_2$, and $U_{0.66}Ce_{0.34}C_2$ contained f.c. cubic dicarbide phases in the arc melted state, the alloy with equal quantities of UC_2 and CeC_2 contained two cubic dicarbides in the arc melted state and after annealing at 1450°C for 300 - 340 h. The structure of the arc melted alloy consisted of graphite, a uranium rich dicarbide and an intergranular phase containing the cerium rich dicarbide. The alloy of composition $U_{0.34}Ce_{0.66}C_2$ contained graphite, a Ce rich dicarbide phase, and an intergranular phase. The dicarbides are not miscible in the cubic state; a eutectic valley is located between 34 and 50 mol.% UC_2.

The lattice parameters of the f.c. cubic dicarbide phases are given in Table VI; these data were obtained only from the low angle lines of the films. The limits of solubility of CeC_2 in UC_2 and UC_2 in CeC_2 are difficult to obtain as there are no data for the room temperature cell sizes of the cubic dicarbides.

The recent studies of McColm et al.[3] have also shown that arc melted alloys in the dicarbide region showed similar behaviour to that found above; two immiscible cubic phases were found for alloys with compositions between $U_{0.5}Ce_{0.5}C_2$ and $U_{0.4}Ce_{0.6}C_2$, and alloys close to the binary edges contained the b.c. tetragonal dicarbides. Whilst the lattice parameter values given by McColm et al. are in reasonable agreement with those obtained in this work, there is a disagreement between our data with respect to the liquidus of this region; evidence of a eutectic indicating that the miscibility gap cuts the solidus-liquidus has been found in this work whilst McColm et al. suggest a region of complete cubic miscibility; the solid-liquidus temperatures were almost constant from UC_2 to CeC_2 in their studies.

We now consider the phase relationships in this system for temperatures less than 1450°C. Lorenzelli and Marcon have suggested that the phase relationships at 1250°C are essentially the same as those at 1600°C. Alloys annealed at 1000°C and 1250°C have been examined and it has been found the phase relationships are much more complex than those at 1450 and 1600°C. Evidence has been found for an additional f.c. cubic phase which contains less than 50 at.% carbon on the uranium rich side of the ternary diagram.

Since the preliminary observations on this system[1] attempts have been to determine the composition of this second cubic phase which has a carbon concentration less than 50 at.% but greater than 45 at.%; and probably contains between 15 and 20 at.% Ce; it is probable that this phase exists over a range of composition which may be dependent on temperature. At 1250°C the lattice parameter varied from 4.972-4.980Å, the lattice parameter increasing with increasing cerium and/or carbon content; at 1000°C the lattice parameters varied from 4.979-4.988Å. The lattice parameter of Ce saturated UC was 4.968Å corresponding to 6 mol.% 'CeC' solubility. A proposed isothermal section for 1250°C is shown in Figure 4.

It is interesting to note that Holleck and Wagner[5] found two coexisting cubic phases after annealing pellets of UC-'CeC' for 16 h at 1500°C and 64 h at 1400°C; because one phase had the same lattice parameter as UC it was assumed this was simply unreacted UC but in fact could have been in equilibrium, the presence of impurity oxygen lowering the lattice parameter of the Ce saturated U monocarbide phase. A lack of equilibrium would have resulted in ill-defined lines in the X-ray powder photographs, but the temperature at which the second cubic phase was observed was somewhat higher than expected from this work.

2.2.3 The uranium-gadolinium-carbon system

This system is considered separately because lanthanides heavier than or equal to Sm (except Eu) form the lanthanide carbide - LnC_x - where $0.35 < x < 0.65$[17] in addition to the sesquicarbide and dicarbide. At temperatures above 800 - 1000°C these compounds have a cubic Fe_4N structure [18] below these temperatures ordering occurs with the formation of a lower symmetry rhombohedral phase[19,20]. The variations of lattice parameter with carbon content are indicative of a range of composition. The lattice parameter of GdC_x varies between 5.02[17] and 5.126[18].

The dicarbide region of the ternary system has been examined by Wallace, Krikorian and Stone[6]; there was a complete range of solid solution in the dicarbide cubic region. Solidus temperatures were measured using DTA techniques and decrease continuously with the addition of GdC_2 to UC_2. Lattice parameter measurements were made on the b.c. tetragonal dicarbide phase which also possessed regions (probably metastable) of continuous solid solution. The cubic-tetragonal transformation temperatures decrease with addition of GdC_2 to UC_2 and UC_2 to GdC_2 as found by McColm et al.[3] for the dicarbide regions of the U-La-C and U-Ce-C systems. Between 1300 and 1500°C the alloys containing > ~ 11 mol.% GdC_2 decomposed into a three-phase mixture of $U_{0.89}Gd_{0.11}C_{1.5}$, C and dicarbide; the lattice parameter of the sesquicarbide was 8.104Å, and assuming a Vegard's law relationship between Gd_2C_3 and U_2C_3, the calculated solubility is only ~ 6.5 mol.% Gd_2C_3.

These studies for the temperature range 1000 - 1450°C have shown that up to ca. 16.5 mol.% hypothetical 'GdC' can dissolve in UC if it is assumed that the solid solution has a constant 50 at.% carbon content. A maximum solubility of 5.5 mol.% Gd_2C_3 in U_2C_3 was found in the presence of GdC_2 and UC (Gd solubility). Another feature of this phase diagram is that the GdC_x phase extends a long way into the uranium rich side of the ternary system. The lattice parameters of this

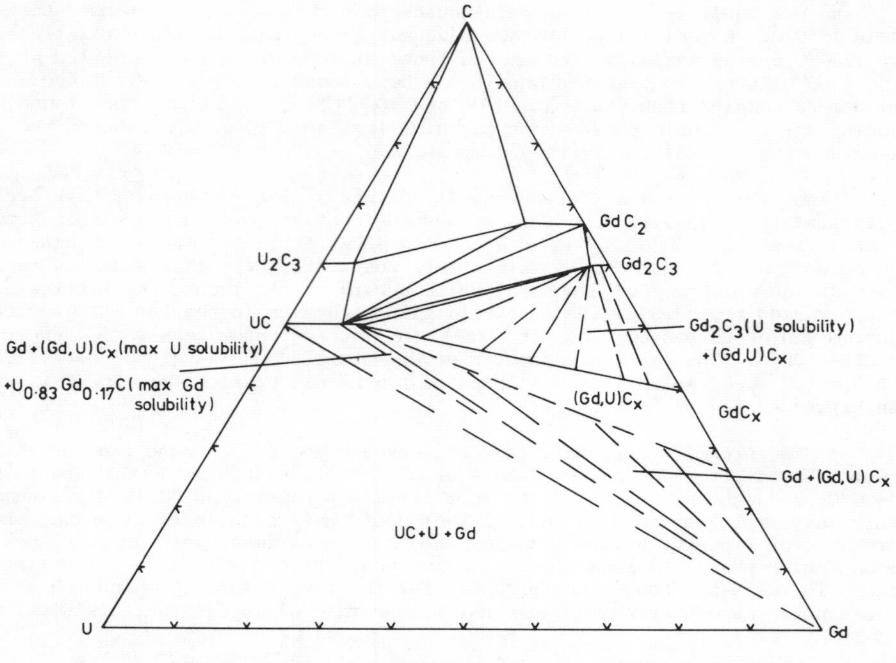

FIG. 5. A possible section of the U-Gd-C phase diagram at 1000-1200°C.

phase decrease with increase of uranium concentrations, variations have been found here from 5.089 to 4.999Å. The lattice parameters of Gd_2C_3 (8.318Å) in the presence of the above phase indicate a maximum solubility of ~ 8 mol.% U_2C_3. The Gd rich region of the phase diagram is at present under investigation. A tentative isothermal section is given in Figure 5.

2.2.4 The uranium-yttrium-carbon system

The binary Y-C system is complex and has been described by Coulson and Paulson[21] and Storms[22]; there are four phases of the system YC_x isomorphous with GdC_xwhich exists over a wide composition range, at 1200°C between 30 and 40 at.% C, Y_5C_6, Y_2C_3 and YC_2. The structures of the compounds Y_5C_6 and Y_2C_3 are unknown. YC_2 is isomorphous with the lanthanide dicarbides.

Studies of the ternary U-Y-C system have been reported by Chubb and Keller[23] which suggested that UC and YC_x were completely miscible as were the dicarbide phases at 1500°C. These authors also suggested that YC_x is in equilibrium with U. Recent studies[24] have shown that these conclusions are essentially correct although there may be a narrow region of immiscibility in the UC-YC_x section as was found in the U-Gd-C system; uranium is in equilibrium with Y and YC_x.

2.2.5 The plutonium - (La, Ce, Pr, Nd and Gd)-C system

The above systems have been examined and all behave rather similarly, the phase relationships can be represented by an isothermal section as in Figure 6.

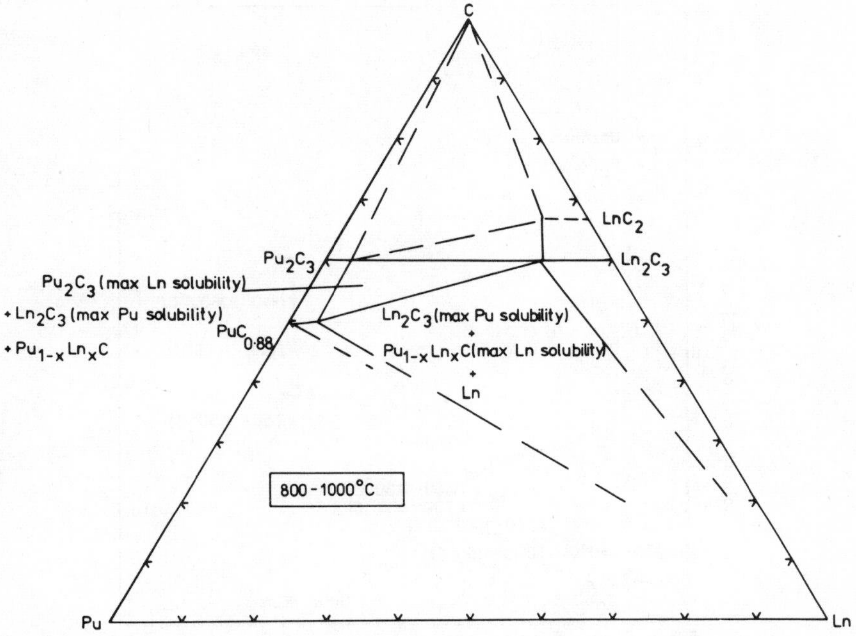

FIG. 6. The form of the section of the Pu-Ln-C phase diagrams (Ln = La, Ce, Pr, Nd).

The binary Pu-C system[14] has four carbide phases. Pu_3C_2 of unknown structure which decomposes peritectoidally at 575°C to -Pu and Pu monocarbide. Pu monocarbide exists over a wide range of composition but is always hypostoichiometric with respect to carbon. 'PuC' is isomorphous with UC; Pu_2C_3 and U_2C_3 are isomorphous but unlike U_2C_3, Pu_2C_3 possesses a small stoichiometry range. The lattice parameter of Pu monocarbide in equilibrium with Pu_2C_3, is 4.972Å and those of Pu_2C_3 are 8.126Å in the presence of $PuC_{0.87}$ and 8.133Å in the presence of carbon[25]. Pu dicarbide is not stable at ambient temperatures and is formed from Pu_2C_3 and carbon at temperatures close to 1660°C and has a cubic structure above this temperature.

Data for the Pu-Lanthanide binary phase diagrams have been collected by Ellinger et al.[26]; all these systems possess limited solid solubilities. There is liquid immiscibility in the Pu-La system; probably up to ca. 1250°C, and in the other systems there is complete miscibility in the liquid state.

2.2.5.1. The Pu-La-C system. For Pu-La-C alloys containing ca. 50 at.% carbon, the primary freezing monocarbide phase contained very little if any La in solution. In alloys containing 55 - 60 at.% carbon, where Pu_2C_3 was the primary freezing phase, the lattice parameters (8.14-8.19Å) indicated up to ~ 5 mol.% La_2C_3 solubility. In the dicarbide region, an arc melted alloy close to the Pu-C edge ($Pu_{0.85}La_{0.15}C_2$) contained sesquicarbide and carbon, and an alloy close to the La-C edge ($Pu_{0.10}La_{0.90}C_2$) contained a primary freezing dicarbide. An alloy of composition $Pu_{0.5}La_{0.5}C_2$ was single-phase and had an f.c. cubic structure with lattice parameter 5.71 ± 0.1Å. As in the uranium systems the high temperature cubic form of the dicarbides could be stabilised by solid solution formation.

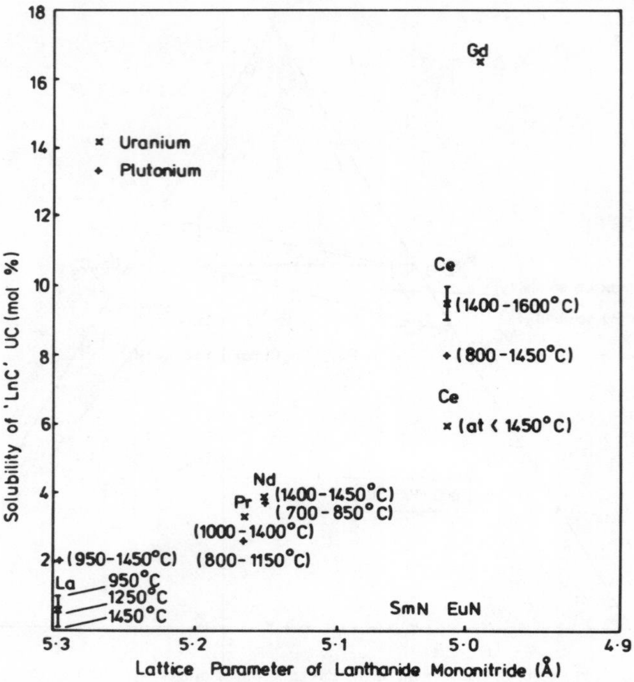

FIG. 7a. The variation of the solubility of the lanthanides (mol% hypothetical monocarbide) in U and Pu monocarbide with size of the lanthanide (expressed as lattice parameter of Ln mononitride).

Examination of alloys annealed at 950, 1250 and 1450°C for varying times from 50 - 500 h indicated a small constant solubility of 'LaC' of ca. 2 mol.% 'LaC'. (a_o = 4.978Å). The La_2C_3 solubility in Pu_2C_3 increases with increase in annealing temperature from 6 - 7 mol.% at 950°C to ca. 8.5 mol.% at 1450°C. The f.c. cubic dicarbide phase for the alloy composition of $Pu_{0.5}La_{0.5}C_2$ remaine after annealing at 1450°C for 50 hours.

We are currently examining the solubility of Pu_2C_3 in La_2C_3.

2.2.5.2. The Pu-Ce-C system. Alloys were prepared with carbon contents of 45 - 67 at.%. In arc-melted alloys which contained the monocarbide the lattice parameter of this phase increased from 4.978Å to a maximum value of 4.979Å; thi value suggesting a maximum solubility of ~ 4 mol.% 'CeC' assuming that 'CeC' dissolves in the 4 valent state as for 'CeC' solubility in UC, or if Ce is in th 3 valent state the solubility of 'CeC' would be ~ 2 mol.%.

Alloys with compositions of 60 at.% carbon indicated appreciable or complet solubility between the sesquicarbides, and arc melted alloys with 67 at.% carbon contained a b.c. tetragonal dicarbide phase; although an alloy containing equal amounts of Pu and Ce probably contained a small quantity of the cubic dicarbide the arc melted alloys gave no indication of immiscibility.

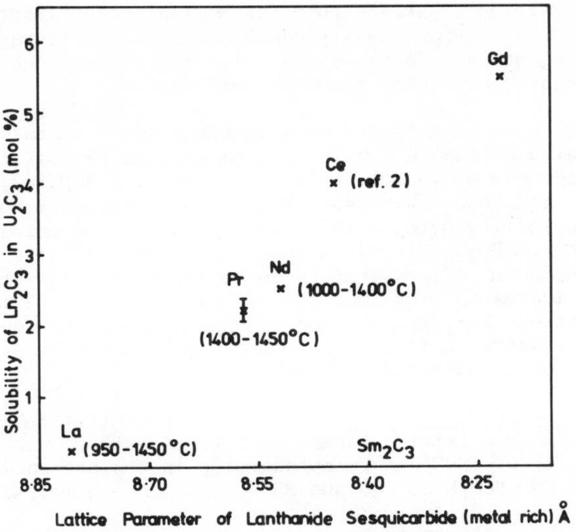

FIG. 7b. The variation of the solubility of the lanthanides (mol% Ln_2C_3) in U sesquicarbide with size of the lanthanide (expressed as lattice parameter of Ln sesquicarbide).

The alloys of this ternary system were annealed at 1450, 1250, 1000 and 800°C. The maximum solubility of 'CeC' in Pu monocarbide increases slightly with temperature, the maximum value of the solubility would be ~ 8 mol.% 'CeC' assuming tetravalent Ce, or half this value for trivalent 'CeC'. For the sesquicarbides, Ce seems to dissolve in the tetravalent state, at ~ 1500°C the sesquicarbides are completely miscible and below this temperature the sesquicarbides separate into two immiscible phases. The critical temperature is just above 1450°C and the critical concentration at ca. 22 mol.% Pu_2C_3. At 800°C the terminal solubility of Pu_2C_3 in Ce_2C_3 was 10 mol.% and at 1000°C, the solubility was 23 mol.%. The solubility of Ce_2C_3 in Pu_2C_3 was 6 mol.% at 1000°C.

The alloys containing 67 at.% carbon were annealed at 1450°C for 50 hours; alloys of composition $Pu_{0.95}Ce_{0.05}C_2$, $Pu_{0.8}Ce_{0.2}C_2$ contained graphite and a sesquicarbide phase, whilst alloys containing equal quantities of Pu and Ce were b.c. tetragonal after annealing; the exact existence range of the b.c. tetragonal region will depend on temperature.

2.2.5.3. The Pu-Pr-C system. The lattice parameter of Pu monocarbide in arc melted alloys containing nominally 50 at.% carbon indicated solubilities of up to 6 mol.% of hypothetical 'PrC' (lattice parameter 4.987Å); this value of solubility may however be high due to the presence of oxygen; the solution of oxygen in the hypostoichiometric Pu monocarbide also has the effect of increasing the lattice parameters[27].

Alloys containing 60 at.% carbon of nominal composition $Pu_{0.75}Pr_{0.25}C_{1.5}$, $Pu_{0.5}Pr_{0.5}C_{1.5}$ and $Pu_{0.25}Pr_{0.75}C_{1.5}$ indicated almost complete miscibility of the sesquicarbides of the frozen alloys, although in the alloy containing equal quantities of Pu and Pr there was some indication of phase separation as the lines of the X-ray powder photograph were rather broad.

Alloys containing 67 at.% carbon, of nominal composition $Pu_{0.55}Pr_{0.45}C_2$, $Pu_{0.3}Pr_{0.7}C_2$ and $Pu_{0.10}Pr_{0.90}C_2$ all contained the b.c. tetragonal dicarbide phas in the arc melted state. There was complete miscibility between the metastable PuC_2 and PrC_2 phases over this range of composition.

The alloys containing 50 and 60 at.% carbon were annealed at 800 and 1150°C for 170 h and also at 1450°C for 75 h. The maximum solubility at 800°C of 'PrC' in Pu monocarbide was ~ 2.5 mol.% (lattice parameter 4.979Å); the other phases present were Pr and Pr_2C_3 (with some Pu_2C_3 solubility (< 5 mol.%)). At 800°C, there was immiscibility between the sesquicarbides; 1.8 mol.% Pr_2C_3 dissolved in Pu_2C_3 and < 5 mol.% Pu_2C_3 dissolved in Pr_2C_3. The solubility of Pr_2C_3 in Pu_2C_3 was almost independent of temperature up to 1450°C, whilst the solubility of Pu_2C_3 in Pr_2C_3 increased with increase of temperature (~ 52 mol.% at 1450°C). The dicarbide alloys $Pu_{0.3}Pr_{0.7}C_2$ and $Pu_{0.10}Pr_{0.9}C_2$ after annealing at 1450°C contained a b.c. tetragonal dicarbide phase, whilst the alloy $Pu_{0.55}Pr_{0.45}C_2$ contained only the sesquicarbide phase.

2.2.5.4. The Pu-Nd-C system. The 'NdC' concentration in Pu monocarbide in arc melted alloys from lattice parameter determinations was ~ 2 mol.%. The maximum solubility of 'NdC' in Pu monocarbide in the presence of Nd and Nd_2C_3 in alloys annealed for 250 h at 700 and 850°C was 3.5 - 4 mol.%.

For the sesquicarbides there is some evidence that there may be immiscibili in the sesquicarbides; there were two sesquicarbide phases in arc melted alloys with compositions close to 20 mol.% Nd_2C_3. However if there is any immiscibilit remaining in alloys annealed at 1450°C it is confined to the region of compositions between 14 and 27 mol.% Nd_2C_3.

There is a complete range of b.c. tetragonal dicarbide solubility; these solid solutions are metastable in the plutonium rich regions. No evidence was found for the f.c. cubic dicarbide phase in the arc melted alloys.

2.2.6 General comments on the uranium and plutonium-lanthanide-carbon systems

The lanthanides considered here do not form a monocarbide phase similar to those found in the actinide carbon system, but such a monocarbide can be stabilised by solubility in either uranium or plutonium monocarbide; the solubility would be expected to increase with decrease of atom size of the lanthanide elements. This is in fact found for the uranium and plutonium systems, and the solubility of the lanthanide (expressed as mol.% hypothetical monocarbide) as a function of size of the lanthanide atom expressed as lattice parameter of the mononitride; mononitride lattice parameters were used to estimate monocarbide lattice parameters. The solubilities are shown in Figure 7a. The magnitude of the solubility of Ce which has been checked using electron microprobe analysis is indicative of Ce dissolving in U monocarbide in a valency state greater than 3; this appears to be the case for Pu monocarbide also.

There is limited solubility of the lanthanides in uranium sesquicarbide, which again increases with increase in lanthanide atomic number. These data are shown in Figure 7b; the solubilities expressed as mol.% Ln_2C_3 are plotted against the lattice parameter of the lanthanide sesquicarbide. There is some disagreement concerning the solubility of Ce_2C_3; the lattice parameter data we obtained and that of Stecher et al.[4] with those of Lorenzelli and Marcon[2]. Our solubility is 1.5 mol.% indicating solubility as Ce^{3+} whilst that of Lorenzelli and Marcon is ~ 4 mol.% suggesting that Ce is in solution with a valency state > 3.

For the sesquicarbide region of the plutonium systems we find evidence of complete solubility in the Ce, Pr and Nd systems. As the temperature is lowered phase separation into two b.c.c. sesquicarbide phases occurs. The critical temperature and concentration for the Ce system is ca. 1450°C and 20 mol.% Ce_2C_3, for the Pr and Nd systems the critical temperature is probably slightly higher than 1450°C, but the critical concentration very similar. No evidence of complete solubility was found in the lanthanum system; there is a much greater difference between the size of the La and Pu atoms.

At present there are no data on the terminal solubilities of uranium and plutonium in the lanthanide sesquicarbides.

The dicarbide region of the uranium and plutonium systems with the lanthanide behaves in a similar way in that for both systems the high temperature form of the actinide or lanthanide dicarbides presumably with the KCN structure[11], has been observed at room temperatures in the UC_2-LaC_2, UC_2-CeC_2, Uc_2-PrC_2, UC_2-NdC_2, PuC_2-LaC_2 and possibly in the PuC_2-PrC_2 systems. For the uranium systems there was immiscibility in the dicarbides; a eutectic alloy at the dicarbide section on the lanthanide dicarbide rich side of an equi-molar composition. For the Pu above systems there was complete miscibility. The immiscibility in the uranium systems resulted in the observation of two cubic phases as was observed by McColm et al. in the UC_2-LaC_2 and UC_2-CeC_2 systems. The f.c.c. dicarbides were only observed in alloys for which the concentrations of actinides and lanthanides were close; alloys with compositions nearer the end members contained the b.c. tetragonal dicarbide phase. The UC_2-GdC_2 system contained only a b.c. tetragonal phase; UC_2-GdC_2 are miscible in all proportions, as is the case for all the plutonium systems examined. The difference in size between the dicarbides of UC_2 and PuC_2 with the LnC_2 decreases as the atomic number of the lanthanide increases; the size difference is less for the Pu systems than for the U systems, and this is reflected in the observation that PuC_2 and LaC_2 are completely miscible.

The tetragonal to cubic transformation temperatures of lanthanide dicarbides increase as the lanthanide atomic number increases, that for LaC_2 is reported as 995 [28] and 1060°C[11] whilst that for GdC_2 is given as 1218 [28] or 1265°C[11]. The transformation temperature for UC_2 is 1765°C[7], and that for PuC_2 is assumed to be coincident with the eutectoid decomposition temperature[14] of 1660°C. The high temperature cubic phase of the pure compounds cannot be quenched to room temperature; however in the solid solutions an overall negative deviation from ideality results in a lowering of the transition temperature; this has been directly observed in the UC_2-LaC_2 and UC_2-CeC_2[3] systems and the minimum transformation temperatures occur near equimolar concentrations of components; the region where the cubic form is observed at room temperature. This same behaviour has been observed in solid solutions of lanthanide dicarbides[28,29].

Finally for the systems with the lanthanides in which a sub-carbide (LnC_x) occurs, U-Gd-C and U-Y-C, it seems that a small area of immiscibility between the lanthanide sub-carbides and the actinide monocarbide may be a feature of these systems; further examination is required.

2.3 The plutonium-group VIII element-carbon systems

Several studies have been made on the analogous systems with uranium[30-39]. The U-Ru-C and U-Rh-C systems are characterised by the formation of ternary compounds U_2RuC_2 and U_2RhC_2; there is also carbon solubility in URu_3 up to a composition $URu_3C_{0.7}$[34], whilst in URh_3 the solubility is less. There is no ternary compound analogous to U_2RuC_2 in the U-Pd-C system, and no carbon solubility in UPd_3.

In some preliminary studies of the plutonium systems[30] we found no evidence for any ternary compounds analogous to U_2RuC_2.

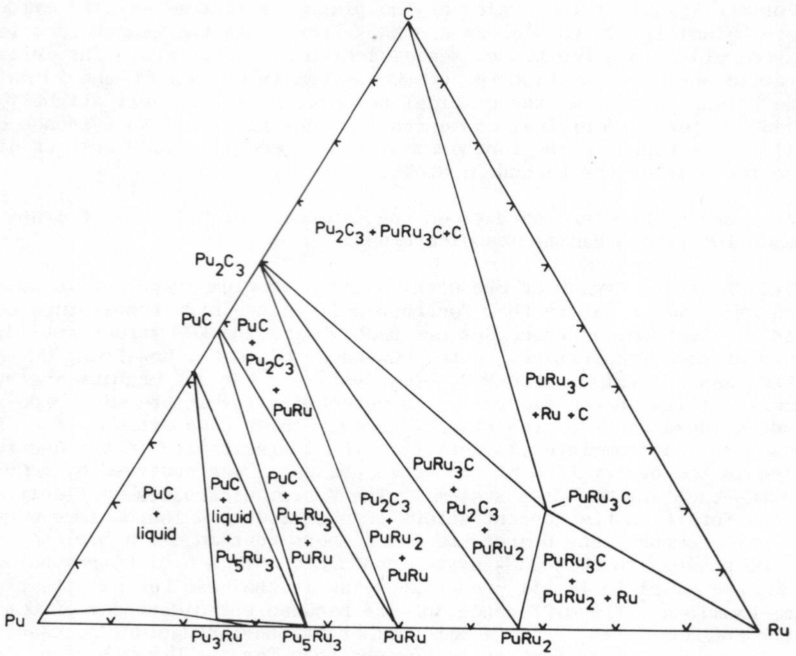

FIG. 8. A tentative section of the Pu-Ru-C phase diagram at 800°C.

2.3.1 The Pu-Ru-C system

Before discussing the ternary system the relevant binary systems are discussed. The details of the Pu-C system have been discussed earlier (section 2.2.5). For the Pu-Ru system[25] the melting behaviour of the alloys with Ru concentrations greater than ca. 20 at.% requires further studies. Five intermetallic compounds have been identified: $Pu_{19}Ru$, Pu_3Ru, Pu_5Ru_3, PuRu and $PuRu_2$. Only the structures of the latter two compounds have been reported. PuRu has a CsCl cubic structure (a_o = 3.363Å) and $PuRu_2$ has the $MgCu_2$ face-centred cubic structure (a_o = 7.472-6Å). No evidence has been found for binary compounds in the Ru-C system[40], although earlier studies suggested the existence of a compound RuC[41,42].

Examination of annealed alloys of Pu, Ru and C was consistent with the phase diagram shown in Figure 8. These further studies confirmed the previous conclusion that there was no ternary compound analogous to the uranium compound, U_2RuC_2. However, a ternary compound of composition $PuRu_3C$ was identified; this has a face-centred cubic structure and appears to be isomorphous with $URu_3C_{0.7}$; URu_3 has the Cu_3Au structure whilst the carbon containing compounds probably have a perovskite-type structure[34]. This filled Cu_3Au or perovskite structured compound for the Pu-Ru-C system was predicted[43] on the basis of empirical rules for the formation of these compounds; these filled structures have been found in the actinide-rhodium-carbon systems and in the lanthanide-rhodium-carbon systems, and in the analogous ruthenium systems. The suggested empirical rules are that the filled An or $LnRh_3C_{1-x}$ (An actinide, Ln lanthanide) would exist if the An or $LnPd_3$ compounds existed, and that the An or $LnRu_3C_{1-x}$ compound would exist if the An or $LnRh_3$ compounds existed. On this basis the predicted formation of the $PuRu_3C_{1-x}$ phase was made. The phase diagram (Figure 8) is somewhat different from the preliminary suggestion and that suggested by Holleck in that the tie lines in the region bounded by Pu_2C_3-$PuRu_2$-Pu are different from those given by Holleck based of the present authors' preliminary data, which suggested that there was a phase field, Pu_2C_3-$PuC_{0.87}$-$PuRu_2$. These further studies indicated that there is a tie line between Pu_2C_3 and PuRu up to the maximum annealing temperature of 1000°C.

Thus:

$$1.5\langle PuC_{0.87}\rangle + 0.63\langle PuRu_2\rangle \rightarrow 1.26\langle PuRu\rangle + 0.87\langle PuC_{1.5}\rangle$$

and $$\Delta G^{o}_{f\,\langle PuRu\rangle} < -18771 + 3.04T \text{ cals.mol}^{-1} \quad (1073\text{-}1273K)$$

using the recommended data for the Pu carbides[43a], and those of Campbell et al.[44] for $PuRu_2$ obtained using a molten chloride E.M.F. cell for the temperature range 935 - 1069K. The upper and lower limits for the stability of the ternary compound $PuRu_3C$ can be obtained and are given below; the upper limit for the free energy of formation (ΔG^o_f) is given by the reaction

$$6\langle Ru\rangle + 2\langle PuC_{1.5}\rangle \rightarrow \langle C\rangle + 2\langle PuRu_3C\rangle$$

and the lower limit by the reaction

$$1.215\langle PuC_{0.88}\rangle + 0.431\langle PuRu_3C\rangle \rightarrow \langle PuC_{1.5}\rangle + 0.646\langle PuRu_2\rangle$$

and $$-21460 + 2.1T > \Delta G^{o}_{f\,\langle PuRu_3C\rangle} > -52223 + 15.36T \text{ cals.mol}^{-1}$$

Further aspects of this ternary compound will be discussed in section 2.3.4.

2.3.2 The Pu-Rh-C system

In the Pu-Rh system six intermetallic compounds have been identified[25], Pu_2Rh, Pu_5Rh_3, Pu_5Rh_4, PuRh, Pu_3Rh_4, $PuRh_2$, $PuRh_3$. The only compounds for which the structures have been determined are $PuRh_2$ which has the $MgCu_2$ f.c. cubic structure (a_o = 7.488Å) and $PuRh_3$ which has the Cu_3Au f.c. cubic structure (a_o = 4.009-4.040Å). Pu_2Rh, Pu_5Rh_3, Pu_5Rh_4, Pu_3Rh_4 and $PuRh_2$ all melt peritectically at 940, 980, 1180, 1310 and 1340°C respectively. PuRh and $PuRh_3$ probably melt congruently.

No compound has been found in the Rh-C system[40] the solubility of carbon in Rh is low and the eutectic temperature is ca. 1690°C. The Pu-C system has been discussed in section 2.2.5.

We have examined alloys of this ternary system which were annealed at temperatures between 800 and 1250°C for times up to 750 h; from the constitution of these alloys a phase diagram as shown in Figure 9 can be drawn. A feature of the system is the high carbon solubility in $PuRh_3$. From the empirical rules outlined in the previous section it would indeed be expected that a perovskite or filled Cu_3Au structured compound - $PuRh_3C$ - would form in this system. The lattice parameter of $PuRh_3C$ (a_o = 4.098Å) was greater than reported for the binary compound.

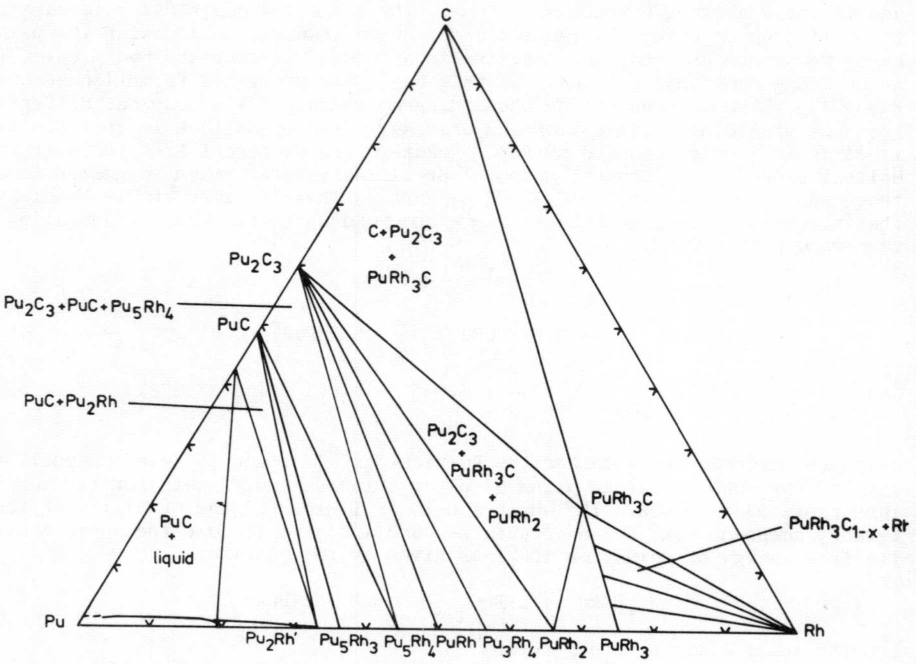

FIG. 9. A tentative section for the Pu-Rh-C phase diagram at 800°C.

The preliminary observations[30] which indicated that no ternary compound analogous to U_2RhC_2 was formed in the system have been confirmed in these later studies; however the preliminary evidence for phase field $PuC_{0.88}$-Pu_2C_3-$PuRh_3C$ has not been confirmed. It is most probable that there is a phase field $PuC_{0.88}$-Pu_2C_3-Pu_5Rh_4.

2.3.3 The Pu-Pd-C system

Four compounds have been reported for the Pu-Pd binary system[25], Pu_3Pd_4 formed peritectically at 970°C, PuPd formed peritectically at 1150°C and stable down to 950°C at which temperature it decomposes into Pu_5Pd_4 and Pu_3Pd_4. Pu_3Pd_4 melts congruently at ~ 1190°C. $PuPd_3$ melts congruently at 1500°C. Pu_3Pd_4 has a rhombohedral structure (a_o = 7.916Å, α = 114.2°), and $PuPd_3$ which exists ove a range of composition (~ 5 at.%) has the Cu_3Au structure with a_o = 4.077-4.199Å.

There are no compounds in the Pd-C system[40], the Pd-C eutectic temperatur is ca. 1500°C. The Pu-C system has been discussed in section 2.2.5.

Little additional data on the ternary system has been obtained since the preliminary studies[30]. No C-filled Cu_3Au structured or perovskite compound ha been found and no ternary compound analogous to U_2RuC_2 and U_2RhC_2 has been found no such ternary compound exists in the U-Pd-C system.

TABLE VII

The Lattice Parameters of Compounds of the Type MX_3 and MX_3C (M = Ce, Pu, X = Ru, Rh, Pd) and the Valency of M

Compound	Lattice parameter	Valence of Ce or Pu	Reference
$PuPd_3$ (variable composition)	4.077 - 4.119	3.5 - 3.25	25, 30
$CePd_3$	4.136	3.1	42
$PuRh_3$ (variable composition)	4.009 - 4.040	4.6 - 4.3	25, 30
$CeRh_3$	4.037	~4	42
$PuRh_3C$	4.098	~3.6	this work
$CeRh_3C$	4.117	~3.5	42
$PuRu_3C$	4.113 - 4.134	4.5 - 4.1	this work
$CeRu_3C$	4.137	~3.9	42

2.3.4 The C-filled or Cu_3Au structured perovskite compounds

By comparing the lattice parameters of the binary lanthanide and actinide - Ru, Rh or Pd compounds and the ternary C-filled Cu_3Au structures with the atomic size of the various lanthanides and actinides, Holleck suggested[44] that the actinides or lanthanides which form binary Cu_3Au type compounds with Pd (i.e. Sc, Y, Ln, Np, Pu and Am) or perovskite type carbides with Rh and C are present predominantly in the trivalent state, whereas the actinide, lanthanide and transition metal elements which form Cu_3Au type compounds with Rh (Th, U, Pu, Ce, Sc, Ti, V, Zr, Nb, Hf and Ta) or the perovskite type carbides with Ru and C (Th, U, Pu, Ce, V, Zr, Hf and Ta) are present predominantly in the tetravalent or pentavalent state.

The lattice parameters of the Pu phases with Ru, Rh and Pd with and without carbon are shown in Table VII, together with the estimated Pu valency and the corresponding data for the analogous Ce compounds. It can readily be seen that the $PuRu_3C$ compound has a Pu valency greater than Pu in the $PuRh_3C$.

These intermetallic compounds with the Cu_3Au or C-filled Cu_3Au structure have been observed in irradiated oxide[45] and carbide[46] fuels in fast reactor conditions. Clearly the phases obtained from oxide irradiations will be of the Cu_3Au structure whilst those in the carbide will have the C-filled Cu_3Au structure, providing that the Pd containing compounds do not exist as separate phases. Phase separation is likely to occur in these alloys; for example an alloy of composition $U_{0.5}Pu_{0.5}(Ru_{0.54}Rh_{0.13}Pd_{0.33})_3$ which contained the transition elements in the ratio of their fission yield from $U_{0.8}Pu_{0.2}$-C (Table I) but with the Pu/U+Pu ratio 0.5 was two phase; the Pu/U+Pu ratio was greater than 0.15 in the intermetallic phase in the carbide irradiations described by Bramman et al.[46]

namely 0.41. The alloy was prepared by arc melting and then annealed at 1250°C for 250 h, and contained two f.c. cubic phases (a_o = 3.987Å and 4.097Å). Immiscibility has been found in the URu_3-URh_3 and URu_3-UPd_3 systems which is being currently investigated, as well as some of the systems containing a lanthanide-GroupVIII-transition element system as an understanding of these systems is important in assessing the distribution of the lanthanide elements between the fuel matrix and fission product phases.

In the Ce-Ru-C system a ternary compound of formula $Ce_3Ru_2C_5$ has been identified[47] which is formed peritectically from CeC_2; this ternary phase was identified after annealing an arc melted alloy at 1000°C for 335 h. This phase had an hexagonal lattice with a_o = 8.792Å and c_o = 5.2983Å. This phase coul form the basis of a new type of phase observable in irradiated carbides.

From the earlier data on the phase relationships of the uranium-Group VIII element-carbon systems together with the plutonium containing systems described here, there is evidence that the phases appearing during burn-up are most likely to be ternary compounds of formulae based on U_2RuC_2, U_2RhC_2 and/or URu_3C, $PuRu_3C$ If separate Pd containing intermetallic phases separate, carbon would not be present in these phases.

Several features of these systems require further examination. For example although U_2RuC_2 and U_2RhC_2 are completely miscible, one needs to know how much P can be accommodated in these ternary compounds before the UX_3C type compounds form (X = $Ru_{1-x-y}Rh_xPd_y$) and how much Pd can be accommodated in the URu_3C type phases.

We have shown that Pu does not form compounds analogous to U_2RuC_2 and U_2RhC_2, but studies on the U-Pu-Ru-C system indicate that U_2RuC_2can accommodate Pu up to $(U_{0.8}Pu_{0.2})_2RuC_2$; above this concentration of Pu, phase separation occurs. The limits of solubility are required for the Rh system as well as data on the effects of the presence of Pd.

The above limits of Pu solubility in the ternary compounds are important in predicting the chemical state of irradiated carbide fuels. Addition of Ru and Rh to $PuC_{0.88}$ in equilibrium with Pu_2C_3 will result in the formation of PuRu and Pu_5Rh_4 respectively; such an addition to UC would result in formation of the ternary U_2RuC_2 and U_2RhC_2 compounds. The constitutional behaviour of the U-Pu-X-C (X = Ru, Rh, Pd) system is thus very important in determining whether intermetallic compounds with An/X ratios less than 3 can occur in carbide fuels, the initial compositions of which are likely to be in the monocarbide-sesquicarbide phase field of the U-Pu-C system, the Pu/U+Pu ratios usually being $\leq$ 0.30.

It seems likely that the following types of compound could form the basis of multicomponent phases which might be found in irradiated carbides.

U_2RuC_2 (b.c. tetragonal), UPd_3 (hexagonal), URu_3C (Cu_3Au f.c. cubic)
PuRu (CsCl cubic), Pu_5Rh_4 (unknown structure), $Ce_3Ru_2C_3$

Ewart and Taylor[48] found evidence for phases containing U, Pu, Ln, Ru, Rh, Pd where the Ru + Rh + Pd/U + Pu + Ln ratio was close to unity in a fuel of initial composition $U_{0.85}Pu_{0.15}C$ irradiated to 9.2% burn up in DFR.

2.4 The other phase relationships

We have described our studies on the lanthanide systems and the Group VIII transition elements, Ru, Rh and Pd. The elements Mo and Tc (paragraph 2, group b) must be considered separately from Ru, Rh and Pd. The phase relationships of U-C and Pu-C with Mo have been determined[30,49]; both systems contain the ternary compounds $UMoC_2$ and $PuMoC_2$ which most probably exist over a range of composition. There is a complete range of solid solution between $UMoC_2$ and $PuMoC_2$[30], and also solubility of Mo in UC. $UTcC_2$ is isomorphous with $UMoC_2$[39], and there is also a ternary compound $PuTcC_2$ for which the structure has not yet been determined.

For Zr and Nb (section 2, group c), both ZrC[50] and NbC[23] form a complete range of solid solution in UC; the solubility of PuC in ZrC with UC is however limited[30] and no data are available for the Pu-Nb-C system. Ba and Sr (section 2, group d) form dicarbides and there will be a tie line between these compounds and U and Pu monocarbides, but there are no data on the terminal solubilities of Ba and Sr in the uranium and plutonium carbides.

The phase relationships of Se and Te (section 2, group e) with U-C have been determined[51], Te forms a ternary compound U_2TeC_2. No data are available on the Pu systems.

Br and I (section 2, group f) will be associated with Cs and Rb, as the bromides and the iodides.

The final group of elements (section 2, group g) Ag, Cd, In, Sn, Sb are present in quite low concentrations at 10% burn-up and have not been considered here.

3. THE CHEMICAL STATE OF AN IRRADIATED CARBIDE FUEL

The constitutional studies of Bramman[46] and Ewart and Taylor[48] on irradiated carbides have already been discussed in the context of the U-C and Pu-C-Group VIII fission product elements in section 2.3.4.

There have also been irradiation experiments and burn-up simulation experiments on uranium carbides described[2b] in the context of determining the constitutional changes as a result of burn-up, Holleck and Smailos[52] found evidence that in an initially hypostoichiometric uranium monocarbide that at simulated burn-ups up to 20% separate phases based on $UMoC_2$ and U_2RuC_2 were formed, zirconium and some of the lanthanides were in solution in the U monocarbide matrix. Additional phases containing lanthanides, Ru, Rh and Pd and probably some carbon were found. Similarly Lorenzelli[2b] found many similar phases in simulation experiments on stoichiometric UC, as in a simulation of a plutonium containing carbide.

An alloy, $(U_{0.8}Pu_{0.2})$-C at 10% burn-up was simulated and its constitution examined. If the initial composition contained about 20 mol.% sesquicarbide, the simulated carbide contained U_2RuC_2, $(UPu)MoC_2$, and $PuRu_2$, but if $C/U+Pu<1$ the U_2RuC_2 was not observed, but the $PuRu_2$ type phase was again present containing lanthanides and rhodium. This behaviour is consistent with the determined phase relationships.

The distribution of the lanthanides between phases expected to appear as burn-up increases is not known; there is, however, evidence in both the simulation and irradiation experiments that lanthanides can be distributed between the fuel matrix and other phases; the exact nature of the distribution will most probably depend on the initial carbon content of the carbide fuel.

Whilst it is not claimed that a knowledge only of the phase relationships of a single fission product element with uranium- and plutonium-carbon can be used to obtain a complete chemical picture of an irradiated carbide, such a knowledge

can however provide a foundation for an understanding of the complex chemistry c
a burn-up.

With a knowledge of the phase relationships described in this paper an attempt has been made to calculate the carbon potential of the two compositions given in Table I at 10% burn-up.

To calculate the carbon potential of the fuel available carbon is distribut amongst the fission product elements as well as uranium and plutonium according their stabilities.

It has been assumed that the lanthanides and yttrium, together with Zr, and Nb dissolve in the matrix of the fuel, although studies are required on the solubilities of the rare earths in U and Pu monocarbides in the presence of Zr. Before the formation of a sesquicarbide phase, the Mo, Tc ternary compound (U,Pu (Mo,Tc)C_2 will form as well as U (Te,Se)C_2. If it is firstly assumed that the Group VIII elements are present as the ternary compound $(UPu)_2(Ru,Rh,Pd)C_2$ then the carbon potential of $U_{0.8}Pu_{0.2}C_{1.0}$ could be buffered by the reaction

$$[Ba, Sr] + 2[C] \rightleftharpoons [Ba, Sr]C_2$$

The band of carbon potentials for this reaction will be less than that for the initial composition if there were a trace quantity of sesquicarbide present[53]. Thus for this case the carbon potential decreases with burn-up. For the initial composition, $U_{0.8}Pu_{0.2}C_{1.09}$ the carbon potential after burn-up will only be fractionally less than for the initial composition[53] because there is not enou Ba and Sr present to buffer the carbon potential to the previous lower value.

If it is assumed that all the Ru, Rh and Pd are present in the form of $(UPu)(RuRh)_3C$ and $(UPu)Pd_3$ type compounds then for the almost single phase monocarbide the carbon potential at 10% burn-up will be buffered by the (Ba Sr)C_2 formation, whilst that for the carbide with the higher carbon content will remai almost unchanged. The same behaviour is predicted if it is assumed that the Pu is associated with Ru, Rh and Pd as compounds with Ru + Rh + Pd/Pd $\approx$ 1, and the uranium forms ternary compounds U_2(Ru, Rh, Pd)C_2.

If Ba and Sr were associated as compounds with the Group VIII elements as suggested by Lorenzelli this would decrease the quantity available to take part in the buffering reaction.

If the carbide fuel is subjected to temperature gradients comparable to oxide fast reactor fuels, material transport within a fuel pin will occur[54] and then the analysis of carbon potential change briefly outlined here must be modified.

4. ACKNOWLEDGEMENTS

The authors are grateful to H. E. Chamberlain, R. Paige, R. H. Smith and A. E. W. Taylor for their experimental help.

REFERENCES

[1] HAINES, H.R., POTTER, P.E., UKAEA Report. AERE-R 6513 (1970).

[2] LORENZELLI, N., MARCON, J.P., (a) J. Less Common Met. 26 (1972) 71. See also (b) LORENZELLI, N., CEA Report R-4465 (1973).

[3] McCOLM, I.J., COLQUHOUN, I., CLARK, N.J., J. Inorg. Nucl. Chem. 34 (1972) 3809.

[4] STECHER, P., NECKEL, A., BENESOVSKY, F., NOWOTNY, H., Planseeber. für Pulvermet. 12 (1964) 181.

[5] HOLLECK, H., WAGNER, W., "Thermodynamics 1967", IAEA Vienna (1968), 667.
[6] WALLACE, T.C., KRIKORIAN, N.H., STONE, P.L., J. Electrochem. Soc. 111 (1964) 1404.
[7] BENZ, R., HOFFMAN, C.G., RUPERT, G.M., High Temp. Science 1 (1969) 342.
[8] BOWMAN, A.L., ARNOLD, G.P., WITTEMAN, W.G., WALLACE, T.C., NERESON, M.G., Acta. Cryst. 21 (1966) 670.
[9] SPEDDING, F.H., GSCHNEIDNER, K.A., Jr., DAANE, A.H., Trans AIME 215 (1959) 192.
[10] POTTER, P.E., HAINES, H.R., ESTALL, J.K., unpublished work, AERE Harwell.
[11] KRIKORIAN, N.H., WALLACE, T.C., BOWMAN, M.G., Colloques. Intern. CNRS Univ. Paris (Orsay) 1965.
[12] GSCHNEIDNER, K.A., Jr., "Rare Earth Alloys" van Nostrand, Princeton (1961) 141.
[13] GSCHNEIDNER, K.A., Jr., ibid 327.
[14] HOLLEY, C.E., Jr., STORMS, E.K., "Thermodynamics 1967", IAEA Vienna (1968) 397.
[15] TENNERY, V.J., BOMAR, E.S., J. Amer. Ceram. Soc. 54 (1971) 247.
[16] HOLLECK, H., SMAILOS, E., THÜMMLER, F., J. Nucl. Mat. 32 (1969) 281.
[17] LALLEMENT, R., VEYSSIE, J.J., "Progress in the Science and Technology of the Rare Earths" 3 (EYRING, L. ed) Pergamon, Oxford (1968) 285.
[18] GSCHNEIDNER, K.A., Jr., "Rare Earth Alloys" van Nostrand, Princeton (1961) 134.
[19] BACHELLA, G.L., MERIEL, P., PINOT, M., LALLEMENT, R., Bull. Soc. Franc. Min. Chem. 89 (1966) 226.
[20] DEAN, G., LALLEMENT, R., LORENZELLI, R., PASCARD, R., Compt. Rend. Acad. Sci. Paris 89 (1966) 226.
[21] CARLSON, N., PAULSON, W.M., Trans. AIME 242 (1968) 846.
[22] STORMS, E.K., High Temp. Sci. 3 (1971) 99.
[23] CHUBB, W., KELLER, D.L., "Carbides in Nuclear Energy" (RUSSELL, L.E. et al, eds.) Macmillan, London (1964) 1 208.
[24] BRETT, N.H., HAINES, H.R., PEATFIELD, M., POTTER, P.E., unpublished work, AERE Harwell and Sheffield University.
[25] MULFORD, R.N.R., ELLINGER, F.H., HENDRIX, G.S., ALBRECHT, E.D., "Plutonium 1960". Cleaver Hume, London (1961) 301.
[26] ELLINGER, F.H., MINER, W.N., O'BOYLE, D.R., SCHONFELD, F.W., USAEC report LA 3870 (1968).
[27] POTTER, P.E., J. Nucl. Mat. 42 (1972) 1.
[28] McCOLM, I.J., QUIGLEY, T.A., CLARK, N.J., J. Inorg. Nucl. Chem. 35 (1973) 1931.
[29] ADACHI, G., KOTANI, H., YOSHIDA, N., SHIOKAWA, J., J. Less Common Metals 22 (1970) 517.
[30] HAINES, H.R., POTTER, P.E., UKAEA report AERE-R 6512 (1970).
[31] HAINES, H.R., POTTER, P.E., UKAEA report AERE-M 2837 (1968).
[32] HAINES, H.R., POTTER, P.E., Nature 221 (1969) 1258.
[33] HOLLECK, H., J. Nucl. Mat. 28 (1968) 339.
[34] HOLLECK, H., KLEYKAMP, H., ibid 34 (1970) 158.
[35] HOLLECK, H., Report KFK 1111 (1969).
[36] NARAINE, M.G., BELL, H.B., J. Nucl. Mat. 30 (1974) 83.
[37] KRIKORIAN, N.H., WALLACE, T.C., KRUPKA, M.C., RADOSEVICH, C.L., J. Nucl. Mat. 21 (1967) 236.
[38] HOLLECK, H., KLEYKAMP, H., J. Nucl. Mat. 45 (1972/3) 47.
[39] FARR, J.D., BOWMAN, M.G., "Carbides in Nuclear Energy" (RUSSELL, L.E. et al, eds.) Macmillan, London (1964) 1 184.
[40] JEANTET, R., KNAPTON, A.G., Planseeber für Pulvermet. 12 (1964) 12.
[41] NADLER, M.B., KEMPTER, C.P., J. Phys. Chem. 64 (1960) 1468.
[42] KEMPTER, C.P., NADLER, M.B., J. Chem. Phys. 33 (1960) 1580.

[43] HOLLECK, H., J. Nucl. Mat. 42 (1972) 278.
[43a] Panel on Thermodynamic Properties of the U-C, Pu-C, and U-Pu-C Systems, IAEA Vienna to appear.
[44] CAMPBELL, G.N., MULLINS, L.J., LEARY, J.A., "Thermodynamics 1967" IAEA Vienna (1968) 75.
[45] BRAMMAN, J.I., SHARPE, R.M., THORN. D., YATES, G., J. Nucl. Mat. 25 (1968) 201.
[46] BRAMMAN, J.I., Proc. 9th Commonwealth Mining and Met. Cong. (London 1969) 4 (1970).
[47] BRETT, N.H., HAINES, H.R., PARNELL, D.G., POTTER, P.E., unpublished work, AERE Harwell and Sheffield University.
[48] EWART, F.T., TAYLOR, R.G., AERE, private communication.
[49] BOUCHER, R., BARTHELEMY, P., MILET, C., "Plutonium 1965" (KAY, A.E., WALDRON, M.B., eds.).
[50] INANC, Ö., Report Jul-516-RW (1968).
[51] BREEZE, E.W., BRETT, N.H., WHITE, J., J. Nucl. Mat. 39 (1971) 157.
[52] HOLLECK, H., SMAILOS, E., "Behaviour and Chemical State of Irradiated Ceramic Fuels". IAEA Vienna (1974) 361.
[53] MARDON, P.G., POTTER, P.E., "Plutonium 1970 and other actinides" (W.N. MINER ed.) Nuclear Metallurgy 17 part 2. Proc. of the 4th International Conf., Sante Fe, New Mexico. Oct. 1970, Met. Soc. Am. Inst. Min. Met. Pet. Eng. New York 842.
[54] HORSPOOL, J.M., PARKINSON, N., FINDLAY, J.R., POTTER, P.E., RAND, M.H., RUSSELL, L.E., BATEY, W., "Fuel and Fuel Elements for Fast Reactors" 1 IAEA Vienna (1974) 3.

DISCUSSION

H. HOLLECK: You propose that a narrow region of immiscibility occurs in the UC-Gd_2C and the UC-YC_x systems. Taking into account the fact that at high temperatures you can fill up the Y_2C lattice in a homogeneous phase up to YC_2 (this means that there must be a great deal of disorder in the carbon lattice), do you believe that a miscibility gap also exists at high temperatures?

P.E. POTTER: At the temperatures of our studies on the U-Gd-C and U-Y-C systems there is definite evidence for a slight immiscibility gap between the monocarbide and the subcarbide solid solutions. We do not know whether these immiscibilities remain at high temperatures, and further studies at such temperatures using X-ray diffraction techniques are required.

H. HOLLECK: In our work relating to the behaviour of fission products in hyperstoichiometric UC (see Reaktortagung Berlin, 1974 or Karlsruhe Rep. KFK-1953) we simulated a burn-up of 10, 20 and 30 at.% in UC + U_2C_3. Without going into the details I may say that we found, even at a simulated burn-up of 10%, rare earths occurring as sesquicarbides or dicarbides — in addition to small amounts in solution. This behaviour can influence the carbon potential in a burnt fuel. You do not find higher carbides in your (U, Pu)C simulation, and I wonder whether you believe this to be due to the influence of the plutonium.

P.E. POTTER: In our paper we do not present data on simulations of burnt fuels containing plutonium, although such studies are indeed a part of our current programme. I would stress that the amount of sesquicarbide present in the initial fuel composition will influence the nature of the phases which could precipitate and that at burn-ups higher than 10% it is possible that lanthanide sesquicarbides or dicarbides could form.

N.A. JAVED: In our studies of uranium monocarbide systems, we have observed that the lattice parameter increased for hyperstoichiometric uranium monocarbides ($C/U > 1.0$) as compared with UC. Do you think that the increased lattice parameter you observed for various solid solutions of UC and lanthanide carbides may be due to the hyperstoichiometry ($C/M > 1.0$) rather than to the formation of solid solutions? What was the atomic percentage of carbon in these solid solutions?

P.E. POTTER: The solid solutions close to uranium and plutonium monocarbide with limited solubility of the lanthanides when in equilibrium with sesquicarbides or dicarbides have carbon concentrations very close to 50 at.% in the alloys which we have examined. The upper boundary of this phase would of course move to higher carbon concentrations with an increase in temperature.

A. NAOUMIDIS: The sesquicarbide of uranium cannot be directly prepared by cooling melted samples. UC_2 is a fairly metastable phase, depending on its preparation history. For the transformation of a melted sample to U_2C_3 quite a long annealing time is necessary. I should therefore like to ask whether you have eliminated the time needed to achieve equilibrium in your study of the various ternary systems and whether the heat treatments were significantly longer than this time.

P.E. POTTER: We have not examined this particular aspect of the equilibria in detail. Solubility of the lanthanides in the sesquicarbide and dicarbide will of course influence the equilibria. It is possible that in the region to which you refer equilibrium may not have been achieved; studies in this region close to the U-C binary are necessary in order to settle the matter. We believe, however, that while modifications in detail may be required, the general form of the phase diagrams is correct.

A.S. PANOV: One of the least studied aspects of ternary systems is the width of the homogeneity region. What is your opinion on this point? Did you study the change in the homogeneity region of $UC_{1\pm x}$ when rare-earth metal carbides were added?

P.E. POTTER: You are probably referring to the large regions of single-phase solid solution, for example in the U-Gd-C system. This particular region in the Gd-rich region of the phase diagram has not been investigated in detail as it is not the primary region of interest for nuclear fuel studies. The exact boundaries of the single phase carbide ternary phases and the variation with temperature require further research.

НЕКОТОРЫЕ НОВЫЕ ДАННЫЕ ЭКСПЕРИМЕНТАЛЬНОГО ИССЛЕДОВАНИЯ ФАЗОВОГО СТРОЕНИЯ СПЛАВОВ И ДИАГРАММ СОСТОЯНИЯ СИСТЕМ U-C-Mo (W, Cr, Re)

З.М.АЛЕКСЕЕВА, О.С.ИВАНОВ
Институт металлургии им.А.А.Байкова
Академии наук СССР,
Москва,
Союз Советских Социалистических Республик

Abstract–Аннотация

SOME NEW DATA FROM THE EXPERIMENTAL INVESTIGATION OF THE PHASE STRUCTURE OF ALLOYS AND THE PHASE DIAGRAMS OF THE U-C-Mo, -W, -Cr, -Re SYSTEMS.

The interaction of uranium carbides with transition metals of groups VI and VII is characterized by the formation of a ternary compound of the $UMeO_2$ type with a rhombic lattice ($UCrC_2$, $UMoC_2$, UWC_2, $UMnC_2$, $UTcC_2$, $UReC_2$). However, the tendency of uranium carbides to form ternary chemical compounds with transition metals of groups VI and VII is not limited to this: carbon-deficient compounds of the $UMeC_{2-x}$type with a monoclinic lattice are formed in the U-Mo-C and U-W-C systems. The authors discovered a similar compound in the U-Re-C system. Ternary compounds with a tetragonal lattice were also discovered in the U-W-C and U-Cr-C systems, in the $UMeC_2$-"MeC" composition region, and in the U-Re-C system, in the $UReC_2$-UC composition region. A comparative analysis was performed of the elementary cell volumes of the ternary compounds. It is possible that not all ternary compounds have been discovered in the ternary systems described. In the region of ternary systems within which the absence of ternary compounds has been established with certainty, a study was made in order to ascertain the nature of the phase equilibria. In the U-Mo-C, U-W-C and U-Cr-C systems, the authors have ascertained the nature of the crystallization of alloys in the U-UC-$UMoC_2$-Mo_2C-Mo, U-UC-UWC_2-W and UC-$UCrC_2$-Cr_7C_3-Cr regions, respectively.

НЕКОТОРЫЕ НОВЫЕ ДАННЫЕ ЭКСПЕРИМЕНТАЛЬНОГО ИССЛЕДОВАНИЯ ФАЗОВОГО СТРОЕНИЯ СПЛАВОВ И ДИАГРАММ СОСТОЯНИЯ СИСТЕМ U-C-Mo(W, Cr, Re).

Взаимодействие карбидов урана с переходными металлами VI-VII групп характеризуется образованием тройного соединения типа $UMeC_2$ с ромбической решеткой ($UCrC_2$, $UMoC_2$, UWC_2, $UMnC_2$, $UTcC_2$, $UReC_2$). Однако склонность карбидов урана образовывать с переходными металлами VI-VII групп тройные химические соединения этим не ограничивается. В системах U-Mo-C и U-W-C образуется дефицитное по углероду соединение типа $UMeC_{2-x}$ с моноклинной решеткой. Авторами обнаружено аналогичное соединение в системе U-Re-C, а также тройные соединения с тетрагональной решеткой в системах U-W-C и U-Cr-C в области составов $UMeC_2$-"MeC" и в системе U-Re-C в области составов $UReC_2$-UC. Проведен сравнительный анализ объемов элементарной ячейки тройных соединений. Возможно, что в описанных тройных системах выявлены еще не все тройные соединения. В области тройных систем, в пределах которых точно установлено отсутствие тройных соединений, проведено исследование с целью определения характера фазовых равновесий. В системах U-Mo-C, U-W-C и U-Cr-C установлен характер кристаллизации сплавов соответственно в областях U-UC-$UMoC_2$-Mo_2C-Mo, U-UC-UWC_2-W, UC-$UCrC_2$-Cr_7C_3-Cr.

I. Введение

Использование тугоплавких соединений урана в качестве ядерного горючего породило интерес к изучению диаграмм состояния тройных систем, содержащих уран, углерод и один из переходных металлов УI-УШ групп, как потенциальных источников новых тугоплавких соединений урана и физико-химических данных по их совместимости с тугоплавкими металлами. Взаимодействие карбидов урана с переходными металлами УI-УП групп характеризуется образованием тройного соединения типа $UMeC_2$ с ромбической решеткой. Однако склонность карбидов урана образовывать с переходными металлами УI-УП групп тройные химические соединения этим не ограничивается. Так в системах $U-Mo-C$, $U-W-C$ и $U-Re-C$ образуется дефицитное по углероду соединение типа $UMeC_{2-x}$ с моноклинной решеткой, где $x \simeq 0{,}12$. В системах $U-W-C$ и $U-Cr-C$ в области составов $UMeC_2$ - „MeC" обнаружено тройное соединение с тетрагональной решеткой. В разрезе $UReC_2 - UC$ системы $U-Re-C$ зафиксировано третье соединение этой системы (Y) также с тетрагональной решеткой.

2. Кристаллографические характеристики тройных соединений.

В системе $U-Mo-C$ определено два тройных соединения $UMoC_2$ [I,2] и $UMoC_{2-x}$ [3,4,5], и с большой степенью достоверности можно сказать, что в этой системе других тройных соединений не образуется. В таблице I приведены кристаллографические характеристики тройных соединений системы $U-Mo-C$.

ТАБЛИЦА I. ХАРАКТЕРИСТИКА ТРОЙНЫХ СОЕДИНЕНИЙ СИСТЕМЫ U-Mo-C

	Константы элементарной ячейки				Объем $Å^3$	Сингония	Источник
	а,Å	в,Å	с,Å	β			
$UMoC_2$	5,625	3,249	I0,990	—	200,85	Ромбическая	(а)
$UMoC_{2-x}$	5,628	3,238	II,655	I09°30'	200,I0	Моноклинная	/5/

(а) Собственные данные

ТАБЛИЦА II. ХАРАКТЕРИСТИКА ТРОЙНЫХ СОЕДИНЕНИЙ СИСТЕМЫ U-W-C

	Константы элементарной ячейки				Объем $Å^3$	Сингония	Источник
	а,Å	в,Å	с,Å	β			
UWC_2	5,633	3,250	I0,990	-	20I,2	Ромбическая	(а)
UWC_{2-x}	5,628	3,238	II,655	I09°30'	200,2	Моноклинная	/5/
Z	8,327	3,I37	-	-	2I7,4	Тетрагональная	/5/

(а) Собственные данные.

В системе $U-W-C$ существует три соединения: UWC_2 /I/ UWC_{2-x} /5,6,7,8,9/ и Z /5/. Два первых соединений аналогичны $UMoC_2$ и $UMoC_{2-x}$. Z - соединение не имеет аналогий в предыдущей системе, его состав приближенно соответствует составу сплава с 50 ат.% W + IO ат.% U + 40 ат.% C. В таблице II представлены кристаллографические характеристики тройных соединений системы $U-W-C$.

Для большей наглядности в случае Z - соединения мы отказались от принципа выбора осей, принятых в тетрагональ-сингонии, и обозначили ось четвертого порядка координатной осью "в".

В системе $U-Cr-C$ надежно определено соединение $UCrC_2$ /I/, авторами обнаружено соединение с составом, приближенно, соответствующим составу сплава с 45 ат.% Cr + IO ат.% U + 45 ат.% C (условное обозначение Cr-45), и по кристаллическому строению похожему на Z - соединение. В литературе имеется упоминание еще об одном соединении (X) системы $U-Cr-C$ /IO/, но авторы этого соединения не обнаружили. В таблице III представлены кристаллографические характеристики тройных соединений системы $U-Cr-C$.

ТАБЛИЦА III. ХАРАКТЕРИСТИКА ТРОЙНЫХ СОЕДИНЕНИЙ СИСТЕМЫ U-Cr-C

	Константы элементарной ячейки			Объем, Å^3	Сингония	Источник
	a, Å	в, Å	c, Å			
$UCrC_2$	5,433	3,23I	IO,536	I86,70	Ромбическая	/I/
Cr45	7,937	3,I48	-	I98,3I	Тетрагональная	(a)
X	-	3,636	I5,74	208,09	Тетрагональная	/I /

(a) Собственные данные

ТАБЛИЦА IV. ХАРАКТЕРИСТИКА ТРОЙНЫХ СОЕДИНЕНИЙ СИСТЕМЫ U-Re-C

	Константы элементарной ячейки				V, Å^3	Сингония	Источник
	a, Å	в, Å	c, Å	β			
$UReC_2$	5,549	3,223	IO,74		I92,08	Ромбическая	/II/
$UReC_{2-x}$	5,569	3,207	II,643	IO9^{o}52'	I95,58	Моноклинная	/I2/
Y	II,306	3,292	-	-	420,80	Тетрагональная	(a)

(a) Собственные данные.

По аналогии с Z в случае $Cr45$ обозначили кристаллографическую ось четвертого порядка координатной осью "в".

В системе $U-Re-C$ существует три соединения $UReC_2$, $UReC_{2-x}$ и Y, состав которого сдвинут относительно $UReC_2$ в сторону урана на 2-3 ат.%. Первые два соединения аналогичны $UMoC_2$ и $UMoC_{2-x}$, Y – соединение не имеет аналогий среди рассмотренных выше соединений.

В таблице IV представлены кристаллографические характеристики тройных соединений системы $U-Re-C$.

Для удобства сравнительного анализа в случае Y обозначили кристаллографическую ось четвертого порядка координатной осью "в".

В пятой колонке каждой из таблиц I-IV указаны объемы элементарных ячеек. Чтобы иметь возможность по этой величине судить об относительной устойчивости соединений, нужно знать количество атомов в элементарной ячейке. Для $UMoC_2$ Кромер /2/ определил число формульных групп (4) в элементарной ячейке. Количество атомов в элементарной ячейке $UMoC_{2-x}$ можно оценить косвенно. Дефицит по углероду в соединениях этого типа, определенный авторами ($x \simeq 0,I2$), составляет в атомном выражении I атом на две элементарных ячейки. Если подсчитать объем атома углерода, взяв за исходную величину кристаллический радиус по Макарову (0,77Å) /I3/, то он окажется равным V_c = I,9Å^3. Разница в объемах между $UMoC_2$ и $UMoC_{2-x}$, UWC_2 и UWC_{2-x}, $UReC_2$ и $UReC_{2-x}$ равна 0,75, I,0, -3,5Å^3, что для первых двух пар соответствует I/2 V_c. Следовательно, уменьшение объема дефицитного по углероду соединения связано просто с уменьшением числа атомов углерода в элементарной ячейке. Количество атомов в элементарных ячейках соединений Z, Y, $Cr45$ пока неизвестно.

Таким образом, показана способность карбидов урана образовывать с переходными металлами УI-УП групп значительное количество тройных соединений.

3. Исследование фазовых равновесий

Изо всех перечисленных тройных соединений только $UMoC_2$, UWC_2 и $UReC_2$ плавятся конгруэнтно, остальные образуются по перитектической реакции, что значительно затрудняет определение фазовых равновесий тройных систем.

3.I. Разрез $U:Me$ = I:I

Авторы попытались установить характер фазовых равновесий между соединениями $UMeC_2$ и $UMeC_{2-x}$ для систем $U-Mo-C$, $U-W-C$ и $U-Re-C$. Для этого сплавы по разрезам $U:Mo$ = I:I, $U:W$ = I:I и $U:Re$ = I:I, приготовленные методом аргонно-дуговой плавки, были подвергнуты закалке соответственно с температур до 2200, 2400 и 2000°С.

На рис.I,а представлены результаты рентгенофазового анализа сплавов разреза $U:Mo$ = I:I. В литых сплавах с содержанием углерода до 46 ат.% присутствует соединение $UMoC_{2-x}$, признаков соединения $UMoC_2$ нет; в сплавах с 46-50 ат.% C преимущественной фазой является соединение $UMoC_2$. На этом основании ранее /4/ соединению

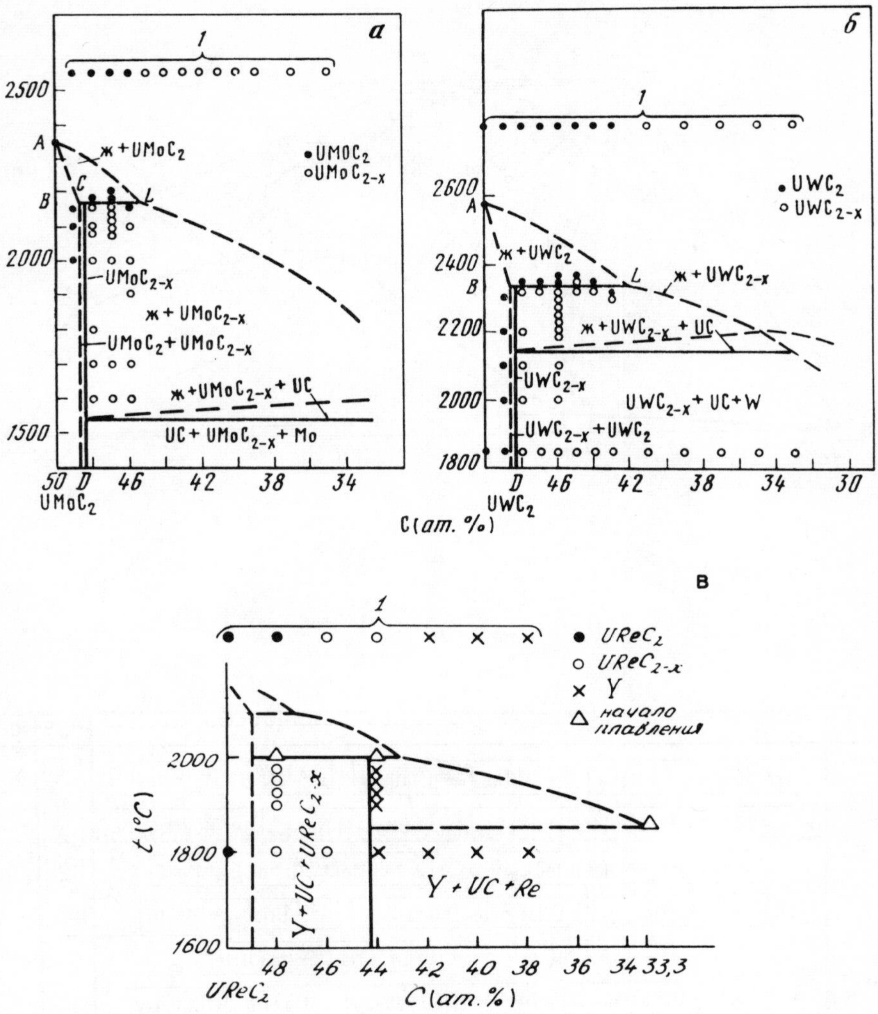

Рис.1. Результаты рентгенофазового анализа сплавов разрезов U : Mo = 1 : 1 (а), U : W = 1 : 1 (б), U : Re = 1 : 1 (в).

$UMoC_{2-x}$ был приписан состав $UMoC_{1,67}$. Как выяснилось в ходе описанного ниже эксперимента, этот вывод был ошибочным, но его не избежали Анзелин и Бартелеми /3/, обозначив исследованное ими соединение $UMoC_{1,7}$.

В закаленном состоянии сплавы, содержащие $UMoC_{2-x}$ локализованы по составу и температуре. $UMoC_{2-x}$ наблюдается в сплавах с содержанием углерода меньше 49 ат.% и при температуре ниже 2170°С. Для разделения областей существования соединений $UMoC_2$ и $UMoC_{2-x}$ достаточно провести горизонталь BL и вертикаль СД. Схематическая линия ликви-

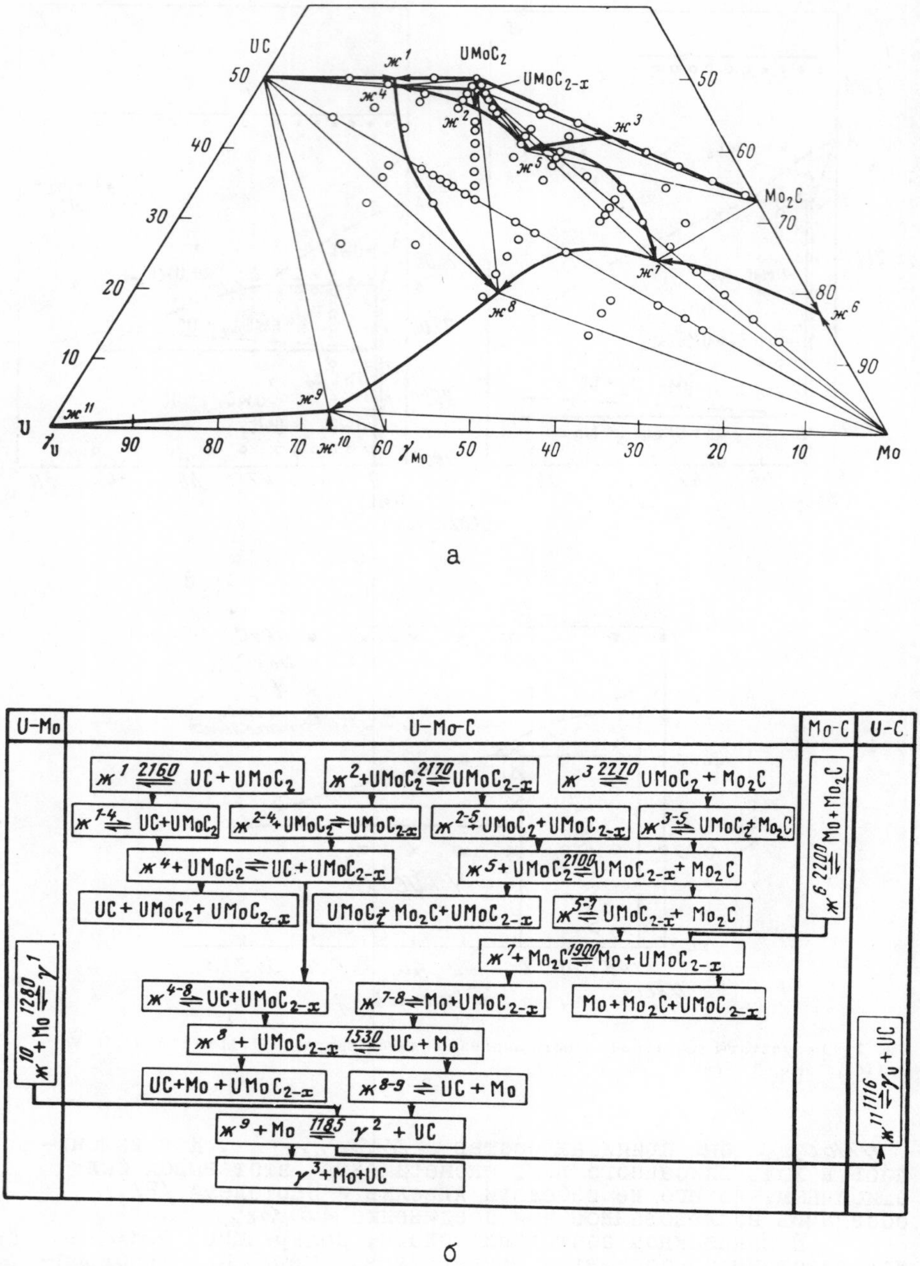

Рис.2. Проекция диаграммы состояния системы U-Mo-C (а) и схема моно- и нонвариантных реакций (б).

дуса (пунктирная кривая) проведена через точку (А) плавления соединения $UMoC_2$ (2350°С), точку L, определенную изотермой 2170°С и распределением фазовых полей в литом состоянии, и точку ликвидуса сплава Mo 33,3 - U 33,3 - - C 33,3 (около 1800°С).

На рис.1,б представлены аналогичные результаты разреза U : W = 1:1. В литых сплавах с содержанием углерода до 42 ат.% преимущественной фазой является соединение UWC_{2-x}, признаков соединения UWC_2 нет; в сплавах с 43-50 ат.% преобладает фаза UWC_2. В закаленном состоянии сплавы, содержащие UWC_{2-x}, локализованы по составу и температуре. Соединение UWC_{2-x} наблюдается в сплавах с содержанием углерода меньше 49 ат.% и при температуре ниже 2330°С. Для разделения областей существования соединений UWC_2 и UWC_{2-x} достаточно провести горизонталь BL и вертикаль СД. Схематическая линия ликвидуса проведена через точку (А) плавления соединения UWC_2 (2575°С), точку L, определенную изотермой 2330°С и распределением фазовых полей в литом состоянии, и точку ликвидуса сплава W 33,3 - U 33,3 - C 33,3 (около 2200°С).

На рис.1,в представлены результаты рентгенофазового анализа сплавов разреза U : Re = 1:1. Сплав с содержанием углерода 50 ат.% является соединением $UReC_2$ с ромбической структурой. На рентгенограмме литого сплава с 48 ат.% преимущественной фазой также является $UReC_2$. В литых сплавах с 46 и 44 ат.% C не обнаружено следов $UReC_2$, преимущественная фаза в них - $UReC_{2-x}$. В литых сплавах с 42,40 и 38 ат.% C преимущественной фазой является Y, признаков присутствия соединений $UReC_2$ и $UReC_{2-x}$ нет.

В результате отжига при 1800°С в течение 10 час. в сплаве с 48 ат.% C фаза $UReC_2$ полностью превратилась в $UReC_{2-x}$; в сплаве с 44 ат.% C фаза $UReC_{2-x}$ полностью превратилась в Y. В остальных сплавах методом качественного рентгенофазового анализа заметных изменений не обнаружено.

Сплавы с 44 и 48 ат.% C подвергались закалке с 1800, 1900, 1925, 1950, 1975 и 2000°С. Оба сплава при 2000°С оплавились. После закалок с остальных температур сплавы сохранили соответственно структуру Y и $UReC_{2-x}$.

На основании изложенных результатов была составлена реакция образования соединения $UMeC_{2-x}$ ($Me = Mo$, W, Re).

$$ж + UMeC_2 \rightleftharpoons UMeC_{2-x}, \text{ где } x \simeq 0{,}12,$$

имеющая место при 2170°С для $UMoC_{2-x}$, при 2330°С для UWC_{2-x}; для $UReC_{2-x}$ температура осталась неопределенной.

В связи с вышеизложенным следует проанализировать эксперимент Югаджина /9/, исследовавшего 4 сплава в области составов $UWC_{2,00}$ - $UWC_{1,90}$ и наблюдавшего увеличение периодов решетки ромбического соединения UWC_2 : для "а" - на 0,001Å и для "в" - на 0,002А°. Следовательно, изъятие атомов углерода сопровождается увеличением объема элементарной ячейки UWC_2, что свидетельствует об ослаблении межатомных связей. На этом участке составов распределение вакансий носит статистический характер. При достижении общего количества вакансий, выражающегося изъятием 1/2 атома углерода на одну элементарную ячейку,

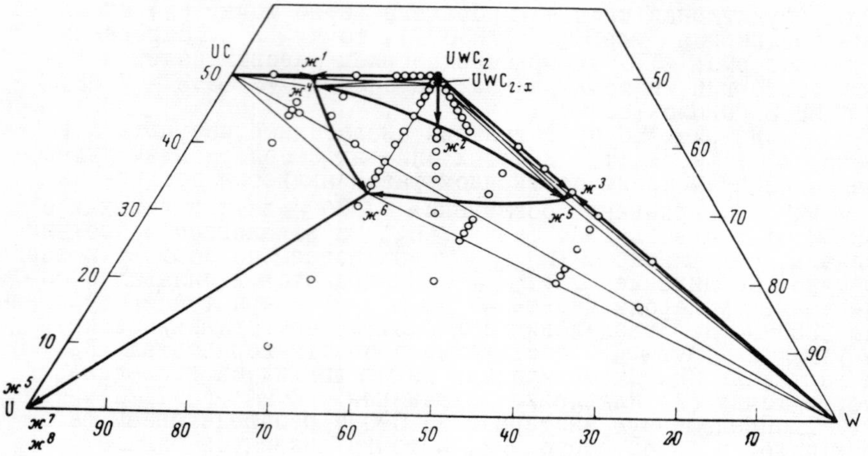

а

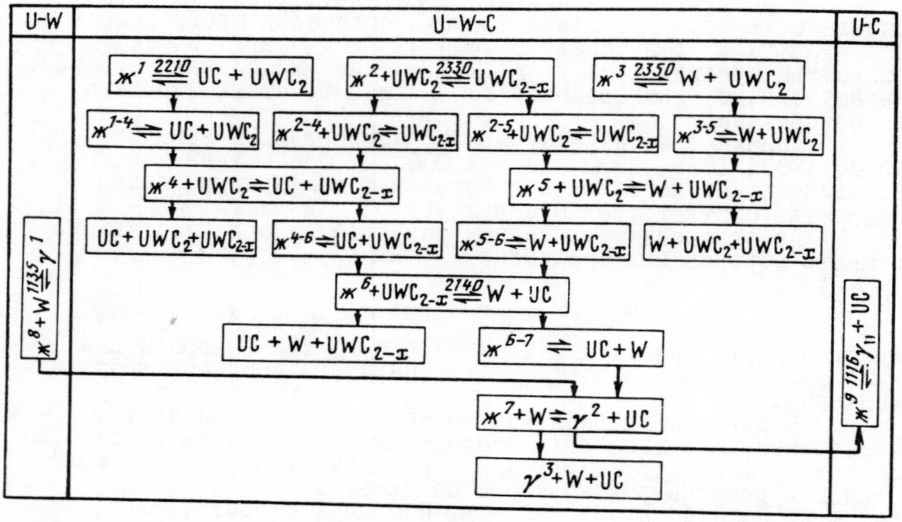

б

Рис.3. Проекция диаграммы состояния системы U-W-C (а) и схема моно- и нонвариантных реакций (б).

ромбическая упаковка перестает быть устойчивой. Атомы перестраиваются в более устойчивую моноклинную конфигурацию, при этом характер связей восстанавливается, о чем свидетельствует подсчет объемов элементарных ячеек соединений UWC_2 и UWC_{2-x}.

К сожалению, Югаджин, комментируя свой эксперимент, исходит из неправильной структуры UWC_2 /I/, исправленной Кромером /2/, показавшим отсутствие ацетиленовых связей в структуре типа $UMeC_2$.

3.2. Система $U - Mo - C$.

Вследствие того, что удалось найти фазовое соотношение между соединениями $UMoC_2$ и $UMoC_{2-x}$, стало возможным составить полную картину фазовых равновесий в области $UC - UMoC_2 - Mo_2C - Mo - U$ (рис.2).

3.3. Система $U - W - C$.

Существование четырехфазного перитектического равновесия,

$$Ж + UWC_{2-x} \rightleftharpoons UC + W$$

так долго обсуждавшееся в литературе /I4/, теперь ни у кого не вызывает сомнения. Обнаружение фазового соотношения, связывающего соединения UWC_2 и UWC_{2-x}, дало возможность проследить его возникновение (рис.3).

3.4. Система $U - Cr - C$

В системе $U - Cr - C$ установлен характер кристаллизации сплавов для области $UC - UCrC_2 - Cr_7C_3 - Cr_{23}C_6 - C$.

Путем микроструктурного и рентгенофазового анализов установлено четырехфазное перитектическое равновесие

$$Ж^1 + UCrC_2 \rightleftharpoons UC + Cr_7C_3 \quad (I)$$

Моновариантное равновесие $Ж^{1-2} \rightleftharpoons UC + Cr_7C_3$, вытекающее из I, пересекаясь с моновариантным равновесием $Ж^{3-2} + Cr_7C_3 \rightleftharpoons Cr_{23}C_6$, берущим начало в системе $Cr - C$, образует четырехфазное перитектическое равновесие

$$Ж^2 + Cr_7C_3 \rightleftharpoons UC + Cr_{23}C_6 \quad (П)$$

От П отходит моновариантное равновесие $Ж^{2-3}$ $UC + Cr_{23}C_6$ которое, пересекаясь с $Ж^{4-3} \rightleftharpoons Cr + Cr_{23}C_6$, берущим начало в системе $Cr - C$, образует четырехфазное перитектическое равновесие

$$Ж^3 + Cr_{23}C_6 \rightleftharpoons UC + Cr \quad (Ш)$$

Следовательно, разрез $UC - Cr$ пересекает области первичной кристаллизации трех фаз: UC, хрома и $Cr_{23}C_6$, и, как аналогичные разрезы в системах $U - Mo - C$ и $U - W - C$, не является бинарным.

ЛИТЕРАТУРА

[1] NOWOTNY, H., KIEFFER, R., BENESOVSKY, F., LAUBE, E., Monatsh. Chem. 89 6 (1958) 692.
[2] CROMER, T., LARSON, A.C., ROOF, R.B., Acta Crystallogr. 17 (1964) 272.
[3] ANSELIN, F., BARTHELEMY, P., Bull. Soc. Fr. Mineral. Crystallogr. 89 1 (1966) 132.
[4] АЛЕКСЕЕВА, З.М., ИВАНОВ, О.С., В сб. Физико-Химия Сплавов и Тугоплавких Соединений с Торием и Ураном, Под ред. О.С.Иванова, "Наука", М., 1968, стр.145.
[5] АЛЕКСЕЕВА, З.М., ИВАНОВ, О.С., В сб. Строение и Свойства Сплавов для Атомной Энергетики, Под ред. О.С.Иванова и Т.А.Бадаевой, "Наука", М., 1973, стр.19.
[6] CHUBB, L.W., J. Nucl. Mater. 23 3 (1967) 336.
[7] UGAJIN, M., TAKAHASHI, I., J. Nucl. Mater. 37 3 (1970) 303.
[8] POLITIS, C., J. Nucl. Mater. 39 3 (1971) 258.
[9] UGAGIN, M., SUZUKI, Y., SHIMOKAWA, J., J. Nucl. Mater. 43 3 (1972) 277.
[10] BRIGGS, G., DUTTA, S.K., WHITE, J., In "Carbides in nuclear energy", v. I. Ed. L.E. Russel, Macmillan and Co. Ltd, London, 1964, p. 231.
[11] FARR, J.D., BOWMAN, M.G., ibid., p.184.
[12] ALEXEYEVA, Z.M., J. Nucl. Mater. 49 3 (1974) 333.
[13] МАКАРОВ, Е.С., Изоморфизм Атомов в Кристаллах, Госатомиздат, М., 1973, стр.138.
[14] АЛЕКСЕЕВА, З.М., ИВАНОВ, О.С., В сб. Строение и Свойства Сплавов для Атомной Энергетики, Под.ред. О.С.Иванова и Т.А.Бадаевой, "Наука", М., 1973, стр.5.

DISCUSSION

P.E. POTTER: It is extremely interesting to see that in the U-C-Mo and U-C-W systems there is immiscibility between $UMoC_{2-x}$ and $UMoC_2$, and between UWC_{2-x} and UWC_2. How do you explain this?

O.S. IVANOV: This may possibly be due to the precipitation of one carbon atom into two elementary cells of the $UMoC_2$ compound and to the subsequent appearance of a new crystal lattice. The same reasons appear to be involved here as in the occurrence of polymorphism. Narrow regions of two-phase states are well known, for example $(\alpha + \gamma)$ in the systems Fe-Ti, Fe-Al, etc.

THE INTERACTION OF CAESIUM WITH UO_2 AT LOW OXYGEN PRESSURES AT TEMPERATURES BETWEEN 600 AND 1000°C

II. Investigations on caesium uranates

E.H.P. CORDFUNKE
Reactor Centrum Nederland,
Petten,
The Netherlands

Abstract

THE INTERACTION OF CAESIUM WITH UO_2 AT LOW OXYGEN PRESSURES AT TEMPERATURES BETWEEN 600 AND 1000°C: II. INVESTIGATIONS ON CAESIUM URANATES.

The phase relationships in the Cs-U-O system are described. The phase which exists at low oxygen pressures and at temperatures between 600 and 1000°C is $Cs_2U_4O_{12}$. Its thermal stability depends on the oxygen potential of the system, and it decomposes via two reversible phase transitions into UO_{2+x}. The structural relationship between $Cs_2U_4O_{12}$ and UO_2 is discussed. The thermochemical stability as a function of temperature was investigated by means of emf measurements.

1. INTRODUCTION

The phase relationships in the Cs-U-O system have been investigated in detail as a function of oxygen pressure and temperature, as part of our programme on fuel chemistry. Our results for a section of the phase diagram in air have been published in an earlier paper [1]. In this paper the phase relationships in the system at low oxygen pressures are dealt with. These are of special interest since caesium is one of the fission products which may play a role in the in-pile performance of oxide fuel pins. Therefore, it is important to have detailed knowledge of the chemical state of the fuel and of the possibility of caesium uranate formation, together with its dependence on the oxygen potential of the fuel.

2. EXPERIMENTAL

$Cs_2U_4O_{12}$ was made by heating of $Cs_2U_4O_{13}$ in an inert atmosphere (e.g. argon) at about 700-800°C. The latter compound could be prepared by ignition in air of the stoichiometric amounts of UO_3 and Cs_2CO_3 at about 600°C, as described previously [1].

The emf measurements were carried out in a quartz apparatus of the type described by Markin and Bones [2]. In our case the electrode compartments were separated by means of a ZrO_2-CaO tube which was closed at one end. A Ni/NiO pellet was used as the reference electrode, and a pellet consisting of a mixture of $Cs_2U_4O_{12}/UO_2$, which was placed inside the tube, was used as the other electrode. The cell was operated in an argon atmosphere. For the Ni/NiO compartment a slow stream of argon was used, this being passed through a tube filled with a mixture of Ni/NiO that was kept at the temperature of the measurements. The $Cs_2U_4O_{12}/UO_2$ compartment was first filled with argon (purified by passing over molecular sieves, BTS catalyst and UH_3) and then closed at the beginning of the measurements. The emf measurements were made with a precision Solartron digital voltmeter (A 210). Temperature measurements were done with a calibrated Pt/Pt-10% Rh thermocouple of which the junction was placed adjacent to the $Cs_2U_4O_{12}/UO_2$ electrode. The differential thermal analyses (DTA) were done with a Bureau de Liaison (Paris) apparatus using a heating rate of 9°C/min.

High temperature X-ray films were taken with a Nonius Guinier-III camera using Cu Kα-radiation.

3. RESULTS

3.1. Phase relationships in the Cs-U-O system

The pseudo-binary phase diagram of the Cs-U-O system at low oxygen pressures ($p_{O_2} \leq 10^{-10}$ atm) is given in Fig. 1. It appears that under these conditions and below 600°C the same ternary compounds are stable which were found previously [1] in the section of the phase diagram in air. The compounds have the Cs/U atomic ratios of 0.125, 0.286, 0.40, 0.5, 0.8, 1.0 (α and β) and 2.0, corresponding with the formulae $Cs_2U_{16}O_{49}$, $Cs_2U_7O_{22}$, $Cs_2U_5O_{16}$, $Cs_2U_4O_{13}$, $Cs_4U_5O_{17}$, $Cs_2U_2O_7$ (α and β) and Cs_2UO_4. In these compounds uranium is in the hexavalent state. The X-ray data of these well-defined phases have been indexed and the relevant unit cell parameters are published in part I of this series [1], whereas a structure study, based on single crystal work, will be published separately as part III of the series [3].

When heated above 600°C in an inert atmosphere at a low oxygen potential ($p_{O_2} \leq 10^{-10}$ atm) all hexavalent caesium uranates slowly decompose, and the only caesium uranate remaining has the formula $Cs_2U_4O_{12}$. This is the same phase that can be obtained by thermal decomposition of $Cs_2U_4O_{13}$, according to:

$$Cs_2U_4O_{13} \rightleftharpoons Cs_2U_4O_{12} + \tfrac{1}{2}O_2 \qquad \text{(I)}$$

This reaction is reversible, indicating that loss of Cs_2O does not occur. The dissociation temperature depends on the oxygen pressure; preliminary experiments indicate that, in air, dissociation occurs at about 1040°C, whereas at 600°C the equilibrium oxygen pressure is $\sim 10^{-13}$ atm.

All other caesium uranates decompose into a mixture of either $UO_2 + Cs_2U_4O_{12}$ or $Cs_2U_4O_{12} + Cs_2O$. It should be noted that this decomposition is very slow (at 600°C about one week is required), and that the temperature at which the decomposition begins depends on the oxygen potential of the system.

It is assumed that in $Cs_2U_4O_{12}$ uranium is both hexavalent and pentavalent; this is based on magnetic studies on related uranium compounds [4]. Since no caesium uranate in which uranium has a lower valency has been found, neither at low oxygen pressures nor in reducing atmospheres, it must be concluded that $Cs_2U_4O_{12}$ is the compound which can be formed – at a given oxygen pressure and temperature – during fission in the cooler regions of the oxide fuel. This is emphasized by the fact that $Cs_2U_4O_{12}$ in turn easily de-

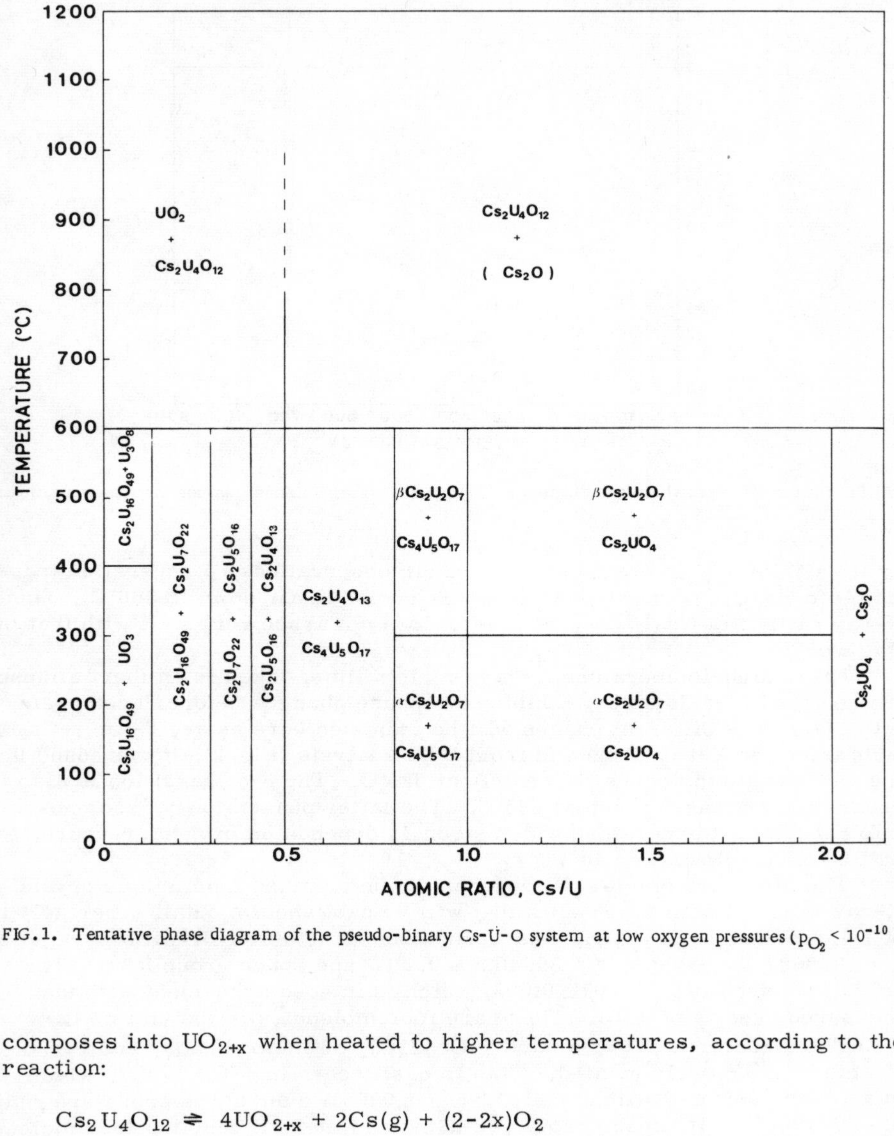

FIG.1. Tentative phase diagram of the pseudo-binary Cs-U-O system at low oxygen pressures ($p_{O_2} < 10^{-10}$ atm).

composes into UO_{2+x} when heated to higher temperatures, according to the reaction:

$$Cs_2U_4O_{12} \rightleftharpoons 4UO_{2+x} + 2Cs(g) + (2-2x)O_2 \qquad (II)$$

The temperature at which the reaction takes place depends strongly on the oxygen potential of the system, as will be demonstrated below.

3.2. Phase transitions in $Cs_2U_4O_{12}$

The formation of $Cs_2U_4O_{12}$ has been observed previously by Efremova et al. [5]. The authors obtained this product as a result of prolonged heating of 'caesium mesouranate' – Cs_4UO_5, a phase not found in our investigations –

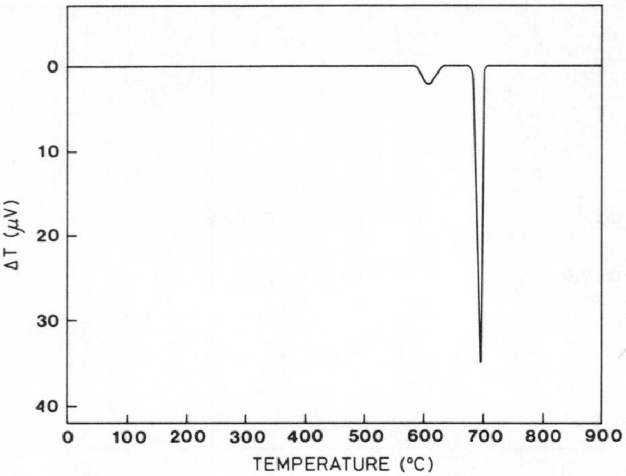

FIG.2. Differential thermal analysis diagram of $Cs_2U_4O_{12}$ (in argon, 9°C/min) showing the phase transitions.

in air at 1200°C. In agreement with their observations, a reversible transition into $Cs_2U_4O_{13}$ was found when it is cooled in air down to 500°C. Their X-ray pattern of $Cs_2U_4O_{12}$, however, does not agree very well with that of this work.

From high-temperature X-ray Guinier films, taken in an inert atmosphere it was found that $Cs_2U_4O_{12}$ exhibits two phase changes before decomposing into UO_2. The different phases will be indicated here as α-, β- or γ-$Cs_2U_4O_{12}$. Both from the X-ray films and from DTA analysis (Fig. 2), it was found that the α-β transition occurs reversibly at 625°C. The β-γ transition is also reversible and takes place at 695°C. The latter phase slowly decomposes into UO_{2+x} at a temperature which strongly depends on oxygen pressure, as will be seen below.

The structure of α-$Cs_2U_4O_{12}$ could be determined from single crystal X-ray work. From the results that will be published in detail separately [3], it follows that α-$Cs_2U_4O_{12}$ has a rhombohedral unit cell with a = 10.9623 ± 0.0006 Å, α = 89.402° ± 0.007° and space group R3m. The calculated density is 7.107 ± 0.002, which is in good agreement with the measured density of 7.16. There are four molecules in the unit cell.

The structures of α-, β- and γ-$Cs_2U_4O_{12}$ on the one hand, and UO_2 on the other are closely related. This is easily seen in γ-$Cs_2U_4O_{12}$, which has a cubic unit cell with a = 11.2295 ± 0.0006 Å over the temperature range of 700-1000°C. Its space group is Fd3m, and the cell contains four molecules, which corresponds with a calculated density of 6.610 ± 0.002. The resemblance of the α and γ-form is evident[1], whereas the relationship with the UO_2 structure (cubic, a = 5.470 Å, space group Fm3m, d = 10.95) follows from consideration of Fig. 3. It can be seen that half of the uranium atoms have to be taken away from the UO_2 structure, being replaced by a caesium atom to yield the $Cs_2U_4O_{12}$ structure.

---

[1] The monoclinic β-phase is also strongly related to this structure (see Ref. [3]).

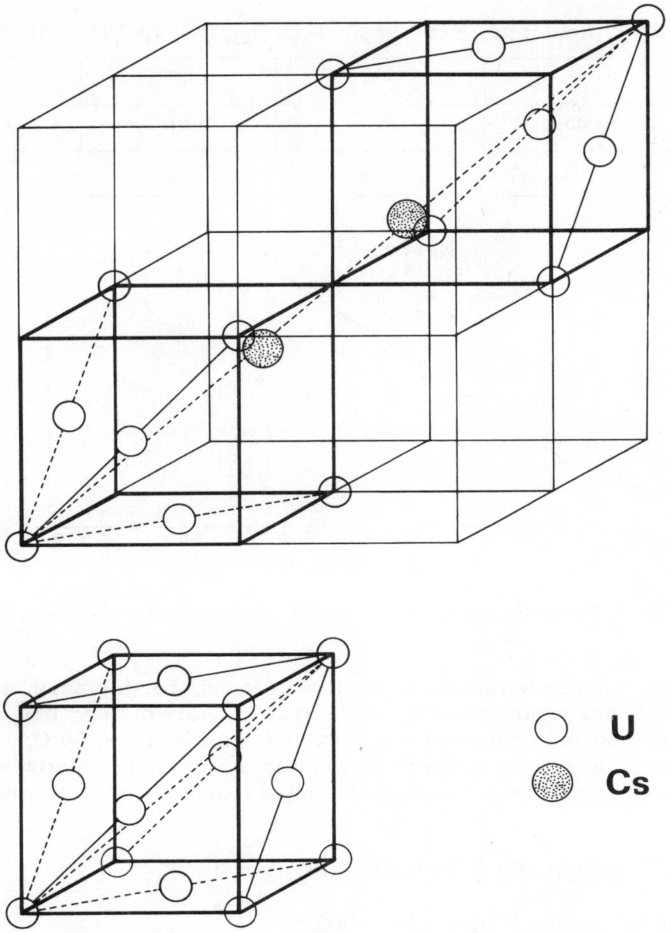

FIG.3. Unit cells of UO_2 (below) and $Cs_2U_4O_{12}$ (top) showing the relationship between their structures. Note that, in $Cs_2U_4O_{12}$, only those uranium atoms are drawn from which the formation from UO_2 can be visualized.

It is evident that formation of $Cs_2U_4O_{12}$ from UO_2 – as may occur in the cooler parts of the oxide fuel pins in the reactor [6] – causes a large increase in volume. This may lead to swelling of the fuel, and even to fuel-pin failure [7].

3.3. Thermochemical stability of $Cs_2U_4O_{12}$

γ-$Cs_2U_4O_{12}$ is the form which is stable above 695°C. Its stability depends strongly on the oxygen potential of the surrounding atmosphere. For instance, in air, $Cs_2U_4O_{12}$ can be heated up to about 1250°C. However, at lower oxygen pressures, decomposition into UO_{2+x} according to:

$$Cs_2U_4O_{12} \rightarrow 4UO_{2+x} + 2Cs(g) + (2-2x)O_2 \qquad \text{(III)}$$

TABLE I. $\Delta\overline{G}_{O_2}$ (kcal/mol) FOR THE $Cs_2U_4O_{12}/UO_2$ EQUILIBRIUM

t (°C)	701	728	751
$-\Delta\overline{G}_{O_2}$ (kcal/mol)	87.3	81.2	74.5

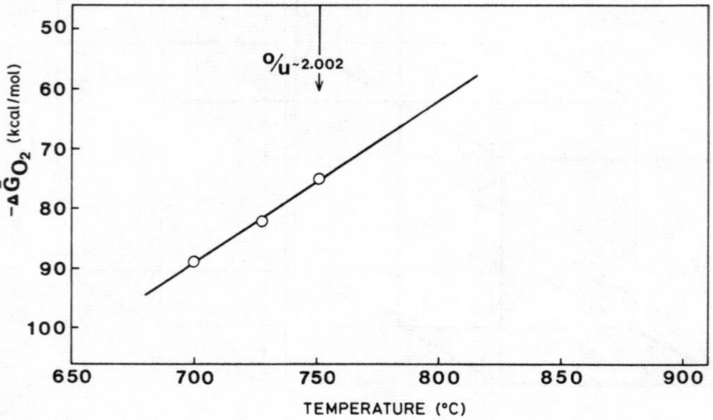

FIG.4. The equilibrium $Cs_2U_4O_{12}/UO_2$; $\Delta\overline{G}_{O_2}$ as a function of temperature.

occurs at much lower temperatures. It was found, for instance, that in an atmosphere of thoroughly purified argon (using finely divided UH_3 as oxygen getter) decomposition occurs at temperatures as low as ~ 650°C.

An attempt has been made to determine the oxygen potential at which $Cs_2U_4O_{12}$ decomposes as a function of temperature, using a reversible emf cell of the type:

$$Ni, NiO \mid ZrO_2(CaO) \mid Cs_2U_4O_{12}, UO_{2+x}$$

as described in Section 2 (Experimental).

In the emf cell experiments, steadiness of the emf reading is taken as an indication that the cell has reached equilibrium. The time required to attain steady emf values is generally long (at 750°C at least 50 h). The emf values are used to calculate $\Delta\overline{G}_{O_2}$, by taking for the Ni/NiO reference electrode the value [8]:

$$\Delta G_T^0 = -111\,940 + 40.67\,T \quad \text{kcal/mol } O_2 \qquad (640 - 1100°C)$$

Using this approach, equilibrium $\Delta\overline{G}_{O_2}$ values have been obtained at the temperatures shown in Table I. The results are shown in Fig. 4.

4. DISCUSSION

In a reactor fuelled with mixed oxide, a steep temperature gradient exists and, as a result, an oxygen gradient exists also. The oxygen potential increases during irradiation. It is evident that the behaviour of fission product caesium is determined by the local oxygen potential. From previous work [6, 10] it appears that, at low oxygen potentials, caesium behaves both as the free element and, in the combined state, as CsI, migrating rapidly to the coolest position in the fuel rod. At the stoichiometric composition caesium appears to combine with other constituents in an order of preference given by I > Te > fuel > Mo [6].

At higher oxygen potentials (O/M > 2, $\Delta\overline{G}_{O_2}$ thus being > 100 kcal/mol at 800°C), it has been observed that caesium strongly associates with the fuel between 500 and 1000°C, thus influencing the behaviour and the in-pile performance. For instance, in hyperstoichiometric fuel (O/M = 2.075) heated in sealed capsules at about 725°C for 75 h, caesium uranate was found to be present [11].

A number of caesium uranates that could be formed have been postulated, i.e. Cs_2UO_4 [9], Cs_2UO_4 and $Cs_2U_2O_7$ [6] and $CsUO_3$ [11]. However, as demonstrated in this paper, only $Cs_2U_4O_{12}$ can be in equilibrium with UO_2 at the low oxygen potentials existing in the fuel rods. Its formation leads to local swelling, as is evident from the large difference in densities between UO_2 and $Cs_2U_4O_{12}$.

At much higher oxygen potentials (> 40 kcal/mol) other caesium uranates, i.e. $Cs_2U_5O_{16}$ and $Cs_2U_4O_{13}$, can be formed [1].

ACKNOWLEDGEMENTS

The experimental assistance of Mrs. G. van Voorst is gratefully acknowledged. The author wishes to thank Dr. G. Prins and Mr. A. B. van Egmond for valuable discussions.

REFERENCES

[1] CORDFUNKE, E.H.P., VAN EGMOND, A.B., VAN VOORST, (Mrs.) G., J. Inorg. Nucl. Chem. (in press, 1974).
[2] BONES, R.J., MARKIN, T.L., WHEELER, V.J., Proc. Br. Ceram. Soc. 8 (1967) 51.
[3] VAN EGMOND, A.B., to be published.
[4] KEMMLER-SACK, S., STUMPP, E., RÜDORFF, W., ERFURTH, H., Z. Anorg. Allg. Chem. 354 (1967) 287.
[5] EFREMOVA, K.M., IPPOLITOVA, E.A., SIMANOV, Yu.P., Vestn. Mosk. Univ., Ser.II. Khim. 24 (1969) 57.
[6] AITKEN, E.A., EVANS, S.K., RUBIN, B.F., Behaviour and Chemical State of Irradiated Ceramic Fuels (Proc. Panel Vienna, 1972), IAEA, Vienna (1974) 269.
[7] NEIMARK, L.A., LAMBERT, J.D., USAEC Rep. ANL-RDP-report 11 (1972).
[8] KELLOGG, H.H., J. Chem. Eng. Data 14 (1969) 41.
[9] HOEKSTRA, H.R., J. Chem. Thermodyn. 6 (1974) 251.
[10] ADAMSON, M.G., AITKEN, E.A., Trans. Am. Nucl. Soc. 17 (1973) 195.
[11] JOHNSON, I., JOHNSON, C.E., Trans. Am. Nucl. Soc. 17 (1973) 194.
[12] MAYA, P.S., USAEC Rep. ANL-7833 (1971) 5.11.

DISCUSSION

M.H. RAND: Have you observed the presence of $Cs_2O(g)$ in the decomposition of $Cs_2U_4O_{12}$?

E.H.P. CORDFUNKE: No, because we have not carried out any mass spectrometric work on this decomposition reaction.

M.H. RAND: Does not the oxygen pressure for this decomposition vary with the total pressure of the caesium-containing species?

E.H.P. CORDFUNKE: Yes, it does. In our investigations, however, p_{O_2} had a fixed value.

M.G. ADAMSON: It is very gratifying to learn that Cs-U-O compounds with a net uranium valency of less than 6+ do indeed exist. As you know, at the Vallecitos Laboratory we have obtained evidence for the existence of such compounds in the Cs-U-O and Cs-U, Pu-O systems (described in paper IAEA-SM-190/52 these Proceedings, Vol. 1, p. 187).

I should first like to ask you whether you know the caesium pressure in your emf cell measurements of p_{O_2} for the stability of $Cs_2U_4O_{12}$ + UO_2 mixtures.

E.H.P. CORDFUNKE: It is not possible to measure caesium pressures in our emf apparatus. Since, however, the decomposition reaction

$$Cs_2U_4O_{12} \rightarrow 2Cs(g) + 2O_2\ (g) + 4UO_2$$

takes place in argon (at 1 atm pressure), and $p_{Cs} = p_{O_2}$, $\Delta\bar{G}_{O_2}$ is only a function of temperature.

M.G. ADAMSON: Have you performed any experiments in which $Cs_2U_4O_{12}$ was heated with liquid caesium and, if so, did you obtain any evidence for the formation of additional ternary compounds (e.g. with a Cs/U ratio greater than 0.5)?

E.H.P. CORDFUNKE: No, we have not. However, we have performed some experiments in closed crucibles, the results of which do not indicate that additional ternary caesium uranates can be formed.

J.H. DAVIES: On the basis of the measurements reported by Mr. Adamson in paper IAEA-SM-190/54 (Vol. 1, p. 187), which showed that the oxygen potential in Zircaloy-clad UO_2 fuel at appreciable burn-up is very low, one would not expect the formation of caesium uranate. However, Mr. Förthmann's work on ceramic additives to fuel indicated that the caesium alumino-silicates are more stable than caesium uranate. It would be interesting and valuable to characterize these compounds in the same way as you have characterized the Cs-U-O system.

Would you care to comment on or predict the stability of the caesium alumino-silicates?

E.H.P. CORDFUNKE: This is an interesting point. In fact, you may be right, since it is not possible to heat the caesium uranates in silica or alumina without severe attack on the crucibles. It seems worth while to compare the relative stabilities, either by estimation or experimentally.

ТЕРМОДИНАМИКА ВЗАИМОДЕЙСТВИЯ ДВУОКИСИ И МОНОНИТРИДА УРАНА С МЕТАЛЛАМИ

В.Н.ЗАГРЯЗКИН, А.С.ПАНОВ, Б.Ф.УШАКОВ,
Е.В.ФИВЕЙСКИЙ, Б.Н.БОБКОВ
Физико-энергетический институт,
Обнинск,
Союз Советских Социалистических Республик

Abstract–Аннотация

THE THERMODYNAMICS OF THE INTERACTION OF URANIUM DIOXIDE AND MONONITRIDE WITH METALS.

The authors show that the interaction of non-stoichiometric UO_{2+x} and UN_{1-x} with molybdenum, tungsten, tantalum and niobium is accompanied by the formation of oxides and solid solutions of oxygen and nitrogen in these metals. The authors determine the maximum permissible deviations from stoichiometric composition at which the interactions set in. They establish the fact that a transfer of components from non-stoichiometric compounds takes place when a variable source is used. With the passage of time the component flows achieve equilibrium and the composition of the compound ceases to vary. The stable compositions calculated for non-stoichiometric uranium compounds lie in the pre-stoichiometric region.

ТЕРМОДИНАМИКА ВЗАИМОДЕЙСТВИЯ ДВУОКИСИ И МОНОНИТРИДА УРАНА С МЕТАЛЛАМИ.

Показано, что взаимодействие нестехиометрических UO_{2+x} и UN_{1-x} с Mo, W, Ta, Nb сопровождается образованием окислов и твердых растворов кислорода и азота в этих металлах. Определены предельно допустимые отклонения от стехиометрических составов, при которых начинаются реакции взаимодействия. Установлено, что перенос компонент из нестехиометрических соединений происходит в условиях переменного источника. С течением времени потоки компонент уравновешиваются и состав соединения перестает меняться. Рассчитанные стабильные составы нестехиометрических соединений урана располагаются в достехиометрической области.

Специфика твердофазных высокотемпературных реакций заключается в значительном влиянии нестехиометричности участвующих в реакции соединений. Для иллюстрации этого факта достаточно вспомнить работы [1-12], в которых было показано значительное влияние отношения ат%O/ат%U на характер и степень взаимодействия UO_{2+x} с молибденом и вольфрамом. Поэтому понимание и описание высокотемпературных твердофазных реакций взаимодействия невозможно без учета областей гомогенности реагирующих соединений. В настоящей работе это будет продемонстрировано на примере систем двух типов UO_{2+x} - Me и UN_{1-x} - Me, где Me – тугоплавкий металл.

Для описания процессов взаимодействия в выбранных для анализа системах потребуются знания парциальных термодинамических функций компонент нестехиометрических соединений, которые можно получить из анализа имеющихся литературных данных по равновесию между нестехиометрическим соединением и газовой фазой. При переходе от термодинамической активности компонент в газовой фазе к активности в твердой фазе можно использовать известные уравнения статистической термодинамики.

1. ХИМИЧЕСКИЙ ПОТЕНЦИАЛ АЗОТА И КИСЛОРОДА В ГАЗОВОЙ ФАЗЕ

Химический потенциал и абсолютную активность одноатомной газовой компоненты А можно вычислить по известным уравнениям статистической термодинамики [13,14].

$$\mu_A = RT\ln\lambda_A = \frac{1}{2}RT\ln\lambda_{A_2} = \frac{1}{2}RT\ln\frac{\overline{P}_{A_2}}{B(T)} \qquad (1)$$

где $\overline{P}_{A_2}$ – парциальное давление молекулы A_2; λ_A и λ_{A_2} – абсолютные активности А и A_2, а величина B(T) определяется уравнением:

$$B(T) = kT\left[\frac{2\pi m_A \cdot kT}{h^2}\right]^{3/2} \cdot \frac{T}{\Theta_2} \cdot \frac{1}{1 - e^{-e_v/T}} \cdot \frac{\rho_{oe}\cdot\rho_{on}}{2} \cdot e^{-\frac{D_{A_2}}{kT}} \qquad (2)$$

в котором Θ_r и Θ_v – характеристические температуры, равные $\Theta_r = h^2/8\pi^2 Jk$ и $\Theta_v = h\nu/k$; m_{A_2} – масса молекулы A_2; J – момент инерции, ν – частота колебаний гармонического осциллятора; ρ_{oe} – вырождение самого низкого электронного уровня; ρ_{on} – вырождение самого низкого ядерного уровня; k и h – постоянные Больцмана и Планка; D – энергия диссоциации молекулы на атомы.

При расчетах были приняты следующие значения констант [13-17].

	Θ_r, °K	Θ_v, °K	ρ_{oe}	ρ_{on}	D, ккал/моль
O_2	2,0802	3356,75	3	1	117,973
N_2	2,8694	2202,4	1	1	225,072

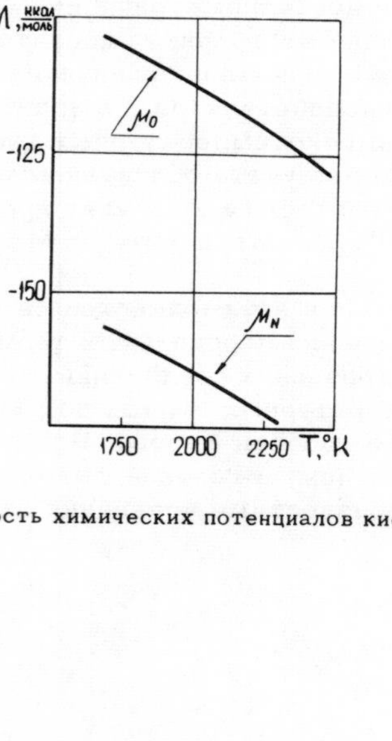

Рис.1. Температурная зависимость химических потенциалов кислорода и азота в газовой фазе при давлении, равном 1 ат.

Проверка правильности выбранных констант осуществлялась сравнением рассчитанных и известных из литературы [16-18] значений констант диссоциации O_2 и N_2. Оказалось, что отличия не превышали 0,5-0,7%. Авторы сочли такие отклонения вполне приемлемыми. На рис.1 приведены результаты расчета температурной зависимости химических потенциалов O и N в газовой фазе при $\bar{P}_{A_2} = 1$ ат с использованием выбранных констант.

2. ТЕРМОДИНАМИЧЕСКАЯ АКТИВНОСТЬ КИСЛОРОДА В ТВЕРДОЙ НЕСТЕХИОМЕТРИЧЕСКОЙ ДВУОКИСИ УРАНА

Для описания зависимости термодинамической активности кислорода от температуры и состава нестехиометрической двуокиси урана привлечем методы статистической термодинамики. Однако прежде оговорим тип структурных дефектов. В работе Виллиса Б. [19] с помощью нейтронографических методов исследований показано, что при избыточном содержании кислорода в двуокиси урана основными типами дефектов являются атомы или ионы кислорода, внедренные в междуузлия анионной подрешетки. Для двуокиси урана с дефицитом по кислороду отсутствуют прямые доказательства того, что основными типами дефектов являются вакансии в анионной подрешетке. Нами с учетом работы [20] будет рассматриваться модель двуокиси урана с бездефектной подрешеткой урана. Тогда часть большой функции состояния, ответственная за отклонения от стехиометрии, будет иметь вид, представленный уравнением (3) [13]:

$$\Gamma(T,\lambda_O) = \sum_{N_O^v N_O^i} \frac{(2N_U)!}{N_O^v!\,(2N_U - N_O^v)!} \cdot \frac{(\alpha N_U)!}{N_O^i!\,(\alpha N_U - N_O^i)!} (\lambda_O)^{(2N_U - N_O^v - N_O^i)} j_O^s(T,N_O^v) \cdot j_O^i(T,N_O^v) \times$$

$$\times \exp\left\{-\left[N_O^v E_O^v + N_O^i E_O^i + \frac{(N_O^v)^2}{2N_U} E^{vv} + \frac{(N_O^i)^2}{\alpha N_U} E^{ii}\right]/kT\right\} \qquad (3)$$

В принятых обозначениях подстрочечный индекс означает принадлежность к компоненте. Суммирование производится по всем возможным значениям числа дефектов N_O^v и N_O^i, которые обозначают, соответственно, число вакансий и число атомов в междуузлиях в подрешетке кислорода. При написании уравнения (3) использованы следующие обозначения: $2N_U$ – число мест в узлах подрешетки и (αN_U) – число мест в междуузлиях, в которых распределяются атомы кислорода; E_O^v и E_O^i – энергии образования соответственно вакансии и атома кислорода в междуузлии; E_O^{vv} и E_O^{ii} – учитывают энергии взаимодействия дефектов между собой; j_O^s и j_O^i – колебательные функции состояний.

При построении большой функции состояний принято, что дефекты (вакансии и атомы в междуузлиях) располагаются произвольно, а изменение в электронном распределении мало и им можно пренебречь.

Замена суммы максимальным членом и его отыскание дают возможность получить следующие уравнения для абсолютной активности кислорода [13]:

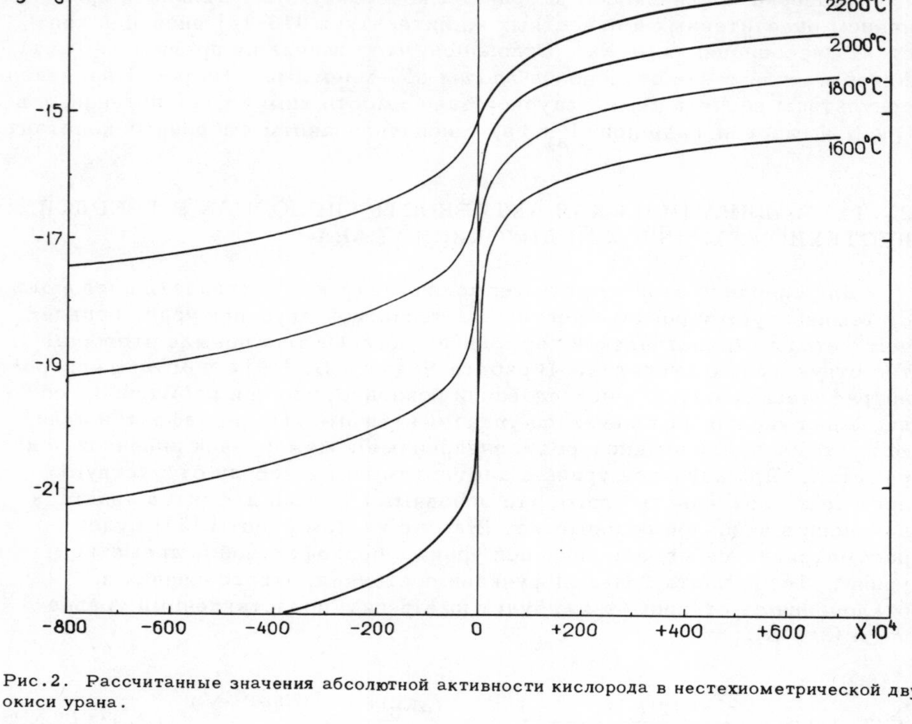

Рис.2. Рассчитанные значения абсолютной активности кислорода в нестехиометрической двуокиси урана.

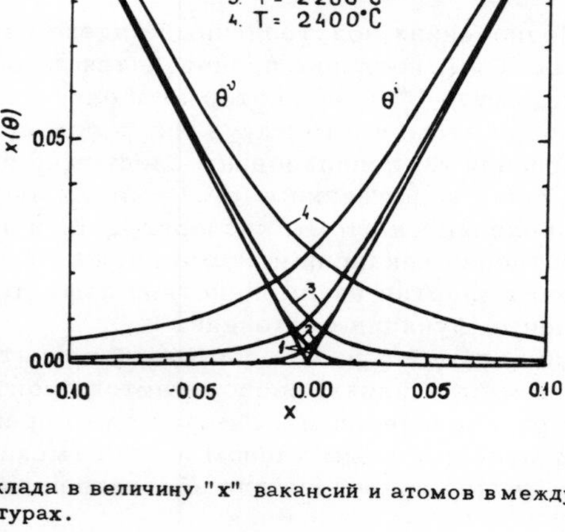

Рис.3. Зависимость вклада в величину "x" вакансий и атомов в междуузлиях от состава UO_{2+x} при различных температурах.

$$\ln\lambda_O = -\ln\frac{\Theta_O^v}{1-\Theta_O^v} - \frac{E_O^v}{kT} - 2\Theta_O^v\frac{E_O^{vv}}{kT} + \frac{\delta \ln j_O^s(T, N_O^v)}{\delta N_O^v} \qquad (4)$$

$$\ln\lambda_O = \ln\frac{\Theta_O^i}{1-\Theta_O^i} + \frac{E_O^i}{kT} + 2\Theta_O^i\frac{E_O^{ii}}{kT} + \frac{\delta \ln j_O^s(T, N_O^i)}{\delta N_O^i} \qquad (5)$$

$$x = \alpha\Theta_O^i - 2\Theta_O^v \qquad (6)$$

В уравнениях (4,5) величины N_O^v и N_O^i заменены долями вакансий и атомов кислорода в междуузлиях, причем соотношения между ними следующие:

$$\Theta_O^v = \frac{N_O^v}{2N_U} \qquad \text{и} \qquad \Theta_O^i = \frac{N_O^i}{\alpha N_U}$$

Введенная величина "x" характеризует отклонение от стехиометрии (ат%O/ат%U = 2 + x) и в соответствии с фазовой диаграммой состояния U - O [20-24] может принимать как положительные, так и отрицательные значения. Число мест в междуузлиях будем считать равным $\alpha = 1$, ограничив тем самым область применимости полученных зависимостей значением $x \leqslant 0{,}08$ [25, 26].

Во всех своих последующих рассуждениях мы будем рассматривать регулярное приближение, учитывающее взаимодействие дефектов между собой (E^{vv} и E^{ii}), различие колебательных функций состояния атомов кислорода в узлах решетки и междуузлиях (j_O^s и j_O^i), взятых в приближении Дебая, и поправку в колебательную функцию состояния за счет наличия вакансий [13, 25].

Вычисление постоянных, входящих в уравнения (3), (4) и (5), производилось путем обработки литературных сведений о температурной и концентрационной зависимости парциального давления кислорода над нестехиометрической двуокисью урана [27-29]. При этом химические потенциалы кислорода в газовой фазе и твердой нестехиометрической UO_{2+x} считались равными и для перехода от $\overline{P}_{O_2}$ к λ_O в UO_{2+x} использовались уравнения (1,2). Рассчитанные с использованием полученных значений констант величины термодинамической активности кислорода в нестехиометрической UO_{2+x} приведены на рис.2.

Следует заметить, что использование уравнений (4)-(6) вблизи стехиометрического состава, когда $x = 0$ и $\Theta_O^i = 2\Theta_O^v$, следует производить с особой тщательностью и осторожностью. Представляет интерес рассмотреть, какой вклад в отклонение от стехиометрии вносят атомы в междуузлиях и вакансии при различных температурах вблизи стехиометрического состава. На рис.3 показана доля вкладов этих дефектов в величину "x" по нашим расчетным данным. Видно, что величина вкладов дефектов противоположного типа становится значительной, начиная с 2000 °C, приводя к необходимости учета разупорядочения $\delta\,(\delta_{x=0} = \sqrt{N_O^v N_O^i})$ для стехиометрической двуокиси урана. При этом область применимости явных зависимостей при $|x| > \delta$ вблизи стехиометрии сокращается.

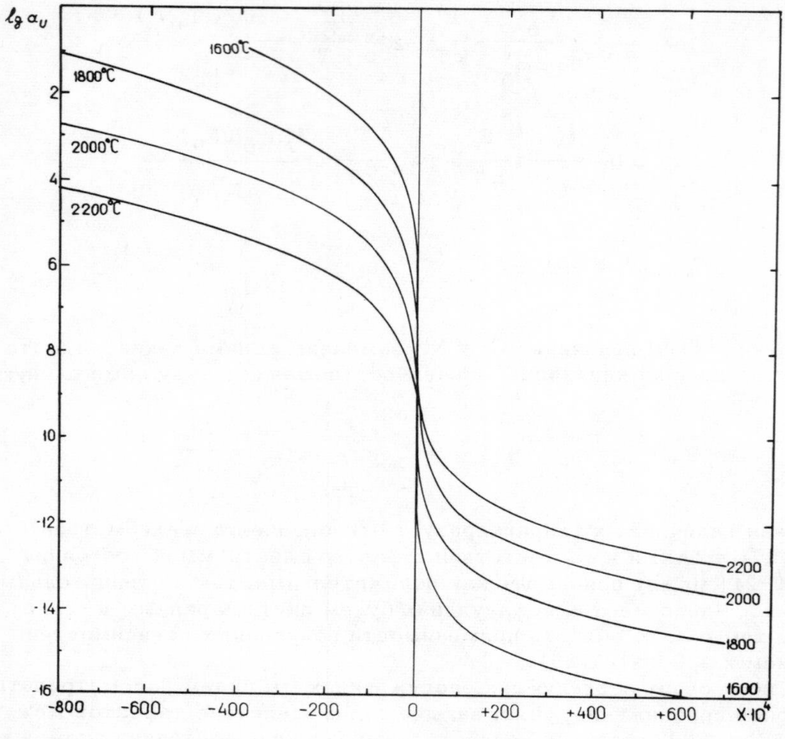

Рис.4. Рассчитанные значения относительной активности урана в нестехиометрической двуокиси урана (стандартное состояние – чистый уран).

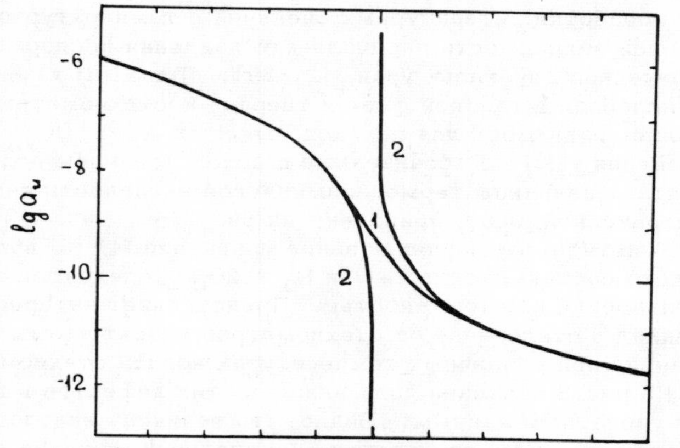

Рис.5. Значение относительной активности урана при 2400 °С, вычисленные по уравнениям (9), (10) (кривые 2), и методом мелкой разбивки участков (кривая 1) (стандартное состояние – чистый уран).

В том случае, когда один тип дефектов преобладает над другим, уравнения (4), (5) можно записать в упрощенном виде. Так, например, для двуокиси с дефицитом по кислороду можно принять $x \simeq -2\Theta_O^v$. Тогда уравнение (4) запишется в виде:

$$\ln\lambda_O = -\ln\frac{-x}{2+x} - a^v(T)x + b^v(T) \tag{7}$$

Для двуокиси урана, имеющей избыточное содержание кислорода, $x \simeq \Theta_O^i$ и уравнение (5) запишется в виде:

$$\ln\lambda_O = \ln\frac{x}{1-x} + a^i(T)x + b^i(T) \tag{8}$$

В уравнениях (7), (8) константы a^v, b^v, a^i и b^i включают энергии образования и взаимодействия дефектов, а также вклад колебательных функций состояния. Границы применимости уравнений (7), (8) видны из рис.3.

3. ТЕРМОДИНАМИЧЕСКАЯ АКТИВНОСТЬ УРАНА В ТВЕРДОЙ НЕСТЕХИОМЕТРИЧЕСКОЙ ДВУОКИСИ УРАНА

Располагая сведениями о зависимости абсолютной активности кислорода от состава при различных температурах, можно вычислить с использованием уравнения Гиббса-Дюгема абсолютную активность урана λ_U. Константа интегрирования определялась путем приравнивания активностей на границах двухфазной области ($U + UO_{2+x}$) для достехиометрической двуокиси урана (C^v) и путем сопоставления интегральных свободных энергий UO_{2+x} по обе стороны от стехиометрического состава при $x \to 0$ для сверхстехиометрической двуокиси урана (C^i). При этом использовались численные значения работ [18,21-24,30]. В общем виде уравнения для термодинамической активности урана в нестехиометрической двуокиси UO_{2+x} имели вид:

$$\ln\lambda_U = 2\ln(-x) + \frac{a^v x(4+x)}{2} + C^v \tag{9}$$

$$\ln\lambda_U = \ln\frac{(1-x)^3}{x^2} - \frac{a^i x(4+x)}{2} + C^i \tag{10}$$

Численные значения термодинамической активности урана в UO_{2+x} приведены на рис.4. Следует отметить, что в области составов, вблизи стехиометрии, когда заметен вклад дефектов другого типа, пользоваться уравнениями (9), (10) надо осторожно. Нами при расчетах активности урана вблизи стехиометрического состава участок при $x \sim 0$ разбивался на небольшие интервалы, в которых изменение λ_O с составом было невелико. Границы применимости уравнений (9)-(10) видны из рис.3,5.

ТАБЛИЦА I. ИЗМЕНЕНИЕ СТАНДАРТНОЙ СВОБОДНОЙ ЭНЕРГИИ ДЛЯ РЕАКЦИИ (11) ПРИ 2000 °K (ккал/г · атом кислорода)

	$1,8\cdot10^{-2}$	$1,7\cdot10^{-2}$	$1,6\cdot10^{-2}$	$1,4\cdot10^{-2}$	$1,3\cdot10^{-2}$	$1,2\cdot10^{-2}$	$1,0\cdot10^{-2}$	$1,0\cdot10^{-3}$	$1,0\cdot10^{-4}$	$-1\cdot10^{-4}$	$-1\cdot10^{-3}$
MoO_2	-2,54	-2,03	-1,62	-0,50	+0,12	+0,73	+2,25	+20,65	+37,72	+186,73	+ 165,87
WO_2	-0,22	+0,29	+0,70	+1,82	+2,44	+3,05	+4,57	+22,97	+40,04	+189,05	+168,19
NbO_2	-50,42	-49,91	-49,50	-48,38	-47,76	-47,15	-45,63	-27,23	-10,16	+138,85	+117,99
Ta_2O_5	-57,38	-56,87	-56,46	-55,34	-54,72	-54,11	-52,59	-34,19	-17,12	+131,89	+111,03

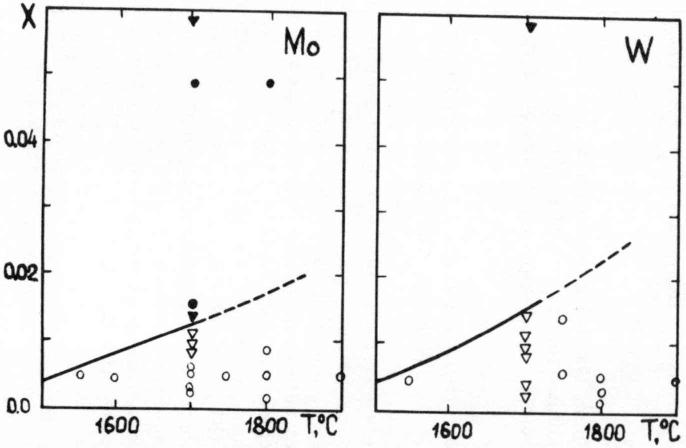

Рис.6. Температурная зависимость предельно допустимых отклонений от стехиометрии двуокиси урана при взаимодействии ее с молибденом и вольфрамом. Зачерненные точки обозначают появление в зоне контакта новых фаз, светлые – их отсутствие.

△▲ – результаты настоящей работы;
○● – [1-12].

Проверка достоверности полученных значений осуществлялась путем сравнения парциального давления урана, полученного в экспериментах работы [31], и рассчитанного по уравнениям (9)-(10). Сходимость результатов была вполне удовлетворительная (отличия не превышали 5÷10%).

4. ТЕРМОДИНАМИЧЕСКИЙ АНАЛИЗ ОБМЕННЫХ РЕАКЦИЙ ВЗАИМОДЕЙСТВИЯ МЕТАЛЛОВ С НЕСТЕХИОМЕТРИЧЕСКОЙ ДВУОКИСЬЮ УРАНА

Реакция взаимодействия различных металлов с нестехиометрической двуокисью урана можно описать уравнением типа:

$$\frac{m}{x} UO_{2+x} + Me \rightleftarrows MeO_m + \frac{m}{x} UO_{2,00} \qquad (11)$$

При такой форме записи обменной реакции были сделаны следующие предположения. Во-первых, считалось, что взаимодействие металлов с UO_{2+x} начинается и заканчивается реакцией (11), т.е. исключалась возможность взаимодействия $UO_{2,00}$ и MeO_m между собой с образованием твердых растворов или сложных тройных окислов. Во-вторых, не учитывалась возможность образования твердых растворов урана и кислорода. В-третьих, предполагалось, что продуктами взаимодействия являются стехиометрические соединения. Возможность прохождения реакции (11) определяется знаком изменения свободной энергии.

Используя полученные ранее зависимости, для изменения свободной энергии реакции (11) были выведены соответствующие уравнения для $x>0$ и $x<0$, на основании которых с использованием данных работы [18] и наших расчетов были вычислены изменения свободной энергии реакции (11) при 1600-2100 °K и $x = (-0,001) \div (+0,030)$. Результаты расчетов (табл.I) показывают, что ни один из всех рассмотренных металлов не должен образовывать окислы при контакте со строго стехиометрической двуокисью урана и двуокисью, имеющей избыток урана. В то же время даже небольшой избыток кислорода ($x = 10^{-4} \div 10^{-5}$) в двуокиси является причиной образования окислов ниобия и тантала. Значения предельно допустимых отклонений от стехиометрического состава, при которых может начинаться образование окислов, для всех металлов различны и определяются свободной энергией образования окислов этих металлов.

Рассчитанные значения предельно допустимых отклонений от стехиометрии двуокиси при взаимодействии ее с молибденом и вольфрамом в области 1600-2100 °K приведены на рис.6. Видно, что границы предельно допустимых отклонений от стехиометрии двуокиси урана симбатно расширяются с температурой. Это является следствием различий температурных зависимостей свободных энергий образования окислов молибдена, вольфрама и урана. Причем для W, как металла, образующего наиболее непрочные с термодинамической точки зрения окислы, допускаются наибольшие предельно допустимые отклонения от стехиометрии.

Для проверки полученных соотношений нами было проведено несколько экспериментов. Образцы двуокиси урана, имеющие формульные составы ат%О/ат%U = 2,002; 2,003; 2,008; 2,011; 2,014 и 2,059, отжигались в контакте с монокристаллическими молибденом и вольфрамом при 1700 °C в течение 1000 часов. В том случае, если в зоне контакта не образовывалось никаких новых соединений, то граница раздела оставалась ровной и вблизи нее не было видно никаких включений. Это было характерно для составов 2,002 (Mo, W); 2,003 (Mo, W); 2,008 (Mo, W); 2,011 (Mo, W) и 2,014 (W). Если же в системе образовывались новые фазы, то граница контакта становилась неровной, а вблизи нее появлялись поры и включения новой фазы, по-видимому, оксидного происхождения [1,2,11,32]. Такое взаимодействие наблюдалось в случае использования двуокиси, имеющей составы 2,014 (Mo) и 2,059 (Mo, W).

Результаты экспериментов вместе с литературными данными нанесены на рис.6. Нетрудно видеть хорошее согласие результатов расчета и экспериментов авторов настоящей работы и работ [1-12].

5. ИЗМЕНЕНИЕ СОСТАВА ДВУОКИСИ УРАНА В ПРОЦЕССЕ ЕЕ ВЗАИМОДЕЙСТВИЯ С МЕТАЛЛАМИ

В том случае, если содержание кислорода в UO_{2+x} будет превышать предельно допустимые значения (рис.6), то в зоне контакта будут образовываться окислы реагирующих металлов. Когда отношение ат%O/ат%U было меньше предельно допустимых значений, то все взаимодействие сводилось к диффузии урана и кислорода в контактирующий с UO_{2+x} металл. В любом случае в процессе взаимодействия будут уходить уран и кислород. Так как скорости их ухода и потребления, как правило, не совпадают, то не исключены случаи, когда состав UO_{2+x} изменяется в процессе взаимодействия.

Для изменения состава двуокиси урана со временем легко получить следующее уравнение:

$$\frac{dx}{dt} = -\frac{S}{n_U} J_O + (2+x)\frac{S}{n_U} \cdot J_U \qquad (12)$$

где S – площадь, через которую уходят U и O из системы; J_O и J_U – потоки соответственно урана и кислорода через площадь S; n_U – количество молей урана в системе.

В том случае, когда состав двуокиси не меняется ($\frac{dx}{dt} = 0$), то выполняется условие (13)

$$J_O = (2+x) J_U \qquad (13)$$

Такой состав двуокиси урана, при котором соотношение между компонентами не меняется, назван нами стабильным. Рассчитать его легко из уравнения (13). Для этого необходимо знать потоки U и O и лимитирующую стадию. Рассматривая стационарный режим переноса компонент (разумность этого предположения проверялась) и пользуясь данными работ [4,33-36] и собственными экспериментами, было установлено, что лимитирующей стадией взаимодействия UO_{2+x} с герметичной оболочкой являются процессы переноса через оболочку.

Для расчетов стабильного состава UO_{2+x} необходимы сведения о диффузионной подвижности атомов урана и кислорода в материале оболочки. Анализ литературных данных [37-40] по диффузии урана в тугоплавких металлах выявил существенный разброс в значениях коэффициентов диффузии, достигающий нескольких порядков по данным разных авторов.

Выполненная авторами обработка известных литературных данных по самодиффузии и диффузии примесей в ОЦК-металлах позволила получить обобщенное корреляционное соотношение, связывающее параметры диффузии с процессом испарения и имеющее следующий вид:

$$D = 0{,}545 \cdot a^2 \cdot \nu \cdot \exp\left\{-0{,}35(\Delta H_{исп} - T \cdot \Delta S_{исп})/RT\right\} \times$$

$$\times \exp\left\{-0{,}25(\Delta H^*_{исп} - L(1-N)^2/RT\right\} \frac{см^2}{с} \qquad (14)$$

Здесь a – период решетки; $\Delta H_{исп}$ и $\Delta S_{исп}$ – теплота и энтропия испарения металла-основы; L – энергия взаимообмена в твердом растворе; $\Delta H^*_{исп}$ – теплота испарения металла-основы в случае самодиффузии или теплота испарения примесного металла в случае диффузии примеси; N – атомная доля примесных атомов. Для численных расчетов принимается, что N = 1 при самодиффузии и N = 0 при диффузии примесных атомов. Величина L определялась из обработки фазовых диаграмм. Для систем U - W, U - Mo, U - Ta величины L соответственно равны 18,9; 11,7; 14,5 ккал/г · атом.

Частота колебаний ν оценивается по формуле:

$$\nu = \left\{0{,}17[\Delta H^*_{исп} - L(1-N)^2]/a^2 m\right\}^{1/2}, \ с^{-1} \qquad (15)$$

где m – масса диффундирующего атома.

Расчеты по формулам (14,15) удовлетворительно согласуются с большинством известных литературных данных по диффузии примесных атомов и самодиффузии в "нормальных" ОЦК-металлах. Для случая диффузии атомов урана в тугоплавких металлах полученные результаты расчета приведены в табл.II. Здесь же для сравнения приведены экспериментальные параметры диффузии урана в монокристаллах W, Mo и Ta, получен-

ТАБЛИЦА II. ПАРАМЕТРЫ ДИФФУЗИИ УРАНА D_0 $см^2/с$ И Q, ккал/г · атом ПО ДАННЫМ РАСЧЕТА И ЭКСПЕРИМЕНТА

Система	Расчет		Эксперимент	
	D_0	Q	D_0	Q
W-U	0,45	95,5	0,30	95,5
Mo-U	0,32	81,6	0,075	81,6
Ta-U	0,34	90,8	0,19	87,7

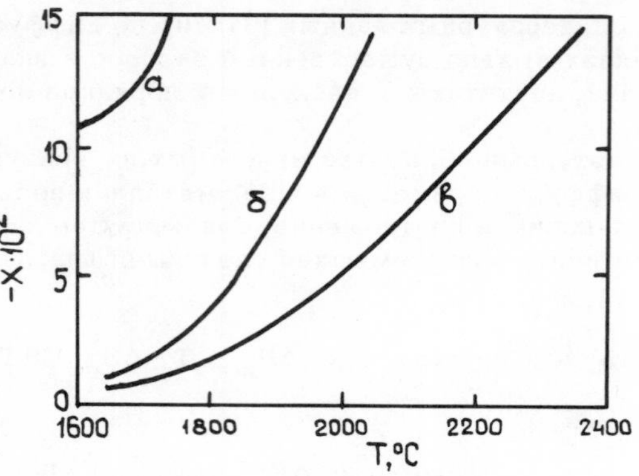

Рис.7. Температурная зависимость стабильных составов двуокиси урана для танталовой (а), вольфрамовой (б) и молибденовой (в) оболочек.

ные авторами в интервале температур 1800-2100 °С с применением изотопа ^{235}U в металлическом виде. Коэффициенты диффузии U, полученные авторами, отличаются от результатов работ [37-40], что может быть объяснено исключением переменного источника диффузии, имевшего место во всех экспериментах, выполненных ранее.

При расчетах стабильного состава UO_{2+x} использовались наши экспериментальные данные по диффузии U через монокристаллы W, Mo и Ta. Для диффузии O брались данные работы [4]. При этом химические потенциалы компонент на границе в оболочке и соединении UO_{2+x} считались одинаковыми.

Рассчитанные значения стабильных составов приведены на рис.7. Видно, что они находятся в достехиометрической области, близки к конгруэнтно испаряющимся составам UO_{2+x} и практически совпадают с результатами работ [1,2,6].

Однако случай, когда состав двуокиси не меняется со временем, является редким. Чаще потоки компонент не уравновешивают друг друга и состав UO_{2+x} с течением времени изменяется. Для избыточного содержания кислорода, когда поток кислорода значительно превышает поток урана, из уравнения (12) с использованием полученных ранее констант после некоторых упрощений можно получить следующее выражение для описания изменения состава UO_{2+x} со временем

$$x = x_0 \exp(-\xi_O t) \qquad (16)$$

где x_0 соответствует значению при $t = 0$, а величина ξ_O определяется уравнением:

$$\xi_O = \frac{S \cdot P_{O_2}}{n_U l} \cdot \exp\left[b^i + \frac{1}{2}\ln B(T)\right] \qquad (17)$$

в котором P_{O_2} — проницаемость кислорода через герметичную стенку толщиной l.

Зависимость изменения состава двуокиси со временем для достехиометрической области составов в условиях, когда поток кислорода значительно меньше потока урана, запишется после ряда упрощений в следующем виде:

$$-\frac{1}{2(x-x_0)} + a^v \cdot \ln\left(\frac{x}{x_0}\right) + (a^v)^2 (x - x_0) = \xi_U t \quad (18)$$

где x_0 соответствует $t = 0$, n_O — число молей кислорода, а величина ξ_U определяется уравнением

$$\xi_U = \frac{2SD_U}{n_O l} \exp\left[C^v - \frac{(L + Z_U^O)}{RT}\right] \quad (19)$$

в котором D_U — коэффициент диффузии урана; Z_U^O — стандартная свободная энергия чистого урана.

Следует отметить, что при получении уравнений (18) и (19) предполагалось, что твердые растворы U-Me подчиняются закономерностям регулярного приближения.

Пользуясь уравнениями (16) и (18), можно оценить время достижения стабильного состава. При этом не следует забывать, что выведенные нами закономерности неточны вблизи строго стехиометрического состава, и при переходе через стехиометрию ($x = 0$) расчеты следует производить приближенными методами, как уже отмечалось ранее.

Оценочные расчеты показывают, что время выхода на установившийся стабильный состав для оболочек 1 мм, например, от состава двуокиси $UO_{2,005}$- $UO_{2,010}$ составляет для Mo и W сотни часов при 2000 °C и тысячи часов при 1700 °C. Таким образом, процесс диффузионного взаимодействия металлов с двуокисью урана, имеющей состав, отличный от стабильного, проходит в условиях переменного источника диффузии (особенно на первой стадии взаимодействия).

Используя известные решения уравнений диффузии из переменного источника [36], нами были получены зависимости для описания концентрационного распределения компонент в оболочке, контактирующей с UO_{2+x} . Так, например, для концентрационного распределения урана в тугоплавком металле в случае взаимодействия его со сверхстехиометрической двуокисью урана, было получено следующее выражение:

$$C(y,t) = C_0 \exp(2\xi_O t) \cdot \exp\left\{\left[-\left(\frac{2\xi_O}{D}\right)^{1/2}\right] y\right\} \quad (20)$$

где C_0 соответствует $t = 0$ и $y = 0$ и y — координата, описывающая распределение урана.

Таким образом, при диффузии из переменного источника должны выполняться линейные зависимости типа $\ln C - y$. Из представленного

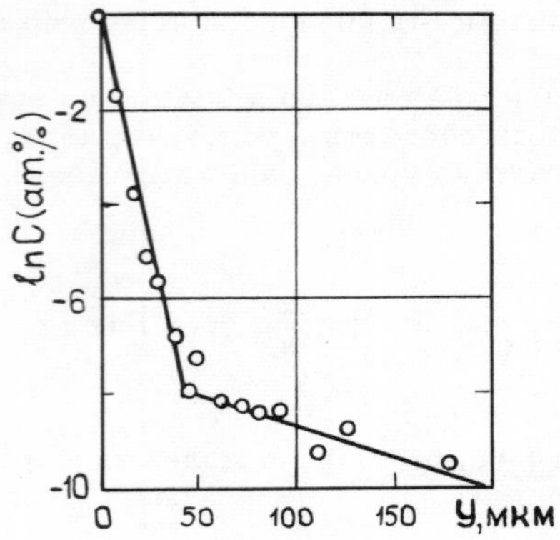

Рис.8. Концентрационное распределение урана в монокристаллическом Mo после взаимодействия с $UO_{2.003}$ при 1900 °С в течение 1000 часов. Граница контакта параллельна плоскости < 100> монокристалла.

нами экспериментального материала (рис.8) видно, что экспериментальные точки действительно неплохо укладываются на прямые в координатах $\ln C - y$.

6. ТЕРМОДИНАМИКА НЕСТЕХИОМЕТРИЧЕСКОГО МОНОНИТРИДА УРАНА

При выводе уравнений для активности компонентов в нестехиометрическом мононитриде урана рассматривались составы от нижней границы области гомогенности до стехиометрического и по аналогии с UC [41] и VN [42] делалось предположение о бездефектности металлической подрешетки.

Кроме того, учитывая, что область гомогенности UN_{1-x} мала, энергиями взаимодействия дефектов (E^{vv}) и вкладом колебательных функций пренебрегалось.

С учетом сделанных допущений с помощью известных приемов статистической термодинамики [13] для зависимости активностей урана и азота в UN_{1-x} были получены следующие соотношения:

$$\ln \lambda_U = \ln x + a^v \tag{21}$$

$$\ln \lambda_N = \ln \frac{1-x}{x} + O^v \tag{22}$$

Для определения активности урана в нестехиометрическом мононитриде на нижнем пределе области гомогенности воспользуемся сведениями о парциальном давлении урана над мононитридом, насыщенным ураном, и над чистым жидким ураном [30,43]. Данные о нижней границе области гомогенности мононитрида возьмем из работ [39,44].

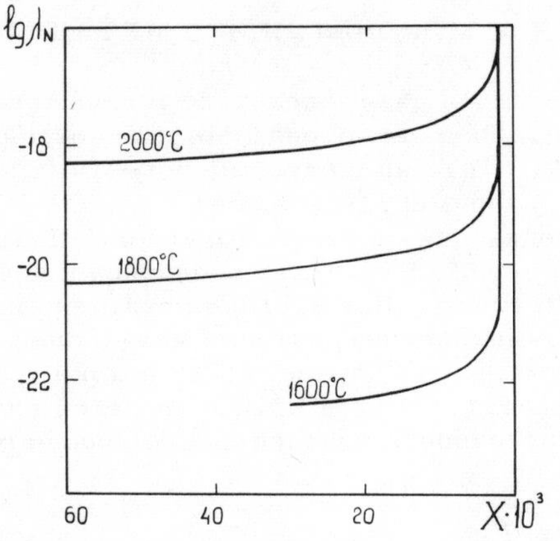

Рис.9. Абсолютная активность азота в UN_{1-x}.

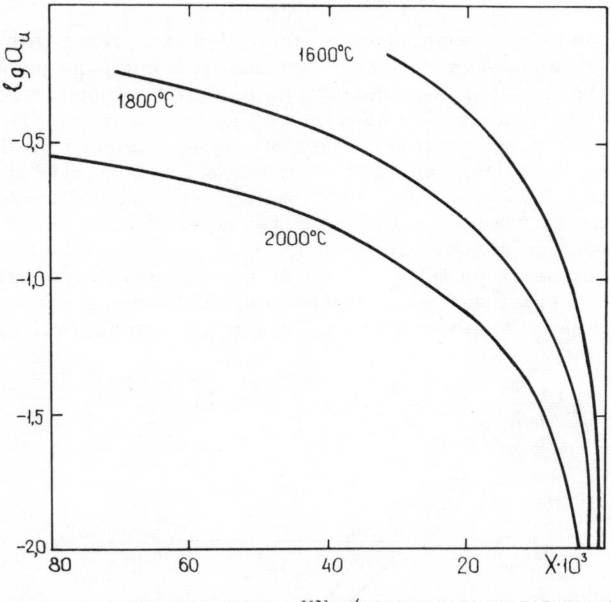

Рис.10. Относительная активность урана в UN_{1-x} (стандартное состояние – чистый уран).

Термодинамическая активность азота рассчитывалась с использованием уравнения Гиббса-Дюгема. Константа интегрирования определялась из интегральной свободной энергии образования UN для стехиометрического состава. При этом использовались результаты работы [45] с учетом данных [18,46] и уравнения (1,2). Результаты расчетов показаны на рис.9 и 10. Проверка достоверности выполненных расчетов проводилась путем сопоставления парциальных давлений азота, полученных из уравнений для активности и измеренных в работе [47]. Сходимость результатов была вполне удовлетворительной (расхождения не превышали 20-30%).

7. ВЗАИМОДЕЙСТВИЕ МОНОНИТРИДА С МЕТАЛЛАМИ

Нами рассматривалось взаимодействие мононитрида урана с молибденом и вольфрамом. Так как нитриды Мо и W нестойки при температурах 1500-2000 °С [17,18], то ни в одной из известных нам экспериментальных работ (см., например, [48]) в зоне контакта UN_{1-x} - Mo(W) никаких новых нитридных соединений не образовывалось. Поэтому весь процесс взаимодействия UN_{1-x} с Мо и W будет сводиться к диффузии U и N из мононитрида в эти металлы. При этом не исключено изменение состава UN_{1-x} в процессе взаимодействия, которое можно описать уравнением, аналогичным (12). В том редком случае, когда потоки урана и азота будут взаимно уравновешиваться, состав UN_{1-x} в процессе взаимодействия меняться не будет. Расчетное уравнение для стабильного состава UN_{1-x} имело вид:

$$x = \left(\frac{P_{N_2}}{D_U}\right)^{1/2} \cdot \exp\left[\frac{(a^v - C^v)RT + L + 1/2 \ln B(T)}{2RT}\right] \qquad (23)$$

Рассчитанные с использованием полученных результатов и данных работ [49,50] значения стабильного состава мононитрида урана в герметичных Мо и W-оболочках приведены на рис.11. Видно, что для W-оболочки они находятся в пределах области гомогенности и близки к стехиометрическому составу. Рассчитанные значения стабильных составов UN_{1-x} в Мо-оболочке попадают в область гомогенности лишь при 1600 °С и ниже. При более высоких температурах поток азота не будет уравновешиваться потоком урана и состав мононитрида будет определяться нижней границей области гомогенности.

При взаимодействии UN_{1-x} с Мо и W нужно учитывать, что исходный состав UN_{1-x} будет меняться, постепенно приближаясь к стабильному. Изменение состава со временем можно рассчитать, пользуясь зависимостью:

$$x = \left[\frac{\xi_N}{\xi_U} + \left(x_0^2 - \frac{\xi_N}{\xi_U}\right) \cdot \exp\left\{-\frac{2t(\xi_U - \xi_N)^2}{(\xi_U + \xi_N)}\right\}\right]^{1/2} \qquad (24)$$

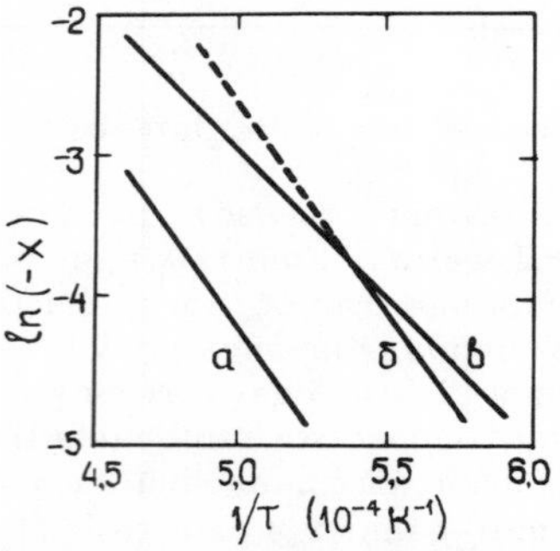

Рис.11. Стабильные составы мононитрида урана в Мо и W – оболочках.

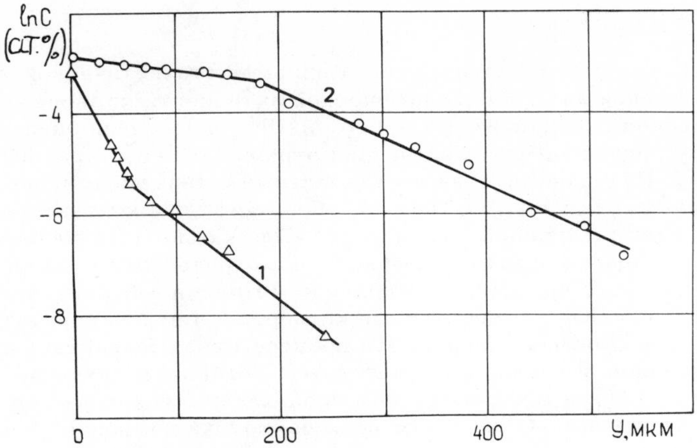

Рис.12. Концентрационное распределение урана в монокристаллическом Mo после взаимодействия с UN_{1-x}. Граница контакта параллельна плоскости <100> монокристалла.
1 – 1700 °C, 1000 ч.
2 – 1900 °C, 1000 ч.

Уравнение (24) получено способом, аналогичным использованным приемам при выводе уравнений (16) и (18). Величина x_0 соответствует $t = 0$, а значения ξ_N и ξ_U определяются уравнениями:

$$\xi_N = \frac{S \cdot P_{N_2}}{n_N l} \cdot \exp\left[a^v + \frac{1}{2} \ln B(T)\right] \qquad (25)$$

$$\xi_U = \frac{S \cdot D_U}{n_N l} \cdot \exp\left[C^v - \frac{L}{RT}\right] \qquad (26)$$

в которых P_{N_2} – проницаемость азота, n_N – число молей азота в системе.

Таким образом, перенос атомов урана и азота из UN_{1-x} в молибден и вольфрам будет проходить в условиях переменного источника. Распределение урана в металлах, контактирующих с UN_{1-x}, будет описываться соотношениями типа (20).

Из рис.12 видно, что экспериментальные точки неплохо укладываются на прямые линии в координатной сетке lnC–y. Обращает на себя внимание разный изгиб прямых на графике 12, что является отражением того факта, что приближение к стабильному составу в данном случае происходит с разных сторон. При 1700°C, когда поток азота превышает поток урана, граничная концентрация урана с течением времени увеличивается. При 1900 °C, когда, наоборот, поток урана превышает поток азота, граничная концентрация урана уменьшается.

ЗАКЛЮЧЕНИЕ

Полученные с помощью известных приемов статистической термодинамики уравнения для термодинамической активности компонент нестехиометрических соединений урана UO_{2+x} и UN_{1-x} позволили описать термодинамику процессов взаимодействия этих соединений с различными металлами. Показано, что менее стабильные с термодинамической точки зрения окислы Mo, W, Ta и Nb образуются при контактировании с UO_{2+x} за счет избыточного кислорода. Для каждого металла существует вполне определенное значение сверхстехиометрического кислорода в UO_{2+x}, начиная с которого происходит окисление металлов. В случае , если состав исходной UO_{2+x} имеет меньше кислорода, чем это требуется для начала образования окислов, все процессы взаимодействия сводятся к диффузии урана и кислорода в металлы. Установлено, что перенос компонент из UO_{2+x} и UN_{1-x} в металлы происходит в условиях переменного источника диффузии. С течением времени потоки компонент уравновешиваются и состав соединения перестает меняться. Рассчитанные стабильные составы UO_{2+x} и UN_{1-x} располагаются в достехиометрической области. Таким образом показано, что при анализе твердофазных реакций взаимодействия необходимо учитывать нестехиометричность участвующих в них соединений.

ЛИТЕРАТУРА

[1] KAZNOFF, A.J. et al., Trans. Am. Nucl. Soc. 7 1 (1964) 100; 8 1 (1965) 32.
[2] GROSSMAN, L.N., High Temp. Nucl. Fuels, N.Y. – L.-P. (1968) 501-15.
[3] MACDONALD, R.D., Rep. AECL-2601, July 1966.
[4] FRYXELL, R.E. et al., High Temp. Nucl. Fuels, N.Y. – L.-P. (1968) 211-21.
[5] VOLLAH, D., Rep. KEK-762, April 1968 (EURFNR-569).
[6] GROSSMAN, L.N. et al., Report of the "Thermionic Conversion Specialists Conf.", L., 1965.
[7] SANDERSON, M.J. et al., Report of the "Thermionic Conversion Specialists Conf.", Schenectedy, N.Y., 1963.
[8] USAEC Rep. GESR-2006, 1962; GESR-2007, 1963.
[9] ЯНГ, Л. и др., Прямое Преобразование Тепловой Энергии в Электрическую, 6 (1969) 173.
[10] SCHUMER, I.F., USAEC Rep. TJD-7687, 1964.
[11] DAYTON, R.W., DICKERSON, R.F., USAEC Rep. BMJ-1650, 1963.
[12] ПАВЛИНОВ, Л.В. и др., Физика Металлов и Металловедение 28 6 (1969) 1129.
[13] ФАУЛЕР, Р., ГУГГЕНГЕЙМ, Е.А., Статистическая Термодинамика, перев.с англ. Изд. ИЛ., М., 1949.
[14] ГОДНЕВ, И.Н., Вычисление Термодинамических Функций по Молекулярным Данным, Гос. изд. техн. теор. лит., М., 1956.
[15] БАРОН, Н.М. и др., Краткий Справочник Физико-Химических Величин, Изд."Химия",1965.
[16] Термодинамические Константы Индивидуальных Веществ, Справочник под ред. В.П.Глушко, Изд. АН СССР, 1962.
[17] КУЛИКОВ, И.С., Термическая Диссоциация Соединений, Изд. Металлургия, М., 1969.
[18] SCHICK, H., Thermodynamics Certain Refractory Compounds, vv. 1, 2, N.Y., 1966.
[19] WILLIS, B.T.M., Nature 197 (1963) 755; J.Phys. 25 (1964) 431.
[20] ACKERMANN, R.J. et al., J. Opt. Soc. Am. 49 (1959) 1107; J.Chem. Fhys. 25 (1956) 1089.
[21] BATES, J.L. et al., Thermodynamics, IAEA, Vienna 2 (1966) 73-88.
[22] EDWARDS, R.K., MARTIN, A.E., ibid, pp. 423-29.
[23] LATTA, R.E., FRYXELL, R.E., Trans. Am. Nucl. Soc. 8 (1965) 375.
[24] ROBERTS, L.E., MARKIN, T.L., Proc. Br. Ceram. Soc. 8 (1967) 201.
[25] THORN, R.J., WINSLOY, G.H., J. Chem. Phys. 44 (1966) 2632.
[26] AITKEN, E.A. et al., Thermodynamics, IAEA, Vienna (1966) 435-53.

[27] PATTORET, A. et al., Thermodynamics of Nuclear Materials, 1967 (Proc. Conf. Vienna, 1967) IAEA, Vienna (1968) 613-34.
[28] AUKRUST, F. et al., Thermodynamics of Nuclear Materials, 1967 (Proc. Conf. Vienna, 1962) IAEA, Vienna (1962) 713-19.
[29] HAGEMARK, K., BROLI, M., J. Am. Ceram. Soc. 50 11 (1967) 563.
[30] ACKERMANN, R.J. et al., J. Phys. Chem. 73 4 (1969) 762, 769.
[31] FOUNTANA, M.H., BAILEY, R.E., Nucl. Sci. Eng. 36 (1969) 268.
[32] КОВБА, Л.М., ТРУНОВ, В.К., Радиохимия 7 3 (1965) 316.
[33] ALTKEN, E.A. et al., Trans. Met. Soc. IAEA 239 (1967) 1565.
[34] MACHLIN, E.S., Trans. ASM 60 (1967) 260.
[35] MARIN, J.F., CONTAMIN, P., J.Nucl.Mater. 30 (1969) 16.
[36] БЭРРЕР, Р., Диффузия в Твердых Телах, перев. с англ., Изд.ИЛ., М., 1948.
[37] ЮРГЕНС, А.В., Автореферат кандидатской диссертации АН Каз.ССР, Алма-Ата, 1970.
[38] SCHWEGLER, E.C., WHITE, F.A., J. Mass Spectrom. Ion. Phys. 1 (1968) 191.
[39] ФЕДОРОВ, Г.В. и др., Атомная Энергия, 31 (1971) 516.
[40] ПАВЛИНОВ, Л.В., НАКОПЕЧНИКОВ, А.И., Атомная Энергия 19 (1965) 521.
[41] СТОРМС, Э., Тугоплавкие Карбиды, перев. с англ., М., Атомиздат, 1970.
[42] AJAMI, F.J., Diss.Abstr. Int.B. (Nuclear Science Abstracts, 1973, 1881).
[43] ALEXANDER, C.A. et al., J. Nucl. Mater. 31 (1969) 13.
[44] BENZ, R., BOWMAN, R.G., J. Am. Chem. Soc. 88 (1966) 264.
[45] RAND, M.H., KUBASCHEWSKI, O., Thermochemical Properties Uranium Compounds, Edinb., L., 1963.
[46] O'HARE, P.A.G. et al., Thermodynamics of Nuclear Materials, 1967 (Proc. Conf. Vienna, 1967), IAEA, Vienna (1968) 265-78.
[47] INOUYE, H., LEITNAKER, J.M., J. Am. Ceram. Soc. 51 1 (1968) 6.
[48] POLITIS, C. et al., J. Nucl. Mater. 32 2 (1969) 181.
[49] CONN, P.K. et al., Trans. Met. Soc. AIME 242 (1968) 626.
[50] FRANENFELDER, R., J. Chem. Phys. 48 (1968) 3966.

DISCUSSION

D. CALAIS: What were the experimental methods you used to determine the concentration-penetration curves of uranium, oxygen and nitrogen in tungsten and molybdenum?

A.S. PANOV: In our paper we present data only on the distribution of uranium in metals. We applied conventional radiometric research methods. We did not determine the distribution of oxygen and nitrogen.

TERNARY PHASE EQUILIBRIA IN THE SYSTEMS ACTINIDE-TRANSITION METAL-CARBON AND ACTINIDE-TRANSITION METAL-NITROGEN

H. HOLLECK
Kernforschungszentrum Karlsruhe,
Karlsruhe,
Federal Republic of Germany

Abstract

TERNARY PHASE EQUILIBRIA IN THE SYSTEMS ACTINIDE-TRANSITION METAL-CARBON AND ACTINIDE-TRANSITION METAL-NITROGEN.

Isothermal sections in the ternary phase diagrams of actinide-transition metal-carbon and actinide-transition metal-nitrogen systems are compiled on the basis of new phase studies, using thermodynamic data of carbides and nitrides, and considering the structure and lattice parameters of binary compounds. The lack of experimental results is compensated by estimating the solution behaviour and the phase stabilities. Phase diagrams are given for the systems: Th-(U or Pu)-C, U-(Th or Pu)-N, (Th or U or Pu)-Y-C, (Th or U or Pu)-Ce-C, (Th or U or Pu)-Y-N, (Th or U or Pu)-Ce-N, U-(La or Ce or Pr or Nd)-N, (Th or U or Pu)-Ti-C, (Th or U or Pu)-Zr-C, (Th or U or Pu)-Nb-C, (Th or U or Pu)-Mo-C, U-(Ti or Zr or Hf or Mo)-N, (Th or Pu)-Zr-N, (Th or U or Pu)-Cr-N and (Th or U or Pu)-(Ru or Rh or Pd)-C. Of particular interest is the occurrence of ternary carbides and nitrides.

1. INTRODUCTION

The reaction behaviour of multicomponent systems containing actinide carbides and nitrides is of great interest for many problems in nuclear technology. Of particular importance in this context are ternary systems involving the transition metals, which are the most frequently occurring fission products, the main constituents of the cladding and structural materials, and potential alloying elements.

The clearest way of surveying the reaction behaviour for specific combinations of elements at high temperatures is to extract information from the appropriate phase diagrams. New experimental work was carried out with the actinide elements thorium, uranium and plutonium. The thermodynamic data of the carbides and nitrides, including some estimations, and, furthermore, structural and lattice parameters of these, as well as the valence states of the actinide elements, form the bases for an evaluation of ternary phase equilibria and of some stability criteria of ternary compounds.

Since thorium, uranium and plutonium are the most important actinide elements in nuclear technology, and since for these three actinide elements details of only the binary carbide and nitride systems are known at present, the evaluation of ternary phase diagrams was restricted to these elements. The binary carbide and nitride systems of the actinides which were considered when preparing the ternary sections are given in Figs 1 and 2; information was drawn from the following sources: Th-C [1]; U-C [2]; Pu-C [3]; Th-N [4,5]; U-N [12]; Pu-N [3].[1]

---

[1] The figures are all to be found at the end of this paper on pages 220-262.

The Gibbs energies of formation of the carbides and nitrides required for calculations and estimations were only rough values in some cases; therefore the estimated and calculated tie lines must be taken as a first approximation (for thermodynamic data see Refs [3] and [6] for the actinide carbides and nitrides, Refs [7] and [8] for the rare-earth carbides and nitrides, Ref. [9] for the IVA to VIA transition metal carbides, and Ref. [8] for the IVA to VIA transition metal nitrides).

It should be pointed out that this work is not a review of the existing data concerning the constitution of ternary carbide and nitride systems, but is the application of new experimental work, thermodynamic data and theoretical estimations for the construction of phase diagrams. Therefore the entire original literature on the constitution of ternary actinide-transition metal-carbon and actinide-transition metal-nitrogen systems is not fully cited in this work. The references [3, 6] and [9] to [15] give reviews of part of the above-mentioned topics, and were taken into account, just as were the three books on the constitution of binary alloys [16-18].

2. TERNARY PHASE EQUILIBRIA

2.1. Actinide-actinide-carbon systems

The systems consisting of two actinide elements and carbon are characterized by a complete miscibility in the monocarbide, sesquicarbide and dicarbide sections at temperatures at which the corresponding isotypic binary compounds exist. Tentative isothermal sections for the Th-U-C system at 1800°C and for the Th-Pu-C and U-Pu-C systems at 1600°C are given in Fig. 3. No tie lines are drawn in the temperature sections. Calculations which were made on the assumption of ideal solution behaviour [19] in the (U, Pu)C + $(U, Pu)_2C_3$ diphase region show satisfying agreement with experimental results. Details in the Th-Pu-C and U-Pu-C systems are reviewed in Ref. [3] and are given for the Th-U-C system in Refs [15] and [20]. Experimental results in the metal-rich part of the Th-Np-C system [21] confirm complete miscibility in the monocarbide section.

2.2. Actinide-actinide-nitrogen systems

It can be accepted that all the actinide mononitrides are mutually miscible, although this is experimentally confirmed only for ThN-UN and UN-PuN. Using thermodynamic data and estimated interaction parameters, ternary phase equilibria were calculated for the Th-U-N and U-Pu-N systems. Figure 4 shows isothermal sections at 1000°C calculated with the specified interaction parameters and the Gibbs energies of formation:

$${}^{f}\Delta G^0_{UN}\ (\text{cal/mol}) = -73\,200 + 22.5\ T$$

$${}^{f}\Delta G^0_{ThN}(\text{cal/mol}) = -76\,000 + 18.5\ T + 2.2 \times 10^{-20}\ T^6$$

$${}^{f}\Delta G^0_{U_2N_{3+X}}(\text{cal/mol}) \ldots\ldots \text{see Ref. [8]}$$

$${}^{f}\Delta G^0_{ThN \rightarrow ThN_{1.33}}\ (\text{cal/mol}) = -12\,100 + 5.4\ T$$

$${}^{f}\Delta G^0_{PuN}\ (\text{cal/mol}) = -70\,600 + 22.1\ T \quad (\text{cf. Ref. [8]})$$

The interaction parameters are estimated for the nitride systems from the lattice parameter differences in connection with a correlation to the carbides, and for the metal systems from the binary phase diagrams (for details see Ref. [8]). Neither a solubility of plutonium in U_2N_3, or of thorium in U_2N_3, or uranium in Th_3N_4 have been considered. Only the invariant points of the three phase equilibria have been calculated for various nitrogen pressures in the nitrogen-rich part of the phase diagrams. α-U_2N_3 and β-U_2N_3 coexist at 1000°C and 1 atm N_2. At 0.1 atm N_2 only β-U_2N_3 occurs [22]. The phase equilibria in the U-Th-N system between metals and mononitrides, investigated experimentally (Ref. [23]), show excellent agreement with the equilibria of the calculated section in Fig. 4. Phase relations in the U-Pu-N system and a description of these in terms of their dependence on various interaction parameters were recently discussed [24].

2.3. Actinide-rare earth-carbon systems

The individual actinide and rare-earth metals show behaviour different to that of carbon. Whereas the light elements of the actinide series (e. g. thorium and uranium) form stable monocarbides and less stable sesquicarbides or dicarbides, the opposite behaviour is observed with the heavier actinide elements (e. g. plutonium and americium). Like the actinide elements in which the trivalent state predominates (plutonium, americium; cf. Ref. [25]), the rare earths form stable sesquicarbides and dicarbides. Only the rare-earth metals with small atomic radii form monocarbides (scandium and yttrium) or subcarbides (samarium to lutetium). The dicarbides of the actinides and rare earths exist with a face-centred cubic high-temperature modification and a tetragonal low-temperature modification. One can accept for the actinide-rare earth-carbon systems with stable monocarbides (ThC, UC) in the high carbon section high-temperature cubic dicarbide solid solutions and two phase equilibria between the rare-earth dicarbides and actinide monocarbides. One can also accept for systems with stable sesquicarbides (Pu_2C_3, Am_2C_3) high-temperature cubic dicarbide solid solutions and solid solutions of the sesquicarbides. Whereas complete solid solutions exist between actinide monocarbides and YC_{1-x}, lanthanide monocarbides (lanthanum, cerium, praseodymium and neodymium) can be stabilized to only some extent by solution in the actinide monocarbides [26, 27].

Typical phase equilibria at 1600°C in these systems are given by tentative ternary sections in Fig. 5 for the Th-Y-C, U-Y-C and Pu-Y-C systems and in Fig. 6 for the Th-Ce-C, U-Ce-C and Pu-Ce-C systems.

Whereas for the cerium-containing systems, experimental data concerning the solution behaviour are available (Th-Ce-C [28], U-Ce-C and Pu-Ce-C [26, 27]) information on yttrium-containing systems are scarce. Preliminary high-temperature X-ray investigations [29] confirm the assumption of a complete solid solution between ThC_{1-x} and YC_{1-x}. At temperatures higher than 1500°C a complete dicarbide solid solution should exist, and a sesquicarbide solid solution could also be obtained at high pressure [30]. Similar to the U-Y-C system [31], extended homogeneity regions of ternary carbides probably exist in the Pu-Y-C-system.

2.4. Actinide-rare earth-nitrogen systems

The experimental work in thorium and uranium-containing ternary nitride systems established the existence of complete solid solutions between ThN and YN, LaN, CeN, PrN and NdN [32], as well as between UN and YN, LaN, CeN, PrN and NdN [33, 34]. Solid solutions between PuN and these rare-earth nitrides are very probable. Figures 7 and 8 show, respectively, tentative isothermal sections of the (Th, U, Pu)-Y-N and (Th, U, Pu)-Ce-N systems. The tie lines in these sections are estimated. However, those of the sections for the U-La-N, U-Ce-N, U-Pr-N and U-Nd-N systems given in Fig. 9 are calculated with the assumption of a regular solution model, with estimated interaction parameters [8]. In agreement with our experimental results the calculation of the decomposition temperature of (U, La)N with ϵ = 8400 cal/mol is about 1700°C. For the calculation the following Gibbs energies of formation were used (cf. Ref. [8]):

$$^{f}\Delta G^{0}_{UN}\ \text{(cal/mol)} = -73\,200 + 22.5\ T$$

$$^{f}\Delta G^{0}_{LaN}\ \text{(cal/mol)} = -72\,100 + 25.0\ T$$

$$^{f}\Delta G^{0}_{CeN}\ \text{(cal/mol)} = -78\,000 + 25.0\ T$$

$$^{f}\Delta^{0}_{PrN}\ \text{(cal/mol)} = -75\,000 + 25.0\ T$$

$$^{f}\Delta G^{0}_{NdN}\ \text{(cal/mol)} = -71\,000 + 25.0\ T$$

2.5. Actinide-IVA, VA, VIA transition metal-carbon systems

New experimental results in the Th-Ti-C, Th-Zr-C, Th-Nb-C and Th-Mo-C systems, together with the known phase equilibria of the corresponding uranium systems, of the Pu-Mo-C system and estimated phase diagrams of further plutonium-containing systems are presented in the ternary sections shown in Figs 10 to 13. The constitution of the actinide (Th or U or Pu)-Ti-C systems seems to be very similar (Fig. 10). The stable monocarbides TiC and ZrC are in equilibrium with the actinide metals, as is also shown for the corresponding zirconium-containing systems (Fig. 11). Only UC is obviously able to form mixed carbides with ZrC and NbC. The reason, therefore, seems to lie in the difference in the atomic radii. Figure 12 shows ternary sections in the Th-Nb-C, U-Nb-C and Pu-Nb-C systems and Fig. 13 gives sections for the Th-Mo-C, U-Mo-C and Pu-Mo-C systems (for references concerning the U-Mo-C and Pu-Mo-C systems see [10] and [13], respectively). Two ternary compounds were observed in the Th-Mo-C system, which could not as yet, however, be exactly characterized as to structure and composition.

2.6. Actinide-IVA, VA, VIA transition metal-nitrogen systems

Extended experimental work on uranium-containing systems [32, 35, 36] and some experimental data in the thorium and plutonium-containing systems (ThN-ZrN [31], PuN-ZrN [37], Th_2CrN_3 [38], Pu-Cr-N [39]) have, together with experimental and estimated thermodynamic data, been used for calculating and estimating ternary phase equilibria in these nitride systems. Calculated isothermal sections for U-Ti-N at 2000°C, U-Zr-N at 1000°C, U-Hf-N at 1200 and 1500°C, U-Nb-N at 1000°C and U-Mo-N at 1200°C are given in Fig. 14. The interaction parameters listed for each section and the following Gibbs energies of formation are used for the calculation:

$$^{f}\Delta G^{0}_{TiN} \text{ (cal/mol)} = -80\,900 + 22.8\,T$$

$$^{f}\Delta G^{0}_{ZrN} \text{ (cal/mol)} = -97\,900 + 23.1\,T$$

$$^{f}\Delta G^{0}_{HfN} \text{ (cal/mol)} = -88\,200 + 23.0\,T$$

$$^{f}\Delta G^{0}_{NbN} \text{ (cal/mol)} = -56\,800 + 20.0\,T$$

For details see Ref. [8].

The calculations result in critical temperatures for the solid solutions of about 4000°C for UN-TiN, 800°C for UN-ZrN, 1300°C for UN-HfN and 2600°C for UN-NbN. These temperatures, as well as the solubility limits, agree well with our experimental results. Thorium and plutonium-containing systems are only presented for Th-Zr-N and Pu-Zr-N (Fig. 15) because the existence of complex nitrides in other systems has not been confirmed as yet. Isothermal sections for thorium, uranium and plutonium are given for the nitride systems with chromium in Fig. 16. Whereas throrium and uranium form complex nitrides [35, 38], no ternary compounds could be found for plutonium [39].

2.7. Actinide-platinum metal-carbon systems

Figures 17, 18 and 19 show ternary sections for the following systems: Th-Ru-C (1200°C) [40], U-Ru-C (1300°C) [41], Pu-Ru-C (1200°C) [42], Th-Rh-C (1200°C) [40], U-Rh-C (1300°C) [43], Pu-Rh-C (1200°C) [42], Th-Pd-C (1100°C) [44], U-Pd-C (1300°C) [44] and Pu-Pd-C (1200°C) [42]. These have resulted from our group's experimental work. The work of Haines and Potter [26] has been considered for obtaining data concerning the existence of three phase equilibria: $PuC_{1-x} + Pu_2C_3 + PuRu$, $Pu_2C_3 + PuRh_2 + PuRh_3C_{1-x}$, and $PuC_{1-x} + Pu_2C_3 + PuPd_3$.

The actinide carbides are not in equilibrium with the platinum metals. Partly ternary complex carbides, partly intermetallic compounds and free carbon are formed during the reaction of binary carbides with platinum metals. The tendency to form ternary carbides decreases from ruthenium to rhodium and palladium. The occurrence of ternary phases is discussed in the next section.

3. TERNARY ACTINIDE-TRANSITION METAL CARBIDES AND NITRIDES

Ternary carbides and nitrides exist as mixed phases of binary compounds and as complex compounds with a structure of their own. Mixed and complex phases are known with two non-metallic components and one metallic component (i.e. carbonitrides) as well as with two metallic components and one non-metallic component. Here only the latter case – with metals from the transition metal series – is described. The ternary carbides and nitrides are listed in Fig. 20. Many of the systems containing thorium or plutonium together with the other actinides (with the exception of uranium) have not yet been investigated. Preliminary results in the carbide systems with thorium indicate the existence of ternary phases with rhenium, osmium, iridium and platinum.

The uranium compounds are the most thoroughly investigated up to now. One can distinguish between four different groups of ternary carbides (Fig. 21):

(i) the cubic mixed carbides of the third, fourth and fifth group;
(ii) the orthorhombic carbides with the transition metals of the sixth and seventh group (and also vanadium);
(iii) the tetragonal complex carbides with the iron-group metals;
(iv) the tetragonal complex carbides with the platinum metals.

Ternary nitrides are observed which are isostructural or similar in structure with transition metals of lower group numbers. Thus, orthorhombic complex nitrides exist with VA transition metals, isostructural with the carbides with VIA and Group VII transition metals. The complex nitrides with chromium and manganese (for structural details see Ref. [38]) are closely related to the complex carbides with the platinum metals (for structural details see Ref. [45]). In Fig. 22, the transition is shown from the face-centred cubic mixed phase to the orthorhombic cell of U_2CrN_3 (or – with some non-metal positions vacant – to U_2OsC_{2+x}) and to the tetragonal cell of U_2RuC_2.

Carbon is able to stabilize ternary actinide-platinum metal carbides [46] with perovskite structure. Actinides (and other transition metals) with a predominant tetravalent or pentavalent state that can be observed as components in ordered Cu_3Au-type phases with rhodium form perovskite carbides with ruthenium (Fig. 23). Actinides (and other transition metals) with a predominant trivalent state that can be observed as components in ordered Cu_3Au-type phases with palladium form perovskite carbides with rhodium (Fig. 24). Tetravalent metals and palladium form MPd_4 compounds with a "defect Cu_3Au-lattice". Metals with trivalent and tetravalent states, such as cerium or plutonium show homogeneous regions between MPd_4 and MPd_3 or between MRh_3 and MRh_3C. Taking into account the above-mentioned observations, the following perovskite carbides should exist: $PaRu_3C$ (a ≃ 4.15 Å); $NpRu_3C$ (a ≃ 4.13 Å), $AmRh_3C$ (a ≃ 4.19 Å) and $CmRh_3C$ (a ≃ 4.19 Å). The existence of the compounds $AmRu_3C$ (a ≃ 4.16 Å) and $CmRu_3C$ (a ≃ 4.16 Å) is possible.

4. DISCUSSION

The thermodynamic information reflected by phase diagrams is only qualitative but is very reliable. An important characteristic feature is consideration of solutions: neglecting these is one of the most frequent errors made when estimating ternary phase equilibria from thermodynamic data of binary phases. Up to now, unfortunately, no well-established method exists for calculating phase diagrams without many assumptions.

Therefore, the experimental investigation of phase equilibria under given conditions still remains the most reliable method of obtaining information on the reaction behaviour in a system. However, it is very valuable to make calculations and estimations, both to extend and to generalize phase equilibria, and to get an approximate impression of the phase behaviour in unknown systems.

The phase diagrams presented in this work are, in most cases, based on experimental work. Estimated and calculated equilibria are characterized by broken lines and have to be considered as tentative equilibria. Their verification needs, however, only a few, controlled experiments. The solution behaviour is of particular importance for defining the phase equilibria in multicomponent systems of transition metal carbides and nitrides. The occurrence of complete solubility between monocarbides, cubic dicarbides and mononitrides is indicated schematically in Fig. 25. Experimental data existing in combinations with thorium, uranium and plutonium, similarities in the electronic structure and in the lattice parameter differences are the bases for the estimation of this miscibility behaviour (UC_2-LaC_2, -CeC_2; cf. Ref. [47]).

Not all the transition metals could be considered in the scope of this work. Typical representatives and systems with experimental results available were discussed. The complex compounds, for instance, probably control the phase equilibria in systems with the iron-group metals, but there are not sufficient data for a discussion of these systems. As actinide carbides and nitrides are of interest in nuclear technology, however, more data concerning the constitutions of multicomponent systems will certainly be available in the near future.

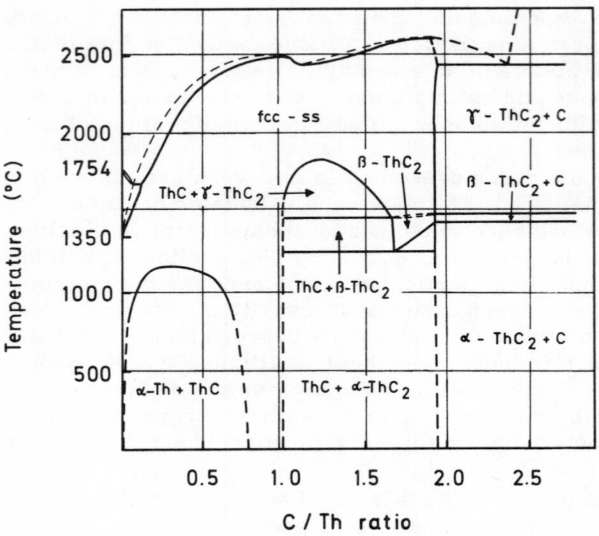
2500
2000
1754
1350
1000
500
Temperature (°C)
fcc - ss
γ - ThC2 + C
ß - ThC2
ß - ThC2 + C
ThC + γ - ThC2
ThC + ß-ThC2
α - ThC2 + C
α-Th + ThC
ThC + α-ThC2
0.5
1.0
1.5
2.0
2.5
C / Th ratio

A

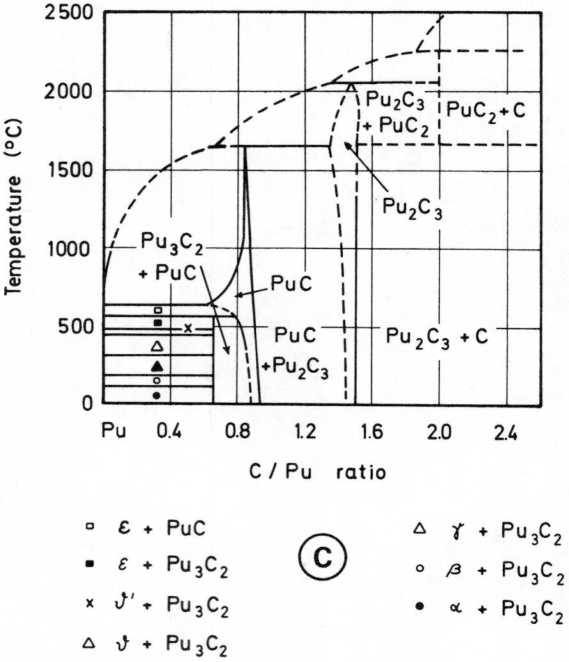
2500
2000
1500
1000
500
0
Temperature (°C)
Pu2C3 + PuC2
PuC2 + C
Pu2C3
Pu3C2 + PuC
PuC
PuC + Pu2C3
Pu2C3 + C
Pu
0.4
0.8
1.2
1.6
2.0
2.4
C / Pu ratio
□ ε + PuC
■ ε + Pu3C2
x ϑ' + Pu3C2
△ ϑ + Pu3C2
▲ γ + Pu3C2
○ β + Pu3C2
● α + Pu3C2

C

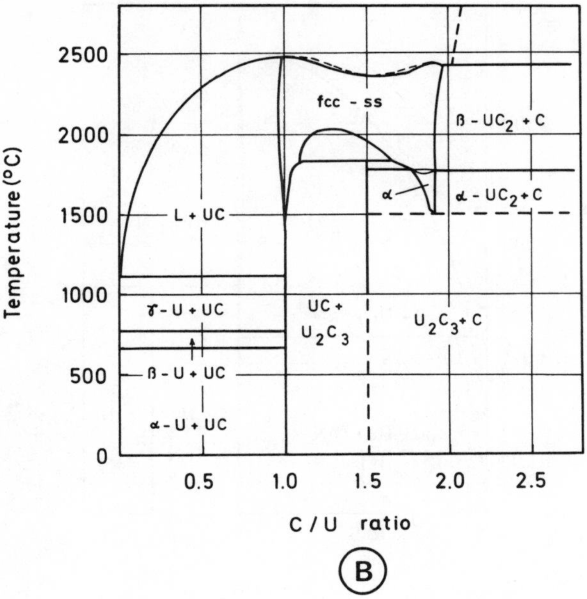

FIG.1. Phase diagrams of binary actinide-carbon systems.

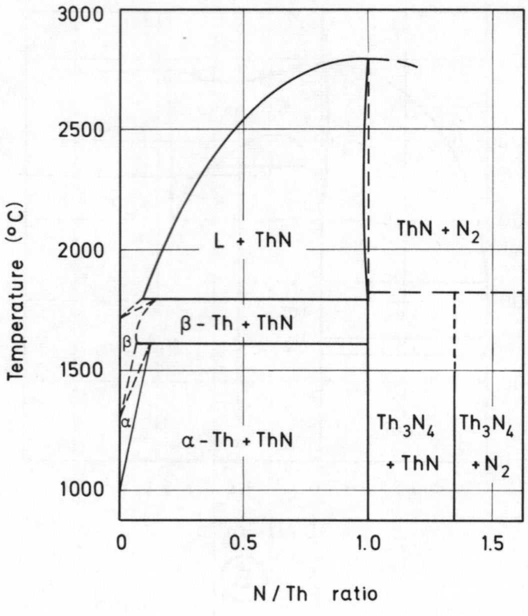
3000
2500
2000
1500
1000
Temperature (°C)
L + ThN
ThN + N$_2$
β - Th + ThN
β
α
α - Th + ThN
Th$_3$N$_4$ + ThN
Th$_3$N$_4$ + N$_2$
0
0.5
1.0
1.5
N / Th ratio

A

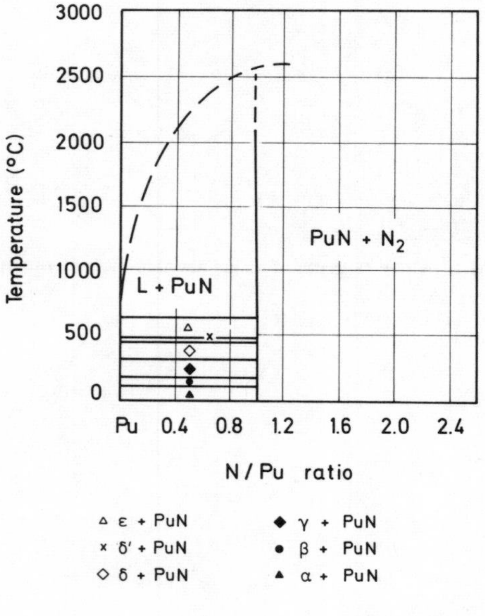
3000
2500
2000
1500
1000
500
0
Temperature (°C)
L + PuN
PuN + N$_2$
Pu
0.4
0.8
1.2
1.6
2.0
2.4
N / Pu ratio
△ ε + PuN
× δ' + PuN
◇ δ + PuN
◆ γ + PuN
● β + PuN
▲ α + PuN

C

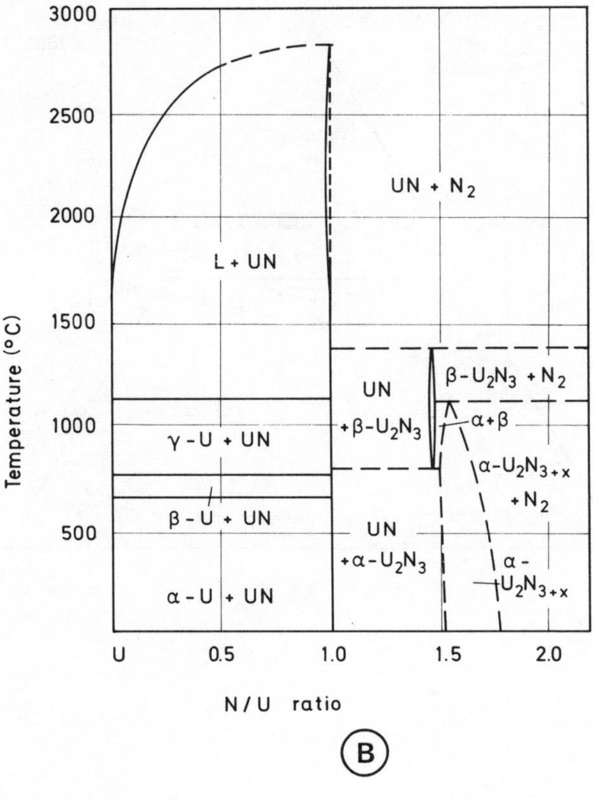

FIG.2. Phase diagrams of binary actinide-nitrogen systems.

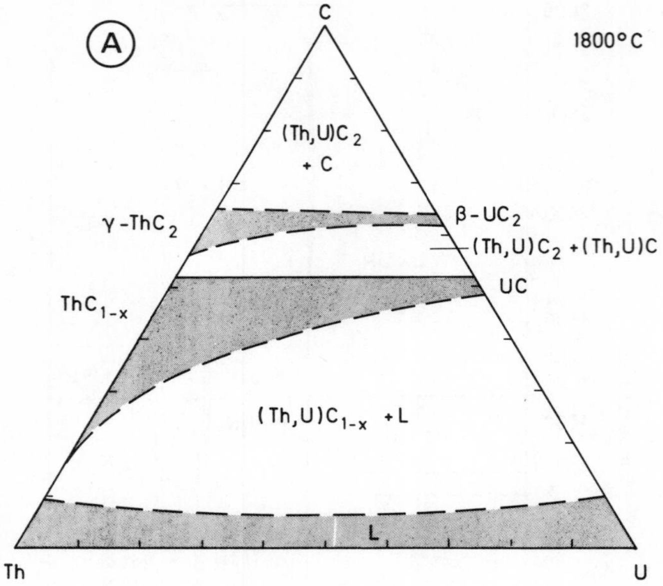

FIG.3. Tentative isothermal sections in the Th-U-C (1800°C), Th-Pu-C (1600°C) and U-Pu-C (1600°C) systems.

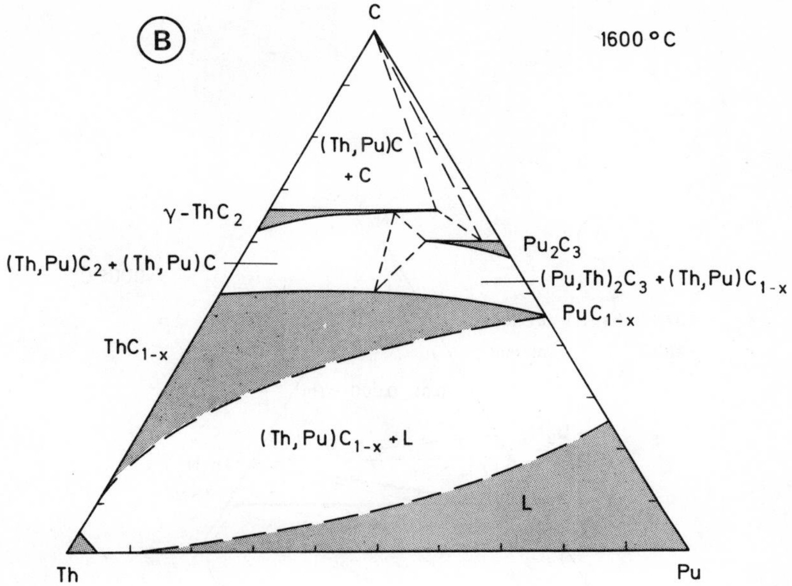
B
C
1600 °C
(Th, Pu)C
+ C
γ - ThC 2
Pu2C3
(Th, Pu)C2 + (Th, Pu)C
(Pu, Th)2C3 + (Th, Pu)C1-x
PuC1-x
ThC1-x
(Th, Pu)C1-x + L
L
Th
Pu

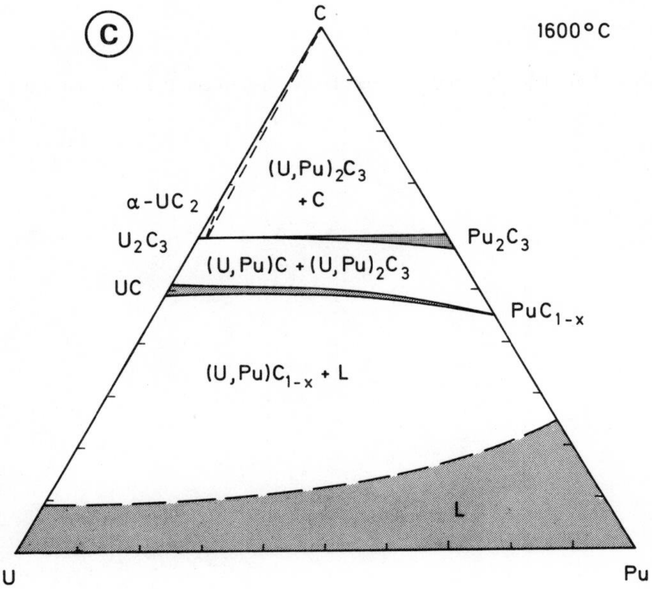
C
C
1600 °C
(U, Pu)2C3
+ C
α - UC 2
U2C3
Pu2C3
(U, Pu)C + (U, Pu)2C3
UC
PuC1-x
(U, Pu)C1-x + L
L
U
Pu

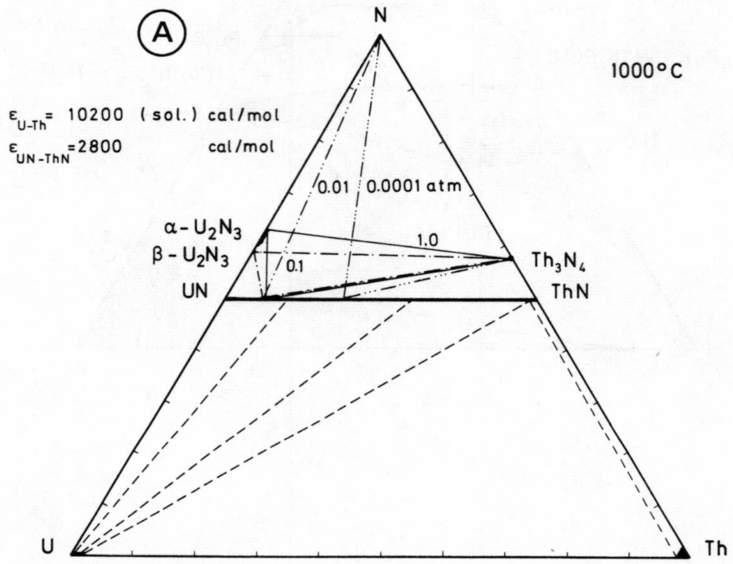

FIG.4. Calculated isothermal sections in the Th-U-N and U-Pu-N systems at 1000°C.

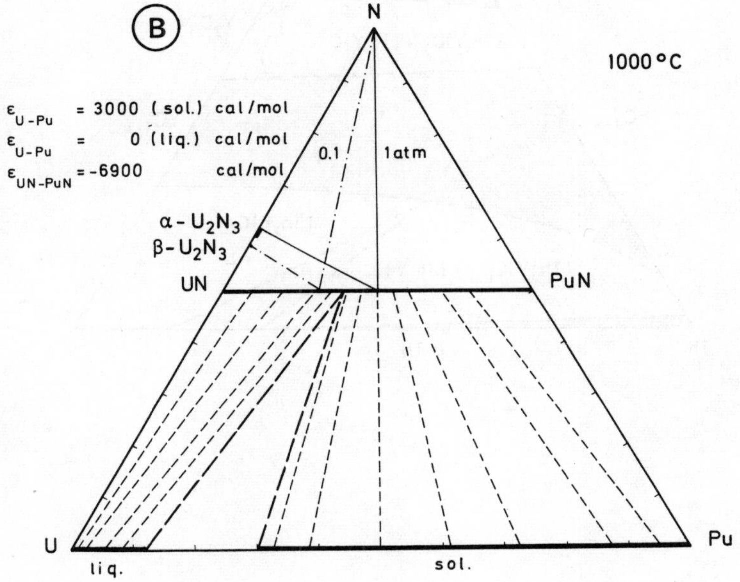
B
N
1000 °C
ε U-Pu = 3000 (sol.) cal/mol
ε U-Pu = 0 (liq.) cal/mol
ε UN-PuN = -6900 cal/mol
0.1
1atm
α-U2N3
β-U2N3
UN
PuN
U
Pu
liq.
sol.

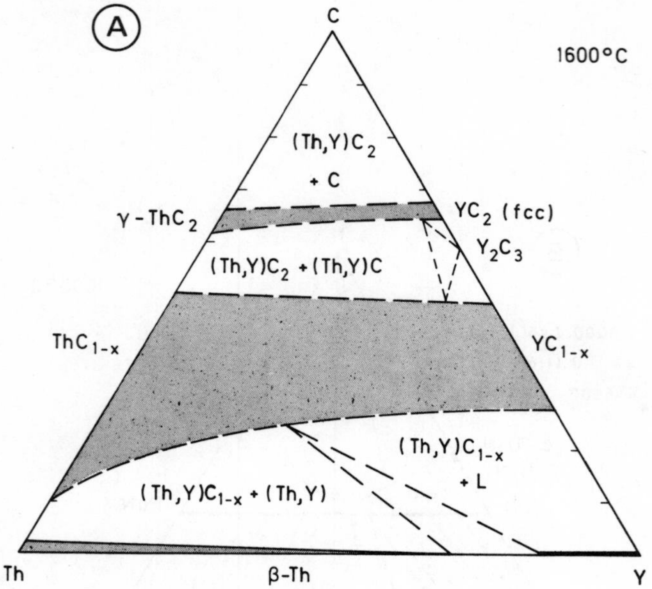

FIG.5. Tentative isothermal sections in the Th-Y-C, U-Y-C and Pu-Y-C systems at 1600°C.

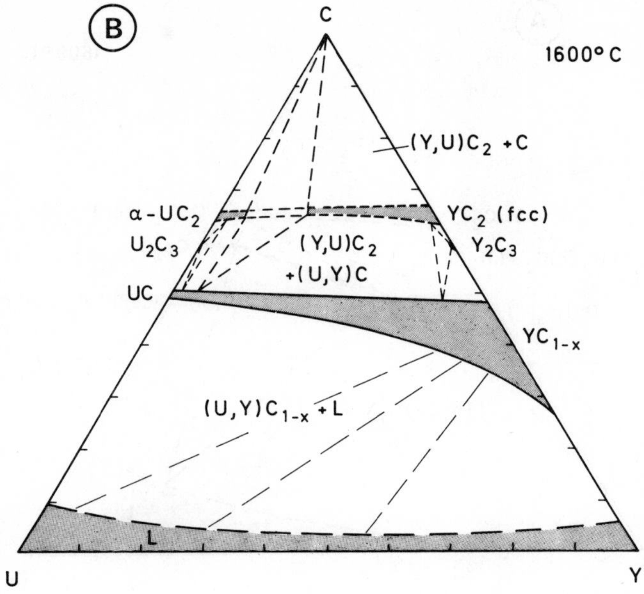
B
C
1600° C
(Y,U)C2 +C
α-UC2
YC2 (fcc)
U2C3
(Y,U)C2
+(U,Y)C
Y2C3
UC
YC1-x
(U,Y)C1-x +L
L
U
Y

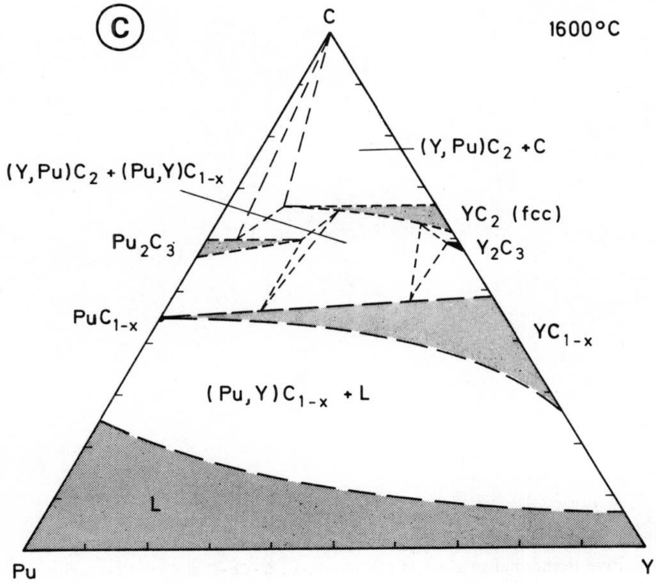
C
C
1600°C
(Y, Pu)C2 +C
(Y,Pu)C2 + (Pu,Y)C1-x
YC2 (fcc)
Pu2C3
Y2C3
PuC1-x
YC1-x
(Pu,Y)C1-x +L
L
Pu
Y

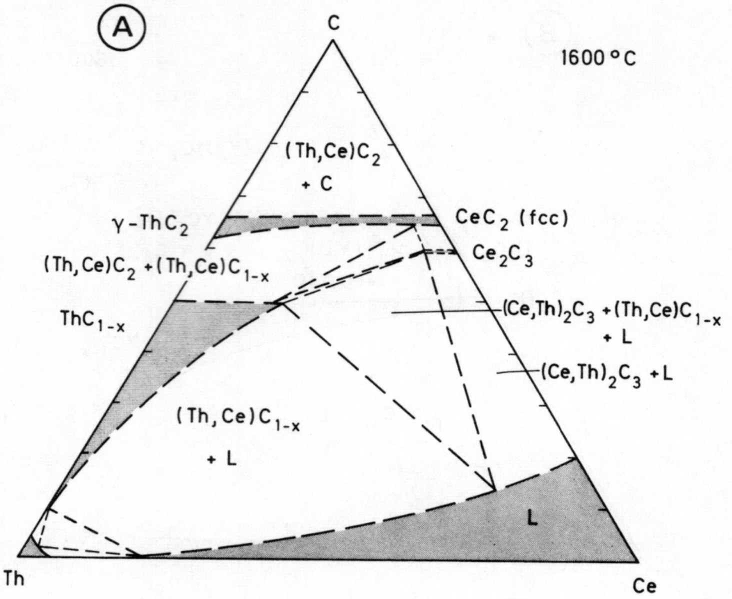

FIG.6. Tentative isothermal sections in the Th-Ce-C, U-Ce-C and Pu-Ce-C systems at 1600°C.

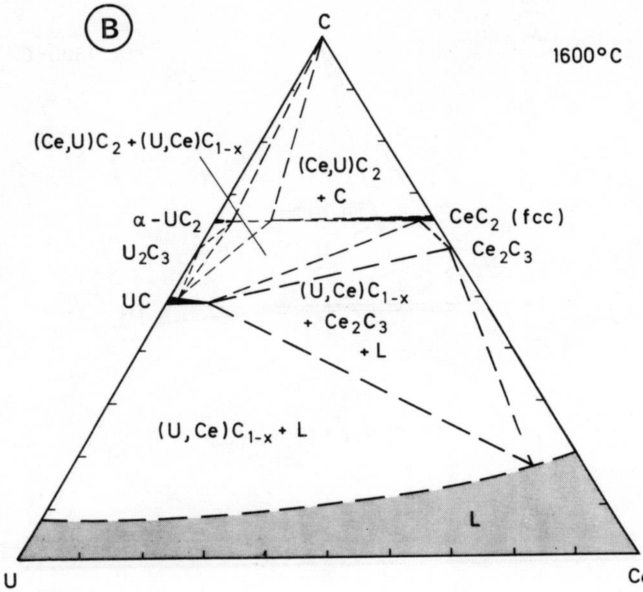
B
C
1600°C
(Ce,U)C₂ + (U,Ce)C₁₋ₓ
(Ce,U)C₂
+ C
α-UC₂
CeC₂ (fcc)
U₂C₃
Ce₂C₃
UC
(U,Ce)C₁₋ₓ
+ Ce₂C₃
+ L
(U,Ce)C₁₋ₓ + L
L
U
Ce

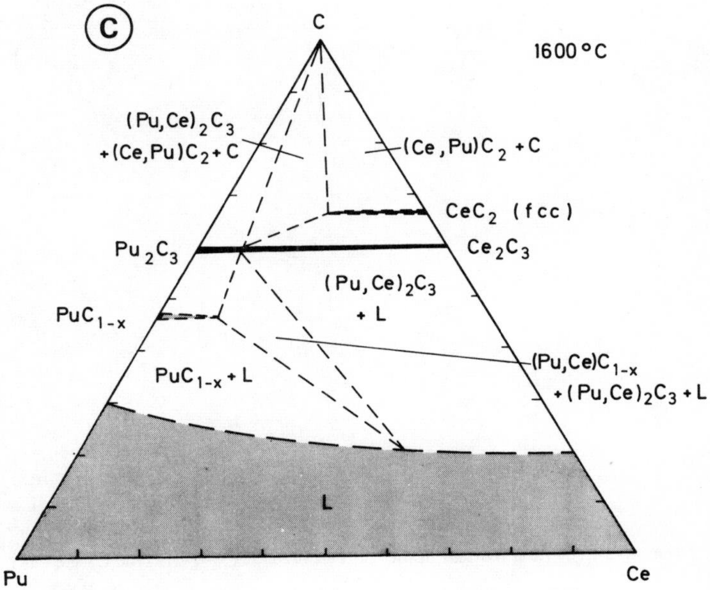
C
C
1600°C
(Pu,Ce)₂C₃
+(Ce,Pu)C₂+ C
(Ce,Pu)C₂ + C
CeC₂ (fcc)
Pu₂C₃
Ce₂C₃
(Pu,Ce)₂C₃
+ L
PuC₁₋ₓ
(Pu,Ce)C₁₋ₓ
+ (Pu,Ce)₂C₃ + L
PuC₁₋ₓ + L
L
Pu
Ce

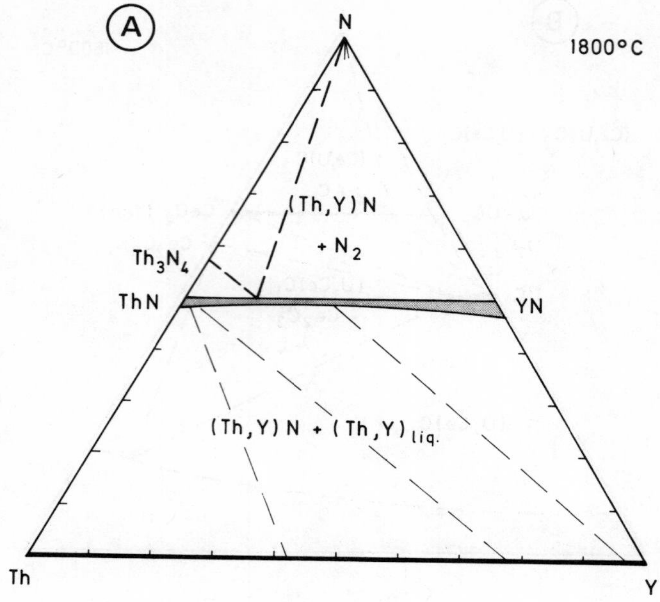

FIG.7. Tentative isothermal sections in the Th-Y-N (1800°C), U-Y-N (1800°C) and Pu-Y-N (1600°C) systems.

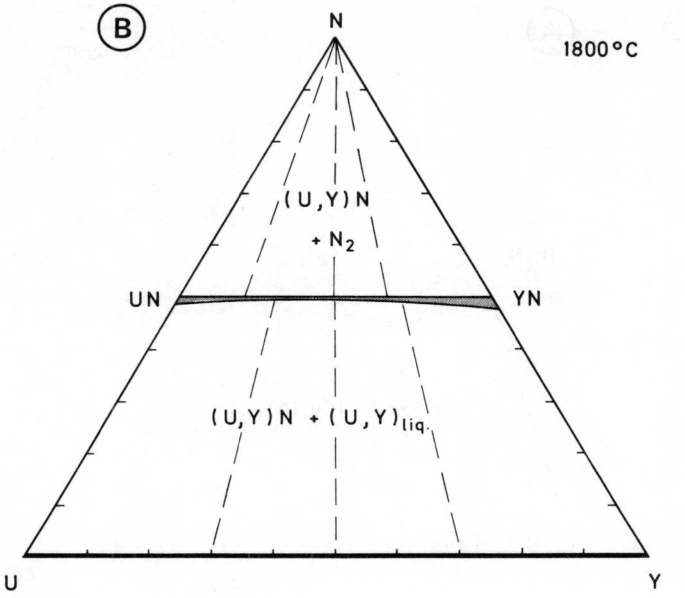
B
N
1800 °C
(U,Y)N
+ N_2
UN
YN
(U,Y)N + $(U,Y)_{liq.}$
U
Y

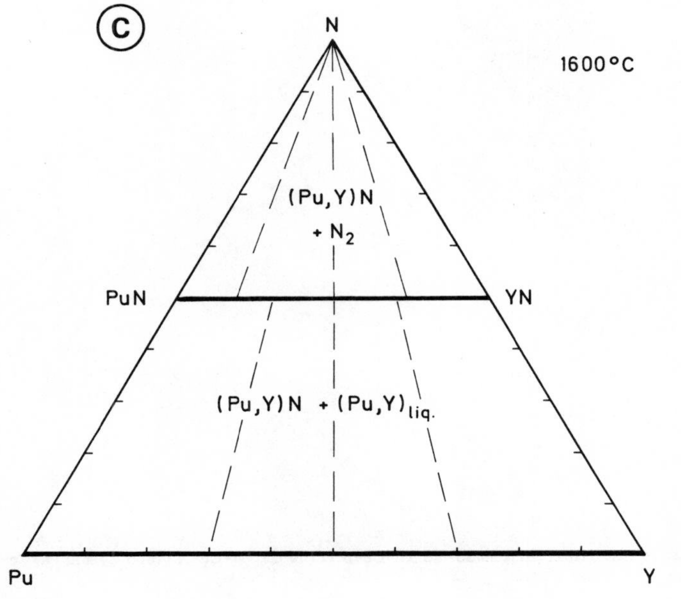
C
N
1600 °C
(Pu,Y)N
+ N_2
PuN
YN
(Pu,Y)N + $(Pu,Y)_{liq.}$
Pu
Y

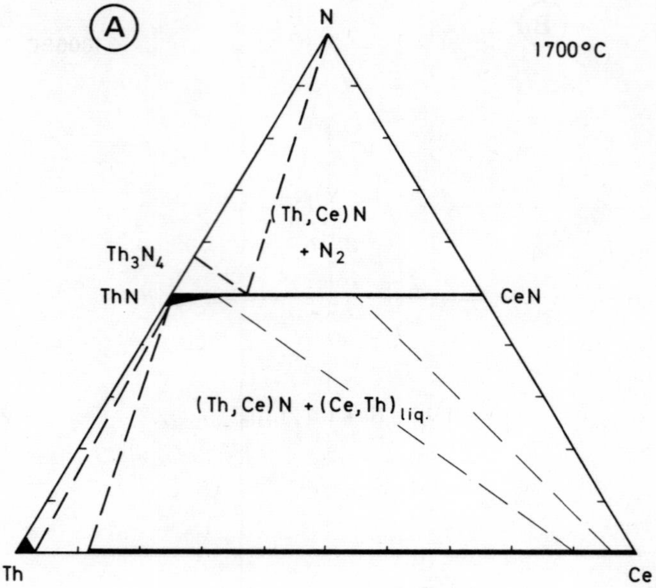

FIG. 8. Tentative isothermal sections in the Th-Ce-N (1700°C), U-Ce-N (1600°C) and Pu-Ce-N (1600°C) systems.

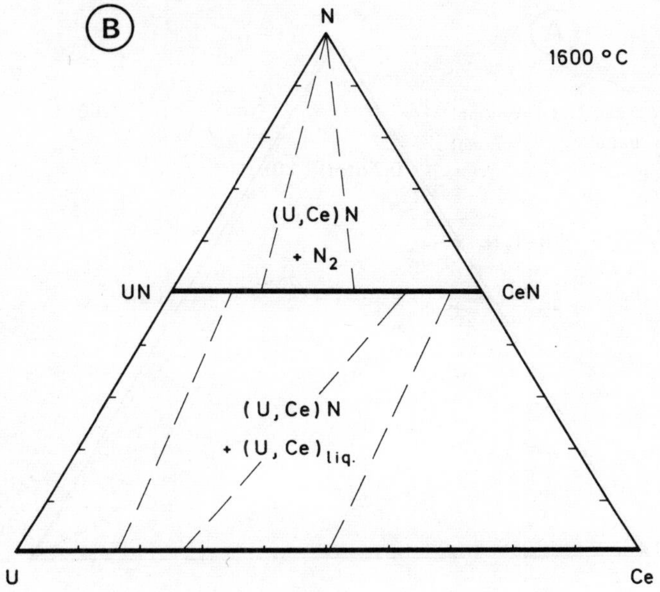
B
N
1600 °C
(U,Ce) N
+ N_2
UN
CeN
(U, Ce) N
+ (U, Ce)$_{liq.}$
U
Ce

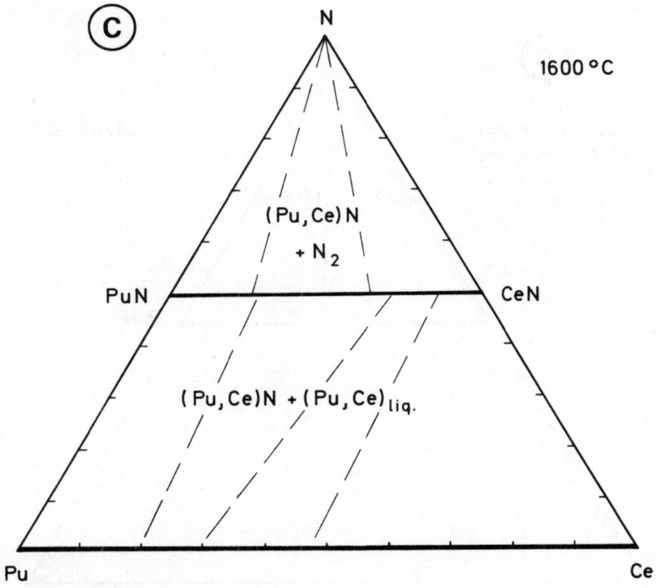
C
N
1600 °C
(Pu, Ce) N
+ N_2
PuN
CeN
(Pu, Ce)N + (Pu, Ce)$_{liq.}$
Pu
Ce

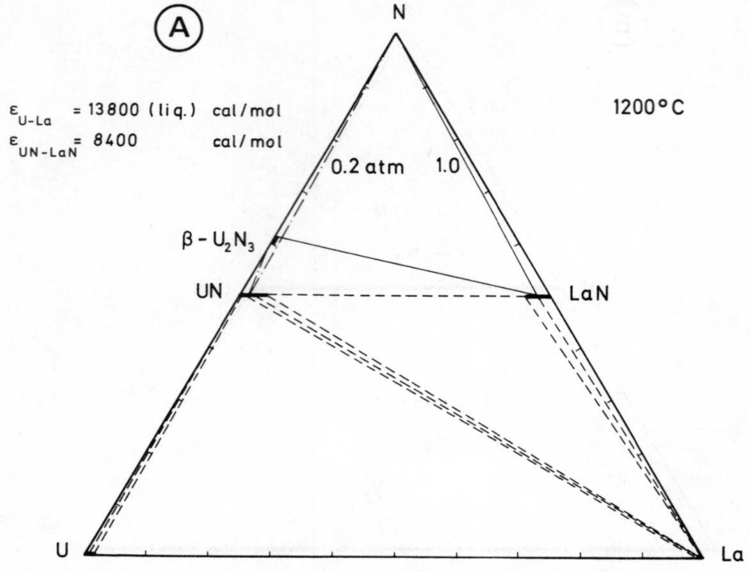
A
N
ε U-La = 13800 (liq.) cal/mol
ε UN-LaN = 8400 cal/mol
1200°C
0.2 atm
1.0
β - U2N3
UN
LaN
U
La

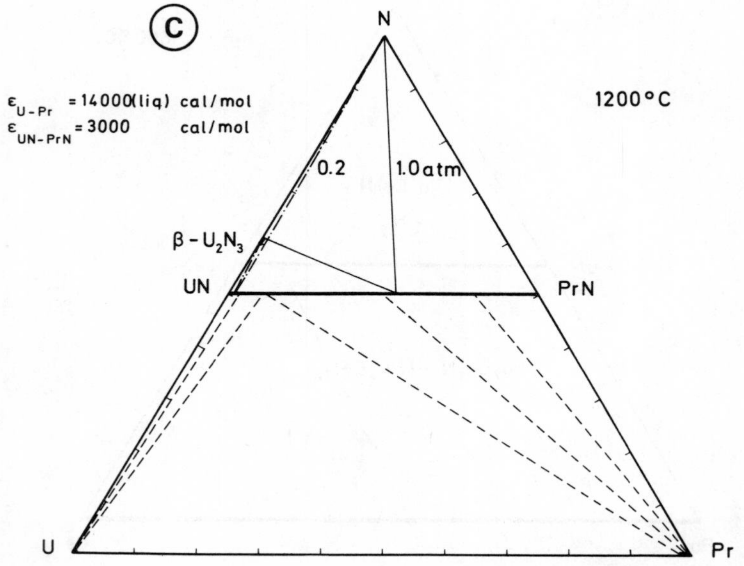
C
N
ε U-Pr = 14000(liq) cal/mol
ε UN-PrN = 3000 cal/mol
1200°C
0.2
1.0atm
β - U2N3
UN
PrN
U
Pr

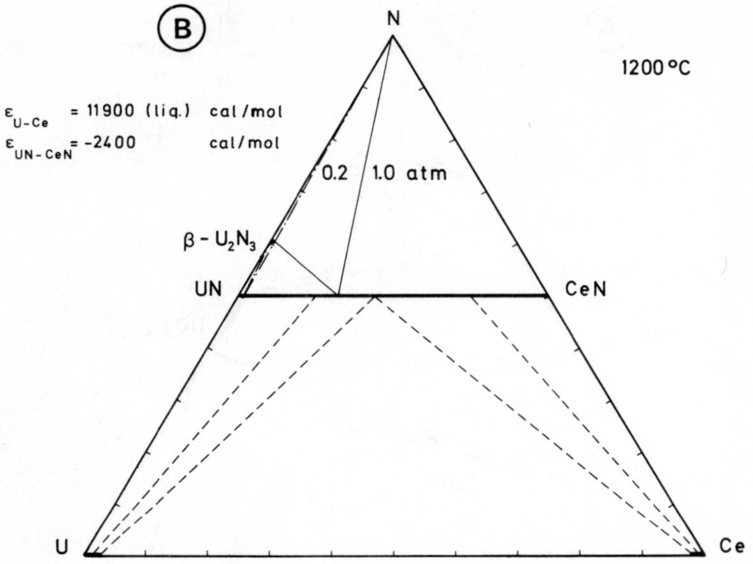

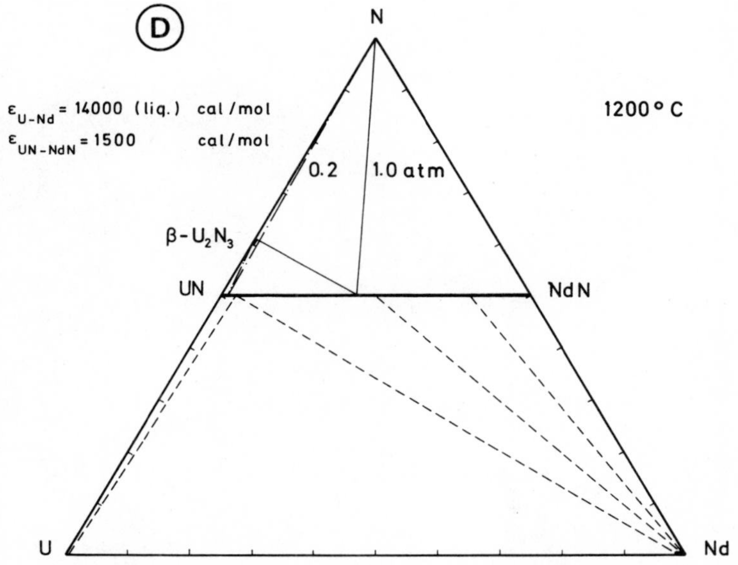

FIG.9. Calculated isothermal sections in the U-La-N, U-Ce-N, U-Pr-N and U-Nd-N systems at 1200°C.

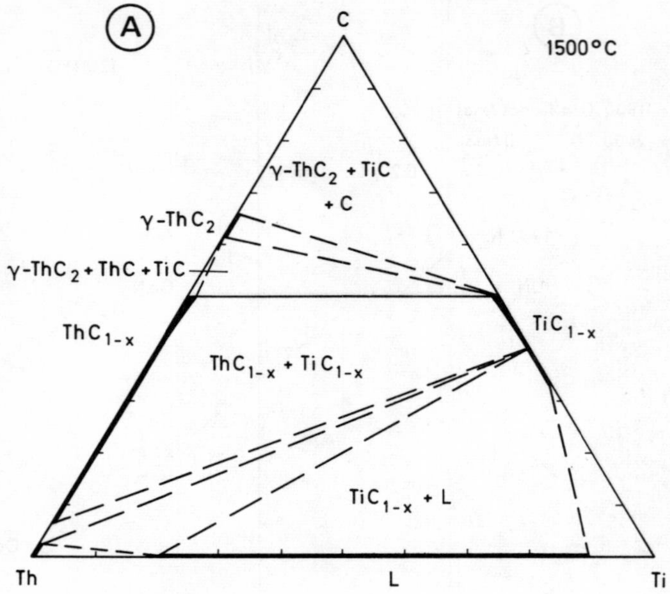

FIG.10. Isothermal sections in the Th-Ti-C (1500°C), U-Ti-C (1500°C) and Pu-Ti-C (1600°C) systems (tentative) at the temperatures indicated.

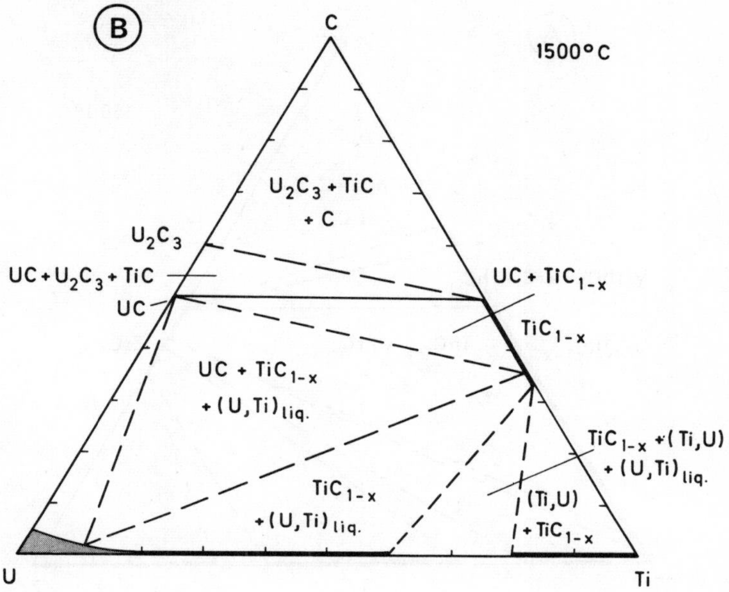
B
C
1500°C
U2C3 + TiC
+ C
U2C3
UC + U2C3 + TiC
UC
UC + TiC1-x
TiC1-x
UC + TiC1-x
+ (U,Ti)liq.
TiC1-x + (Ti,U)
+ (U,Ti)liq.
TiC1-x
+ (U,Ti)liq.
(Ti,U)
+ TiC1-x
U
Ti

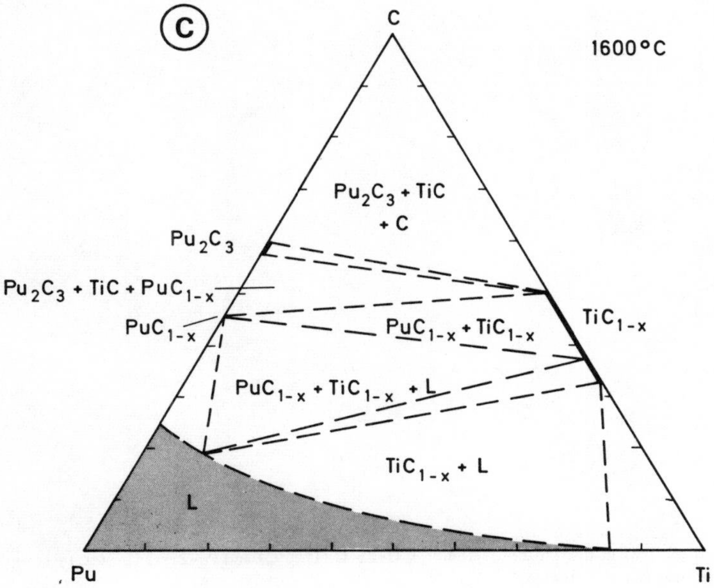
C
C
1600°C
Pu2C3 + TiC
+ C
Pu2C3
Pu2C3 + TiC + PuC1-x
PuC1-x
PuC1-x + TiC1-x
TiC1-x
PuC1-x + TiC1-x + L
TiC1-x + L
L
Pu
Ti

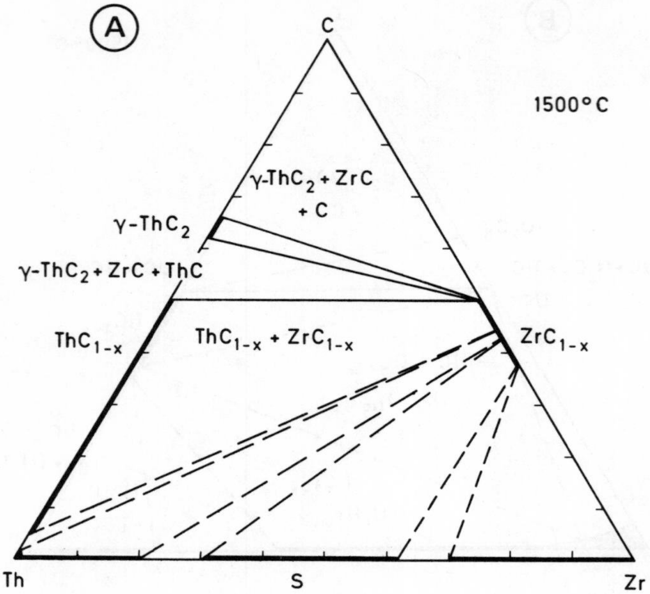

FIG.11. Isothermal sections in the Th-Zr-C (1500°C), U-Zr-C (1700°C) and Pu-Zr-C (1600°C) systems (tentative) at the temperatures indicated.

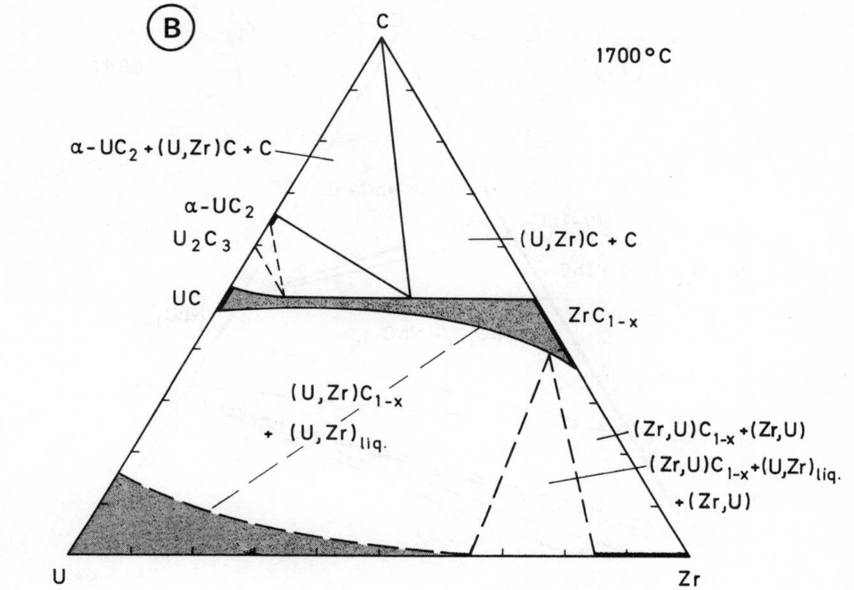
B
C
1700°C
α-UC$_2$ + (U,Zr)C + C
α-UC$_2$
U$_2$C$_3$
(U,Zr)C + C
UC
ZrC$_{1-x}$
(U,Zr)C$_{1-x}$
+ (U,Zr)$_{liq.}$
(Zr,U)C$_{1-x}$ + (Zr,U)
(Zr,U)C$_{1-x}$ + (U,Zr)$_{liq.}$
+ (Zr,U)
U
Zr

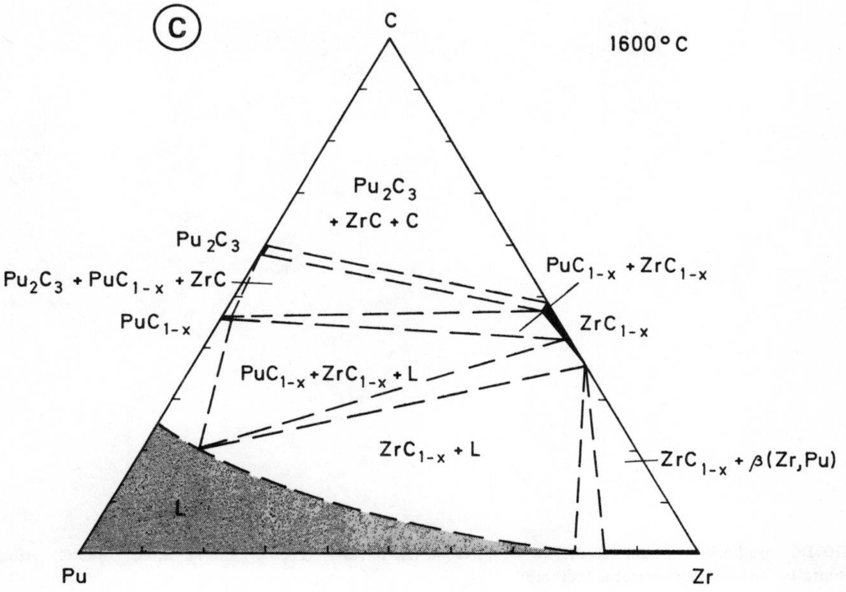
C
C
1600°C
Pu$_2$C$_3$
+ ZrC + C
Pu$_2$C$_3$
Pu$_2$C$_3$ + PuC$_{1-x}$ + ZrC
PuC$_{1-x}$ + ZrC$_{1-x}$
PuC$_{1-x}$
ZrC$_{1-x}$
PuC$_{1-x}$ + ZrC$_{1-x}$ + L
ZrC$_{1-x}$ + L
ZrC$_{1-x}$ + β(Zr,Pu)
L
Pu
Zr

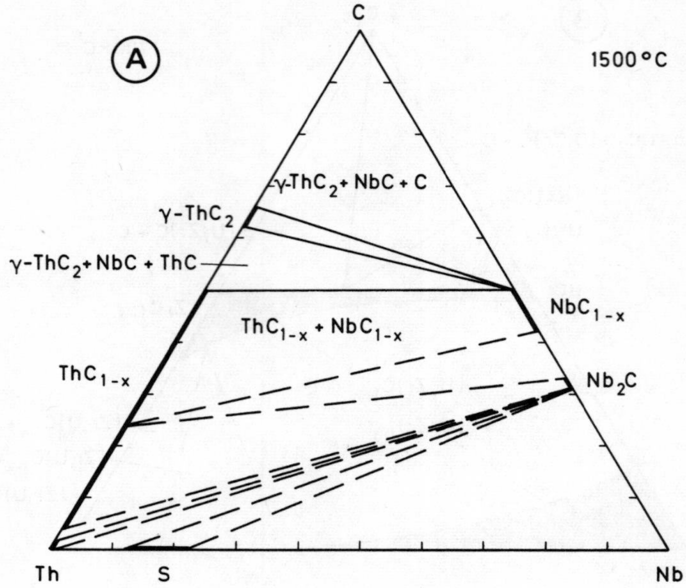

FIG.12. Isothermal sections in the Th-Nb-C (1500°C), U-Nb-C (1700°C) and Pu-Nb-C (1600°C) systems (tentative) at the temperatures indicated.

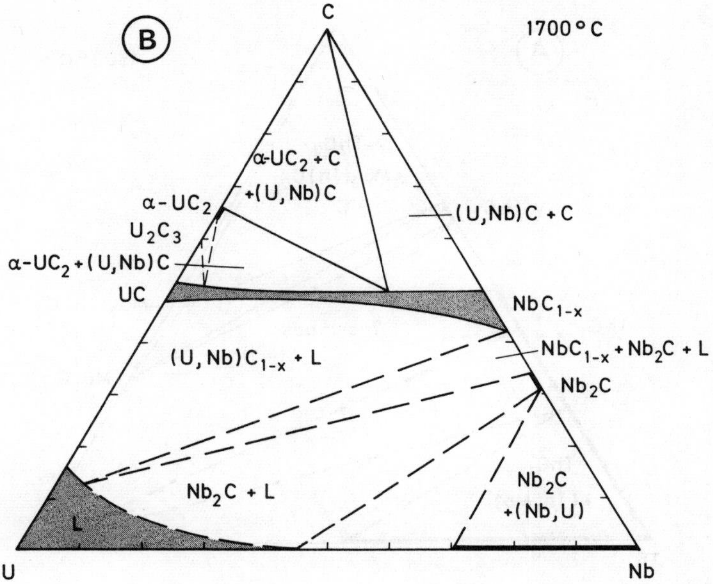
B
C
1700 °C
α-UC2 + C
+(U, Nb)C
α - UC2
(U,Nb)C + C
U2C3
α-UC2 + (U,Nb)C
UC
NbC1-x
(U, Nb)C1-x + L
NbC1-x + Nb2C + L
Nb2C
Nb2C + L
Nb2C
+(Nb, U)
L
U
Nb

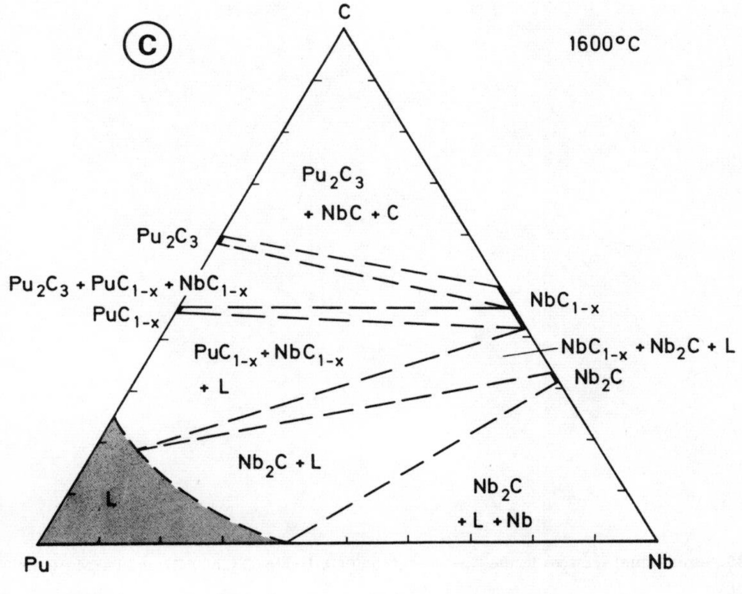
C
C
1600°C
Pu2C3
+ NbC + C
Pu2C3
Pu2C3 + PuC1-x + NbC1-x
NbC1-x
PuC1-x
PuC1-x + NbC1-x
+ L
NbC1-x + Nb2C + L
Nb2C
Nb2C + L
Nb2C
+ L + Nb
L
Pu
Nb

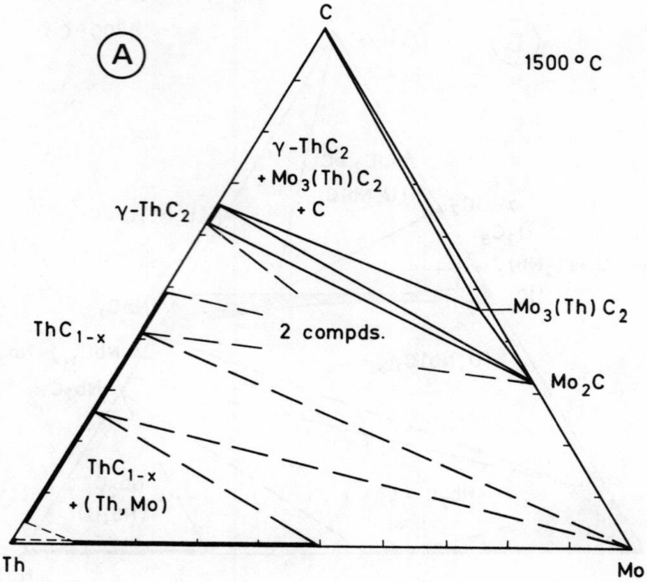

FIG.13. Isothermal sections in the Th-Mo-C (1500°C), U-Mo-C (1500°C) and Pu-Mo-C (900°C) systems.

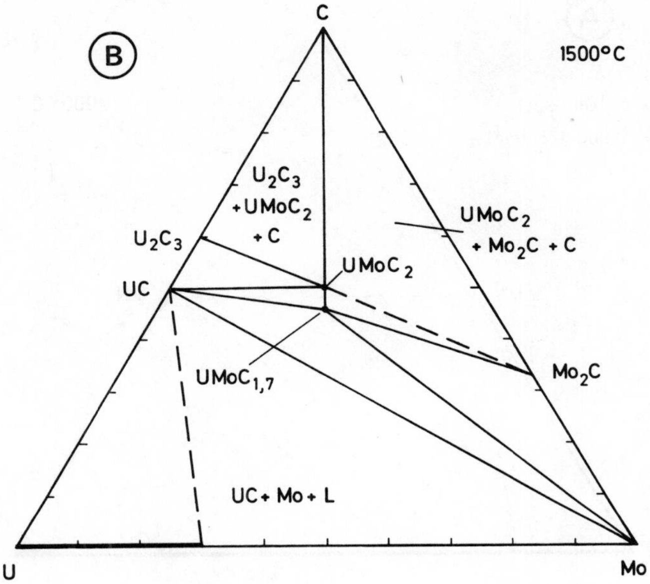
B
1500°C
C
U_2C_3 + $UMoC_2$ + C
$UMoC_2$ + Mo_2C + C
U_2C_3
$UMoC_2$
UC
$UMoC_{1,7}$
Mo_2C
UC + Mo + L
U
Mo

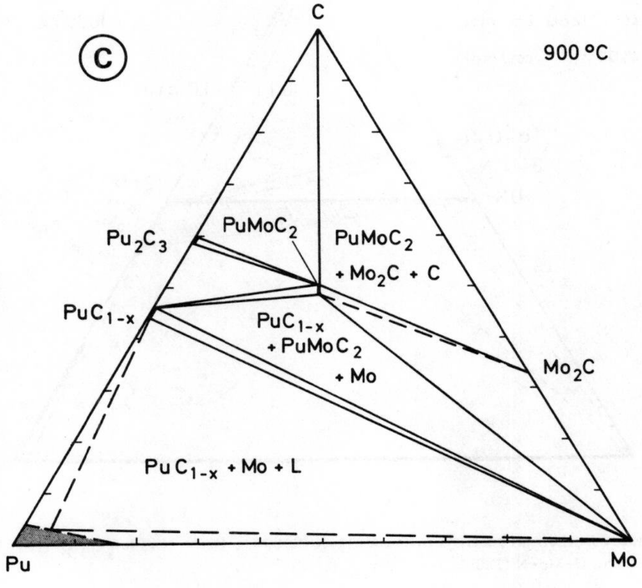
C
900 °C
C
Pu_2C_3
$PuMoC_2$
$PuMoC_2$ + Mo_2C + C
PuC_{1-x}
PuC_{1-x} + $PuMoC_2$ + Mo
Mo_2C
PuC_{1-x} + Mo + L
Pu
Mo

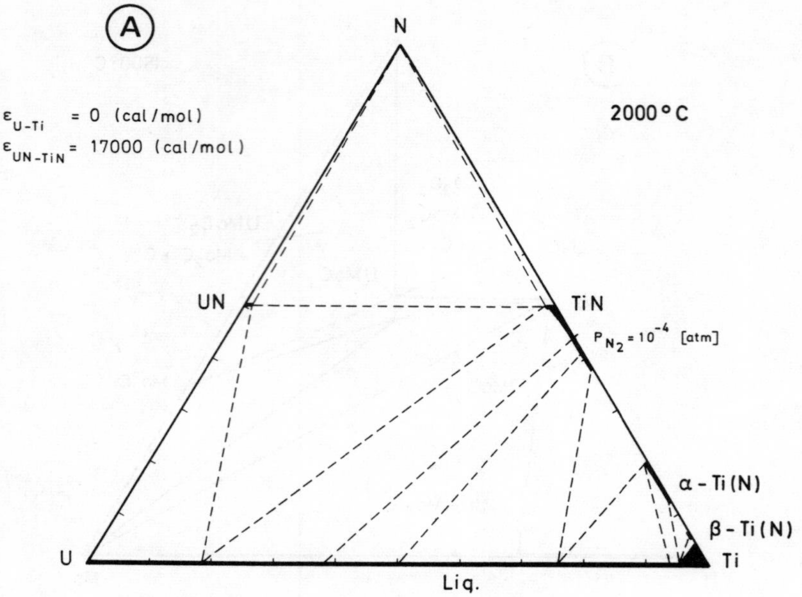

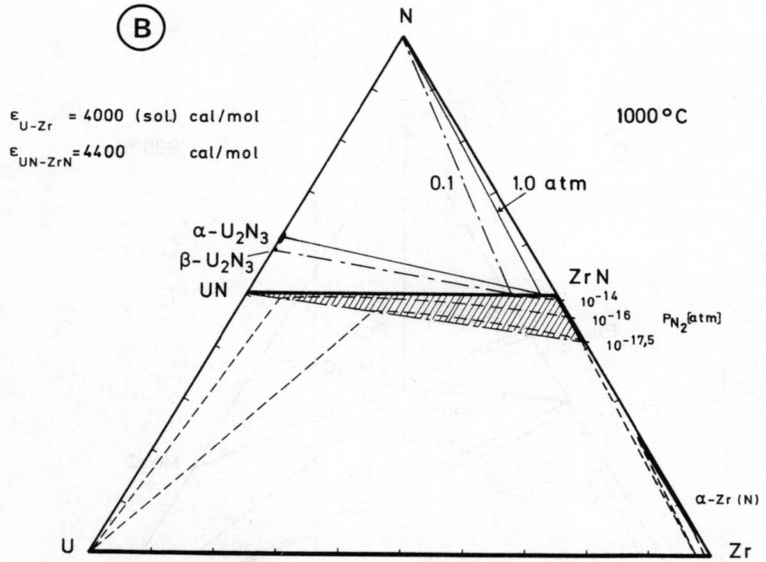

FIG.14. Calculated isothermal sections in the U-Ti-N (2000°C), U-Zr-N (1000°C), U-Hf-N (1200 and 1500°C), U-Nb-N (1000°C) and U-Mo-N (1200°C) systems.

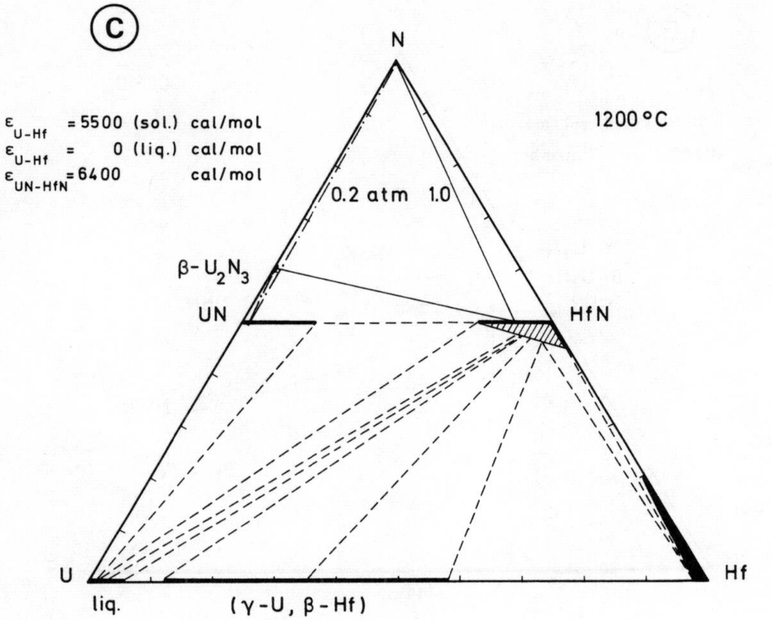
C
N
ε_{U-Hf} = 5500 (sol.) cal/mol
ε_{U-Hf} = 0 (liq.) cal/mol
ε_{UN-HfN} = 6400 cal/mol
1200 °C
0.2 atm 1.0
β-U_2N_3
UN
HfN
U
Hf
liq.
(γ-U, β-Hf)

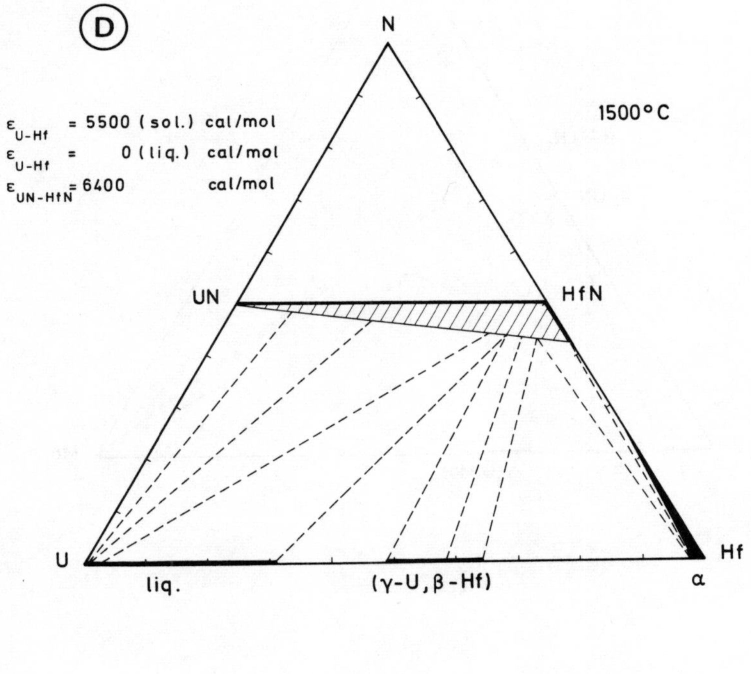
D
N
ε_{U-Hf} = 5500 (sol.) cal/mol
ε_{U-Hf} = 0 (liq.) cal/mol
ε_{UN-HfN} = 6400 cal/mol
1500 °C
UN
HfN
U
Hf
liq.
(γ-U, β-Hf)
α

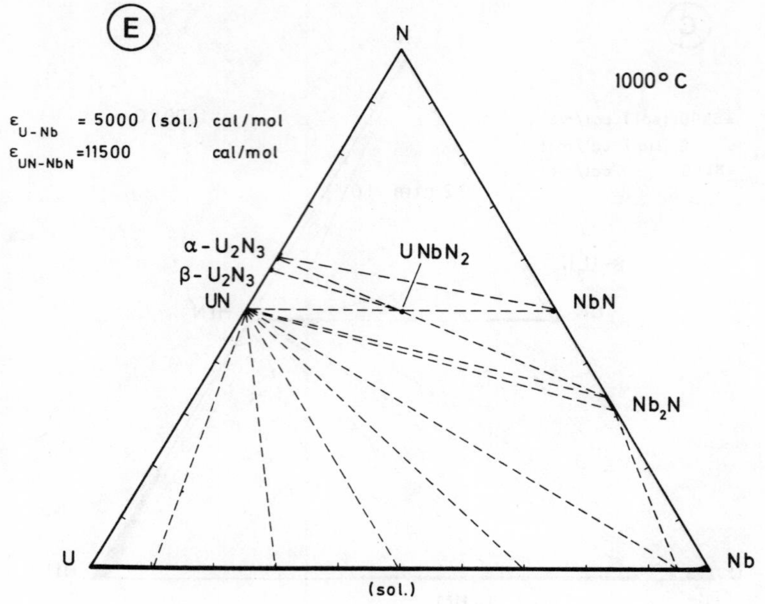

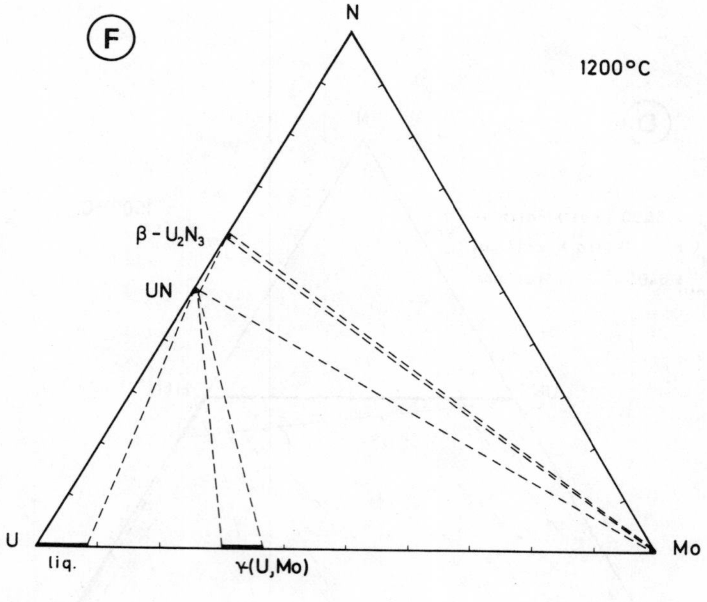

Fig. 14. Continued.

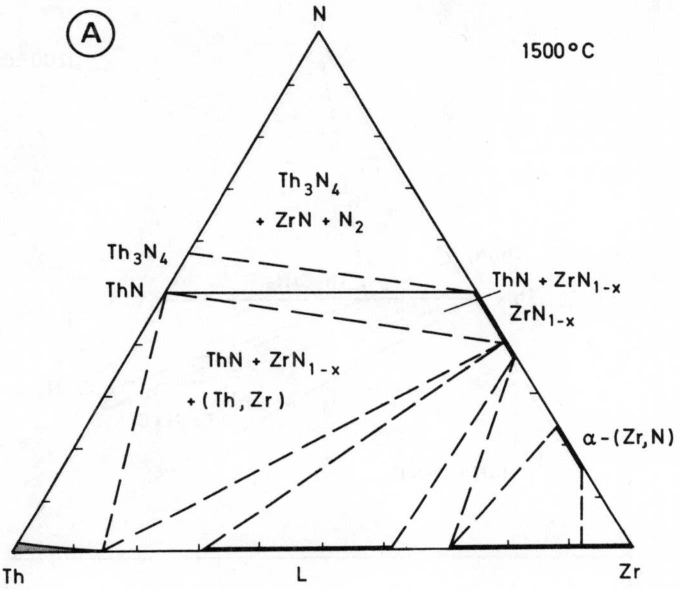

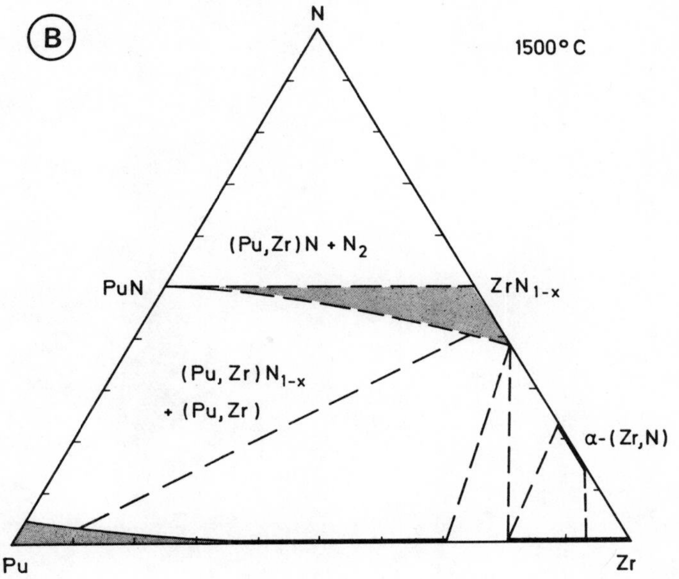

FIG.15. Tentative isothermal sections in the Th-Zr-N and Pu-Zr-N systems at 1500°C.

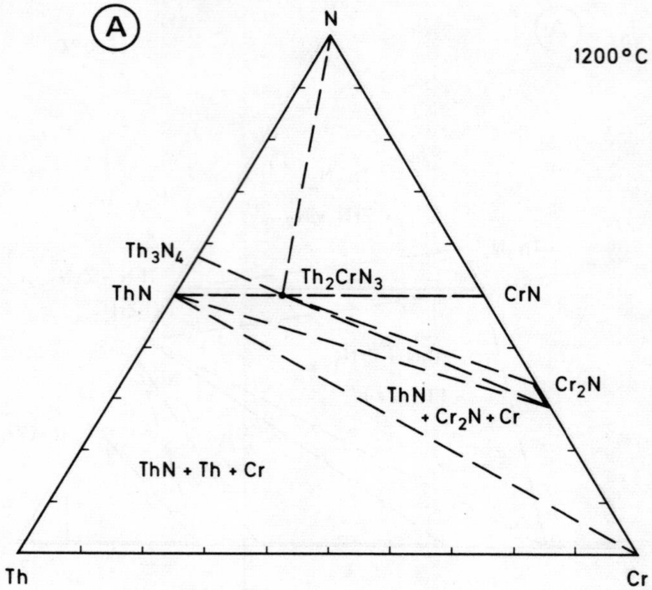

FIG.16. Tentative isothermal sections in the Th-Cr-N, U-Cr-N and Pu-Cr-N systems at 1200°C.

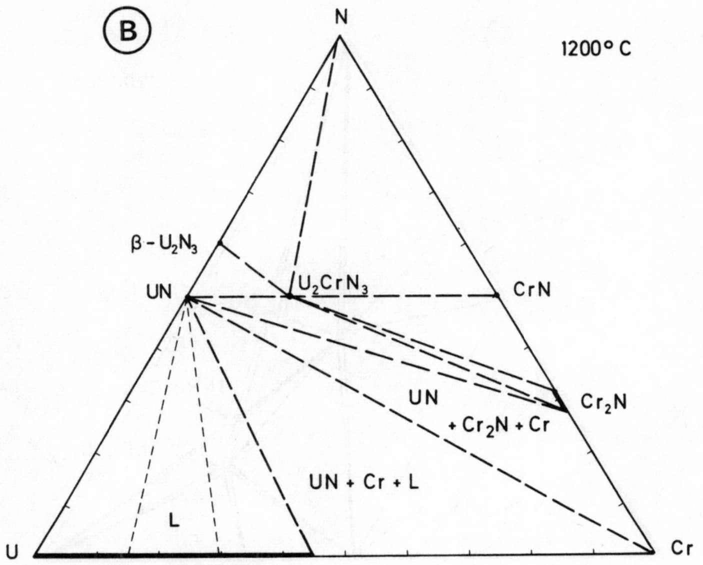
B
N
1200° C
β - U2N3
UN
U2CrN3
CrN
Cr2N
UN
+ Cr2N + Cr
UN + Cr + L
L
U
Cr

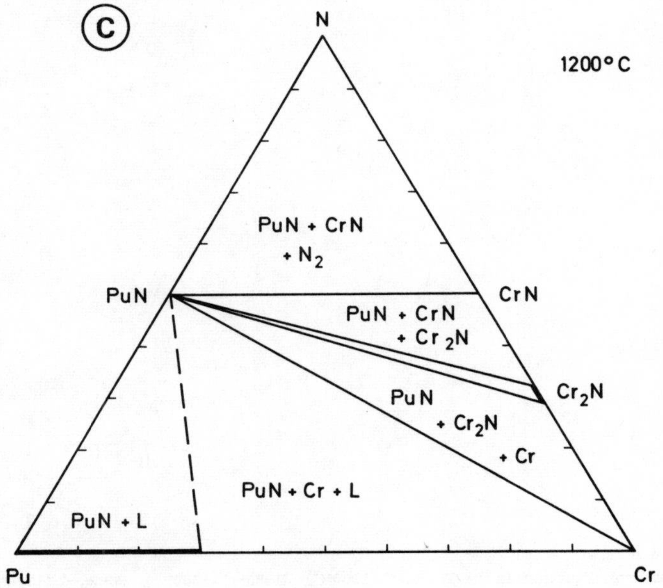
C
N
1200° C
PuN + CrN
+ N2
PuN
CrN
PuN + CrN
+ Cr2N
Cr2N
PuN
+ Cr2N
+ Cr
PuN + Cr + L
PuN + L
Pu
Cr

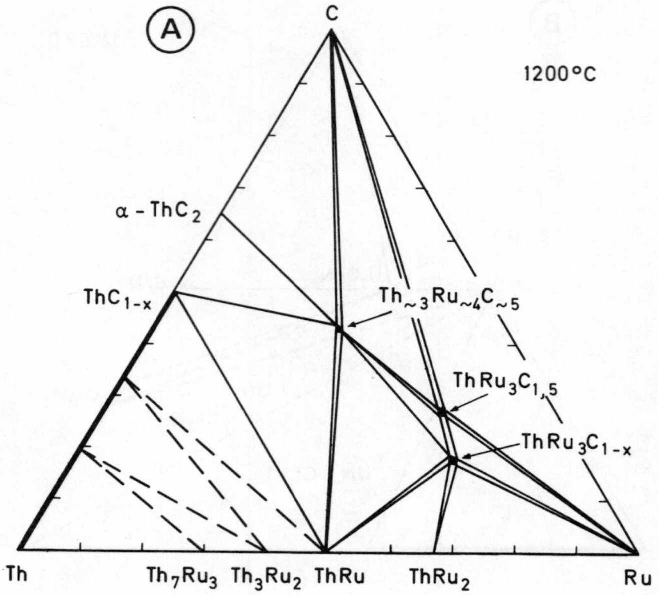

FIG.17. Isothermal sections in the Th-Ru-C (1200°C), U-Ru-C (1300°C), Pu-Ru-C (1200°C) systems.

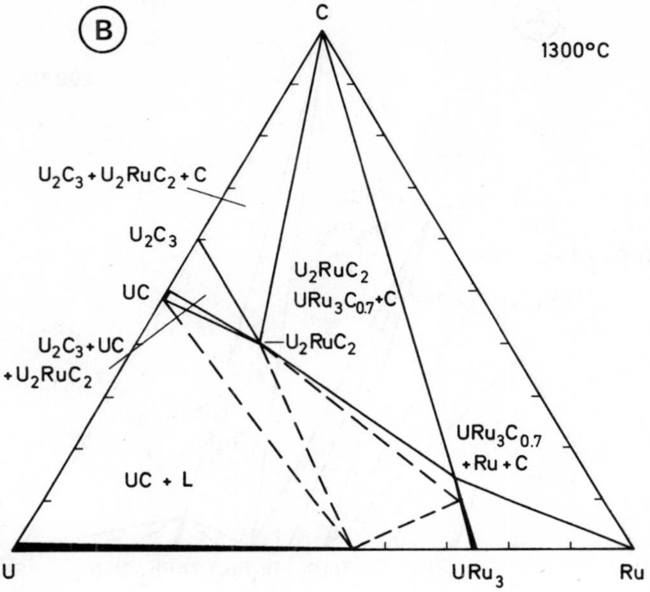
B
C
1300°C
$U_2C_3 + U_2RuC_2 + C$
U_2C_3
U_2RuC_2
$URu_3C_{0.7}$+C
UC
U_2RuC_2
U_2C_3+UC
+U_2RuC_2
$URu_3C_{0.7}$
+Ru+C
UC + L
U
URu_3
Ru

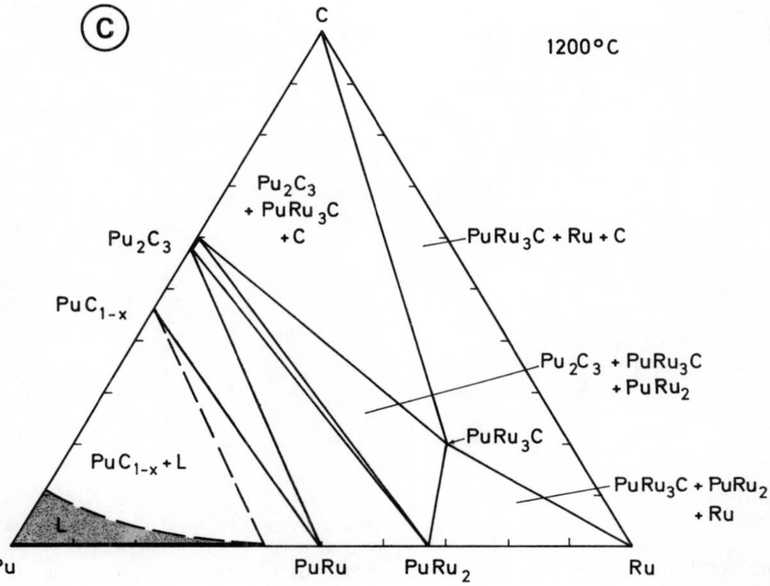
C
C
1200°C
Pu_2C_3
+ $PuRu_3C$
+C
Pu_2C_3
$PuRu_3C$ + Ru + C
PuC_{1-x}
Pu_2C_3 + $PuRu_3C$
+ $PuRu_2$
$PuRu_3C$
PuC_{1-x} + L
$PuRu_3C$ + $PuRu_2$
+ Ru
L
Pu
PuRu
$PuRu_2$
Ru

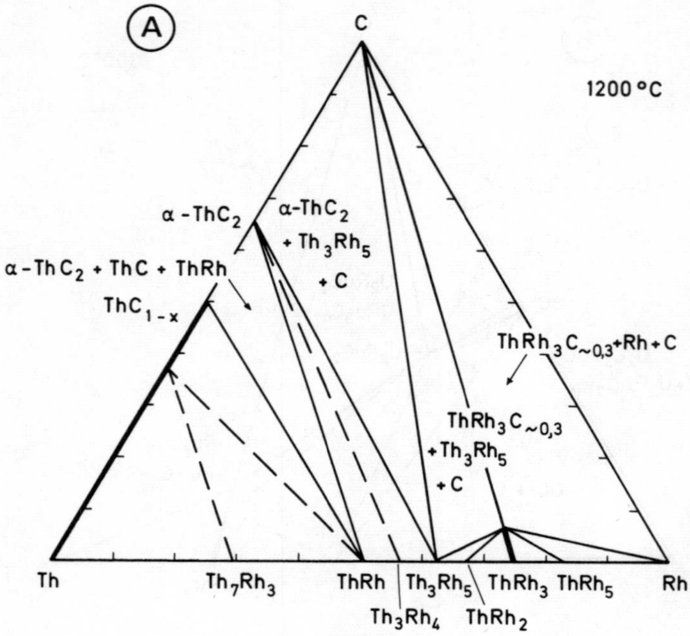

FIG.18. Isothermal sections in the Th-Rh-C (1200°C), U-Rh-C (1300°C), Pu-Rh-C (1200°C) systems.

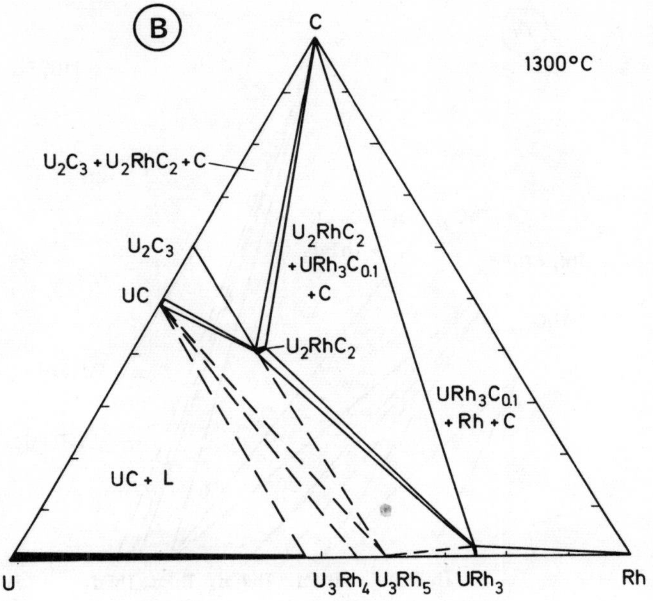
B
C
1300°C
U2C3 + U2RhC2 + C
U2RhC2
+ URh3C0.1
+ C
U2C3
UC
U2RhC2
URh3C0.1
+ Rh + C
UC + L
U
U3Rh4 U3Rh5
URh3
Rh

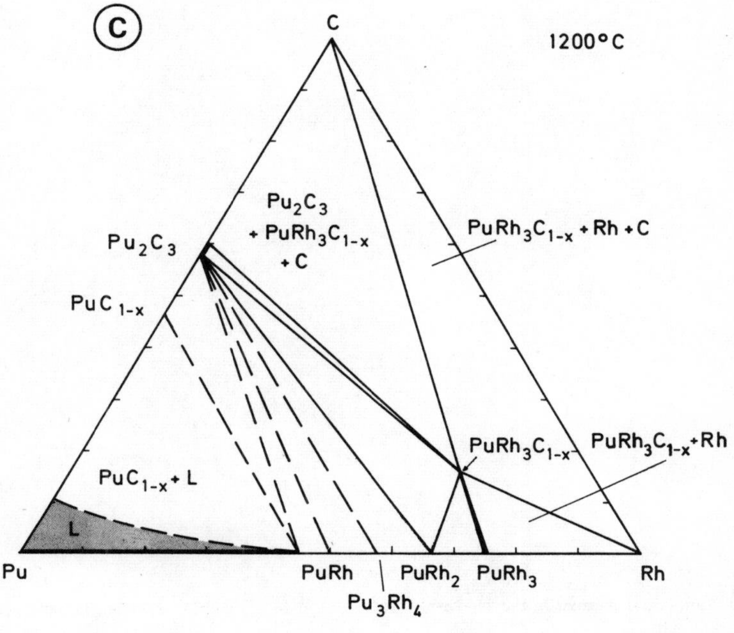
C
C
1200°C
Pu2C3
+ PuRh3C1-x
+ C
PuRh3C1-x + Rh + C
Pu2C3
PuC 1-x
PuRh3C1-x
PuRh3C1-x+Rh
PuC1-x+ L
L
Pu
PuRh
PuRh2
PuRh3
Rh
Pu3Rh4

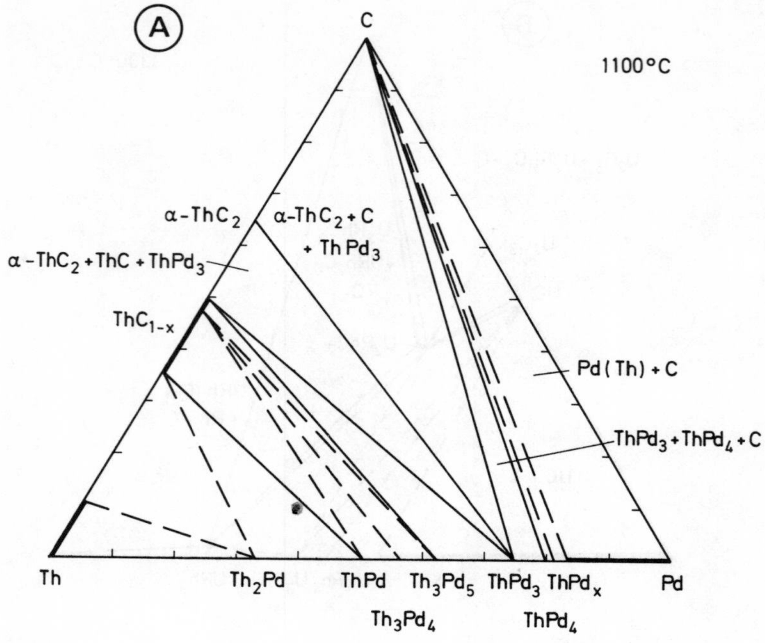

FIG.19. Isothermal sections in the Th-Pd-C (1100°C), U-Pd-C (1300°C), Pu-Pd-C (1200°C) systems.

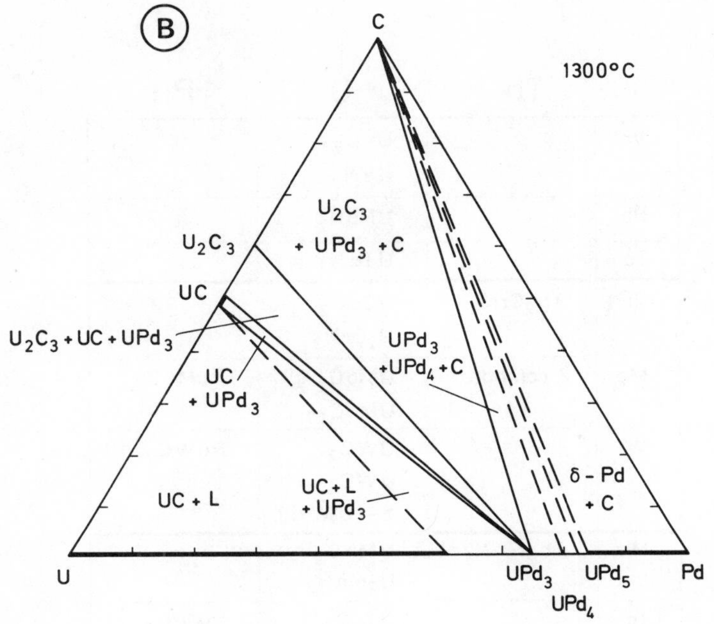
B
C
1300°C
U2C3
+ UPd3 + C
U2C3
UC
U2C3 + UC + UPd3
UPd3
+UPd4 + C
UC
+ UPd3
δ - Pd
+ C
UC + L
UC + L
+ UPd3
U
UPd3
UPd5
Pd
UPd4

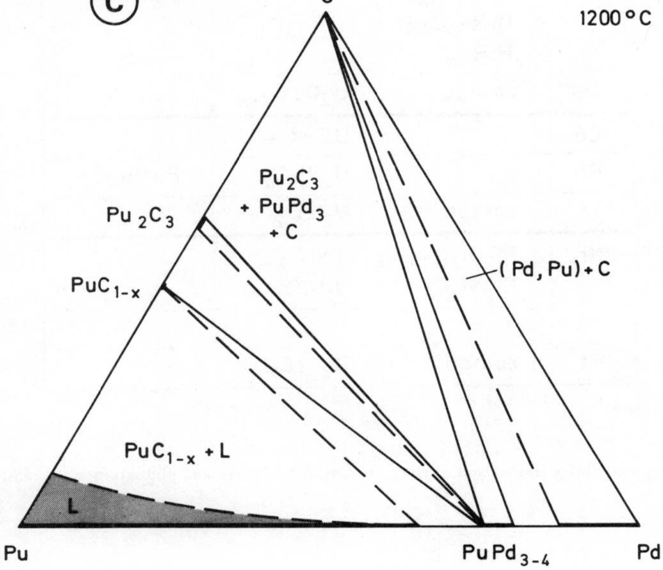
C
C
1200°C
Pu2C3
+ PuPd3
+ C
Pu 2C3
(Pd, Pu) + C
PuC1-x
PuC1-x + L
L
Pu
PuPd3-4
Pd

	Th	U	Pu
V		UVC_2 UVN_2	
Nb		$UNbN_2$	
Ta		$UTaN_2$	
Cr	Th_2CrN_3	$UCrC_2$ U_2CrN_3	
Mo	2 compds.	$UMoC_2$ $UMoC_{1.7}$	$PuMoC_2$
W		UWC_2 $UWC_{1.7}$ z-(U,W,C)	$PuWC_2$
Mn	Th_2MnN_3	$UMnC_2$ U_2MnN_3	
Tc		$UTcC_2$	$PuTcC_2$
Re	compd.	$UReC_2$ $UReC_{1.7}$	
Fe		$UFeC_2$	$PuFeC_2$ $Pu_3Fe_4C_5$
Ru	$Th_3Ru_4C_5$ $ThRu_3C_{1.5}$ $ThRu_3C$	U_2RuC_2 URu_3C_{1-x}	$PuRu_3C$
Os	compd.	U_2OsC_{2+x}	
Co		$UCoC_2$	
Rh		U_2RhC_2	$PuRh_3C_{1-x}$
Ir	compd.	U_2IrC_2	
Ni	$Th_2Ni_3C_2$ $Th_2Ni_3C_5$	$UNiC_2$ U_2NiC_3	
Pd			
Pt	compd.	U_2PtC_2	

FIG.20. Ternary complex carbides and nitrides of thorium, uranium and plutonium with transition metals.

IVA	VA	VIA	VIIA	VIII		
—	UVN_2	U_2CrN_{2+x}	U_2MnN_3	—	—	—
(U,Zr)N	$UNbN_2$	—	—	—	—	—
(U,Hf)N	$UTaN_2$	—	—	—	—	—

IVA	VA	VIA	VIIA	VIII		
—	UVC_2	$UCrC_2$	$UMnC_2$	$UFeC_2$	$UCoC_2$	$UNiC_2$
(U,Zr)C	(U,Nb)C	$UMoC_2$	$UTcC_2$	U_2RuC_2	U_2RhC_2	—
(U,Hf)C	(U,Ta)C	UWC_2	$UReC_2$	U_2OsC_{2+x}	U_2IrC_2	U_2PtC_2

□ f.c.c ▒ orthorh. ▨ tetrag. or orthorh. ▤ tetrag.

FIG.21. Ternary uranium-transition metal carbides and nitrides.

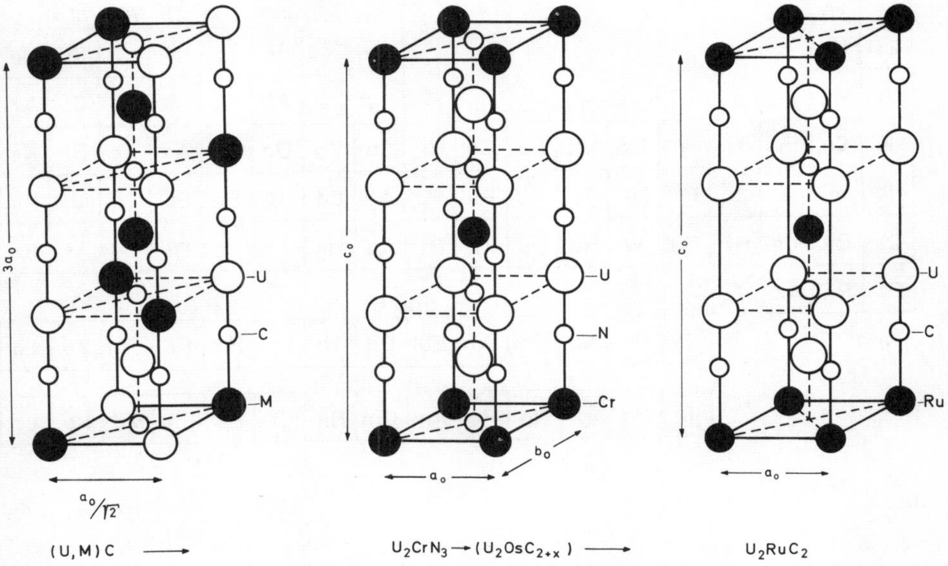

FIG.22. Structures of the (U,M)C mixed crystal (U and M are statistically distributed), orthorhombic U_2CrN_3 and tetragonal U_2RuC_2.

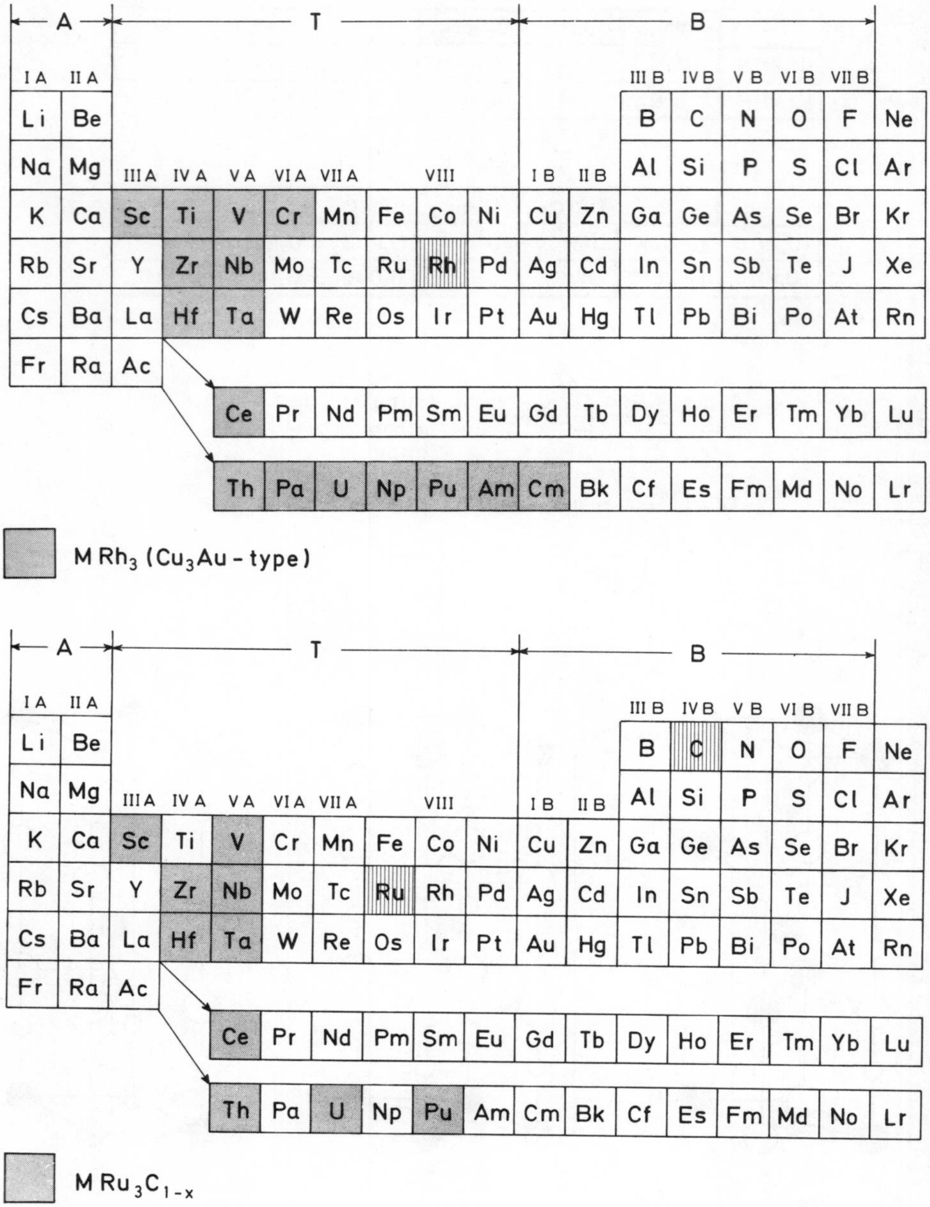

FIG.23. The occurrence of binary MRh_3 phases and ternary MRu_3C_{1-x} carbides.

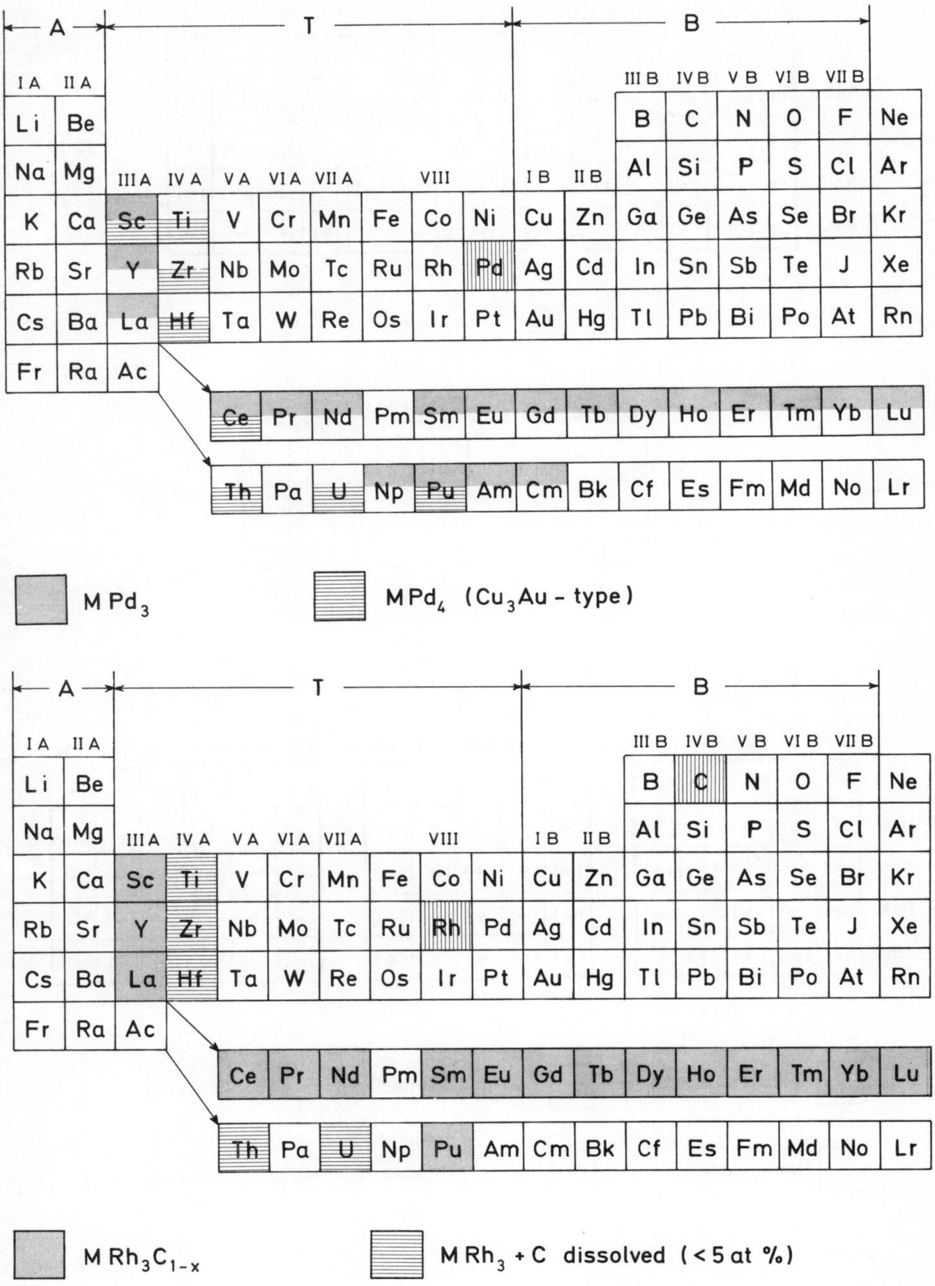

FIG.24. The occurrence of binary MPd_3 and MPd_4 phases and ternary MRh_3C_{1-x} carbides.

(A)

	ThC	PaC	UC	NpC	PuC	YC	TiC	ZrC	HfC	VC	NbC	TaC
ThC		(+)	+	+	+	(+)	−	−	−	−	−	−
PaC	(+)		(+)	(+)	(+)	(+)	(−)	(−)	(−)	(−)	(−)	(−)
UC	+	(+)		(+)	+	+	−	+	+	− comp.	+	+
NpC	+	(+)	(+)		(+)	(+)	(−)	(−)	(−)	(−)	(−)	(−)
PuC	+	(+)	+	(+)		(+)	−	−	(−)	(−)	(−)	(−)

(B)

	ThC_2	UC_2	PuC_2	YC_2	LaC_2	CeC_2	PrC_2	NdC_2
ThC_2		+	+	(+)	(+)	(+)	(+)	(+)
UC_2	+		+	+	+	+	(+)	(+)
PuC_2	+	+		(+)	(+)	(+)	(+)	(+)

(C)

	ThN	UN	NpN	PuN	AmN	YN	LaN	CeN	PrN	NdN	TiN	ZrN	HfN	VN	NbN	TaN
ThN		+	(+)	(+)	(+)	+	+	+	+	+	−	−	(−)	(−)	(−)	(−)
UN	+		(+)	+	(+)	+	+	+	+	+	−	+	+	− comp.	− comp.	− comp.
NpN	(+)	(+)		(+)	(+)	(+)	(+)	(+)	(+)	(+)	(−)	(+)	(+)	(−)	(−)	(−)
PuN	(+)	+	(+)		(+)	(+)	(+)	(+)	(+)	(+)	(−)	(+)	(+)	(−)	(−)	(−)
AmN	(+)	(+)	(+)	(+)		(+)	(+)	(+)	(+)	(+)	(−)	(−)	(−)	(−)	(−)	(−)

+	miscible
(+)	not investigated but probably miscible
−	immiscible
(−)	not investigated but probably immiscible

FIG.25. The occurrence of complete miscibility in face-centered cubic (A) monocarbide, (B) dicarbide and (C) mononitride systems of the actinides.

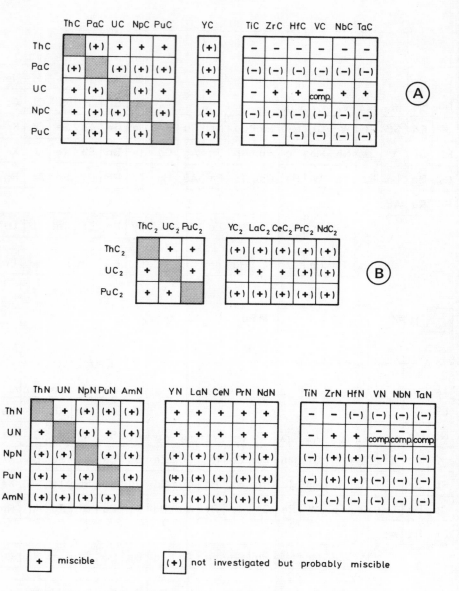

REFERENCES

[1] BENZ, R., STONE, P.L., High Temp. Sci. 1 (1969) 114.

[2] BENZ, R., HOFFMANN, C.G., RUBERT, G.N., High Temp. Sci. 1 (1969) 342.

[3] HOLLECK, H., KLEYKAMP, H., "Compounds with carbon", and "Compounds with nitrogen", in Gmelins Handbuch der anorganischen Chemie, Vol.4, Transurane, Teil C: Verbindungen, Verlag Chemie, Weinheim (1972).

[4] BENZ, R., HOFFMANN, C.G., RUBERT, G.N., J. Am. Chem. Soc. 89 (1967) 191.

[5] BENZ, R., J. Nucl. Mater. 31 (1969) 93.

[6] HOLLEY, C.E., Thermodynamic properties of actinide carbides, J. Nucl. Mater. 51 (1974) 36.

[7] GSCHNEIDER, K.A., KIPPENHAN, N., Thermochemistry of the Rare Earth Carbides, Nitrides and Sulfides for Steelmaking, Iowa State University (Ames) Rep. IS-RIC-5 (1971).

[8] HOLLECK, H., ISHII, T., Calculation of Phase Equilibria in Ternary Systems: Uranium-Transition Metal-Nitrogen, Kernforschungszentrum Karlsruhe Rep. KFK 1754 (1973).

[9] STORMS, E.K., The Refractory Carbides, Academic Press, New York (1967).

[10] HOLLECK, H., KLEYKAMP, H., Thermodynamics of multi-component systems containing UC and PuC. A review, J. Nucl. Mater. 32 (1969) 1.

[11] HOLLECK, H., Ternary Carbides and Nitrides of the Actinides and Lanthanides with other Transition Metals, Kernforschungszentrum Karlsruhe Rep. KFK 1726 (1972).

[12] TAGAWA, H., Phase relations and thermodynamic properties of the uranium-nitrogen system, J. Nucl. Mater. 51 (1974) 78.

[13] TAGAWA, H., Phase behaviour and crystal structure of actinide nitrides, Nippon Genshiryoku Gakkai-Shi 13 (1971) 267.

[14] PETERSON, S., CURTIS, C.E., Thorium Ceramics Data Manual, Vol.2, Nitrides, USAEC Rep. ORNL-4503 (1973).

[15] PETERSON, S., CURTIS, C.E., Thorium Ceramics Data Manual, Vol.3, Carbides, USAEC Rep. ORNL-4503 (1971).

[16] HANSEN, M., Constitution of Binary Alloys, McGraw-Hill, New York (1958).

[17] ELLIOT, R.P., Constitution of Binary Alloys, First Supplement, McGraw-Hill, New York (1965).

[18] SHUNK, F.A., Constitution of Binary Alloys, Second Supplement, McGraw-Hill, New York (1969).

[19] BROWNING, P., PHILLIPS, B.A., POTTER, P.E., RAND, H.M., Third Int. Conf. Chem. Thermodynamics (Proc. Conf. Baden, Austria, 1973) 6, (1974) 100.

[20] HENNEY, J., JONES, J.W., Trans. Br. Ceram. Soc. 65 (1966) 613.

[21] NEVITT, M.V., USAEC Rep. ANL-6868 (1963).

[22] HOLLECK, H., ISHII, T., Thermal Analysis (Proc. Int. Conf. Davos, 1971) 2, (1971) 137-49.

[23] VENARD, J.T., SPRUIELL, J.E., J. Nucl. Mater. 27 (1968) 257.

[24] POTTER, P.E., J. Nucl. Mater. 47 (1973) 7.

[25] HOLLECK, H., J. Nucl. Mater. 42 (1972) 278.

[26] HAINES, H.R., POTTER, P.E., UKAEA Rep. AERE-R-6512 (1970).

[27] LORENZELLI, J.N., MARCON, J.P., J. Less-Common Met. 26 (1972) 71.

[28] STECHER, P., NECKEL, A., BENESOVSKY, F., NOWOTNY, H., Planseeber. Pulvermet. 12 (1964) 181.

[29] HOLLECK, H., POLITIS, C., unpublished results.

[30] KRUPKA, M.C., GIORGI, A.L., KRIKORIAN, N.H., SKLARZ, E.G., J. Less-Common Met. 19 (1969) 113.

[31] CHUBB, W., KELLER, D.L., Battelle Memorial Institute Rep. BMI-1685 (1964).

[32] HOLLECK, H., SMAILOS, E., unpublished results.

[33] HOLLECK, H., SMAILOS, E., THÜMMLER, F., J. Nucl. Mater. 28 (1968) 105.

[34] HOLLECK, H., SMAILOS, E., THÜMMLER, F., J. Nucl. Mater. 32 (1969) 281.

[35] HOLLECK, H., Kernforschungszentrum Karlsruhe Rep. KFK 1011 (1969).

[36] HOLLECK, H., SMAILOS, E., THÜMMLER, F., Monatsh. Chem. 99 (1968) 985.

[37] CAROLL, D.F., USAEC Rep. HW-81603 (1965).

[38] BENZ, R., ZACHARIASEN, W.H., J. Nucl. Mater. 37 (1970) 109.

[39] DeLUCA, J.P., LEITNAKER, J.M., J. Am. Ceram. Soc. 55 (1972) 273.

[40] HOLLECK, H., J. Nucl. Mat. (in press, 1974).

[41] HOLLECK, H., KLEYKAMP, H., J. Nucl. Mater. 35 (1970) 158.

[42] HOLLECK, H., BENEDICT, U., KLEYKAMP, H., SARI, C., in preparation.

[43] HOLLECK, H., KLEYKAMP, H., J. Nucl. Mater. 45 (1972) 47.

[44] HOLLECK, H., Monatsh. Chem. 102 (1971) 1699.

[45] BOWMAN, A.L., Acta Crystallogr., Sect. B 27 (1971) 1067.

[46] HOLLECK, H., J. Nucl. Mater. 39 (1971) 226.

[47] McCOLM, I.J., COLQUHOUN, I., CLARK, N.J., J. Inorg. Nucl. Chem. 34 (1972) 3809.

DISCUSSION

M. KATSURA: This question is addressed to both Mr. Holleck and Mr. Potter. Can the structure of the carbon change the form of the phase diagrams in the regions in which carbon is present?

P.E. POTTER: Yes, this could occur for example in a system such as U-Zr-C, where the composition of the (U, Zr)C phase in equilibrium with carbon and UC_2 would be influenced by the different thermodynamic properties with some departure from the graphite structure.

A.S. PANOV: Mr. Holleck, your paper has to some extent supplied the answer to the question I asked of Mr. Potter at the beginning of this session. Now I should like to ask you whether the phase boundaries of the homogeneity regions shown in your diagram of, for example, $U_y Me_{1-y} C_{1\pm x}$ were obtained experimentally or whether they were derived from estimates.

H. HOLLECK: Most of the extended homogeneity regions are bounded by broken lines. This means, as I pointed out in my paper, that the phase boundaries are estimates. In many cases, however, there is also some experimental evidence for these estimates.

ДИАГРАММЫ СОСТОЯНИЯ ТРОЙНЫХ СИСТЕМ U-Th-Mn, Fe, Ni В ЧАСТИ, БОГАТОЙ U И Th

Т.А.БАДАЕВА, Л.Н.АЛЕКСАНДРОВА
Институт металлургии им.А.А.Байкова
Академии наук СССР,
Москва,
Союз Советских Социалистических Республик

Доклад представлен О.С. Ивановым

Abstract–Аннотация

PHASE DIAGRAMS OF THE TERNARY SYSTEMS U-Th-Mn, -Fe AND -Ni IN A PART RICH IN URANIUM AND THORIUM.

The report contains the results of experimental studies of phase diagrams of the ternary systems U-Th-Mn, -Fe and -Ni. The authors carried out thermal and microstructural analyses of the alloys and measured their hardness and microhardness. Projections of the liquidus surfaces and phase diagrams are constructed and reaction diagrams drawn up. A characteristic feature of the systems studied is the existence of zones of stratification in the liquid phase which spread to the ternary systems from the binary U-Th system and of four-phase eutectic equilibria in the vicinity of binary, uranium-rich eutectics of uranium with manganese, iron and nickel. In the U-UMn_2-Th system, four-phase eutectic equilibrium exists at 730°C; there are also two four-phase peritectic equilibria, at 775°C and 740°C, and a peritectoid and a eutectoid equilibrium. The zone of stratification in the liquid phase spreads to the ternary system at manganese concentrations of up to approximately 18 at.%Mn. In the U-UFe_2-Th system, four-phase eutectic equilibrium is established at 720°C; there are two peritectic equilibria, at 805°C and 800°C; in the solid phase two eutectoid equilibria are observed. There is no zone of stratification in the liquid phase, even at 2 at.%Fe. In the system U-U_7Ni_9-ThNi-Th, four-phase eutectic equilibrium exists at 735°C. In this system there are also two monotectoid four-phase equilibria, at 1650°C and 1050°C, and five peritectic equilibria, at 1040°C, 1020°C, 800°C, 785°C and 780°C; in the solid phase, eutectoid conversions are observed. The zone of stratification in the liquid phase occupies a significant part of the system concentration range studied by the authors, reaching 36 at.%Ni and 64 at.%Th.

ДИАГРАММЫ СОСТОЯНИЯ ТРОЙНЫХ СИСТЕМ U-Th-Mn, Fe, Ni В ЧАСТИ, БОГАТОЙ U И Th.

В докладе излагаются результаты экспериментальных исследований диаграмм состояния указанных систем. Строение сплавов было изучено методами термического и микроструктурного анализа сплавов, а также измерения твердости и микротвердости. Построены проекции поверхностей ликвидуса и диаграмм состояния, составлены схемы реакций. Характерной чертой изученных систем является наличие областей расслоения в жидком состоянии, распространяющихся в тройные системы от двойной системы U-Th, а также четырехфазных эвтектических равновесий вблизи двойных эвтектик урана с Mn, Fe, Ni, богатых ураном. В системе U-UMn_2-Th четырехфазное эвтектическое равновесие существует при 730°C; имеются также 2 четырехфазных перитектических равновесия при 775 и 740°C, а также перитектоидное и эвтектоидное равновесие. Область расслоения в жидком состоянии распространяется в тройную систему до ~18 ат% Mn. В системе U-UFe_2-Th четырехфазное эвтектическое равновесие установлено при 720°C; 2 перитектических равновесия существуют при 805 и 800°C; в твердом состоянии наблюдаются 2 эвтектоидных равновесия. Область расслоения в жидком состоянии не существует уже при 2 ат% Fe. В системе U-U_7-Ni_9 -ThNi-Th четырехфазное эвтектическое равновесие существует при 735°C. В системе установлены также 2 монотектоидных четырехфазных равновесия при 1650°C и 1050°C, 5 перитектических равновесий при 1040, 1020, 800, 785 и 780°C; в твердом состоянии наблюдаются эвтектоидные превращения. Область расслоения в жидком состоянии занимает значительную часть изученной области концентраций системы, достигая 36 ат% Ni и 64 ат% Th.

В докладе излагаются результаты экспериментальных исследований диаграмм состояния тройных систем U – Th – Mn, Fe, Ni в части, богатой U и Th. В качестве исходных данных по строению двойных систем, ограничивающих тройные изученные системы, были использованы диаграммы состояния, опубликованные в литературе: U – Th [1]; U – Mn [2]; U – Fe [3], [4]; U – Ni [5],[6]; Th – Ni [7]. Критические температуры двойных систем приводятся по результатам собственных исследований, полученных при записи кривых нагрева сплавов в отожженном состоянии. Строение сплавов было изучено методами термического и микроструктурного анализа, а также измерения твердости и микротвердости. Исходными материалами для приготовления сплавов служили компактные-уран (99,8%), торий (99,6%), электролитические-марганец (99,98%), никель (99,97%) и Армко железо. Сплавы готовили прямым сплавлением исходных металлов в дуговой печи на медном водоохлаждаемом поддоне в атмосфере химически чистого аргона. Предварительно торий и марганец переплавляли в тех же условиях. Для получения однородных слитков сплавы плавили 4 раза с переворачиванием слитков после каждого расплавления. Сплавы, вес которых после плавки изменился по сравнению с шихтовым, были подвергнуты химическому анализу, состав остальных сплавов принимался по шихте. Гомогенизирующий отжиг всех сплавов проводили при 600° в течение 12 суток, после отжига сплавы охлаждали вместе с печью.

Термический анализ отожженных сплавов с записью кривых нагрева и охлаждения проводили в специальной вакуумной печи с вольфрамовым нагревателем в атмосфере аргона чистоты, содержащим не более 0,005% N и 0,001% O ; образцы весом около 3г помещали в тигли из окиси бериллия. Запись термограмм осуществляли с помощью пирометра Курнакова типа ПК-52 с использованием вольфрам-рениевой термопары; скорость нагрева составляла ~ 40 град/мин; градуировочная кривая была построена по температурам плавления железа, никеля, меди и алюминия; расчет критических температур сплавов производили по кривым нагрева.

Закалку образцов с высоких температур проводили в печи для термического анализа после выдержки в течение 1 часа с выключением нагрева, при этом скорость охлаждения составляла ~ 200 град/мин.

Микроструктуру сплавов выявляли электролитическим травлением в электролите состава: 50 мл H_3PO_4 + 80 мл. C_2H_5OH + 50 мл. $CH_2OH \cdot CH_2OH$. Твердость сплавов мерили на приборе типа ТП алмазной пирамидкой с нагрузкой 10 кг, микротвердость - на приборе ПМТ с нагрузкой 50г.

В результате исследований построены проекции поверхностей ликвидуса и диаграмм состояния, составлены схемы реакций изученных систем.

Характерной чертой всех изученных систем является наличие четырехфазных низкотемпературных эвтектических равновесий вблизи двойных эвтектик урана с Mn , Fe , Ni .

При изучении данных термического анализа и микроструктуры сплавов в закаленном состоянии и после термического анализа сплавов системы U – Mn – Th было обнаружено, что между UMn_2 и Th существует квазибинарный разрез эвтектического характера. Эвтектическая точка отвечает ~ 42ат.% Th при ~ 940°С. В сплавах вблизи этого разреза первично кристаллизуется UMn_2 или Th в зависимости от состава, вторично-эвтектика (UMn_2 + Th).

В системе $U - Mn - Th$ при высоких температурах установлена область несмешиваемости в жидком состоянии, исходящая из двойной системы $U - Th$ и распространяющаяся в тройную до ~ 18 ат.% Mn. Границы распространения этой области определялись микроструктурным методом сплавов, закаленных в интервале температур 1200-1400°С. В сплавах, составы которых находятся в области akb перед первичной кристаллизацией β_{Th} происходит расслоение жидкости на два слоя: один - богатый ураном другой - торием.

На рис.1 представлена проекция поверхности ликвидуса диаграммы состояния $U - Mn - Th$ в области концентраций $U - UMn_2 - Th$. Поверхность ликвидуса имеет 5 областей первичной кристаллизации фаз: γ_U, β_U, U_6Mn, UMn_2, β_{Th}. В системе $U - UMn_2 - Th$ полученные данные позволили построить проекцию диаграммы состояния и составить схему реакций (рис.2 а,б).

В изученной области концентраций тройной системы $U - Mn - Th$ установлено два четырехфазных перитектических и одно четырехфазное эвтектическое равновесия. Трехфазные равновесия, исходящие из двойной системы $U - Th$, $ж \rightleftharpoons \gamma_U + \alpha_{Th}$ и $\gamma_U \rightleftharpoons \beta_U + \alpha_{Th}$ встречаясь в точке , при 775°С образуют перитектическое равновесие $\gamma_U + \alpha_{Th} \rightleftharpoons \beta_U + ж$. С понижением температуры в тройную систему отходит равновесие $ж \rightleftharpoons \beta_U + \alpha_{Th}$, которое, накладываясь с равновесием $ж + \beta_U \rightleftharpoons U_6Mn$, образует перитектическое четырехфазное равновесие $ж + \beta_U \rightleftharpoons U_6Mn + \alpha_{Th}$ при 740°. При 730° в изученной области концентраций установлено четырехфазное эвтектическое равновесие $ж \rightleftharpoons U_6Mn + UMn_2 + \alpha_{Th}$ в результате наложения на эвтектическое

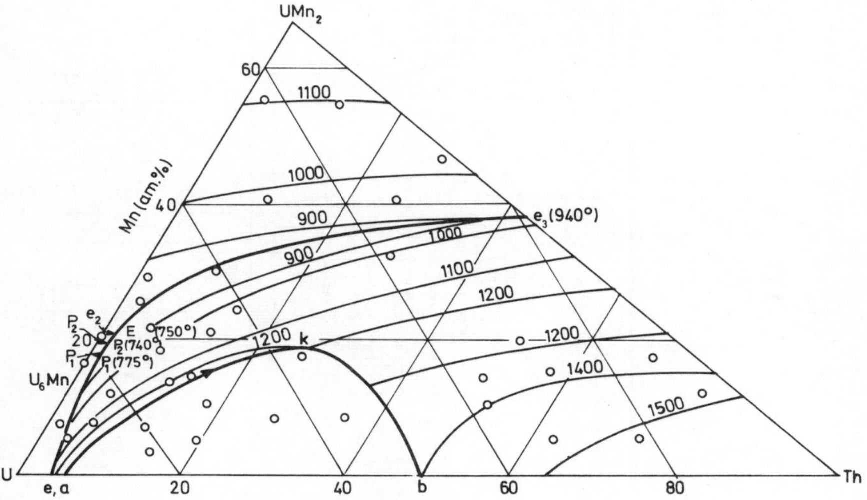

Рис.1. Проекция поверхности ликвидуса диаграммы состояния системы U-UMn_2-Th.

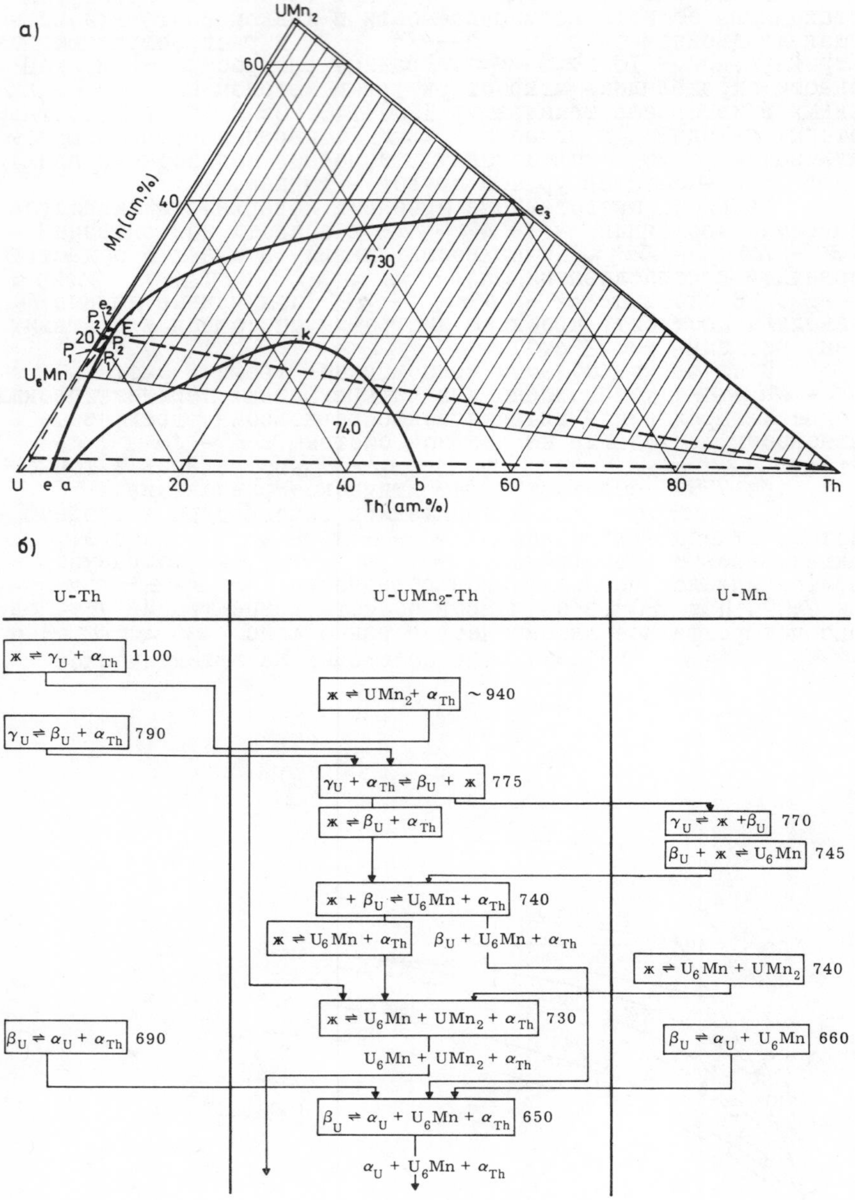

Рис.2. Проекция диаграммы состояния (а) и схема реакций (б) системы $U\text{-}UMn_2\text{-}Th$.

равновесие $Ж \rightleftharpoons U_6Mn + UMn_2$, исходящее из двойной системы $U - Mn$ двух равновесий $Ж \rightleftharpoons U_6Mn + \alpha_{Th}$ и $Ж \rightleftharpoons UMn_2 + \alpha_{Th}$. Температуры этих равновесий снижаются от P_2 и e_3 к E. Плоскость этого четырехфазного равновесия представлена заштрихованным треугольником U_6Mn – UMn_2 – Th. Все сплавы этого треугольника заканчивают кристаллизацию при 730°С.

В результате изучения температур превращений в сплавах системы $U - Fe - Th$ при нагревании и микроструктуры этих сплавов в закаленном состоянии и после термического анализа, установлено наличие квазибинарного разреза UFe_2 – Th, в котором по перитектической реакции $Ж + \alpha_{Th} \rightleftharpoons T$ при~1045° образуется тройная фаза Т, имеющая некоторую протяженность по составу. Фаза Т находится в эвтектическом равновесии с UFe_2 при ~965°С; эвтектическая точка отвечает 33 ат.% Th.

Область расслоения в системе, изученная методом микроструктурного анализа сплавов, закаленных в интервале температур 1200-1500°С, не была обнаружена в сплавах содержащих уже 2 ат.% Fe.

На рис.3 представлена проекция поверхности ликвидуса системы U - Fe - Th. Поверхность ликвидуса имеет 5 областей первичной кристаллизации фаз: γ_U, U_6Fe, UFe_2, Т, β_{Th}, разделенных линиями моновариантных равновесий, температуры которых снижаются от двойных в тройную систему с образованием критических точек четырехфазных равновесий.

Как показано на рис.4 а,б, где представлены проекция диаграммы состояния и схема реакций, перитектическое и эвтектическое равновесия в двойных системах $U - Fe$ и $U - Th$, существующие при 820 и 1100°С, соответственно, встречаясь в тройной системе образуют четырехфазное перитек-

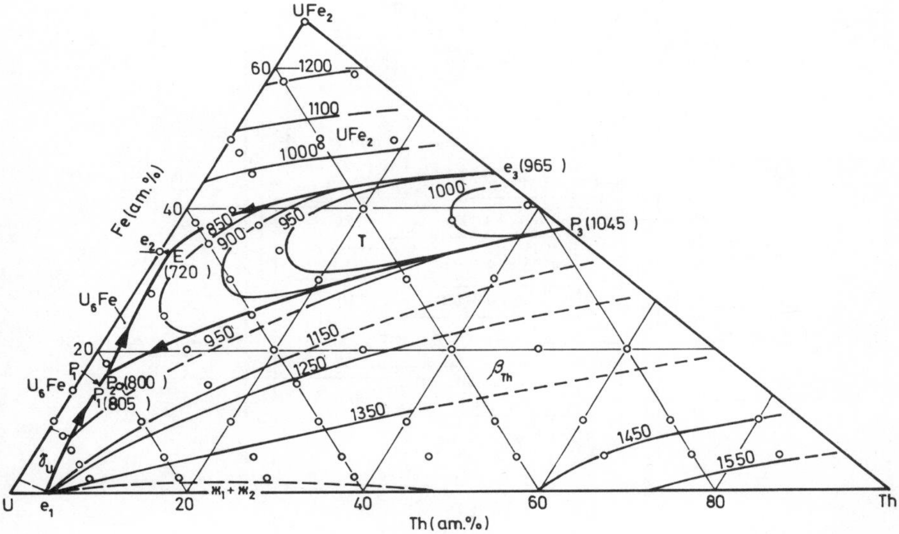

Рис.3. Проекция поверхности ликвидуса диаграммы состояния системы U-UFe_2-Th.

а)

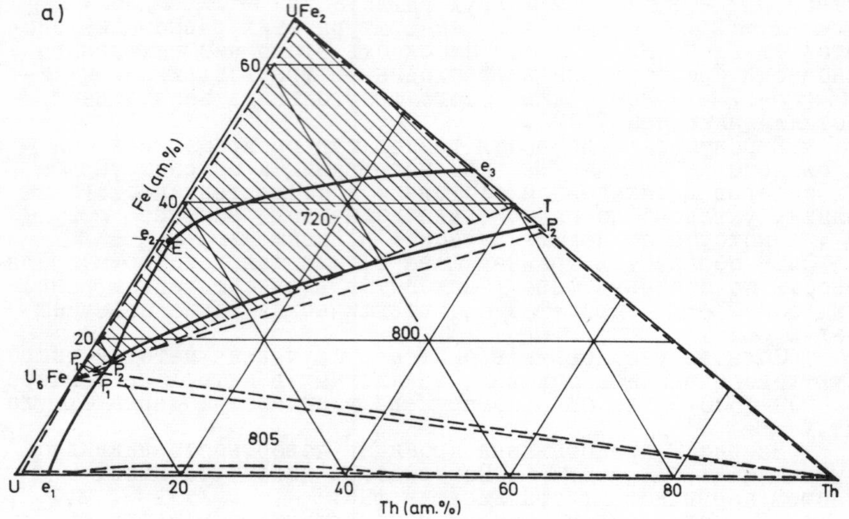

б)

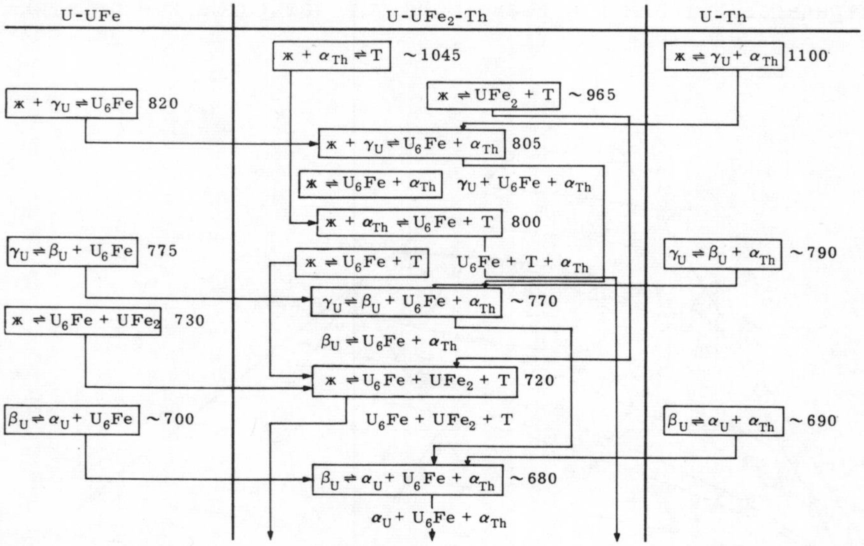

Рис.4. Проекция диаграммы состояния (а) и схема реакций (б) системы U-UFe_2–Th.

тическое равновесие ж + $\gamma_U \rightleftharpoons U_6Fe + \alpha_{Th}$ при 805°C; плоскость этого четырехфазного равновесия очерчена на рис.6а пунктирными линиями $U - U_6Fe - P_1 - Th$. При понижении температуры от этого равновесия отходит равновесие ж $\rightleftharpoons U_6Fe + \alpha_{Th}$, которое, встречаясь с перитектическим равновесием, исходящим из квазибинарного разреза $UFe_2 - Th$ при ~1045°C, образует четырехфазное перитектическое равновесие ж + $\alpha_{Th} \rightleftharpoons U_6Fe + T$ при 800°C; плоскость его очерчена пунктирными линиями $U_6Fe - P_2 - T - Th$. При 720°C в тройной системе устанавливается четырехфазное эвтектическое равновесие ж $\rightleftharpoons U_6Fe + UFe_2 + T$, как результат наложения на равновесие ж $\rightleftharpoons U_6Fe + T$ эвтектических равновесий, исходящих из двойных систем $U - UFe_2$ ж $\rightleftharpoons U_6Fe + UFe_2$ при 730°C и $UFe_2 - Th$ ж $\rightleftharpoons UFe_2 + T$ при ~965°C. Плоскость этого четырехфазного равновесия представлена на рис.4а заштрихованным треугольником $U_6Fe - UFe_2 - T$ Все сплавы, составы которых находятся в этом треугольнике, заканчивают свою кристаллизацию при 720°C.

В тройной системе $U - Ni - Th$ при высоких температурах установлена обширная область несмешиваемости в жидком состоянии, исходящая из двойной системы $U - Th$ и распространяющаяся в тройную систему до 36 ат.% Ni и 64 ат.% Th. Границы этой области были определены в результате изучения микроструктуры сплавов после термического анализа и закалки сплавов в интервале температур 1200-1500°C. На рис.5 приведена проекция поверхности ликвидуса системы $U - Ni - Th$ в области концентраций $U - U_7Ni_9 - ThNi - Th$. Поверхность ликвидуса изученной части диаграммы имеет 8 областей первичной кристаллизации фаз: γ_U, U_6Ni, U_7Ni_9, U_5Ni_7, UNi_2, $ThNi$, Th_7Ni_3, β_{Th}. В сплавах, находящихся в области akb, перед первичной кристаллизацией

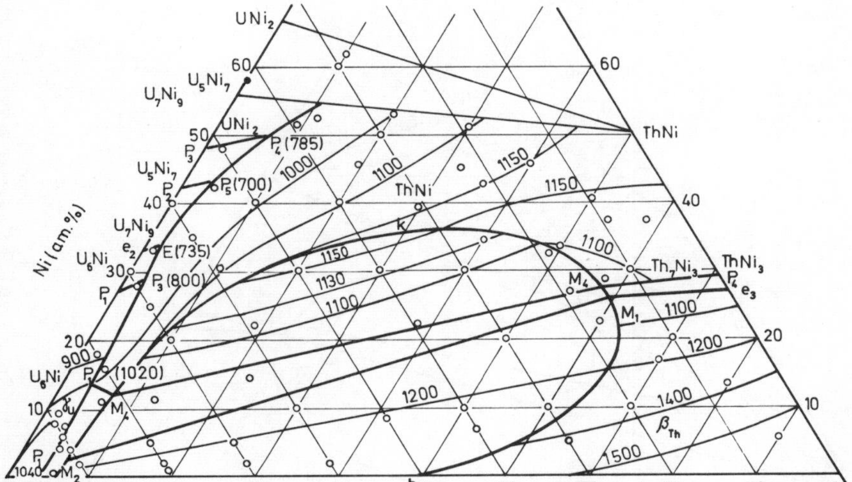

Рис.5. Проекция поверхности ликвидуса диаграммы состояния системы $U-U_7Ni_9-ThNi-Th$.

происходит разделение сплава на две жидкости. На микроструктурах этих сплавов после закалки с высоких температур и после термического анализа наблюдается расслоение.

Как видно на рис.6 а,б, где представлены проекция диаграммы состояния и схема реакций, в изученной области концентраций установлено наличие 8 четырехфазных равновесий, из которых одно эвтектическое, два монотектических и пять перитектических реакций. Точки монотектических равновесий $ж_1 + \alpha_{Th} \rightleftharpoons ж_2 + Th_7Ni_3$ (M_1, M_2) и $ж_3 \rightleftharpoons ж_4 + ThNi + Th_7Ni_3$ (M_3, M_4) находятся на пересечении моновариантных линий с областью несмешиваемости. Прямые M_IM_2 и M_3M_4 являются изотермами, соединяющими обе жидкие фазы, находящиеся в равновесии при 1065 и 1050°С. От точки M_2 и с понижением температуры отходит равновесие $ж_2 \rightleftharpoons Th_7Ni_3 + \alpha_{Th}$, которое встречается с равновесием $ж \rightleftharpoons \gamma_U + \alpha_{Th}$ и при 1040°С образует четырехфазное перитектическое равновесие $ж_2 + \alpha_{Th} \rightleftharpoons \gamma_U + Th_7Ni_3$ в точке P_I. От этого равновесия с понижением температуры отходит равновесие $ж_2 \rightleftharpoons \gamma_U + Th_7Ni_3$. Это равновесие накладывается на равновесие $ж_4 \rightleftharpoons ThNi + Th_7Ni_3$ и при 1020° в точке P_2 устанавливается четырехфазное перитектическое равновесие $ж + Th_7Ni_3 \rightleftharpoons \gamma_U + ThNi$ от которого отходит с понижением температуры равновесие $ж \rightleftharpoons \gamma_U + ThNi$ и при температуре 800°С в точке P_3 образует четырехфазное перитектическое равновесие $ж + \gamma_U + ThNi \rightleftharpoons U_6Ni$. При 785°С в точке P_4 в тройной системе установлено четырехфазное перитектическое равновесие $ж + UNi_2 \rightleftharpoons U_5Ni_7 + ThNi$, от него отходит трехфазное равновесие $ж \rightleftharpoons U_5Ni_7 + ThNi$, которое, встречаясь с перитектическим трехфазным равновесием $ж + U_5Ni_7 \rightleftharpoons U_7Ni_9$, исходящим из двойной системы $U - Ni$, образует четырехфазное перитектическое равновесие $ж + U_5Ni_7 \rightleftharpoons U_7Ni_9 + ThNi$ при 780°С в точке P_5. При наложении на эвтектическое равновесие $ж \rightleftharpoons U_6Ni + U_7Ni_9$, исходящее из двойной системы $U - Ni$, двух трехфазных равновесий $ж \rightleftharpoons U_6Ni + ThNi$ и $ж \rightleftharpoons U_7Ni_9 + ThNi$,

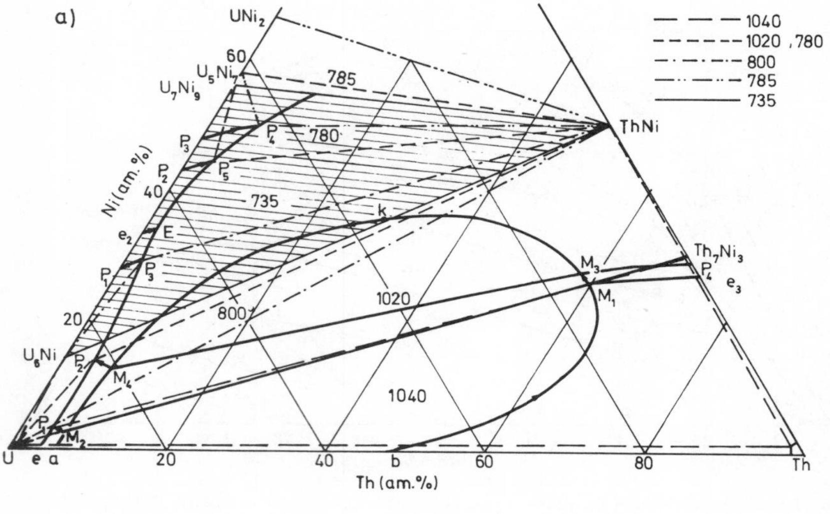

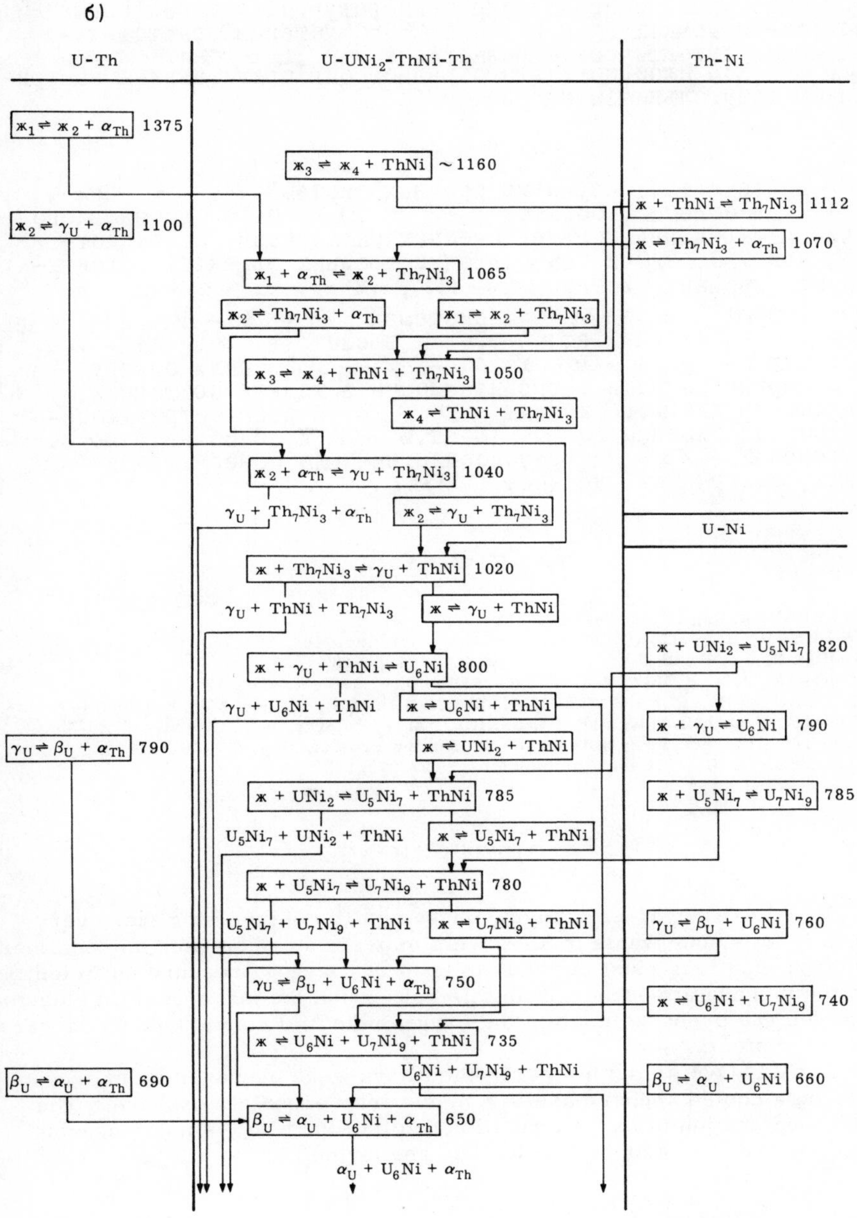

Рис.6. Проекция диаграммы состояния (а) и схема реакций (б) системы $U-UNi_2-ThNi-Th$.

которые протекают с понижением температур от четырехфазных равновесий в точках P_3 и P_5, при 735°С устанавливается четырехфазное эвтектическое равновесие ж ⇌ $U_6Ni + U_7Ni_9$ + + $ThNi$ (Е). Плоскость этого равновесия представлена заштрихованным треугольником на рис.6а.

Выводы

В результате изучения тройных систем U - Th - Mn, Fe, Ni в области, богатой U и Th, было установлено наличие четырехфазных эвтектических равновесий, соответственно, при 730, 720 и 735°С. Точки тройных эвтектик находятся вблизи двойных систем U - Mn, U - Fe, U - Ni и соответствуют I ат.% Th в системах U - Th - Mn и U - Th - Fe и 2 ат.% Th в системе U - Th - Ni. В системах U - Th - Mn и U - Th - Ni были обнаружены обширные области несмешиваемости в жидком состоянии, исходящие из двойной системы U - Th и распространяющиеся в тройные системы до ~ I8 ат.% Mn и до 36 ат.% Ni. В системе U - Th - Fe расслоение не было обнаружено в сплавах, содержащих уже 2 ат.% Fe.

ЛИТЕРАТУРА

[1] MURRAY, J.R., J. Inst. Met. 87 3 (1958) 94.
[2] WILHELM, H.A., CARLSON, O.N., Trans. Am. Soc. Met. 42 (1950) 1311.
[3] GORDON, P., KAUFMAN, A.R., J. Met. 2 1 (1950) 182.
[4] GROGAN, J.D., J. Inst. Met. 77 (1950) 571.
[5] GROGAN, J.D., PLEASANCE, R.J., (B.E. WILLIAMS), J. Inst. Met. 82 (1953) 141.
[6] FOOTE, F.C., CLARK, J.R., CIESLICKI, M., NELSON, B.J., LANE, T.R., The Tuballey-Nickel Phase Diagram, Manhatten Project. Rep. CT-3013 (1945).
[7] THOMSON, J.R., J.Less-Common Met. 29 2 (2972) 183.

DISCUSSION

C. B. ALCOCK: When studying phase equilibria involving these very reactive metals, one wants to know what materials can be used as containers, and what precautions must be taken to keep the oxygen pressure sufficiently low in the furnace atmosphere. In this connection the use of a solid electrolyte oxygen gas probe to monitor the oxygen potential should help to characterize the system.

O.S. IVANOV: In our research the alloys were prepared in an arc furnace on a cooled copper base in a high-purity argon atmosphere. The possible penetration of oxygen and nitrogen into the alloy is not dangerous since ThO_2 and UO_2 with high solubility are formed in the alloy.

MODIFICATIONS D'EQUILIBRE DE PHASES DANS LE SYSTEME UO_2-SiO_2 PAR IRRADIATIONS A TAUX DE COMBUSTION ELEVES

S. LUNGU, O. RADULESCU
Institut de physique atomique,
Bucarest,
Roumanie

Abstract–Résumé

PHASE EQUILIBRIUM CHANGES IN THE SYSTEM UO_2-SiO_2 DUE TO IRRADIATION UNDER CONDITIONS OF HIGH BURN-UP.

Samples of UO_2-SiO_2 irradiated with a burn-up of 20 000 MWd/t at 200 - 1500°C were examined microscopically and observations made of the following: 1. changes in the dimensions and the volume fraction of the UO_2 crystals and in the structure of the "spinodal mass"; 2. fission gas precipitation; 3. phase stability during irradiation. The growth of the grains and the smallness of the variation in the dimensions of the crystals and the fission gas bubbles suggest irradiation-induced diffusion. The single jump diffusion model was used to explain the value of the diffusion coefficients. The decreasing solubility of the fission gases with temperature may explain the threshold for the appearance of the fission gas bubbles. The possibility is noted of swelling which decreases with increasing hydrostatic pressure.

MODIFICATIONS D'EQUILIBRE DE PHASES DANS LE SYSTEME UO_2-SiO_2 PAR IRRADIATIONS A TAUX DE COMBUSTION ELEVES.

Des examens microscopiques ont été faits sur des échantillons de UO_2-SiO_2 irradiés à 20 000 MWj/t et 200 à 1500°C. On étudie les changements observés concernant: 1) la modification des dimensions et de la fraction volumique des cristaux de UO_2 et celle de la structure de la «masse spinodale»; 2) la précipitation des gaz de fission; 3) la stabilité des phases pendant l'irradiation. La croissance des grains et la faible variation des dimensions des cristaux et des bulles de gaz de fission indiquent un processus de diffusion induit par l'irradiation. On a utilisé le modèle «single jump diffusion» pour expliquer la valeur des coefficients de diffusion. La diminution de la solubilité des gaz de fission avec la température peut expliquer le seuil d'apparition des bulles de gaz de fission. On constate la possibilité d'avoir un gonflement qui diminue avec l'augmentation de la pression hydrostatique.

INTRODUCTION

Les échantillons de UO_2-SiO_2 (teneur en UO_2 jusqu'à 80%) sont obtenus par fusion à 2300-2400°C et injection dans des moules métalliques fusibles.

La structure obtenue de cette manière est biphasée: cristaux de UO_2 + verre ($SiO_2 > 99$ mol% $UO_2 < 1$ mol%). La phase cristalline se trouve sous deux formes en raison du mode d'élaboration: cristaux «primaires» (> 1 μm) formés par nucléation homogène dans la masse fondue au passage de la température du liquidus et cristaux «secondaires» (< 1 μm) formés après la décomposition spinodale (fig. 1). Une description de la succession des processus et de l'influence de la vitesse de refroidissement a déjà été présentée [1]. La phase vitreuse SiO_2, formée à la suite de la décomposition spinodale, est mêlée aux cristaux secondaires de UO_2, ce qui constitue la «masse spinodale». A une très grande vitesse de refroidissement (cas

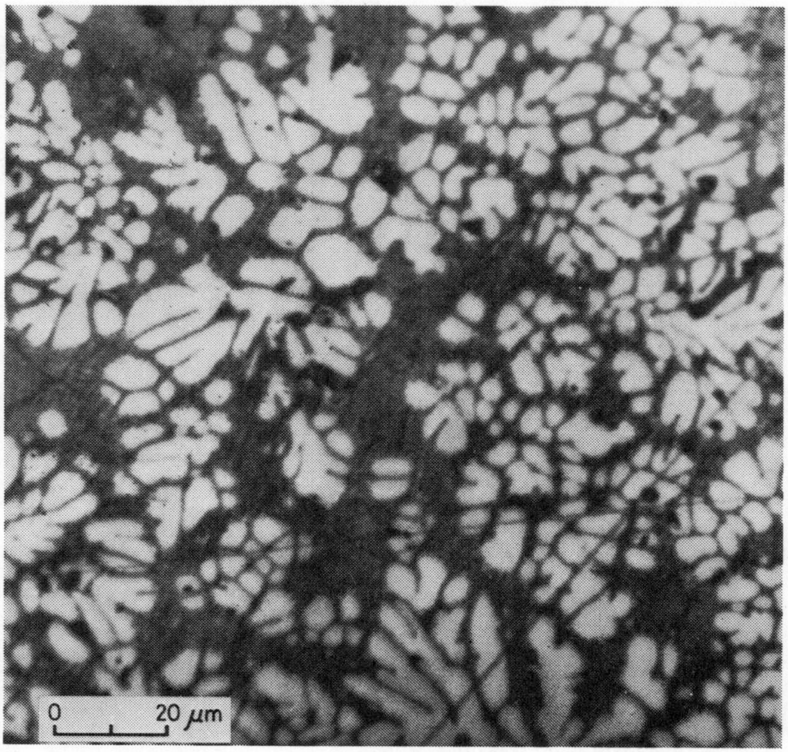

FIG. 1. Microphotographie d'une zone centrale d'un crayon de UO_2-SiO_2 avant irradiation; partie claire: cristaux primaires de UO_2; partie foncée: «masse spinodale».

des couches périphériques des crayons obtenus par injection), les dimensions des cristaux primaires sont inférieures à 1 μm. On ne peut alors plus les distinguer de ceux formés par la décomposition spinodale (fig. 2).

Les essais d'irradiation aux températures de 200 à 1500°C ont permis de suivre, dans des capsules instrumentées, l'évolution des déformations axiales ϵ_z en fonction du taux d'irradiation sous diverses charges axiales. Les résultats obtenus [2] ont permis d'aboutir à une relation qui donne la variation avec la température du taux d'irradiation limite d'apparition du gonflement sous une pression hydrostatique de 1 atm(abs) et une charge axiale de 0,4 kgf/cm^2.

Dans le présent mémoire, sur la base des examens microscopiques après irradiation, les changements observés concernant: 1) la variation des dimensions et de la fraction volumique des cristaux primaires de UO_2 ainsi que la modification de la structure de la «masse spinodale»; 2) la précipitation des gaz de fission; 3) la stabilité des phases, sont présentés.

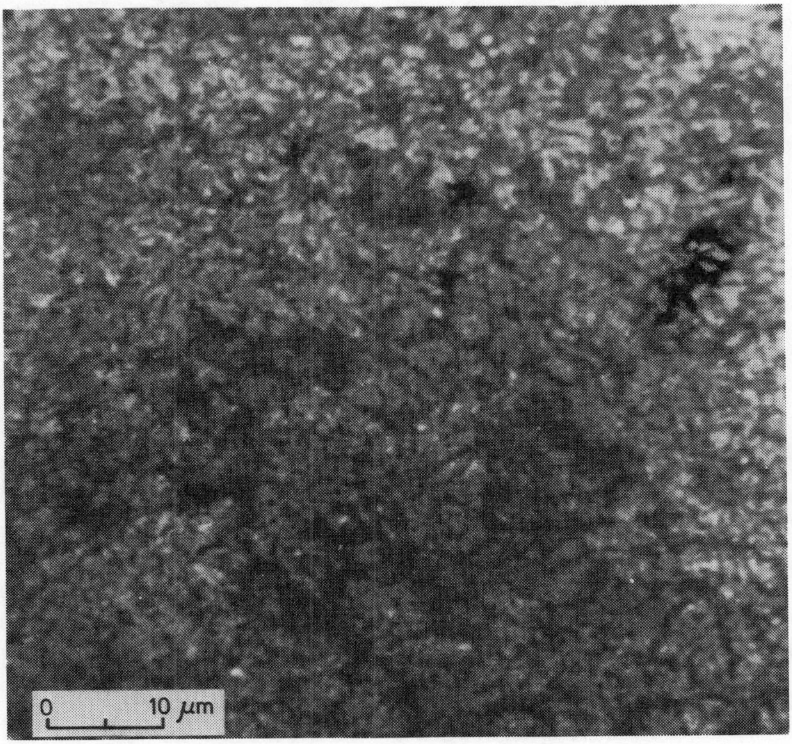

FIG. 2. Microphotographie d'une zone périphérique d'un crayon de UO_2-SiO_2 avant irradiation.

1. MODIFICATION DE LA DISTRIBUTION DES PHASES PENDANT L'IRRADIATION

L'examen microscopique a été fait sur des échantillons irradiés à des températures entre 200 et 1500°C, jusqu'à 20 000 MWj/t. On a examiné des régions qui, avant irradiation, avaient des structures microscopiques différentes. La figure 3 représente une microphotographie typique d'une zone centrale irradiée à 820°C dont la structure avant irradiation est celle de la figure 1; la figure 4 montre une microphotographie typique d'une zone périphérique irradiée à 650°C dont la structure avant irradiation est caractérisée par des cristaux primaires beaucoup plus fins (voir fig. 2).

Les résultats de l'examen microscopique ont conduit principalement aux observations suivantes:

— Grossissement des grains primaires de UO_2 indépendant de leur distribution de taille avant irradiation (deux premières colonnes du tableau I), ainsi que de la température. Ceci implique des coefficients de diffusion de l'uranium et du silicium importants et quasi indépendants de la température. En conséquence, des processus induits par l'irradiation

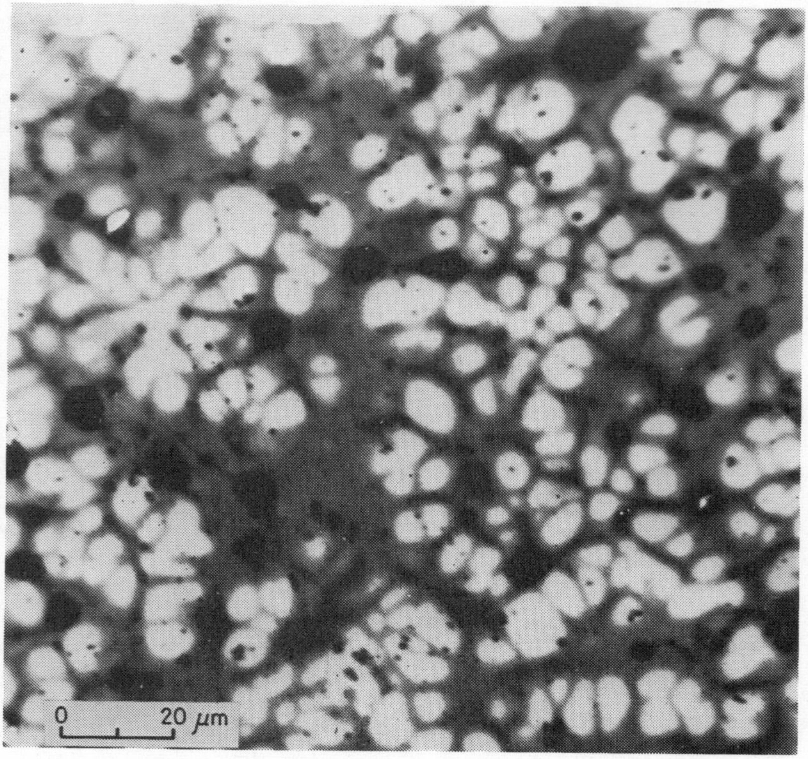

FIG. 3. Microphotographie d'une zone centrale d'un crayon de UO_2-SiO_2 irradiée à 820°C; partie noire: bulles de gaz de fission.

sont certainement mis en cause. Cette croissance du grain est beaucoup plus spectaculaire dans la région périphérique de l'échantillon.

— Augmentation de la fraction volumique des cristaux de $UO_2 > 1\ \mu m$ (à partir des cristaux primaires d'origine) aux dépens des cristaux inférieurs à 1 μm existant dans la «masse spinodale» (troisième colonne du tableau I.A).

— Modification de la structure de la «masse spinodale» indiquée par l'aspect vitreux de celle-ci, ainsi que par la délimitation beaucoup moins nette entre les cristaux primaires et la «masse spinodale» provoquée par une éventuelle transparence de cette dernière. Cette transparence peut exagérer à l'examen la croissance des grains et l'augmentation de la fraction volumique des cristaux primaires.

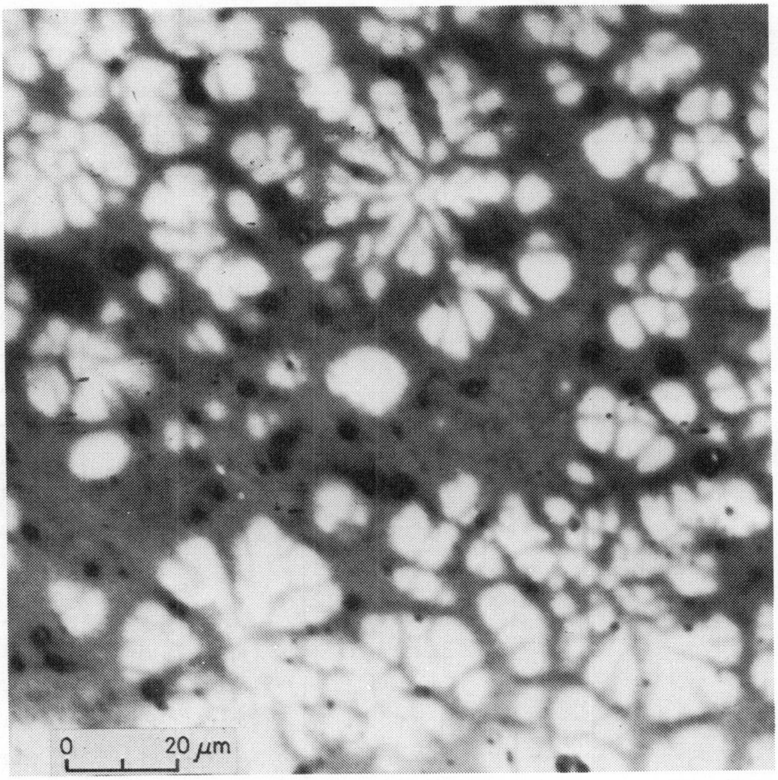

FIG. 4. Microphotographie d'une zone périphérique d'un crayon de UO_2-SiO_2 irradiée à 650°C.

2. PRECIPITATION DES GAZ DE FISSION

L'examen après irradiation concernant les gaz de fission a conduit aux données de la dernière colonne du tableau I, représentées dans la figure 5. Les gaz de fission précipitent de façon importante à une température dépassant 650°C; celle-ci correspond à la température d'apparition du gonflement comme l'ont montré des mesures de déformations axiales rapportées par Lungu [2] (voir fig.6). Dans la figure 7, les distributions réelles du diamètre des bulles calculées à partir des distributions apparentes sont indiquées. Toutes ces distributions présentent deux maximums correspondant, le premier, aux petites bulles dont le développement a été limité dans les zones étroites de «masse spinodale», situées entre les cristaux de UO_2 et, le second, aux bulles formées dans les zones plus étendues de «masse spinodale».

TABLEAU I. RESULTATS DE L'ANALYSE DE LA MICROSTRUCTURE DES ECHANTILLONS IRRADIES OU NON

Température d'irradiation (en °C)	$(\bar{L}_3)\,UO_2$ prim.[a] (en μm)	$(V_V)\,UO_2$ prim.[b]	$(V_V)\,g$[c]
A. Zone centrale			
non irradié	4,1	0,43 ± 0,03	-
410	6,6	0,54 ± 0,05	-
470	6,7	0,57 ± 0,05	-
530	6,7	0,59 ± 0,04	-
580	6,5	0,51 ± 0,05	-
630	5,6	0,52 ± 0,04	-
750	6,5	0,52[d] ± 0,04	0,046
820	6,7	0,48[d] ± 0,03	0,072
1200	5,8	0,51[d] ± 0,06	0,424
B. Zone périphérique			
non irradié	0,3	?	-
230	4,2	0,35 ± 0,03	-
270	5,3	0,49 ± 0,04	-
490	3,9	0,41 ± 0,03	-
650	5,3	0,53[d] ± 0,05	0,027
1050	6,4	0,46[d] ± 0,04	0,118
	moyenne	0,45	

[a] «Mean intercept value» pour les cristaux primaires de UO_2 (distance moyenne entre 2 intersections correspondant aux limites de la phase considérée).
[b] Fraction volumique des cristaux primaires.
[c] Fraction volumique des gaz.
[d] On ne considère que les phases solides.

3. STABILITE DES PHASES PENDANT L'IRRADIATION

Dans les échantillons de UO_2-SiO_2 on observe hors pile, à une température de 1150°C, après une demi-heure, une cristallisation du SiO_2 sous forme de cristobalite. Celle-ci se manifeste par une fragilisation très avancée due à la transformation allotropique de la cristobalite à 250°C. Cette durée de recuit peut être reliée à la vitesse de cristallisation du SiO_2 pur massif. Ceci, en considérant le processus de nucléation de la cristobalite assez rapide à l'interface cristal UO_2-verre et aussi le $(\bar{L}_3)^1$ du verre qui est de l'ordre de grandeur de quelques dixièmes de micron. On

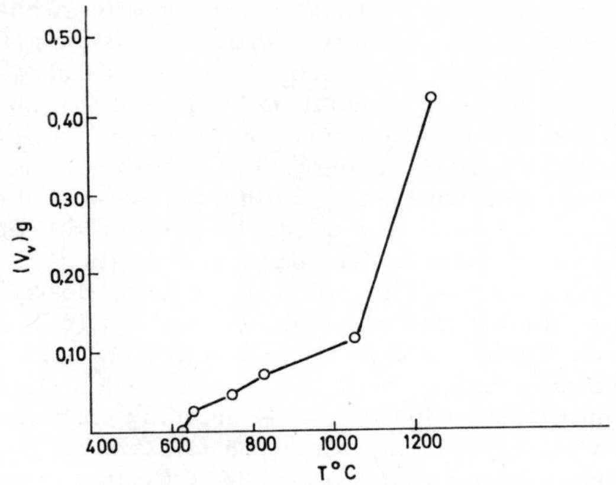

FIG. 5. Fraction volumique de gaz de fission $\big((V_V)g\big)$, en fonction de la température.

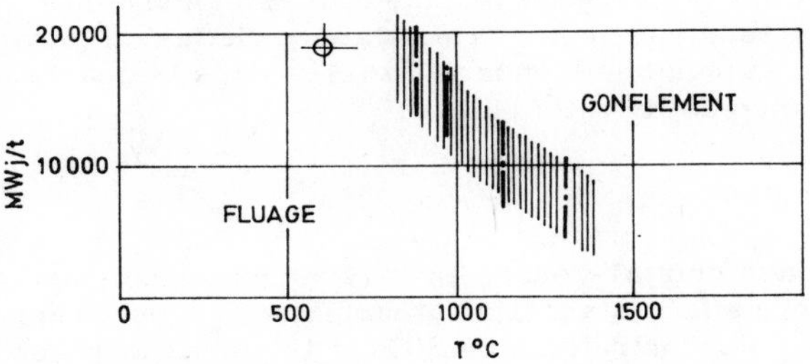

FIG. 6. Taux d'apparition du gonflement en fonction de la température. -⊕-: indique le point correspondant aux résultats du présent mémoire.

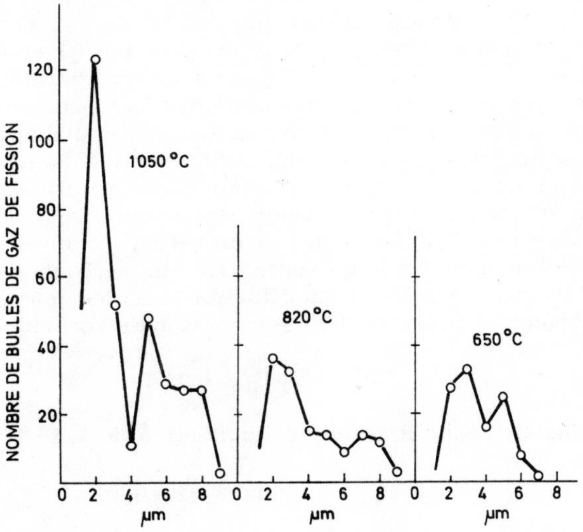

FIG. 7. Distribution de diamètre des bulles de gaz de fission à trois températures différentes.

n'observe pas la présence de cristaux de cristobalite sur les microphotographies des zones irradiées à des températures élevées (1050 à 1200°C). De plus, les enregistrements de déformations axiales au début de l'irradiation (quelques dizaines d'heures d'irradiation [2]) montrent un comportement pratiquement indépendant de la température dans une gamme de 800 à 1500°C, avec le maintien de l'intégrité. Les échantillons, retrouvés intacts après le cyclage thermique dû au démarrage et à l'arrêt du réacteur, confirment la conclusion suggérée par les microphotographies concernant l'absence de cristobalite, donc de fragilisation.

On ne peut pas expliquer ce fait par la destruction des noyaux de cristobalite par les fragments de fission, comme dans le cas de la redissolution des bulles de gaz de fission dans le UO_2. En effet, le caractère statistique du processus implique l'existence d'une concentration stationnaire de germes qui peuvent croître sous forme de cristaux suffisamment stables de cristobalite pendant l'irradiation. Ceci, du fait que la vitesse de croissance dépasse, aux températures supérieures à 1000°C, la vitesse de redissolution. On peut expliquer le phénomène observé si on prend en considération la baisse éventuelle de la température de fusion de la cristobalite sous irradiation. Ceci est suggéré par les faibles valeurs de l'énergie stockée dans le SiO_2 vitreux [3] et la chaleur latente de fusion (environ 1 kcal/mol). L'absence des valeurs de l'énergie stockée dans le quartz nous empêche de calculer le niveau de cette baisse.

4. DISCUSSION

A l'interface cristal-verre, en raison des déplacements induits par l'irradiation et de la restauration athermique, on peut représenter la distribution des concentrations de SiO_2 et UO_2 dans la phase cristalline de UO_2 et dans la phase vitreuse comme cela est indiqué sur la figure 8.

En fonction des surfaces spécifiques, entre les régions occupées par des cristaux plus grands que 1 μm et les régions occupées par des cristaux de taille inférieure, l'apparition d'un gradient de concentration moyenne de UO_2 dans la phase vitreuse, et du SiO_2 dans la phase cristalline, est possible. L'accélération de la diffusion sous irradiation peut conduire à la croissance des grands cristaux aux dépens des petits cristaux.

Une croissance des cristaux par diffusion jusqu'à un rayon $R_1 (= \bar{L}_3/2) = 2\ \mu m$ (dans la zone périphérique de l'échantillon) pour une fraction volumique moyenne de 0,45 (voir tableau I.B), implique une couronne sphérique de « masse spinodale » comme source de UO_2 avec un rayon extérieur $R_2 = 2,6\ \mu m$. La concentration moyenne de UO_2 dans cette couronne correspondrait à une fraction volumique équivalente de 0,27. En utilisant l'approximation de la diffusion dans les plaques (jugée assez bonne pour la diffusion dans les couronnes sphériques), on obtient:

$$D_{eq} t / (R_2 - R_1)^2 = 0,25$$

valeur qui conduit, pour le temps d'irradiation de $1,3 \cdot 10^7$ s, à un coefficient de diffusion:

$$D_{eq} > 0,7 \cdot 10^{-16}\ cm^2/s$$

---

[1] Voir renvoi a du tableau I.

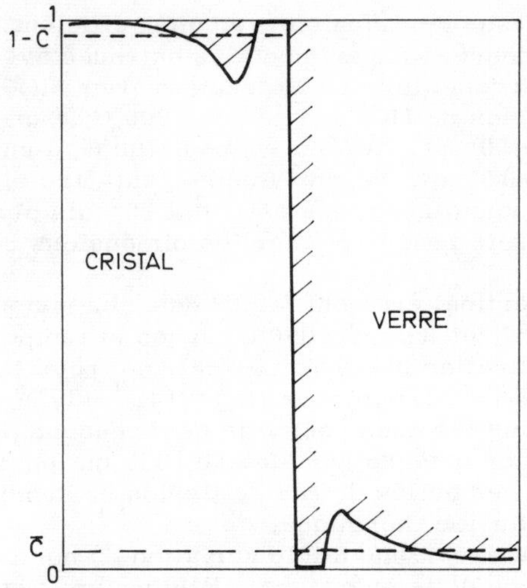

FIG. 8. Concentrations de UO_2 et SiO_2 à l'interface cristal-verre pendant l'irradiation.

Le coefficient de diffusion réel revient à:

$$D = D_{eq}/\overline{c} \gg 0,7 \cdot 10^{-16}\ cm^2/s$$

où $\overline{c}$ est la concentration moyenne de UO_2 dans le verre donnée dans la figure 8.

La contribution de l'irradiation à la diffusion est caractérisée par:

$$D = \frac{1}{3} \cdot \overline{\lambda}_d^2 \cdot \frac{1}{\tau} = 4 \cdot 10^{-18}\ cm^2/s$$

où $1/\tau$ est la fraction d'atomes déplacés par seconde et par fission et $\overline{\lambda}_d^2$ le parcours quadratique moyen des atomes déplacés ($\overline{\lambda}_d^2 = 10^{-13}\ cm^2$). Ce coefficient de diffusion est plus de 10 fois plus petit que le coefficient de diffusion équivalent ce qui implique que les parcours des atomes déplacés sont beaucoup plus grands. Cette possibilité ($\overline{\lambda}^2 > 10^{-11}\ cm^2$) a été suggérée par Matzke [4] pour la diffusion du xénon dans le verre SiO_2. Ce dernier suppose le dégagement d'un atome de xénon implanté à une profondeur de 270 Å, par un seul processus élémentaire d'activation thermique (single jump diffusion).

La modification de la fraction volumique des cristaux de UO_2 plus grands que 1 μm et l'augmentation de la transparence de la «masse spinodale» peuvent être le résultat direct de la croissance des grands cristaux aux dépens des cristaux dont la taille est inférieure à 1 μm.

L'existence d'un seuil de température, pour l'apparition des bulles de gaz de fission visibles au microscope, est un fait normal pour les combustibles solides. Mais pour le UO_2, ce seuil se situe au-delà de 1000°C et la formation des bulles se produit à la limite des grains. Le seuil de 650°C peut être considéré comme lié à la présence de la phase vitreuse dans la «masse spinodale». Dans ce cas, l'accumulation des gaz de fission nécessite un processus de diffusion avec un coefficient apparent de 10^{-16} à 10^{-15} cm^2/s. Les données expérimentales obtenues avec les traitements thermiques après irradiation par Matzke [4] pour SiO_2 correspondent à un coefficient de diffusion de 10^{-18} cm^2/s pour 900°C, à un taux d'irradiation correspondant à $8 \cdot 10^{10}$ at. Xe/cm^2 et, pour 700°C, à un taux d'irradiation correspondant à $4 \cdot 10^{12}$ at. Xe/cm^2 (même peut-être plus élevé). Cela suggère que la diffusion thermique est 10^2 à 10^3 fois plus faible que celle dont nous avons besoin pour expliquer les dimensions des bulles dans notre cas.

La variation particulièrement faible des dimensions des grandes bulles (5 μm à 650°C, et 8 μm à 1050°C), avec la température, correspondrait à une énergie d'activation d'environ 5 kcal/mol pour $(\overline{L}_3) \propto 1/D^2$ (diffusion en volume) et 10 à 13 kcal/mol pour $(\overline{L}_3) \propto 1/D^4 \div 1/D^5$ (migration de très petites bulles). Dans les deux cas, elle est beaucoup plus faible que l'énergie d'activation obtenue par Matzke [4], qui est de 74 ± 7 kcal/mol.

La croissance des bulles de gaz de fission ne semble donc pas être contrôlée par la diffusion thermique.

On observe également une faible variation du nombre de grandes bulles dans la même gamme de température. Elle pourrait être due à un coefficient de diffusion élevé.

Ces faits suggèrent comme probable un processus de diffusion similaire à celui proposé par Matzke dans son modèle « single jump diffusion ». On remplace alors l'activation thermique du saut par l'activation induite par l'irradiation, ceci comme dans l'explication donnée à propos du grossissement du grain et en prenant un déplacement moyen plus grand que 10^{-11} cm^2 au lieu de 10^{-13} cm^2. Dans ce modèle, le seuil de température d'apparition des bulles pourrait être expliqué comme dans [2] par l'existence d'une haute solubilité du gaz de fission dans le verre, solubilité qui diminue quand la température augmente, comme cela est montré dans la courbe de la figure 6.

Aux températures plus hautes (par exemple 1150°C), la diffusion thermique normale pourrait intervenir (avec $10^{-1} < D_0 < 10^0$ et $E = 74$ kcal/mol). De plus, à ces températures, le dégagement des gaz de fission des cristaux de UO_2 peut participer à l'augmentation du gaz qui précipite en bulles.

Ce modèle suggère une relation entre le gonflement et la pression hydrostatique conduisant à un déplacement proportionnel de la courbe de la figure 6, représentant le cas p = 1 atm(abs), vers les taux d'irradiation plus élevés pour les pressions caractéristiques du gaz de refroidissement dans les réacteurs de puissance (60 à 80 atm).

REFERENCES

[1] LUNGU, S., et al., J. Nucl. Mater. 48 (1973) 165-71.
[2] LUNGU, S., J. Nucl. Mater., à paraître.
[3] PRIMAK, W., Phys. Rev. 98 (1955) 1854.
[4] MATZKE, Hj., Phys. Status Solidi 18 (1966) 285.

ТЕРМОДИНАМИЧЕСКИЕ РАСЧЕТЫ ГРАНИЦ ФАЗОВЫХ РАВНОВЕСИЙ ДВУХ- И ТРЕХКОМПОНЕНТНЫХ ОЦК-РАСТВОРОВ УРАНА

А.Л.УДОВСКИЙ, О.С.ИВАНОВ
Институт металлургии им.А.А.Байкова
Академии наук СССР,
Москва,
Союз Советских Социалистических Республик

Abstract–Аннотация

THERMODYNAMIC CALCULATIONS OF THE PHASE EQUILIBRIUM BOUNDARIES FOR TWO- AND THREE-COMPONENT BODY-CENTRED CUBIC SOLID SOLUTIONS OF URANIUM.

The authors put forward models and give basic approximations for a thermodynamic description of two-component and three-component isomorphous solid solutions. They consider the various contributions made to the free energy of solid solutions formed by isomorphous components, namely the chemical contribution (using Becker's model), the contribution made by the energy of elastic deformation of the crystal lattice, the vibrational contribution, and that of thermally excited electrons. Further, the authors demonstrate the predominant part played in shaping the decomposition curves by the elastic component of the free energy, and analyse the effect of the concentration and temperature dependence of the modulus of elasticity on the shape of the decomposition curves for the solid solutions. Results are given for the calculation of: (a) decomposition curves (solubility curves) for eight systems of γ-uranium (with β-Zr, β-Hf, V, Nb, Ta, Cr, Mo and W) and also for the system β-Zr-Nb; and (b) phase boundaries for the separation regions of body-centred cubic solutions in the ternary systems U-Nb-Mo and U-Nb-Zr. A comparison of the results of calculations with experimental phase diagrams indicates, as a whole, that there is satisfactory quantitative concordance. The authors establish the inter-relationship between the shape of the phase equilibrium boundaries for isomorphous solid solutions in two-component and three-component systems, on the one hand, and the concentration relationships for some of the simplest physical characteristics, such as the coefficient of electron thermal capacity, Debye temperature, moduluo of elasticity, and anharmonic characteristics, on the other. The difference in principle between the present calculations and the numerous computer calculations of phase diagrams is the maximum generality of the problem as formulated and its solution in the mathematical sense. Indeed, the phase equilibrium curves were calculated by direct computer plotting of the general tangent to the isothermal sections of the free energy surface, and computation of the points of tangency. The main factors responsible for the shape of the asymmetric curves computed for decomposition and solubility are, for the systems U-Zr, U-Hf, U-Nb and U-Ta, as follows: the elastic component of the free energy, which decreases as the temperature rises, the mixing energy at 0K, and the configuration-entropy term. A further contribution is made by the vibration component of the free energy for the systems U-V, U-Cr, U-Mo, U-W and Zr-Nb; the system U-Mo exhibits an additional factor of electron origin.

ТЕРМОДИНАМИЧЕСКИЕ РАСЧЕТЫ ГРАНИЦ ФАЗОВЫХ РАВНОВЕСИЙ ДВУХ- И ТРЕХКОМПОНЕНТНЫХ ОЦК-РАСТВОРОВ УРАНА.

Изложены модельные представления и основные приближения для термодинамического описания двух- и трехкомпонентных изоморфных твердых растворов. Рассмотрены различные вклады в свободную энергию твердых растворов, образованных изоморфными компонентами: химический (в модели Беккера), энергии упругих искажений кристаллической решетки, колебательный, термически возбужденных электронов. Показана преобладающая роль при формировании кривых распада упругой составляющей свободной энергии. Проанализировано влияние концентрационной и температурной зависимостей модуля Юнга на форму кривых распада твердых растворов. Приведены результаты расчетов а) кривых распада (кривых растворимости) 8 систем γ-урана с β-Zr, β-Hf, V, Nb, Ta, Cr, Mo, W, а также системы β-Zr-Nb; б) фазовых границ областей расслоения оцк-растворов в тройных системах U-Nb-Mo, U-Nb-Zr. Сопоставление результатов расчетов с экспериментальными диаграммами состояния свидетельствует, в целом, об удовлетворительном количественном согласии. Установлена взаимосвязь между формой границ областей фазового равновесия изоморфных твердых растворов двух- и трехкомпонентных систем, с одной стороны, и концентрационными зависимостями некоторых простейших физических характеристик (коэффициентом электронной теплоемкости, температурой Дебая, модулем Юнга, характеристик ангармонизма), с другой стороны. Принципиальным отличием настоящих расчетов в сравнении с многочисленными работами по расчету диаграмм состояния с помощью ЭВМ является максимальная общность постановки задачи и ее решения в математическом смысле. Именно, расчет кривых фазового равновесия осуществлялся путем непосредственного машинного построения общей касательной к изотермическим сечениям поверхности свободной энергии и вычисления точек касания. Основные факторы, ответственные за форму рассчитанных асимметричных кривых распада и растворимости: для систем U-Zr, U-Hf, U-Nb, U-Ta — упругая компонента свободной энергии, уменьшающаяся с ростом температуры, энергия смешения при 0 °K, конфигурационно-энтропийное слагаемое; дополнительный вклад даст вибрационная компонента свободной энергии для систем U-V, U-Cr, U-Mo, U-W, Zr-Nb; в системе U-Mo проявляется дополнительный фактор электронного происхождения.

I. Проблема термодинамического расчета диаграмм состояния

Термодинамический расчет диаграммы состояния двухкомпонентной системы по существу сводится к наложению расчетных границ различных пар фаз, находящихся в равновесии /I/. Проблема термодинамического расчета фазовых границ двух фаз i и j, находящихся в равновесии состоит в решении системы

$$\left.\begin{aligned} RT\ln\frac{1-x}{1-y} &= \Delta G^{i\to j}(0,T,P) + \left[G^{j}_{изб}(y,T,P) - G^{i}_{изб}(x,T,P)\right] - \\ &\quad - \left[\frac{\partial G^{j}_{изб}(y,T,P)}{\partial \ln y} - \frac{\partial G^{i}_{изб}(x,T,P)}{\partial \ln x}\right] \\ RT\ln\frac{x}{y} &= \Delta G^{i\to j}(1,T,P) + \left[G^{j}_{изб}(y,T,P) - G^{i}_{изб}(x,T,P)\right] - \\ &\quad - \left[\frac{\partial G^{j}_{изб}(y,T,P)}{\partial \ln (1-y)} - \frac{\partial G^{i}_{изб}(x,T,P)}{\partial \ln (1-x)}\right] \end{aligned}\right\} \quad (I.I)$$

Здесь x, y - концентрации второго компонента в фазах i и j, соответственно;

$$\begin{aligned} \Delta G^{i\to j}(0,T,P) &= G^{j}(0,T,P) - G^{i}(0,T,P) \\ \Delta G^{i\to j}(1,T,P) &= G^{j}(1,T,P) - G^{i}(1,T,P), \end{aligned} \quad (I.2)$$

$G^{k}(0,T,P), G^{k}(1,T,P)$ - термодинамические потенциалы k - ой фазы $(k = i, j)$ чистых компонентов ($x = 0$ и $x = I$, соответственно) в зависимости от температуры Т и давления Р; $G^{i}_{изб}$ и $G^{j}_{изб}$ - избыточные термодинамические потенциалы сплавов i и j фаз, соответственно.

Система (I.I) получается после применения условий фазового равновесия к термодинамическим потенциалам i - ой

$$G^{i}(x,T,P) = (1-x)G^{i}(0,T,P) + xG^{i}(1,T,P) + RT\left[x\ln x + (1-x)\ln(1-x)\right] + G^{i}_{изб}(x,T,P) \quad (I.3)$$

и j -ой фаз.

При изобарических расчетах диаграмм состояния аргумент Р в (I.I) - (I.3) обычно опускают.

Анализ уравнений системы (I.I) показывает, что для расчета границ $i-j$ фазового равновесия необходимо решить две проблемы. Во-первых, нужно термодинамически описать фазовые переходы (действительные или виртуальные) чистых компонентов - соотношения (I.2). Во-вторых, необходимо количественное описание избыточных термодинамических потенциалов i и j фаз в зависимости от концентрации и температуры (при $p = const$). В решении первой проблемы достигнут определенный прогресс как в термодинамических расчетах Кауфмана /I-4/, так и в работах, базирующихся на теории псевдопотенциала /5-I2/. Однако, проблема количественного описания избыточных термодинамических функций, являясь более сложной, еще далеко не разрешена. Так например, в модели регулярных растворов, на основании которой Кауфманом произведены расчеты диаграмм состояния около I00 двойных и 4 трехкомпонентных систем /I/, параметры B ($G_{изб} = Bx(1-x)$) для оцк-растворов систем $Nb-Mo$, $Ta-Mo$, $Ta-W$ равны 93I; 957 и 2026 кал/г.атом соответственно. Однако экспериментальные данные по энтальпии смешения для сплавов эквиатомных составов при I200°K равны - 2200 для $Nb-Mo$, - I600 для $Ta-Mo$ и - 2600 кал/г-атом для $Ta-W$ /I3/. Теория псевдопотенциалов применительно к сплавам делает лишь первые шаги /5,7, 9,I4-24/.

I.I. Расчеты диаграмм состояния двойных металлических систем на основе урана

Первые расчеты диаграмм состояния двойных систем на основе урана исходя из экспериментальных термохимических данных, провел Кубашевский для систем $U-Zr$ и $U-Fe$ /25-26/. Для расчета границы области несмешиваемости оцк-растворов и кривых солидуса и ликвидуса в системе $U-Zr$ он принял; I) $G_{изб}(x) = \Delta H(x)$; 2) в системе $U-Zr$ $\Delta H(x)$ равно энтальпии смешения в системе $Cr-Mo$, определенной экспериментально. Позднее Чаттерьи /27/ рассчитал кривые солидус и ликвидус в системах $U-Zr$ и $U-Hf$ в приближении идеальных растворов. Недавно авторы привели результаты расчетов кривых расслоения оцк-растворов для систем $U-Zr$, $U-Hf$, $U-Nb$ и $U-Mo$ /28/.

2. Расчеты границ фазовых равновесий двухкомпонентных оцк-растворов урана

2.I. Физические приближения

Содержание модельных представлений и приближения описаны в /28,29/. Свободная энергия двухкомпонентного раствора

$$F(x,T) = U^{\circ}(x) + F_{упр}(x,T) + F_{вибр}(x,T) + F_{эл}(x,T) - TS_{к}(x). \quad (2.I)$$

Анализ этого выражения для свободной энергии неупорядоченного раствора в сравнении с выражением для свободной энергии чистого компонента показывает, что второе и последнее слагаемые отличные от нуля для твердого раствора "вырождаются", т.е. становятся равными нулю, в выражении для свободной энергии чистого компонента. Оставшиеся слагаемые описывают внутреннюю энергию, энергии фононного и термически возбужденного электронного газов чистого компонента. Следовательно, в формальном отношении при переходе от термодинамического описания чистого компонента к термодинамическому опи-

ТАБЛИЦА I. ФИЗИЧЕСКИЕ ХАРАКТЕРИСТИКИ ЧИСТЫХ КОМПОНЕНТОВ

Физическая характеристика	Размерность	γ-U	β-Zr	β-Hf	V	Nb	Ta	Cr	Mo	W
a_{20^oC}	Å	3,474 /35,36/	3,6069 /37/	3,545 /4I/	3,024 /46/	3,300 /46/	3,3025 /4I/	2,8846 /49/	3,I47 /4I/	3,I6475 /49/
$E_{20^oC} \cdot IO^{-3}$	кг/мм^2	I4,92 /30/	9,82^a /37/	I4,3^a /42/	I2,8I /40/	I2,40 /40/	I8,984 /37/	29,53 /37/	33,35 /30/	4I,483 /37/
θ	град,К	I68 /30/	292^а /38/	260^а /43/	399 /38/	277 /38/	258 /38/	630 /38/	4I5 /30/	388 /38/
$\gamma \cdot IO^4$	кал/моль.гр^2	32,8I /30/	6,9^а /38/	6,24^а /43/	2I,2 /38/	I8,44 /38/	I4,I /38/	3,5 /38/	4,778 /30/	2,39 /47/
$\beta \cdot IO^4$	град$^{-I}$	5,2I /36/	5,63 /39/	3,0^б	2,I9 /40/	I,25 /40/	I,3 /48/	I,0 /50/	I,48 /30/	I,0 /48/

[а] Было принято, что физическая характеристика для оцк–структуры равна соответствующей приведенной в таблице характеристике для гпу структуры.

[б] Значение $\beta \simeq 3.IO^{-4}$град$^{-I}$ для Hf было вычислено из температурной зависимости модулей упругости монокристалла Hf $c_{ij}(T)$ /44/, используя переход Фехта /45/ от C_{ij} к модулю Юнга.

санию твердого раствора происходит своеобразное снятие "вырождения" в виде появления двух новых слагаемых в выражении для свободной энергии. Первое ($F_{упр}$) возникает благодаря разности атомов компонентов, второе $(TS_{к})$ - из-за разносортности атомов, образующих раствор.

В более конкретном виде свободная энергия изоморфных растворов

$$F(x,T) = Nz_0\left[x(1-x)E^0_{см} + (1-x)\frac{E^0_{AA}}{2} + x\frac{E^0_{BB}}{2}\right] +$$

$$+\frac{3\pi}{16}Nz_0 E(x,T)d(x,T)\cdot\left[\Delta d(T)\right]^2 x(1-x) +$$

$$+RT\left\{3\ln\left[1-\exp\left(-\frac{\theta(x,T)}{T}\right)\right]-D\left[\frac{\theta(x,T)}{T}\right]\right\}+\frac{9}{8}R\theta(x,0)-$$

$$-\frac{\gamma(x)}{2}\cdot T^2 + RT\left[x\ln x + (1-x)\ln(1-x)\right] \qquad (2.2)$$

Здесь z_0 - координационное число. Остальные обозначения описаны в /29/. Общее число параметров теории равно мести: I) энергия смешения при 0°К $Nz_0E^0_{см}$; концентрационные зависимости при $0 \leq T \leq 0$ 2) периода решетки „a" ; 3) модуля Юнга Е, 4) температуры Дебая θ , 5) коэффициента электронной теплоемкости γ ; 6) температурного коэффициента модуля Юнга при $T \geq \theta$. Эти зависимости, исключая $Nz_0E^0_{см}$, являются исходной информацией для проведения расчетов избыточных термодинамических функций /29-3I/, и, включая $Nz_0E^0_{см}$, для расчетов границ фазовых равновесий и спинодалей изоморфных твердых растворов /28,32-34/. Температурные зависимости $E(x,T)$, $d(x,T)$, $\Delta d(T)$ и $\theta(x,T)$ приведены в /29/.

2.2. Принципы расчета фазовых границ

Вычисление величин $Nz_0E^0_{см}$ производилось двумя способами: либо "привязкой" расчетной энтальпии смешения к экспериментальному значению энтальпии смешения для какого-либо одного состава, либо "привязкой" к температуре максимума кривой распада или к точке максимальной растворимости. Первым способом были рассчитаны фазовые границы оцк-растворов в системах $V - Mo$ /28,32/ и $Zr - Nb$ /28,33/. Все остальные расчеты были проведены вторым способом /28,34/.

Принципиальным отличием настоящих расчетов в сравнении с многочисленными работами по расчету диаграмм состояния с помощью ЭВМ является максимальная общность постановки задачи и ее решения в математическом смысле. Именно, расчет кривых фазового равновесия осуществлялся непосредственным машинным построением общей касательной к изотермическим сечениям поверхности свободной энергии и вычислением точек касания и перегиба при различных фиксированных температурах. Алгоритм описан в /34/. Тогда как обычно расчеты границ фазовых равновесий в двухкомпонентных системах сводят к решению системы двух трансцендентных уравнений типа (I.I).

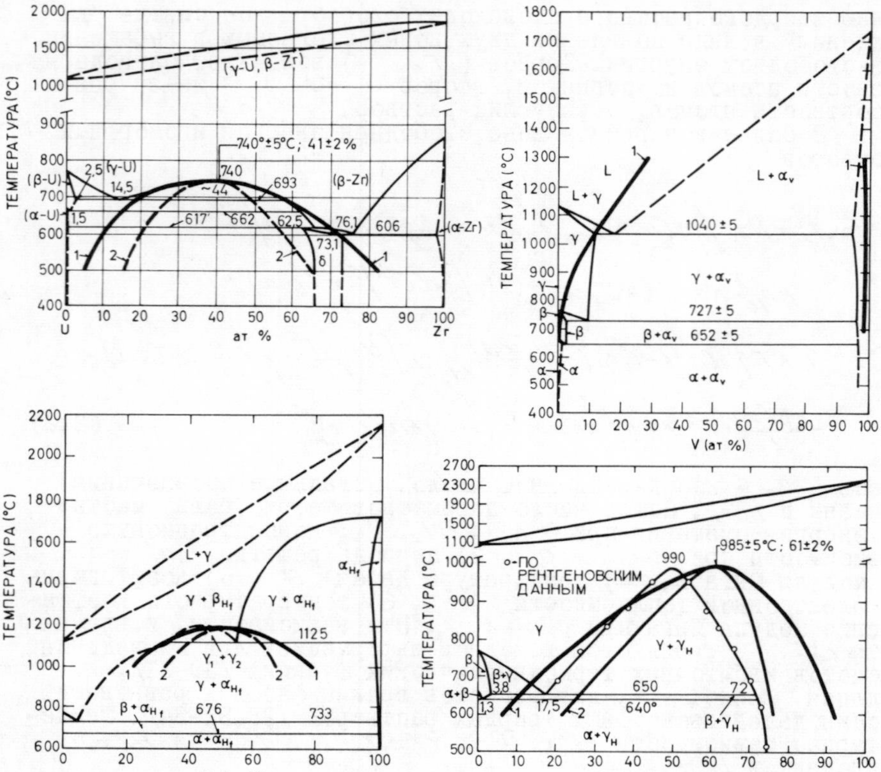

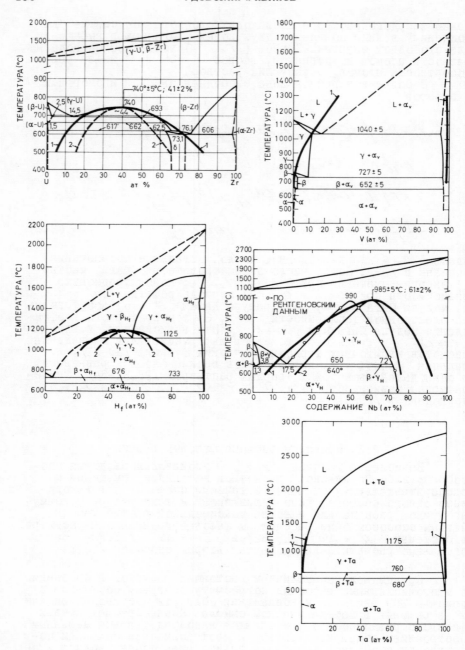

Рис.1. Сопоставление расчетных кривых распада (1) (2-спинодаль) с экспериментальными диаграммами состояния.

2.3. Входная информация

Физические характеристики для оцк структуры чистых компонентов представлены в таблице I. Физические характеристики для γ - U были собраны, проанализированы и оценены в работе /30/. Концентрационные зависимости основных физических характеристик, исключая систему уран-молибден, для оцк растворов были приняты линейными ввиду полного отсутствия экспериментальных данных.

Очевидно, что когда разность соответствующих физических характеристик чистых компонентов по порядку величины равна средней величине этой физической характеристики для системы, то ограничение линейной зависимостью представляет собой учет лишь первых членов разложения в ряд по степеням концентраций полиномиальной зависимости, аппроксимирующей экспериментальную зависимость данной физической характеристики от состава.

2.4. Результаты расчетов

Расчеты кривых распада (ограниченной растворимости) были произведены с помощью ЭВМ "Минск-22М" без учета температурной зависимости размерного фактора, т.е. принималось, что $\Delta d(T) \simeq \Delta d(0)$. А также было принято, что $t(x) = T - \theta(x,0) \simeq T$ для $T \gg \theta(x,0)$ /28/.

2.4.I. Система уран-цирконий.

Величина энергии смешения $B = N z_0 E^0_{CM}$ = +500кал/моль получена "привязкой" к температуре максимума кривой распада To = 740°C /46/. Рассчитанные значения координат точки максимума кривой расслоения: To = 740±5°C и x_0 = 40±2 ат.% Zr, а также расчетные кривая распада (I) и спинодаль (2) нанесены на экспериментальную диаграмму состояния системы уран-цирконий /46/ (рис.I). Видно хорошее количественное согласие расчетной и экспериментальной кривых распада.

2.4.2. Система уран-гафний

Величина B = +4550 кал/моль была вычислена по значению T_0 = I200°C /5I/. Результаты расчета нанесены на экспериментальную диаграмму состояния системы уран-гафний /5I/ рис.I.

2.4.3. Система уран-ванадий

Точкой "привязки", используя которую была вычислена величина B = -I2375 кал/моль, послужила точка максимальной растворимости ванадия в γ - уране (x = 0,I2; T = I040±5°C) экспериментальной диаграммы состояния уран-ванадий /5I/. На рис.I результаты расчета кривых растворимости оцк-растворов (кривые I) сопоставлены с экспериментальными соответствующими кривыми. Следует отметить, что такую асимметричную форму кривых растворимости ванадия в γ - уране и γ - урана в ванадии не в состоянии описать ни одна модель. Изложенные выше представления даже в предположении простейших концентрационных зависимостей физических характеристик позволяют описать указанные кривые в удовлетворительном количественном соответствии с экспериментом.

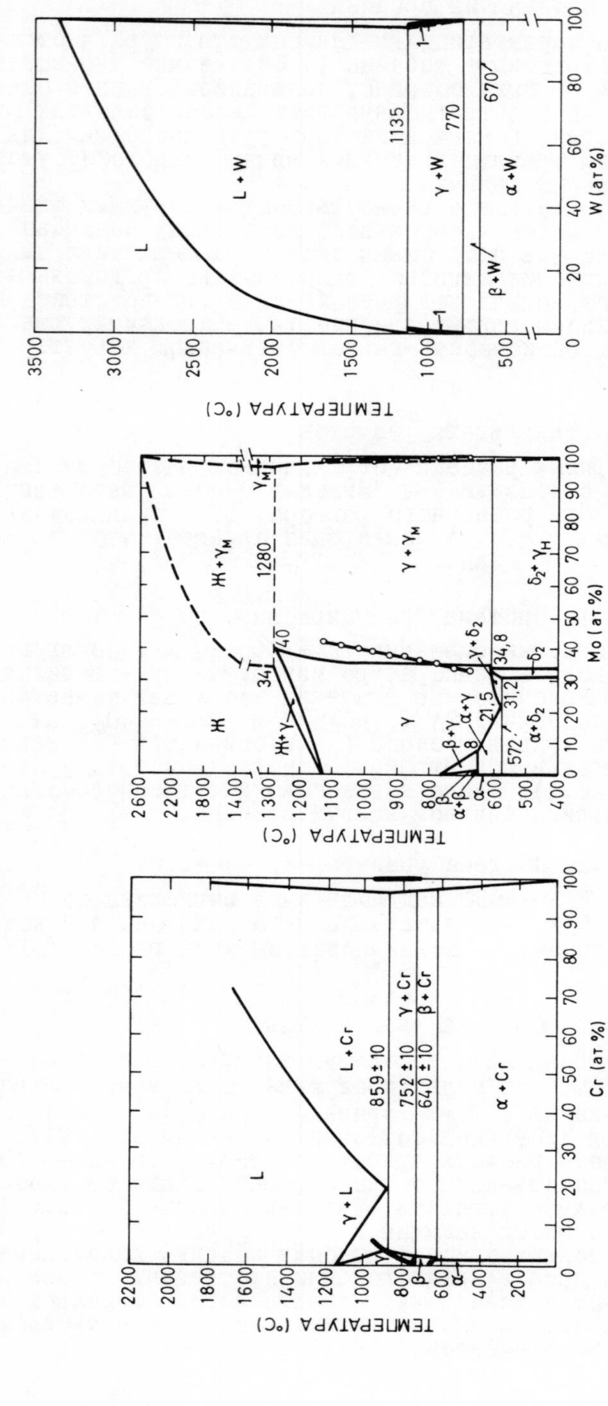

Рис.2. Сопоставление расчетных (1 или кружки) кривых распада с экспериментальными диаграммами состояния.

2.4.4. Система уран-ниобий

Собранные значения модуля Юнга для закаленных оцк-сплавов уран-ниобий указывают на знакопеременное поведение $E(x,0) \simeq E$ (x,300) относительно аддитивной прямой /34/. Поэтому в первом приближении можно взять линейную зависимость $E(x,0)$, принимая во внимание тот факт, что из четырех экспериментальных точек две (x= 0,15 и x = 0,20) не выходят за пределы экспериментально установленной границы γ^{o}/γ /52/, лежащей вблизи x = 0,225. Анализ экспериментальных данных по линейному коэффициенту термического расширения ($\alpha_л$) и температурному коэффициенту модуля Юнга (β) для закаленных сплавов показывает, что линейные зависимости $\alpha_л(x)$ и $\beta(x)$ достаточно хорошо воспроизводят экспериментальные результаты /34/. Зависимость $\theta(x,0)$ была принята линейной. По значению температуры максимума кривой распада T_0 = 990°C /53/ была вычислена энергия смешения B = - 400 кал/моль. Последняя величина находится в соответствии с рентгеновскими данными по измерению диффузного рассеяния на сплавах эквиатомного состава, закаленных с различных температур из γ-фазы, и определенных параметрах ближнего порядка /54/.

Рассчитанные значения координат точки максимума кривой распада оцк-растворов To = 985±5°C и x_o = 61±2 ат.% Nb а также расчетные кривые распада и спинодаль нанесены на экспериментальную диаграмму состояния системы уран-ниобий /54/ рис.I. Следует отметить бóльшую асимметрию расчетной кривой расслоения в сравнении с экспериментальной.

2.4.5. Система уран-тантал

Координаты точки "привязки" взяты согласно экспериментальной диаграммы состояния x = 0,02 и T = 1175°C /51/. Вычисленная энергия смешения B = +9500 кал/моль. Результаты расчета (кривые I) сопоставлены с экспериментальной диаграммой состояния системы уран-тантал (рис.I). Максимальная растворимость $\gamma - U$ в Ta при 1175°C 0,45 ат.% U (расчет) и ~ 2 ат.% (эксперимент) [51].

2.4.6. Система уран-хром

Координаты точки "привязки" равны: x = 0,04, T = 859±10°C [51]. B = - 30 ккал/моль. На рис.2 результаты расчета кривых растворимости Cr в $\gamma - U$ и $\gamma - U$ в Cr сопоставлены с экспериментальной диаграммой состояния /51/. Следует отметить, что экспериментальная кривая растворимости $\gamma - U$ в Cr установлена достаточно неопределенно.

2.4.7. Система уран-молибден

Экспериментальная входная информация для зависимости физических характеристик от состава описывалась линейным приближением для $\theta(x,0)$, $E(x,0)$, $\beta(x)$, тогда как $\gamma(x)$ описывалась аппроксимирующим полиномом /32/. Анализ влияния различных приближений для зависимостей некоторых физических характеристик от состава на форму границ фазовых равновесий оцк растворов $U - Mo$ /28/ показал, что принятое приближение для концентрационных зависи-

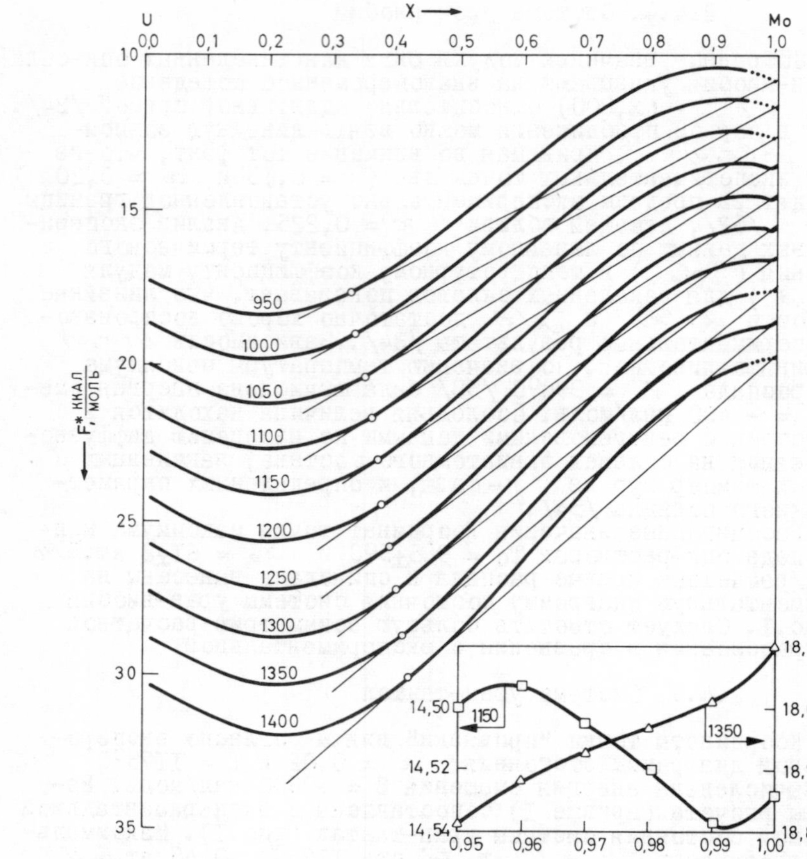

Рис.3. Концентрационные зависимости свободной энергии сплавов U-Mo в температурном диапазоне 950-1400°K.

мостей физических характеристик достаточно для описания основного характера фазовых границ $\gamma/\gamma+\gamma_M$ и $\gamma+\gamma_M/\gamma_M$. Энергия смешения была вычислена из "привязки" к термодинамическим данным по энтальпии смешения ΔH (0,3; II00) = I,3 ккал/моль /32/.
B = - 50,4 ккал/моль.

Результаты расчета свободной энергии F^* (линейные по концентрации члены, не влияющие на фазовые границы, были опущены) в зависимости от концентрации и температуры представлены на рис.3. Кружками отмечены точки соприкосновения общей касательной к изотермам $F^*(x,T)$. В правом нижнем углу этого рисунка в увеличенном масштабе показаны расчетные значения $F^*(x,T)$ в концентрационном диапазоне 95-I00 ат.% Mo при температурах II50 и I350°K для иллюстрации наличия на кривых $F^*(x,T)$ вторых точек перегиба. На рис.2 на экспериментальную диаграмму состояния системы уран-молибден

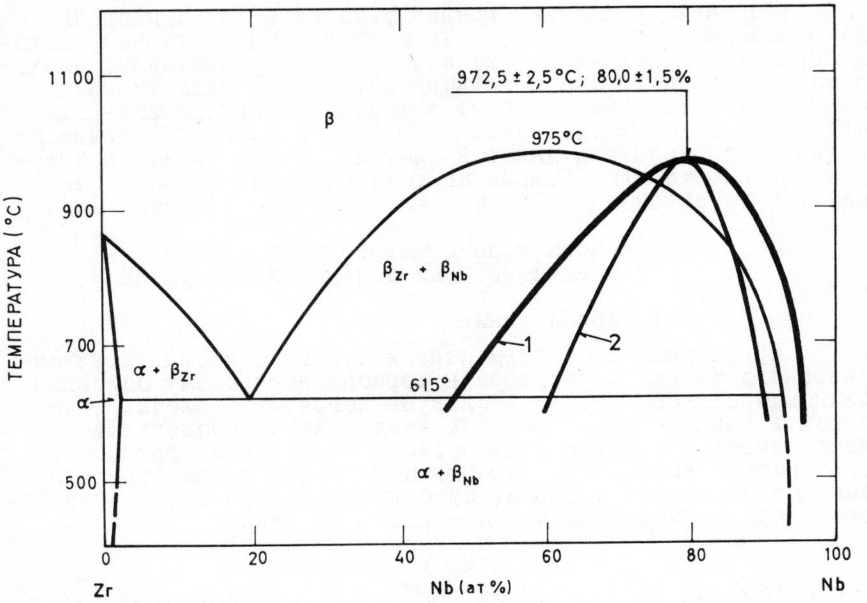

Рис.4. Сопоставление расчетной кривой распада (1) (2-спинодаль) с экспериментальной диаграммой состояния системы Zr-Nb.

/55/ нанесены участки рассчитанной кривой распада (отмечены кружками). Можно видеть хорошее количественное соответствие расчетной и экспериментальной кривых распада. Остальные детали расчетов и анализ результатов изложены в работе /28/. Наиболее существенно то обстоятельство, что предложенная физическая концепция позволили рассчитать в хорошем количественном согласии с экспериментом резко асимметричную кривую распада системы γ - уран-молибден без использования какой-либо экспериментальной точки диаграммы состояния.

2.4.8. Система уран-вольфрам

Координаты точки "привязки" взяты согласно экспериментальной диаграммы состояния уран-вольфрам /5I/: x=0,0II; T = II35°C. Вычисленная энергия смешения B = 4500 кал/моль. Результаты расчета (кривые I) сопоставлены с экспериментальной диаграммой состояния (рис.2). При I000°C растворимость W в $\gamma - U$ составляет 0,2-0,5 ат.% /46/ или 0,9ат% /56/, а U в W 0,4-0,7 ат.% /46/ или 0,I ат.% /56/. Расчетные значения при I000°C: растворимость W в γ-U равна 0,222 ат.% и U в W 0,I2I ат.%.

2.4.9. Система $Zr - Nb$

В настоящей работе осуществлена "привязка" к температуре максимума кривой распада T_0=975°C /57/. B = -II550 кал/моль. Рассчитанные значения координат точки максимума кривой расслоения: T_0 = 972,5±2,5°C и x_o = 80,0±I,5 ат.%

Nb, а также расчетные кривая распада (I) и спинодаль (2) нанесены на экспериментальную диаграмму состояния системы цирконий-ниобий /57/ (рис.4). Причины сравнительно плохого количественного соответствия расчетной и экспериментальной кривых расслоения состоят в том, что принятые линейные концентрационные зависимости $\beta(x)$ и $E(x,0)$ являются чрезмерно упрощенными для этой системы, о чем свидетельствуют экспериментальные данные по температурной зависимости модуля Юнга сплавов Zr с Nb в β - области /58/.

3. Расчеты границ фазовых равновесий трехкомпонентных оцк-растворов урана

3.I. Приближения

При переходе от термодинамического описания двухкомпонентного твердого раствора к термодинамическому описанию трехкомпонентного раствора следуем логической схеме, изложенной в разделе 2.I. При этом каждый двухкомпонентный сплав, имеющий определенный состав, может рассматриваться как "чистый" компонент, который легируется новым (третьим) компонентом. Таким образом, свободная энергия трехкомпонентного твердого неупорядоченного раствора равна:

$$F(x,y,T)=Nz_0\left[x\frac{E^0_{BB}}{2}+y\frac{E^0_{CC}}{2}+z\frac{E^0_{AA}}{2}+xzK_{AB}+xyK_{BC}+yzK_{CA}\right]+$$
$$+\varphi\cdot d(x,T)\cdot\left[\Delta d(x=0,x=1,T)\right]^2\cdot E(x,T)\cdot x\cdot z+$$
$$+\varphi\cdot d(x,y,T)\cdot\left[\Delta d(x,y=1,T)\right]^2\cdot E(x,y,T)\cdot y(x+z)+$$
$$+\frac{9}{8}R\theta(x,y,0)+RT\left\{3\ln\left[1-\exp\left(-\frac{\theta(x,y,T)}{T}\right)\right]-D\left[\frac{\theta(x,y,T)}{T}\right]\right\}-$$
$$-\frac{1}{2}\gamma(x,y)\cdot T^2+RT\left[x\ln x+y\ln y+z\ln z\right] \qquad (3.I)$$

Здесь x,y,z - атомные доли компонентов B,C,A соответственно, связанные соотношением $x+y+z=1$; K_{AB}, K_{BC}, K_{CA} - энергии смешения при 0°К в двойных подсистемах A-B, B-C, C-A; $\varphi=\frac{3\pi}{16}Nz_0$;

$$\Delta d(x=0,x=1,T)=d(0;0;T)-d(1;0;T); \qquad (3.2)$$

$$\Delta d(x,y=1,T)=d(x,0;T)-d(0;1;T); \qquad (3.3)$$

$$E(x,y,T)=E(x,y,0)\left[1-\beta(x,y)\cdot t(x,y)\right], \qquad (3.4)$$

где $t(x,y)=T-\theta(x,y,0)$;

$$\theta(x,y,T)=\theta(x,y,0)\left[1-\alpha_V(x,y)\cdot\gamma_r(x,y)\cdot t(x,y)\right], \qquad (3.5)$$

причем: $2\alpha_V(x,y)\cdot\gamma_r(x,y)\simeq\beta(x,y)$;

$$d(x,y,T)=d(x,y,0)\left[1+\alpha_л(x,y)\cdot t(x,y)\right]; \qquad (3.6)$$

$D[\xi]$ - интеграл Дебая

Сопоставление соотношений (3.I) и (2.2) показывает, что роль члена $U_0(x,y)$ играет первая строка выражения (3.I). Упругая энергия трехкомпонентного твердого раствора состоит из двух слагаемых: вторая и третья строка в формуле (3.I). Первое из них описывает энергию упругих искажений двухкомпонентного раствора, например, на грани А-В концентрационного треугольника А-В-С. Второе слагаемое описывает дополнительный вклад в упругую энергию при добавлении третьего компонента, например, компонента С к двухкомпонентному раствору, который в свою очередь может иметь непрерывный состав. Анализ этих двух слагаемых приведен в /34/. Четвертая строка в (3.I) представляет собой аналог третьей строки в (2.2), т.е. соответствует вибрационной компоненте свободной энергии в дебаевском приближении. Пятая строка в (3.I) состоит из электронной компоненты и конфигурационно-энтропийного слагаемого свободной энергии. Остальные обозначения приведены в /29/.

Общее число параметров, необходимое для расчета свободной энергии трехкомпонентного раствора по формуле (3.I), равно восьми: K_{AB}, K_{BC}, K_{CA}, зависимости от состава периода решетки (или $d(x,y,0)$), $E(x,y,0)$, $\Theta(x,y,0)$, $\beta(x,y)$, $\gamma(x,y)$.

3.2. Принципы расчета

На первом шаге вычислений принимается, что атомная доля третьего компонента, например С, равна нулю ($y=0$) Тогда выражение (3.I) переходит в соотношение (2.2),в дальнейшем поиск точек касания общей касательной к изотермическим сечениям поверхности $F(x,0,T)$ полностью идентичен методу, описанному в разделе 2.2. На втором этапе добавляется определенный шаг по y, например Δy = 0,02, и производится вычисление $F(x,0{,}02;T)$ в пределах $0 \leq x \leq$ 0,98. Поиск точек касания общей касательной к изотермическим сечениям поверхности $F(x,$ 0,02; $T)$ сводится к предыдущему. Таким образом, рассчитывается изоконцентрационное сечение по y = 0,02, которое параллельно грани А-В(y = 0). Увеличивая y от 0 до некоторого конечного значения y_K, при котором заканчиваются вычисления, производится расчет поверхности расслоения, примыкающей к грани А-В. Затем проводим замену переменных $x \rightarrow y$, $y \rightarrow Z$ и $Z \rightarrow x$ и производим расчет,начиная от грани В-С с последовательным приращением y в направлении компонента А . Аналогично проводится расчет и для оставшейся двойной грани С-А вдоль направления возрастания y от грани С-А к компоненту В. Более детальное описание алгоритма расчета фазовых границ областей расслоения трехкомпонентных растворов будет описано в последующей работе.

3.3. Результаты расчета

В настоящем докладе приводятся некоторые результаты расчетов границ фазовых равновесий трехкомпонентных оцк-растворов урана в простейшем линейном приближении для зависимостей физических характеристик $d(x,y,0)$, $E(x,y,0)$, $\Theta(x,y,0)$, $\beta(x,y)$, $\gamma(x,y)$ от состава. Аналитическое описание реальных зависимостей свойство-состав для многокомпонентной системы является сложной задачей. Недавно авторами совместно

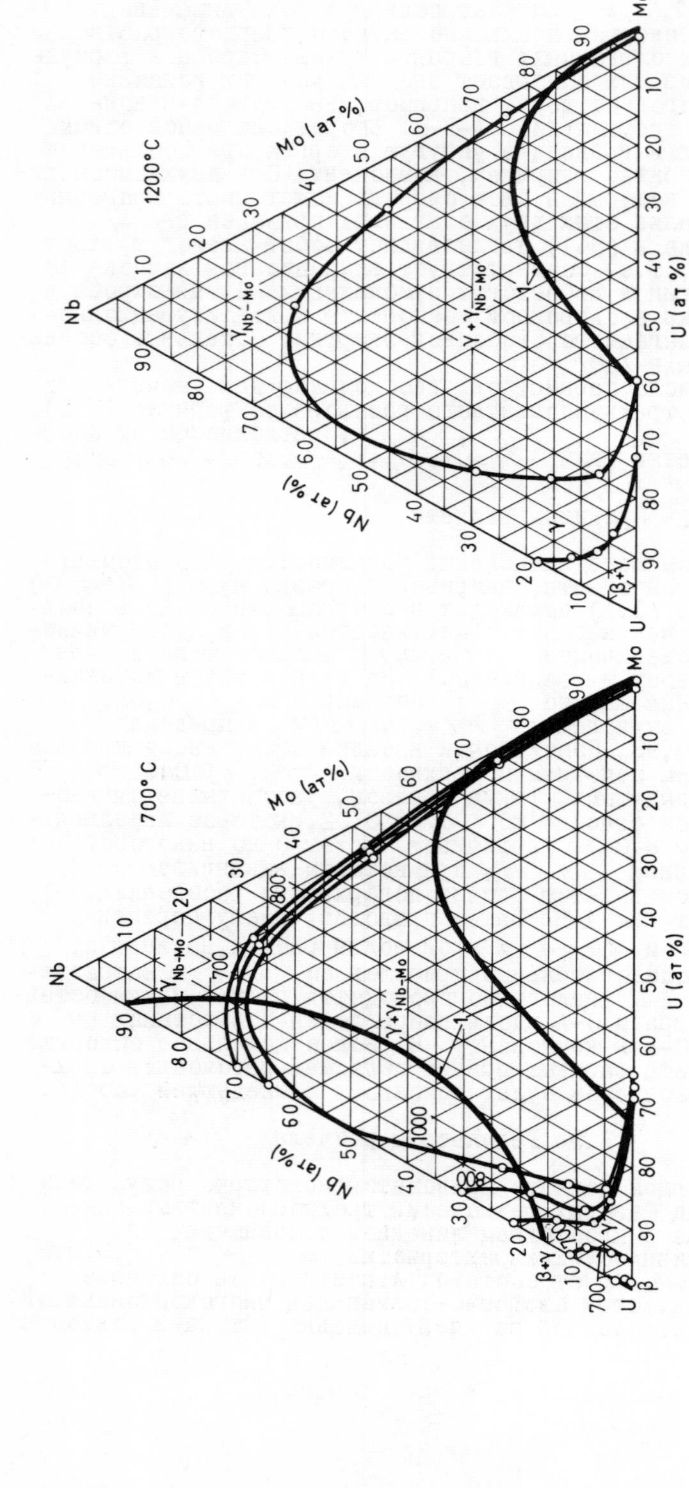

Рис.5. Сопоставление расчетных (1) и экспериментальных изотермических сечений диаграммы состояния U-Mo-Nb.

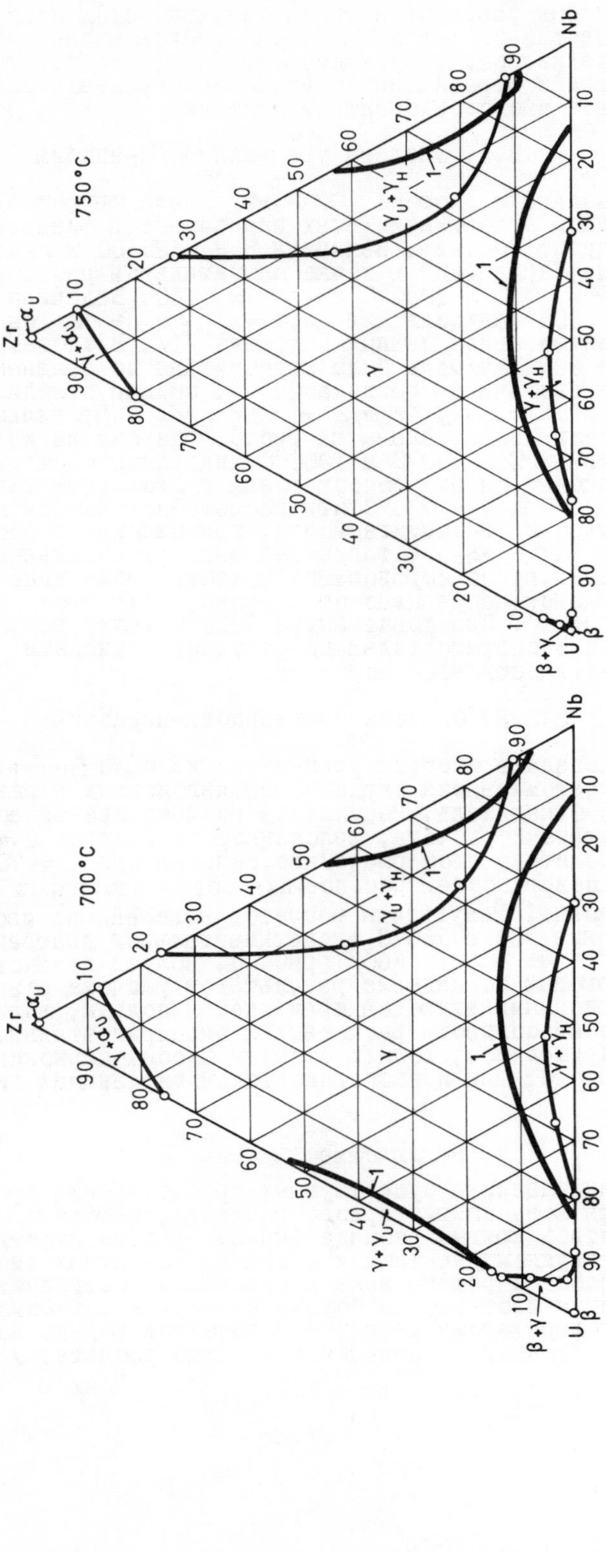

Рис.6. Сопоставление расчетных (1) и экспериментальных изотермических сечений диаграммы состояния U-Nb-Zr.

с А.М.Гайдуковым был разработан аналитический метод описания зависимостей свойство-состав для любых многокомпонентных систем, исходя из известных зависимостей свойство-состав бинарных подсистем и ограниченного числа экспериментальных данных для сплавов многокомпонентной системы.

3.3.I. Система уран-молибден-ниобий

Кривая распада в системе $U-Mo$ была рассчитана повторно в линейном приближении всех зависимостей физических характеристик от состава. Величина В = - 25500 кал/моль была вычислена из "привязки" к точке максимальной растворимости Мо в $\gamma-U$: Т = 1280°С и x = 0,40. Величина $B = Nz_0 K_{BC}$ = - 3008 кал/моль для системы $Mo - Nb$ была вычислена с использованием данных Кауфмана /I/. В системе $Nb - U$ В = - 625 кал/моль была пересчитана по сравнению с соответствующей величиной раздела 2.4.4, снимая приближения $t(x) \simeq T$ и $d(x,0) \simeq \frac{1}{2}[d(0;0) + d(1;0)]$. Частично результаты расчетов представлены на рис.5, где они на изотермических сечениях Т = 700°С и 1200°С диаграммы состояния системы уран-молибден-ниобий сопоставлены с экспериментальными данными /59/. При Т = 700°С при расчете получаются две изолированные области несмешиваемости, примыкающие к сторонам $U - Mo$ и $U - Nb$, тогда как экспериментальные данные свидетельствуют об образовании области несмешиваемости, простирающейся непрерывно от стороны $U - Mo$ к стороне $U - Nb$. При температуре 1200°С между результатами расчетов и экспериментальными фазовыми границами наблюдается лишь качественное согласие.

3.3.2. Система уран-ниобий-цирконий

Кривая распада в системе уран-цирконий была рассчитана повторно, с учетом снятия ограничений, изложенных в разделе 3.3.I. В = +850 кал/моль. Результаты расчета практически совпали с результатами расчета, изложенного в разделе 2.4.I.

Были рассчитаны изотермические сечения при Т = 700, 750 и 800°С/34/ поверхностей расслоения оцк- растворов уран-ниобий-цирконий. Результаты расчетов нанесены на изотермические сечения 700 и 750°С экспериментальной диаграммы состояния $U - Nb - Zr$ /60/ (рис.6). Вблизи двойной системы уран-цирконий совпадение результатов расчета с экспериментальными данными является практически полным, вблизи стороны уран-ниобий согласие расчетных и экспериментальных границ удовлетворительное, вблизи стороны ниобий-цирконий согласие между расчетными и экспериментальными данными лишь качественное.

4. Заключение

На основании анализа существующих представлений предложена физическая концепция твердого раствора, которая позволила рассчитать кривые распада (кривые растворимости) оцк растворов 8 двухкомпонентных и 2 трехкомпонентных систем с ураном. Поставлена и решена задача машинного построения общей касательной к изотерме свободной энергии в зависимости от состава. Благодаря этому развитие физической модели может проводиться без коренного изменения алгоритма расчета. Установлена взаимосвязь между формой границ фазового равновесия изоморфных твердых растворов двух- и трехкомпонентных систем с одной стороны и концентрационными зависимостями некоторых физических характеристик сплавов (коэффициентом электронной теплоемкости, температурой Дебая, модулем Юнга, характеристик ангармонизма), с другой стороны.

Основные факторы, ответственные за форму рассчитанных кривых распада и растворимости, следующие: для систем $U-Zr$, $U-Hf$, $U-Nb$, $U-Ta$, а также $U-Nb-Zr$ и $U-Nb-Mo$ – энергия упругих искажений кристаллической решетки, уменьшающаяся с ростом температуры, энергия смешения при $0^{\circ}K$, конфигурационноэнтропийное слагаемое; дополнительный вклад для систем $U-V$, $U-Cr$, $U-Mo$, $U-W$ дает вибрационная компонента свободной энергии, а в системе $U-Mo$ – проявляется дополнительный фактор электронного происхождения.

За проведение численных расчетов на ЭВМ "Минск-22М" для систем $Zr-Nb$ и $U-Nb-Mo$ авторы выражают благодарность И.В.Семеновской.

ЛИТЕРАТУРА

[1] KAUFMAN, L., BERNSTEIN, H., Computer Calculation of Phase Diagrams, Acad. Press, New York and London, 1970.

[2] KAUFMAN, L., Phase Stability in Metals and Alloys, eds. P.S. Rudman et al., McGraw-Hill, New York, 1967, p. 125.

[3] KAUFMAN, L., Acta Metall. 7 (1959) 575.

[4] KAUFMAN, L., "Calculations of binary phase diagrams". Metallurgical Chemistry, Proc. Symp. at Brunel Univ. and NPL, July 1971, ed. O. Kubaschewski, NPL, London, 1972, p. 373.

[5] INGLESFIELD, J.E., "The physics of metallic cohesion". ibid., p. 191.

[6] PETTIFOR, D.G., "Theory of the crystal structure of transition metals at absolute zero". ibid., p. 157.

[7] HEINE, V., WEAIRE, D., Solid State Physics, eds. F. Zeitz and D. Turnbull, Acad. Press, New York 24 (1970).

[8] HEINE, V., The Physics of Metals, ed. J.M. Ziman, Cambridge Univ. Press, 1969.

[9] HARRISON, W.A., Pseudopotentials in the Theory of Metals, Benjamin, New York, 1966.

[10] ГУРСКИЙ, З.А., КРАСКО, Г.Л., Физ.Тв.Тела 11 10 (1969) 3016.

[11] PETTIFOR, D.G., J.Phys.C., ser. 2 2 (1969) 1051.

[12] PETTIFOR, D.G., J. Phys.C., 3 2 (1970) 367.

[13] SINGHAL, S.C. and WORRELL, W.L., "An emf study of the thermodynamic properties of solid tantalum-molybdenum alloys". Proc. Symp. at Brunel Univ. and NPL, July, 1971, ed. O. Kubaschewski, NPL, London, 1972, p. 65; 149.

[14] INGLESFIELD, J.E., J.Phys.C., ser. 2 2 (1969) 1285; 1293.

[15] HODGES, C.H., STOTT, M.J., Philos. Mag. 26 (1972) 375.

[16] SINGH, S.P. and YOUNG, W.H., J. Phys. F. 2 4 (1972) 672.

[17] БЕЛЕНЬКИЙ, А.Я., ГУРСКИЙ, З.А., КРАСКО, Г.Л., Физика Твердого Тела 15 11 (1973) 3473.

[18] КРАСКО, Г.Л., ГУРСКИЙ, З.А., Физ.Тв. Тела 14 2 (1972) 321.

[19] TANIGAWA, S., DOYAMA, M., Solid State Commun. 11 6 (1972) 787.

[20] TANIGAWA, S., DOYAMA, M., J.Phys. F. 2 5 (1972) L95.

[21] TANIGAWA, S., DOYAMA, M., Phys. Lett. 43A 1 (1973) 17.

[22] TANIGAWA, S., DOYAMA, M., J.Phys. F. 3 5 (1973) 977.

[23] BROEK, J.J., "Calculation of the formation volume of alloys", 3rd Int. Conf. on Chemical Thermodynamics, Sept. 1973, Baden near Vienna, Austria, 9/4a, p. 52.

[24] BROEK, J.J., Phys. Lett. 40A (1972) 219.

[25] KUBASCHEWSKI, O., "Free-Energy and Phase Diagrams", Thermodynamics of Nuclear Materials (Proc. Symp., Vienna, May 1962) IAEA, Vienna (1962) 219.

[26] RAND, M.H., KUBASCHEWSKI, O., The Thermochemical Properties of Uranium Compounds, Oliver and Boyd, Edinburg, 1963, p. 89.

[27] CHATERRYI, D., Met. Trans. 2 10 (1971) 2939.
[28] UDOVSKY, A.L., IVANOV, O.S., J. Nucl. Mater. 49 (1973/74) 309.
[29] ВАМБЕРСКИЙ, Ю.В., УДОВСКИЙ, А.Л., ИВАНОВ, О.С., "Опыт экспериментального и теоретического исследований термодинамических свойств оцк-растворов урана". Доклад на данном Симпозиуме.
[30] VAMBERSKIY, Yu.V., UDOVSKIY, A.L., IVANOV, O.S., J.Nucl.Mater. 46 2(1973)192.
[31] УДОВСКИЙ, А.Л., ВАМБЕРСКИЙ, Ю.В., ИВАНОВ, О.С., Доклады АН СССР, 209 6 (1973) 1377.
[32] ИВАНОВ, О.С., УДОВСКИЙ, А.Л., ВАМБЕРСКИЙ, Ю.В., Доклады АН СССР, 203 5 (1972) 1107.
[33] ИВАНОВ, О.С., УДОВСКИЙ, А.Л., Изв.АН СССР, Металлы, 5 (1973) 309.
[34] УДОВСКИЙ, А.Л., Диссертация, М., 1973.
[35] ИВАНОВ, О.С., ТЕРЕХОВ, Г.И., В сб. Строение Сплавов Некоторых Систем с Ураном и Торием, Госатомиздат, М., 1961, стр.20.
[36] ХОЛДЕН, А.Н., Физическое Металловедение Урана, М., 1962.
[37] Справочник по Редким Металлам, "Мир", М., 1965.
[38] КОЭН, М., ГЛЭДСТОУН, Г., ЙЕНСЕН, М., ШРИФФЕР, Дж., Сверхпроводимость Полупроводников и Переходных Металлов, "Мир", М., 1972, стр.91.
[39] БУГРОВ, В.А. и др., Заводская Лаборатория 34 8 (1968) 994.
[40] ТИТЦ, Т., УИЛСОН, Дж., Тугоплавкие Металлы и Сплавы, Металлургиздат, М., 1969.
[41] СТОРМС, Э., Тугоплавкие Карбиды, Атомиздат, М., 1965.
[42] CHANG, D.H., BUESSEM, W.R., Anisotropy in Single-Crystal Refractory Compounds, v. 2, eds. F.W.Vahlien, S.A.Mensol, Plenum Press, New-York, 1968, p.235-245.
[43] БЛАТТ, Ф., Физика Электронной Проводимости в Твердых Телах, Мир, М., 1971.
[44] FISCHER, E.S., RENKEN, C.J., Phys.Rev., 135 2A (1964) 482.
[45] АНДЕРСОН, О., Физическая Акустика, ред. У.Менон, 3, часть Б, Динамика Решетки, "Мир", М., 1968, стр.62.
[46] ЭЛЛИОТ, Р.П., Структуры Двойных Сплавов, Металлургия, М., 1970.
[47] ISHIKAWA, M., TOTH, L.E., Phys. Rev., B3 6 (1971) 1856.
[48] ФРИДЕЛЬ, Ж., Дислокации, "Мир", М., 1967.
[49] ВЕРЯТИН, У.В. и др., Термодинамические Свойства Неорганических Веществ, Атомиздат, М., 1965.
[50] ГАССНЕР, Р.Х., Некоторые Проблемы Тугоплавких Металлов и Сплавов, ИЛ, 1963.
[51] КУТАЙЦЕВ, В.И., Сплавы Тория, Урана и Плутония, Госатомиздат, М., 1962.
[52] ТЕРЕХОВ, Г.И., ТАГИРОВА, Р.Х., ИВАНОВ, О.С., В сб. Физико-химия Сплавов и Тугоплавких Соединений с Торием и Ураном, "Наука", М., 1968, стр.37.
[53] ИВАНОВ, О.С., ТЕРЕХОВ, Г.И., В сб. Строение Сплавов Некоторых Систем с Ураном и Торием, Госатомиздат, М., 1961, стр.20.
[54] STRELOVA, S.V., UMANSKY, Ya.S., IVANOV, O.S., J.Nucl. Mater. 34 2 (1970) 160.
[55] ИВАНОВ, О.С., СЕМЕНЧЕНКОВ, А.Т., КОЗЛОВА, Н.И., В сб. Строение Сплавов Некоторых Систем с Ураном и Торием, Госатомиздат, М., 1961, стр.68.
[56] SUMMER-SMITH, J. Inst. Met. 83 8 (1954/55) 383.
[57] ИВАНОВ, О.С., РАЕВСКИЙ, И.И., СТЕПАНОВ, Н.В., В сб. Диаграммы Состояния Металлических Систем, Наука, М., 1968, стр.52.
[58] БЫЧКОВ, Ю.Ф., РОЗАНОВ, А.Н., СКОРОВ, Д.М., Атомная Энергия 2 2 (1957) 152.
[59] ИВАНОВ, О.С., ТЕРЕХОВ, Г.И., В сб. Строение Сплавов Некоторых Систем с Ураном и Торием, Госатомиздат, М., 1961, стр.228.
[60] ИВАНОВ, О.С., ГОМОЗОВ, Л.И., там же, стр.107.

DISCUSSION

C.B. ALCOCK: When we consider the fact that a relatively small error in the integral free energy of the mixing curve can substantially alter the appearance of the calculated phase diagram, the results which are presented in this paper are encouraging evidence for the underlying validity of the models which are used. A number of laboratories are at present engaged in computer calculations of phase diagrams and it will be very interesting to compare the results of these calculations for important ternary metallic systems.

Session VIII
BASIC THERMODYNAMIC PROPERTIES

Chairmen: E.H.P. CORDFUNKE
THE NETHERLANDS
and
O.S. IVANOV
UNION OF SOVIET SOCIALIST REPUBLICS

HIGH-TEMPERATURE VAPORIZATION BEHAVIOUR OF HYPOSTOICHIOMETRIC U-Pu-O AND U-Nd-O SOLID SOLUTIONS

M. TETENBAUM
Argonne National Laboratory,
Argonne, Ill.,
United States of America

Abstract

HIGH-TEMPERATURE VAPORIZATION BEHAVIOUR OF HYPOSTOICHIOMETRIC U-Pu-O AND U-Nd-O SOLID SOLUTIONS.

Oxygen potentials over the U-Pu-O system, where Pu/(U + Pu) = 0.2, have been established over the temperature range 2150-2550 K by means of a transpiration technique. These data are of particular importance since no data for the U-Pu-O system were previously available for this temperature range. Our experimental oxygen potentials are approximately 10-15 kcal more positive than the previously accepted values obtained by extrapolation of lower temperature galvanic cell measurements. Values for the relative partial molar enthalpies and entropies of solution of oxygen are presented. Values of the partial pressure of actinide-bearing species have been calculated as functions of temperature and O/M ratio. Oxygen potentials for the U-Nd-O system, where Nd/(U + Nd) = 0.1 and 0.2, are higher than those obtained with unalloyed urania, but lower than those indicated for the U-Pu-O system. The introduction of neodymium into the fluorite lattice appears to have a strong effect on the magnitude of the partial molar entropy and enthalpy of solution of oxygen. X-ray analysis of several U-Nd-O compositions showed only a single face-centred-cubic structure and a decrease in lattice parameter with increasing oxygen content.

Introduction

The performance of fast breeder reactor fuel systems is influenced by the variation of oxygen potential of the fuel pin during burnup. However, a dearth of data exists for the variation of oxygen potential as a function of fuel composition, particularly at fuel operating temperatures above 2000 K. Oxygen potentials derived from the galvanic cell measurements (1073-1373 K) of Markin and McIver [1] and the calculations of Rand and Markin [2] have served as a basis for predicting fuel performance such as oxygen and actinide redistribution across fuel pins, as well as fuel-cladding interactions. Oxygen potential measurements over U-Pu-O compositions have been made by Woodley [3] from 1223-1673 K, and recently by Javed and Roberts [4,5] from 1273 to 1973 K. The agreement between these measurements is poor.

Studies are underway at Argonne of the high temperature thermodynamic behavior of solid solutions of actinide oxides containing rare-earth oxides. This paper will deal with the results of measurements of oxygen potentials over U-Pu-O and U-Nd-O systems. It should be noted that neodymium occurs as a major rare-earth fission product during burnup. The transpiration method was chosen for this study because, for a given temperature, the composition of the condensed phase can be fixed by appropriate choice of the oxygen potential of the carrier gas.

Experimental

Solid solutions were prepared by standard techniques from mechanically blended powders which involved a final sintering step at approximately 1900 K in a helium atmosphere. The resultant sintered pellets were broken up into small irregular granules, approximately 1-2 mm in diameter, for subsequent use in the vaporization studies.

A high-temperature glovebox transpiration complex was constructed to study the vaporization behavior of plutonium-bearing fuel materials. The apparatus and general experimental procedure [6,7] previously used to investigate the vaporization behavior of the uranium-oxygen and uranium-carbon systems were used for the U-Pu-O and U-Nd-O study in the following manner. A small tapered tungsten crucible having a perforated bottom was charged with approximately one gram of sintered granules of sample and press-fitted into an open-ended tungsten condenser tube. The bottom of the charge-support tube was closed. The flow pattern was as follows: The carrier gas, preheated by passage down the annular space between the charge-support tube and condenser tube, flowed up through the powder bed in the tungsten crucible, and up through the condenser tube. Helium was circulated in the annular space between the muffle tube and charge-support tube. With U-Pu-O compositions, the carrier gas consisted of helium containing 6% H_2 and varying amounts of moisture; the total pressure of carrier gas was approximately one atmosphere. With U-Nd-O compositions, the carrier gas consisted of H_2-H_2O mixtures at approximately one atmosphere.

The moisture content of the carrier gas was measured by means of an electrolytic-type moisture monitor which was calibrated by the manufacturer (Beckman Instruments) to read the moisture content directly in parts per million by volume for a sample flow rate of 100 ml per minute at approximately 21°C and one atmosphere pressure. Considerable care was exercised by us in measuring the moisture content of the carrier gas. It was our practice during experimental runs to measure the hygrometer response as a function of carrier gas flow rate through the electrolytic cell. The flow rate of carrier gas passing through the electrolytic cell was measured by means of a calibrated mass flow meter. It should be noted that a properly working electrolytic cell yields a linear relationship when the hygrometer response is plotted as a function of flow rate and extrapolates to zero moisture content at zero flow. The moisture content of the carrier gas can then be obtained from such a plot as the prescribed flow rate of 100 ml/min. Dilution of the exiting carrier gas from the transpiration apparatus with a known amount of ultra-high-purity helium gas was used when the moisture content of the carrier gas exceeded the range of the moisture monitor.

Depending on the sample temperature, oxygen potential, and flow rate of carrier gas, the duration of runs varied from approximately 3 to 10 hours. The carrier gas flow rate was generally about 200 ml per minute. Upon completion of an experiment, the residue in the tungsten crucible was removed and analyzed to determine the oxygen-to-metal ratio. The method of McNeilly and Chikalla [8] was used for U-Pu-O compositions. This is a one-step gas-equilibration method in which the mixed oxide was heated at 850°C in a gas stream which consisted of helium containing 6% H_2 and saturated with water vapor at 0°C. It should be emphasized that the method depends on the assumption that an O/M ratio of 2.000 is reached under these conditions. The precision of the method was estimated to be ±0.003 O/M units with our samples. With the U-Nd-O system, gas fusion analysis was used to determine oxygen content; the precision of the method was estimated to be about ±0.008 O/M units.

Results and Discussion

1. U-Pu-O System

Figure 1 shows a plot of oxygen potential *vs.* O/M ratio for isotherms in the temperature range 2150 to 2550 K; these data are of particular importance because no data for the U-Pu-O system were previously available in this temperature range. The oxygen potential (relative partial molar free energy of diatomic oxygen) associated with the U-Pu-O phase was calculated from the expression

$$\Delta\bar{G}(O_2) = R'T \log p(O_2) \tag{1}$$

Oxygen partial pressures were calculated from the moisture content of the carrier gas by means of the equation

$$1/2 \log p(O_2) = -\frac{12863}{T} + 2.88 - \log \frac{p_{H_2}}{p_{H_2O}} \tag{2}$$

which is derivable from well-known thermodynamic data between water vapor and its elements.

In Fig. 2, our data at 2350 K are compared with (1) oxygen potential data for $U_{0.8}Pu_{0.2}O_{2-x}$ obtained by extrapolation of the low-temperature (1073-1373 K) galvanic-cell measurements of Markin *et al.* [2,3] and of recent transpiration measurements (1273-1973 K) of Javed and Roberts [4,5] and (2) transpiration measurements at 2350 K over UO_{2-x} by Tetenbaum and Hunt [6]. Consistent with expectations, our experimental oxygen-potential values over the U-Pu-O system, where Pu/(U +Pu) = 0.2, are considerably higher (and therefore more oxidizing) than those obtained over unalloyed urania. Our values are also higher than those derived by extrapolation from the measurements of Markin *et al.* and Javed and Roberts. Of particular interest is the fact that our oxygen potentials are approximately 15 kcal more positive than the extrapolated values from the recent measurements of Javed and Roberts. It should be noted that Javed [9] also studied the UO_{2-x} system and obtained more negative values at each O/M ratio than those of Tetenbaum and Hunt. Javed's more negative values may have resulted from high values obtained for the O/M ratios, owing to oxidation of sample residues during cooling in the relatively high-volume system used in his experiments.

It should also be noted that the kinetic model developed by Lindemar and Bradley [10] for the synthesis of hypostoichiometric mixed-oxide compositions by hydrogen reduction at temperatures ranging from approximately 1700 to 2100 K yields results which indicate that the extrapolated oxygen potentials based on the work of Markin *et al.* are too negative by about 5-8 kcal for a given O/M ratio. It follows, therefore, that the measurements of Javed and Roberts may be suspect. Woodley [3] measured oxygen potentials (circulating system, thermogravimetric technique) over $U_{0.75}Pu_{0.25}O_{2-x}$ compositions at temperatures ranging from 1223 to 1673 K. Temperature coefficients are given in Woodley's paper for O/M ratios from 1.95 to 1.99. A comparison can be made with our measurements if we accept the assumption of Markin *et al.* that the oxygen potential is a function only of the valency of the metal whose oxidation state is changing. In the temperature range from 2150 to

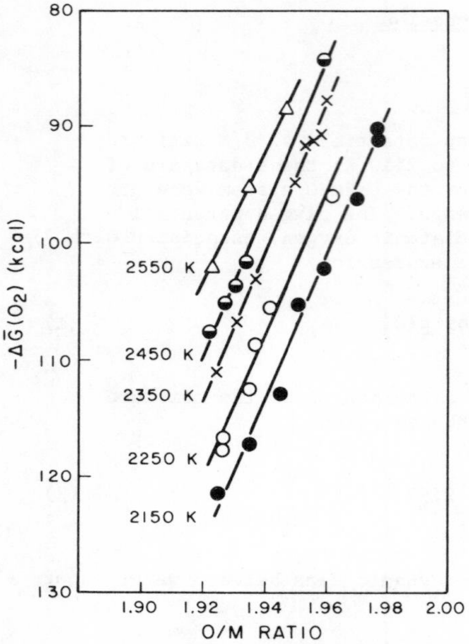

FIG.1. Oxygen potentials over the $(U_{0.8}Pu_{0.2})O_{2-x}$ system as a function of O/M ratio and temperature.

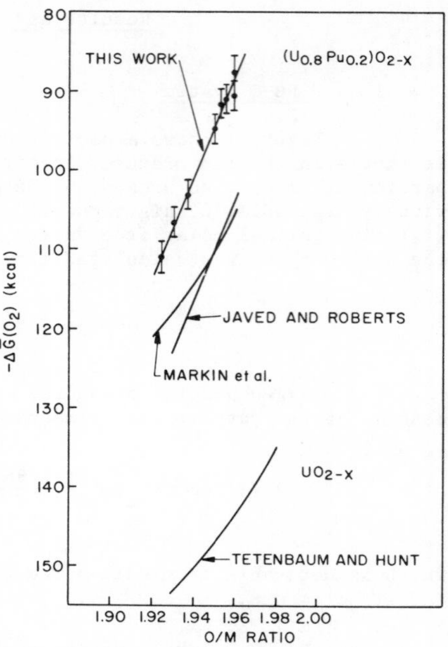

FIG.2. Oxygen potentials as a function of O/M ratio at 2350 K.

2550 K, values estimated from Woodley's temperature coefficient corresponding to the composition $U_{0.8}Pu_{0.2}O_{1.96}$ are approximated 8-16 kcal more positive than our experimental oxygen potentials.

Based on a least square analysis (linear fit) of the data points for the various isotherms investigated, we have estimated the partial molar entropy and partial molar enthalpy of solution of oxygen for selected compositions by means of the thermodynamic relationships

$$\Delta\bar{S}(O_2) = \frac{\partial\bar{G}(O_2)}{\partial T} \tag{3}$$

and

$$\Delta\bar{H}(O_2) = \Delta\bar{G}(O_2) + T\Delta\bar{S}(O_2) \tag{4}$$

The results are summarized in Table I. For comparison, the partial thermodynamic quantities obtained from the measurements of Javed and Roberts and Markin *et al.* on the U-Pu-O system, and of Tetenbaum and Hunt on the UO_{2-x} system are included in the table.

X-ray analysis of several U-Pu-O solid solution compositions from our transpiration runs showed only a single-phase face-centered-cubic structure and a decrease in lattice parameter with increasing oxygen content. The contraction in lattice parameter is in accord with expectations. To maintain electrical neutrality (assuming that uranium retains a valence of +4 in the hypostoichiometric region), the plutonium valence

TABLE I. Partial Molar Enthalpies and Entropies of Solution of Oxygen as Functions of Composition in the U-Pu-O and U-O systems

	$(U_{0.8}Pu_{0.2})O_{2-x}$						UO_{2-x}	
	This Work		Javed & Roberts		Markin *et al.*		Tetenbaum & Hunt	
O/M	$-\Delta\bar{H}_{O_2}$, kcal	$-\Delta\bar{S}_{O_2}$, eu	$-\Delta\bar{H}_{O_2}$, kcal	$-\Delta\bar{S}_{O_2}$, eu	$-\Delta\bar{H}_{O_2}$, kcal	$-\Delta\bar{S}_{O_2}$, eu	$-\Delta\bar{H}_{O_2}$, kcal	$-\Delta\bar{S}_{O_2}$, eu
1.92	243.9	55.0	-	-	193.1	30.6	308.8	65.2
1.93	235.7	54.1	237.8	48.9	191.3	31.4	306.1	65.1
1.94	228.1	53.5	237.1	50.1	189.7	32.2	302.8	64.9
1.95	220.3	52.8	234.7	53.0	185.0	32.2	301.0	65.4
1.96	212.3	52.0	229.1	53.1	180.4	32.1	299.2	66.2

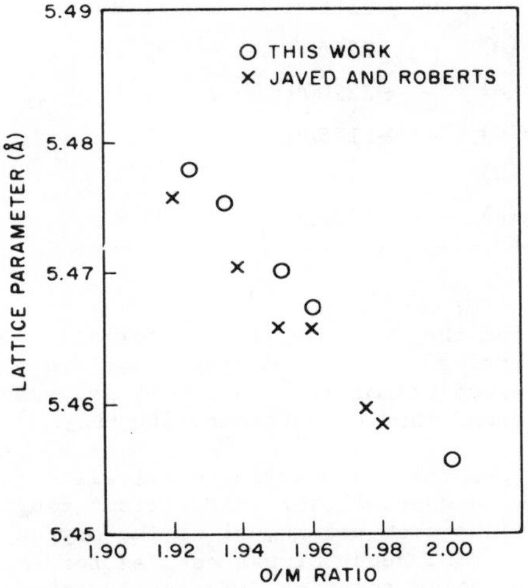

FIG.3. Lattice parameters for the $(U_{0.8}Pu_{0.2})O_{2-x}$ system as a function of O/M ratio.

must increase from a valence of +3 at the composition O/M = 1.90 when the condensed phase is oxidized towards the stoichiometric composition. The results of the x-ray analysis are shown in Fig. 3. Also included in the figure for comparison are lattice-parameter measurements on quenched samples obtained by Javed and Roberts in their transpiration study (1273-1973 K). The two sets of data agree within the limits of accuracy of the lattice-parameter method cited by Schnizlein [11], whose correlation of the data of a number of investigators showed that a 1% change in the O/M ratio produces a change of 0.0005 Å in the lattice parameter.

The total pressure of actinide-bearing species over U-Pu-O compositions, where Pu/(U + Pu) = 0.2, can be estimated from our experimental oxygen potentials and the known or estimated free energies of formation of the gaseous species and condensed phase. The actinide-bearing species over the U-Pu-O system are UO_3, UO_2, UO, U, PuO_2, PuO, and Pu.

TABLE II. Free Energies of Formation of Species

$\Delta G_f^\circ = A + BT$ (cal/mol)

Species	A	B	Ref.
$UO_3(g)$	-200000	19.4	13
$UO_2(g)$	-121500	5.4_5	14
$UO(g)$	-8800	-10.3	15
$U(g)$	117500	-27.0	16
$PuO_2(g)$	-112600	6.6	12
$PuO(g)$	-28500	-9.7	12
$Pu(g)$	78300	-21.0	17
$UO_2(s)$	-258600	41.3	12
$PuO_{1.80}(s)$	-226600	35.2	12
$PuO_{1.75}(s)$	-222000	34.4	12
$PuO_{1.70}(s)$	-217500	33.8	12
$PuO_{1.65}(s)$	-213100	33.3	12
$PuO_{1.60}(s)$	-208800	32.8	12

Equations for the free energies of formation of the gaseous species and those relating to the condensed phase are given in Table II. These equations are based primarily on a recent assessment of actinide oxide systems by Ackermann and Chandrasekhariah [12].

A computer program was written to calculate the partial pressures of the gaseous species over compositions ranging from O/M = 1.92 to 1.96, the compositions over which most of the oxygen potential measurements were obtained. Ideal behavior was assumed for the urania and plutonia components in solid solution. Typically, the results calculated at 2350 K are shown in Fig. 4. The calculated results show that for compositions ranging from O/M = 1.93 to 1.95 $UO_2(g)$ is the predominant species above the condensed phase. It is also apparent that $UO_3(g)$ will begin to make a large contribution to the total pressure at about O/M = 1.97; the total pressure of actinide-bearing species will rise sharply, therefore, at compositions approaching the stoichiometric composition. In the vicinity of O/M = 1.92, the PuO(g) pressure becomes equal to the $UO_2(g)$ pressure. Somewhat similar trends were obtained by Rand and Markin based on their calculations with the $U_{0.85}Pu_{0.15}O_{2-x}$ system.

Figure 5 shows calculated values of the total pressure of actinide-bearing species over U-Pu-O compositions for temperatures ranging from 2150 to 2550 K. It is apparent that the total pressure passes through a minimum, and that the O/M ratio corresponding to the minimum pressure decreases with increasing temperature. The results are summarized in Table III. For comparison, from mass effusion measurements and mass spectrometer measurements in the temperature range 1905-2411 K, Battles *et al.* [18] found a broad-based minimum in the total vapor pressure at an O/M ratio of about 1.92 or 1.93. Ohse and Olsen [19] studied the vaporization of the U-Pu-O system, where Pu/(U + Pu) = 0.15,

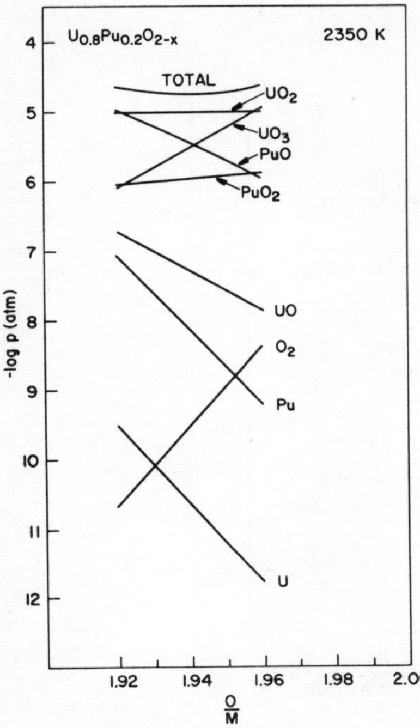

FIG.4. Calculated values of partial pressures of actinide-bearing species over $(U_{0.8}Pu_{0.2})O_{2-x}$ at 2350 K.

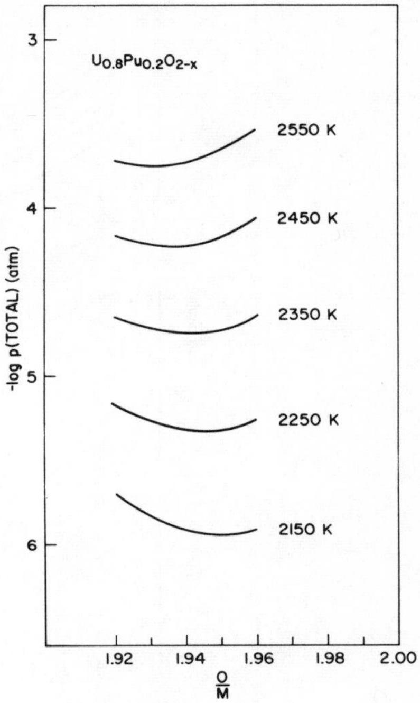

FIG.5. Total pressure of actinide-bearing species over $(U_{0.8}Pu_{0.2})O_{2-x}$ (calculated).

TABLE III

Quasi-Congruently-Vaporizing Compositions for the U-Pu-O System, Where Pu/(U+Pu) = 0.2 (Calculated)

T°K	O/M
2150	1.95_0
2250	1.94_5
2350	1.94_0
2450	1.93_7
2550	1.93_0

between 1800 and 2300 K by means of mass effusion and mass spectrometric techniques. A quasi-congruently-vaporizing composition was found at O/M = 1.97 when these investigators followed the change of composition in the solid at 2108 K starting at O/M = 1.94 and 2.08. It should be emphasized, however, that Battles *et al.* and Ohse and Olson did not establish in their studies the composition dependency with temperature of the total pressure minimum. The trend of decreasing O/M ratio with increasing temperature corresponding to the total pressure minimum, derived

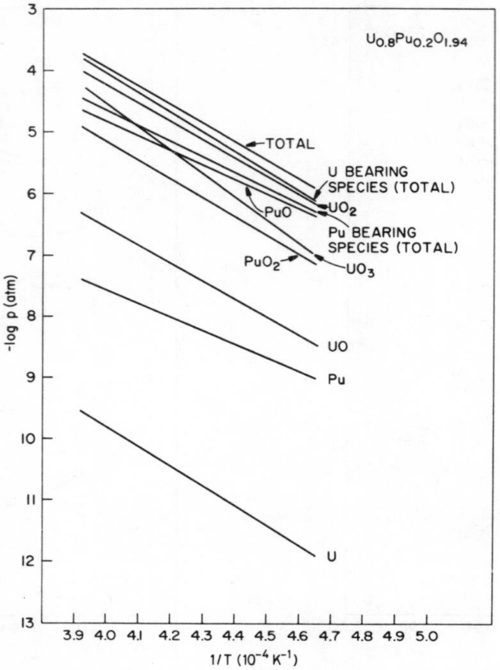

FIG.6. Calculated values for the temperature dependence of actinide-bearing species over $U_{0.8}Pu_{0.2}O_{1.94}$.

from our calculations, has also been observed with the binary actinide oxide systems UO_{2-x} [20,21] and PuO_{2-x}. [22] With the binary systems, these compositions are congruently-vaporizing compositions.

Figure 6 shows calculated values of the partial pressures of actinide-bearing species as a function of the reciprocal temperature over the composition $U_{0.8}Pu_{0.2}O_{1.94}$. Linear equations for the temperature dependency of these actinide-bearing species are summarized in Table IV for O/M ratios 1.92 to 1.96.

2. U-Nd-O System

Oxygen potentials over the U-Nd-O system, where Nd/(U + Nd) = 0.1 and 0.2, have been established at 2250 and 2350 K over a wide range of oxygen-to-metal ratios. The results are shown in Figs. 7 and 8. The shapes of the isotherms show a rapid increase in oxygen potential as the composition of the condensed phase approaches the stoichiometric composition, defined as O/M = 2.000. It is also seen that the variation of oxygen potential as a function of Nd/(U + Nd) ratio is consistent with expectations; that is, with the neodymium in the trivalent state, the oxygen potential increases with increasing Nd/(U + Nd) ratio at a fixed O/M ratio. The higher oxygen potential obtained for Nd/(U + Nd) = 0.2 at a fixed O/M ratio can be attributed to an increase in the average valence of uranium. It is apparent from the results shown in Figs. 7 and 8 that oxygen potentials for the U-Nd-O system are higher (and therefore more oxidizing) than those obtained with unalloyed urania.

TABLE IV. Equations for Temperature Dependency of Actinide-Bearing Species Over $U_{0.8}Pu_{0.2}O_{2-x}$ Compositions

$$\log p = -\frac{A \times 10^4}{T} + B$$

	$UO_3(g)$		$UO_2(g)$		UO(g)		U(g)		ΣU(g)	
	A	B	A	B	A	B	A	B	A	B
1.92	3.95	10.70	3.00	7.74	2.79	5.17	2.89	2.82	3.07	8.09
1.93	3.86	10.60	3.00	7.74	2.90	5.27	3.07	3.01	3.11	8.32
1.94	3.77	10.54	3.00	7.74	2.97	5.34	3.24	3.14	3.18	8.65
1.95	3.69	10.46	3.00	7.74	3.05	5.41	3.41	3.29	3.25	9.04
1.96	3.60	10.37	3.00	7.74	3.14	5.50	3.58	3.47	3.31	9.44
	$PuO_2(g)$		PuO(g)		Pu(g)		ΣPu(g)		Σ[U(g)+Pu(g)]	
1.92	3.17	7.43	2.34	4.99	2.01	1.45	2.40	5.29	2.72	6.94
1.93	3.10	7.20	2.36	4.86	2.12	1.42	2.46	5.36	2.85	7.41
1.94	3.04	6.90	2.39	4.72	2.23	1.34	2.54	5.49	3.00	8.02
1.95	2.99	6.81	2.42	4.61	2.35	1.31	2.63	5.71	3.14	8.66
1.96	2.96	6.68	2.47	4.57	2.49	1.36	2.73	5.99	3.25	9.22

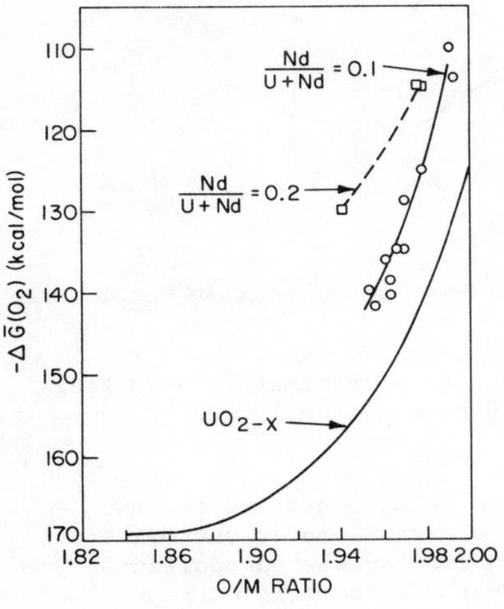

FIG.7. Oxygen potentials as a function of O/M ratio for the U-Nd-O and UO_{2-x} systems at 2250 K.

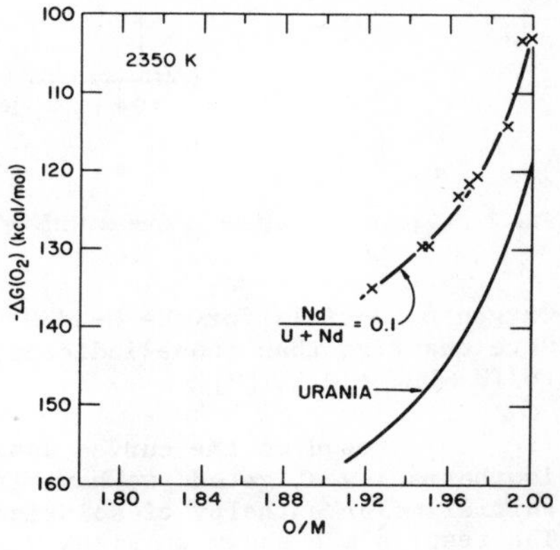

FIG.8. Oxygen potentials as a function of O/M ratio for the U-Nd-O and UO_{2-x} systems at 2350 K.

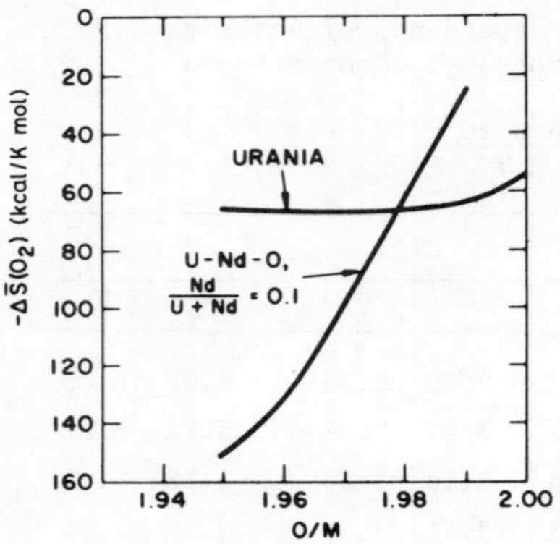

FIG.9. Partial molar entropy of solution of oxygen as a function of O/M ratio for the U-Nd-O and UO_{2-x} systems.

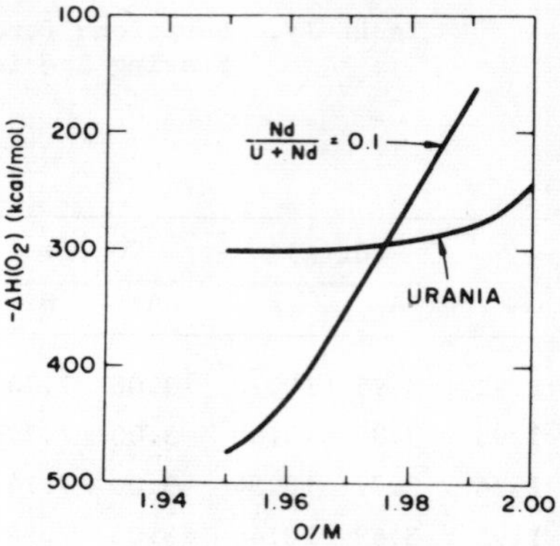

FIG.10. Partial molar enthalpy of solution of oxygen as a function of O/M ratio for the U-Nd-O and UO_{2-x} systems.

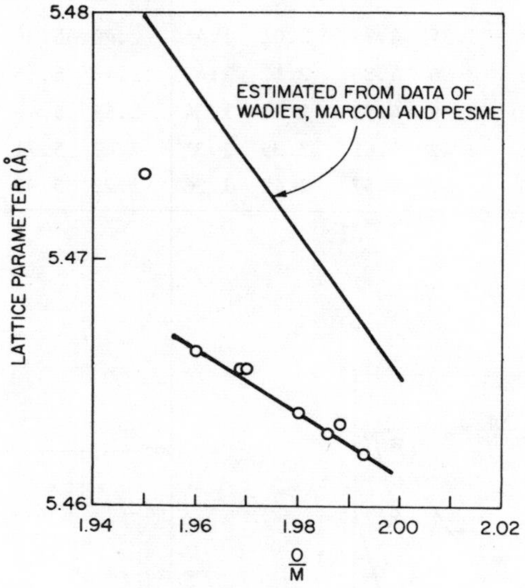

FIG.11. Variation of lattice parameter with O/M ratio for the U-Nd-O system, where Nd/(U + Nd) = 0.1.

Oxygen potentials for the U-Nd-O system are also approximately 5-20 kcal more negative than those indicated for the U-Pu-O system, where Pu/(U + Pu) = 0.1. [2]

Based on the curves assigned to the data points for the two isotherms investigated, we have estimated the partial molar entropy and partial molar enthalpy of solution of oxygen for various compositions. The results are shown in Figs. 9 and 10. Over the composition range in which the O/M ratio is 1.95 to 1.99, a very rapid increase in $-\Delta\bar{S}(O_2)$ and $-\Delta\bar{H}(O_2)$ occurs with increasing oxygen deficiency as the composition of U-Nd-O solid solution approaches O/M = 1.95. The rapid change in these partial thermodynamic quantities probably reflects significant changes in bonding energy due to either the addition or removal of oxygen in the defect lattice. It is apparent that the introduction of neodymium into the fluorite lattice has a strong effect on the magnitude of the partial entropy and enthalpy of solution of oxygen in these compositions.

X-ray analysis of several U-Nd-O solid-solution compositions, where Nd/(U + Nd) = 0.1, showed only a single-phase, face-centered-cubic structure, and a decrease in lattice parameter with increasing oxygen content. The results are given in Fig. 11. The contraction in the lattice parameter is in accord with expectations because, to maintain electrical neutrality, the uranium valence must increase from the value of +4 at the composition O/M = 1.95, when the condensed phase is oxidized toward the stoichiometric compositions. The U-Nd-O system has also been investigated by Wadier *et al.* [23] at 1123 K with Nd/(U + Nd) ratios of 0.2, 0.3, 0.4, and 0.5. Based on the limited lattice-parameter measurements given in their paper at these compositions, we have estimated the variation of lattice parameter with O/M ratio at Nd/(U + Nd) = 0.1, and for comparison we have included these results in Fig. 11.

Chemical analysis for uranium and neodymium content of a residue after several oxygen potential runs with the U-Nd-O system, where Nd/(U + Nd) = 0.2, showed that the Nd/(U + Nd) ratio remained unchanged. X-ray examination of two residues having oxygen-to-metal ratios of 1.97_7 and 1.98_0, respectively, showed a face-centered-cubic structure and a lattice parameter of a_o = 5.459 ±0.005 Å. This value is lower than the lattice parameter value obtained at the corresponding O/M ratio, when Nd/(U + Nd) = 0.1. The decrease in lattice parameter is in accord with expectations, because at a fixed O/M ratio, an increase in the average valence of uranium would be expected in order to compensate for the increase in the concentration of lower valent neodymium.

ACKNOWLEDGMENTS

The author wishes to thank Dr. M. S. Foster for writing the computer programs used for the calculation of partial pressures of actinide-bearing species. He is also indebted to M. I. Homa for x-ray analysis and to Dr. B. D. Holt and R. E. Telford for the determination of oxygen-to-metal ratios in the residues from the transpiration runs.

REFERENCES

[1] Markin, T.L., McIver, E.J., Plutonium 1965, Chapman and Hall, London (1967) 845-857.

[2] Rand, M.H., Markin, T.L., Thermodynamics of Nuclear Materials, IAEA, Vienna (1968) 637-650.

[3] Woodley, R.E., HEDL-TME-72-85 (1972).

[4] Javed, N.A., Roberts, J.T.A., ANL-7901 (1972).

[5] Javed, N.A., J. Nucl. Mater. 47 (1973) 336.

[6] Tetenbaum, M., Hunt, P.D., J. Chem. Phys. 49 (1968) 4739.

[7] Tetenbaum, M., Hunt, P.D., J. Nucl. Mater. 40 (1971) 104.

[8] McNeilly, C.E., Chikalla, T.D., J. Nucl. Mater. 39 (1971) 77.

[9] Javed, N.A., J. Nucl. Mater. 43 (1971) 219.

[10] Lindemar, T.B., Bradley, R.A., J. Nucl. Mater. 41 (1971) 293.

[11] Schnizlein, G.J., Chemical Engineering Division Fuel Cycle Quarterly Report, April-June 1970, ANL-7735, 76.

[12] Ackermann, R.J., Chandrasekhariah, M.S., Paper IAEA-SM-190/44, this symposium.

[13] Ackermann, R.J., Chang, A.T., J. Chem. Thermodynamics 5 (1973) 873.

[14] Ackermann, R.J., Gilles, P.W., Thorn, R.J., J. Chem. Phys. 25 (1956) 1089.

[15] Blackburn, P.E., J. Nucl. Mater. 46 (1973) 244.

[16] IAEA Panel for the Assessment of the Thermodynamic Properties of the U-C, Pu-C, and U-Pu-C Systems, Technical Report Series, to be published (1975).

[17] Ackermann, R.J., Rauh, E.G., J. Chem. Thermodynamics, to be published.

[18] Battles, J.E., Shinn, W.A., Blackburn, P.E., Edwards, R.K., Nuclear Metallurgy 17, Part II (1970) 733.

[19] Ohse, R.W. Olson, W.M., Nuclear Metallurgy 17, Part II (1970) 743.

[20] Edwards, R.K., Chandrasekhariah, M.J., Danielson, P.M., High Temp. Sci. 1 (1969) 98.

[21] Tetenbaum, M., Hunt, P.D., J. Nucl. Mater. 34 (1970) 86.

[22] Ohse, R.W., Ciani, V., Thermodynamics of Nuclear Materials, IAEA, Vienna (1968) 545.

[23] Wadier, J.F., Marcon, J.P., Pesme, O., Panel on the Behavior and Chemical State of Fission Products in Irradiated Fuel, Vienna, Paper 9E (1972).

DISCUSSION

N. GOLDENBERG: Did you look for any tungsten or molybdenum transport in the cooler region of your apparatus? In our PuO_2 compatibility experiments with Ir-0.3% W we have consistently observed tungsten transport together with the other cations referred to in our own paper. The tungsten transport may be a result of the formation of a complex species, but we have been very concerned about the compatibility of tungsten with PuO_2 at temperatures around 1400 - 1500°C.

M. TETENBAUM: We did not observe tungsten contamination in our residues. Since the material used in the cooler region of the transpiration apparatus was itself tungsten, we did not look for tungsten transport. However, the oxygen potentials used in our experiments were low enough to render tungsten contamination unlikely.

N. GOLDENBERG: What material did Battles use for his cell in the mass spectrometry experiments you referred to?

M. TETENBAUM: I believe it was iridium.

I. JOHNSON: I should first like to suggest that, now that we have a fairly consistent set of experimental data on the mixed urania-plutonia system, we should use one generally accepted method for interpolation and extrapolation. I would urge the use of the quasichemical model developed by P. Blackburn; the most recent parameters have been reported in paper IAEA-SM-190/50 presented at this Symposium (these Proceedings, Vol.1, p.17). The settof equations reported can be readily solved using a small programmable desk calculator. The computed oxygen potentials agree to within experimental error with the data reported by Tetenbaum and the low-temperature galvanic cell data of Markin and McIver.

My second comment is that the results reported for urania-neodymia as shown in Figs 7 and 8 illustrate an important aspect of the change in the oxygen potential of oxide fuel with burn-up. The substitution of rare-earth fission product elements for the actinide elements that have fissioned increases the oxygen potential even if the O/M ratio remains the same. Thus the effect of burn-up on the oxygen potential is due to at least two factors, the replacement of the actinide atoms fissioned with fewer atoms of rare-earth (and zirconium) atoms, and the replacement of actinides by the rare earths as reported in this paper.

G. DÉAN: You have compared your oxygen potential results with those of various other authors. In particular you find that the extrapolated values of Markin and of Woodley are too low and too high, respectively, in comparison with your own. This disagreement is based on the hypothesis of a linear variation of $\Delta\overline{G}_{O_2}$ as a function of temperature. We have measured $\Delta\overline{H}_{O_2}$ by microcalorimetry at 1100°C. At this temperature the $\Delta\overline{G}_{O_2}$ values of Woodley and of Markin are in agreement. It is therefore possible on the basis of our $\Delta\overline{H}_{O_2}$ and $\Delta\overline{G}_{O_2}$ figures at 1100°C to calculate the $\Delta\overline{S}_{O_2}$ values. Finally, I should like to mention that, taking account of the variation in the specific heats of mixed oxides as a function of the O/M and Pu/M ratios, we have calculated a complete set of $\Delta\overline{G}_{O_2}$ = f(T, O/M and Pu/M). The results obtained show that the experimental values determined by various authors may be considered valid in their respective temperature ranges, the function $\Delta\overline{G}_{O_2}$ = f(T) not being linear.

M. TETENBAUM: Thank you for your comments, which I find most interesting.

M.G. ADAMSON: Mr. I Johnson has just made a strong plea for use of the Blackburn curve-fitting 'model' to correlate ΔG_{O_2} data for mixed (U,Pu) oxides over a wide temperature range. May I ask Mr. Tetenbaum why he excluded the recent Blackburn and Battles vapour pressure measurements of metal-bearing species as a source of high temperature $\Delta\overline{G}_{O_2}$ data from his comparison of results, and how these measurements in fact compare with his own? I would stress that whatever model is used to represent the available data it should be applied to <u>all</u> of the latter, not just to selected sets of results.

M. TETENBAUM: Oxygen potentials were derived from the vapour pressure measurements of Battles et al. and the ionization cross-section measurements for UO_2(g) and UO_3(g) of Blackburn and Danielson. The oxygen potentials obtained from these measurements were approximately 5-10 kcal higher than those yielded by transpiration measurements as reported in my paper. Additional ionization cross-section measurements of gaseous species are necessary in order to define the oxygen potential using mass spectrometric techniques. Mr. Blackburn has normalized his model to the oxygen potential data reported in my paper, because these appear to be the best measurements available at present, at least in the temperature range 2150-2550 K.

M.H. RAND: I should first like to take up Mr. I. Johnson's comments on the use of the Markin and McIver model for the interpolation and extrapolation of data for (U,Pu) oxides. Mr. R. Chilton at Windscale has recently performed CO/CO_2 microbalance experiments on superstoichiometric oxides up to 1750 K, which show that the oxygen potentials in this regime are not simply a function of the uranium valency. The values for $U_{0.9}Pu_{0.1}O_{2+x}$ were a reasonable linear extrapolation of the Markin and McIver values, but those for $U_{0.7}Pu_{0.3}O_{2+x}$ were distinctly higher (i.e. less negative) than the Markin and McIver values.

My second remark concerns the effect of the addition of neodymium to uranium oxides. While it is clear that the replacement of uranium in UO_2 by neodymium, cerium (or even plutonium) undoubtedly increases the oxygen potential at the same temperature and O/M ratio, we must be more cautious in predicting the change when neodymium or other lanthanides replace plutonium in $(U,Pu)O_{2-x}$. This is in fact what happens (to a close approximation) in a nuclear fuel.

I. JOHNSON: I believe Paul Blackburn has found that when his simple model is applied to hyperstoichiometric urania with an O/M ratio $\gtrsim$ 2.1, deviation from the experimental data is found. He has therefore applied an empirical correction.

With regard to Mr. Rand's second point, I understand that Mr. Tetenbaum plans to determine experimentally whether the replacement of plutonium and uranium by the rare-earth metals will produce as great a change as is found when uranium is replaced by neodymium in urania. I would not expect the effect to be as great.

M. TETENBAUM: Yes, we are indeed planning to measure the oxygen potentials and total pressures of actinide and rare-earth-bearing species over the U-Pu-Ce-O and U-Pu-Nd-O systems. I agree entirely with Mr. Rand that oxygen potential measurements over the U-Pu-O system as a function of the Pu/(U+Pu) ratio are necessary in order to verify the assumption of Markin et al. that the oxidation potential is a function only of the valency of the metal whose operational state changes.

T.B. LINDEMER: I should like to comment briefly that, at present, measurements of oxygen potential and oxygen release (as CO and CO_2) per fission are being made as a function of temperature and burn-up in the oxide HTGR particles. This information may subsequently be of use in connection with the more fundamental studies presented at this Symposium. The HTGR work is being done by Strigl at the Oesterreichische Studiengesellschaft für Atomenergie, Seibersdorf, by McMillan at the Atomic Energy Research Establishment, Harwell, by Pluchery and Pattoret at the Centre d'études nucléaires, Grenoble, and by myself at Oak Ridge National Laboratory. Assuming the experimental techniques prove to be valid, these measurements will provide oxygen potential data on the actual fissioned system from burn-ups ranging from a few per cent to 70% FIMA and at temperatures from 1000 to 2000°C.

R.W. OHSE: May I refer to the total pressure minimum of the oxide systems as given by Mr. Tetenbaum. This minimum and its temperature dependence are clearly defined in the binary systems as shown in the paper on plutonium oxide (by Ciani and myself) presented at the IAEA Symposium on Thermodynamics of Nuclear Materials held in 1967. The total pressure minimum over the ternary U-Pu-O oxide, reported by Olson and myself at the Conference entitled 'Plutonium 1970 and Other Actinides' in Santa Fe, is necessarily very flat since UO_2, the most predominant species in the so-called quasi-equilibrium range, is almost constant. Because of this uncertainty of almost ± 0.01 in the O/M ratio, experimental measurements are less important than indirect methods of calculation.

M.H. RAND: I believe that before we can calculate the position of the minimum in the vapour pressure versus O/M ($M = U_{0.8}Pu_{0.2}$) to a greater degree of accuracy than is possible using an absolute pressure method such as that put forward by Mr. Tetenbaum, we must first obtain more accurate data on all the thermodynamic functions of the gaseous oxides of uranium and plutonium.

R.J. ACKERMANN: From an engineering point of view, the fact that the total pressure is so insensitive to the composition (O/M, $M = U_{0.8}Pu_{0.2}$) makes it easier to predict the total fuel loss and means that one does not have to be concerned with any change in composition. However, as Mr. Rand prompts me to say, the U/Pu ratio does change rather significantly in the composition range O/M = 1.93 to 1.97, and therefore the O/M ratio has a considerable influence on the actinide redistribution. The thermodynamic data for UO_2(g), UO_2(g) and PuO(g) are therefore the most important values in models predicting actinide redistribution.

R.W. OHSE: Let me briefly point out one of the main reasons for the discrepancies in pressure measurements over the ternary U-Pu-O system, apart from the fact that there are no congruent compositions. The appearance and fragmentation potentials of UO_3 in the range of 13 eV are close together. The various techniques proposed for the evaluation of the ion intensity measurements of the nine species found over the mixed oxides cannot be used because of an insufficient range of linearity of the ionization curve, and the interference of fragmentation. The technique of normalizing the various ion currents using a normalization factor obtained by measuring one gaseous species, e.g. UO_2, at two different electron energies, or that of applying a constant difference in electron energy to the appearance potential, can only be used in the linear range of the ionization curve, if no fragmentation occurs. Thus a direct calibration of the mass spectrometer, as reported by Olson and myself at the Plutonium 1970 Conference is preferable.

A.S. PANOV: I agree with Mr. Tetenbaum when he says, in connection with substoichiometric dioxide, that an influence of tungsten is unlikely. However, when neodymium oxide is added and the chemical potential of oxygen thereby increased, oxidation of, for example, tungsten may occur, particularly when one is dealing with compositions close to stoichiometric. This needs to be borne in mind.

M. TETENBAUM: Certainly one must watch for tungsten contamination with U-Pu-O compositions close to stoichiometric, because of the high oxidation potentials required to achieve O/M ratios of around 1.99 to 2.00.

R.W. OHSE: I also consider Mr. Panov to be right in stressing the need to control possible interactions between the samples and the cell materials. Our liquidus/solidus measurements of the uranium plutonium oxides showed that there is no serious interaction in the quasi-congruent composition range 1.95 to 1.97 and below. However, we found strong interactions of the stoichiometric PuO_2 with tungsten, showing clearly the transport of tungsten to the collection targets.

C.B. ALCOCK: I should like to know whether there have been any studies carried out on the uranium-oxygen system as a result of which oxygen potentials obtained from solid-state electrochemistry and mass spectrometry in the same temperature range can be directly compared?

M.H. RAND: As far as I am aware there are no measurements on the uranium-oxygen system using these two techniques which can be directly compared, mainly because different temperature ranges are normally used. Many years ago, however, P. Blackburn made some mass-loss measurements on UO_{2+x} solid solutions which are in good agreement with the emf measurements. For the U-Pu-O system there are mass-spectrometric measurements by Battles et al. which could be compared with oxygen potential measurements. However, it is very difficult to interpret these mass-spectrometric values, and further measurements on isomolecular reactions involving elements such as lanthanum, yttrium and silicon are therefore necessary.

A. PATTORET: May I add a few remarks on the UO_{2+x}-W interaction. From mass spectrometric observations it is quite easy to determine the deviation from stoichiometry of UO_{2+x} that allows tungsten to oxidize to form $WO_2(s)$. When $T \simeq 1400°C$ and the O/U ratio $\lesssim$ 2.005, no oxidized species of tungsten is observed. At higher O/U ratios the UO_2, UO_3, WO_2 and WO_3 species may be measured simultaneously; the polymer forms of the tungsten oxides, of course, represent a considerable part of the gas phase (W_3O_9, W_3O_6.... etc.). For uranium oxides such as O/U < 2.005, the chemical interaction of the tungsten forming the effusion cell may be disregarded. On the other hand, diffusion of the oxygen through the tungsten walls may not be negligible and may, through the action of a parasitic flux, disturb the establishment of the evaporation congruence.

C.B. ALCOCK: The upper temperature limit of emf studies with solid-oxide electrolytes has until recently been set mainly by the lack of satisfactory reference electrode systems. As a result of work on steelmaking oxygen probes, the electrode $Mo-MoO_2$ has been established with a good degree of precision up to 1650°C. It is therefore possible that the emf mass spectrometry 'temperature gap' could be closed in the near future.

ЭКСПЕРИМЕНТАЛЬНОЕ ОПРЕДЕЛЕНИЕ И ТЕОРЕТИЧЕСКИЙ РАСЧЕТ ТЕРМОДИНАМИЧЕСКИХ СВОЙСТВ ОЦК-РАСТВОРОВ УРАНА

Ю.В. ВАМБЕРСКИЙ, А.Л. УДОВСКИЙ,
О.С. ИВАНОВ
Институт металлургии им. А.А. Байкова
Академии наук СССР,
Москва,
Союз Советских Социалистических Республик

Abstract – Аннотация

EXPERIMENTAL DETERMINATION AND THEORETICAL CALCULATION OF THE THERMODYNAMIC PROPERTIES OF BODY-CENTRED CUBIC SOLID SOLUTIONS OF URANIUM.

Data on the thermodynamic properties of solid solutions of uranium are extremely limited. The paper describes the procedure and results of the experimental and theoretical study of the thermodynamic properties of U-Mo and U-Nb body-centred cubic solutions. The experimental studies were carried out using an emf method with a liquid electrolyte. The emf of the following Galvanic cells were measured in the 1048-1173 K temperature range:

⊖ U solid	$NaCl + KCl + UCl_3$ (equimolar)	U - Me ⊕ solid solution

where Me = Mo, Nb.

After the experiments, the electrolyte was tested systematically to detect molybdenum and niobium by means of chemical analysis; passage of these elements from alloy to electrolyte was not observed. Particular care was taken in dehydrating the chlorides before preparing the electrolyte. The effect of the corrosion of the samples in the electrolyte on the emf at $T \leq 1173$ K was insignificant. The excess thermodynamic values calculated from the emf values in both systems were of complex and alternating nature, depending on composition. ΔS^E in the U-Mo system changed its sign twice. A description is given of the model which was used to obtain the equations for calculating the thermodynamic functions on the basis of the physical properties of solid solutions. The results of calculations for 1100 K are compared with experimental data. In the U-Mo system, the calculated dependence of ΔS^E practically reproduces the experimental curve; the calculated dependence of the enthalpy of the mixture is similar in slope to the experimental one. Calculation of the excess thermodynamic functions in the U-Nb system made it possible to predict the experimental concentration dependence of the excess entropy of the mixture (with change of sign) with satisfactory quantitative agreement. The physical reasons have been found for the sign changes of ΔS^E in both systems and for the sign change of ΔH in the U-Mo system. It was noted that the vibration component of the enthalpy of the mixture, excluding the zero-vibration energy, was different from zero only if anharmonicity was taken into account. The negative sign, obtained in the calculations, in the case of the elastic component of the excess entropy in the U-Mo system and for a niobium content higher than 40 at.% in the U-Nb system demonstrates the growth of elastic energy with increasing temperature, and is due to the predominant effect of the increase of difference in the lattice periods of pure components over the decrease in Young's modulus of the solid solutions.

ЭКСПЕРИМЕНТАЛЬНОЕ ОПРЕДЕЛЕНИЕ И ТЕОРЕТИЧЕСКИЙ РАСЧЕТ ТЕРМОДИНАМИЧЕСКИХ СВОЙСТВ ОЦК-РАСТВОРОВ УРАНА.

Количество данных о термодинамических свойствах твердых растворов урана весьма ограничено. В работе изложены методика и результаты экспериментального и теоретического исследований термодинамических свойств оцк-растворов U-Mo и U-Nb. Экспериментальные исследования выполнены методом электродвижущих сил (ЭДС) с жидким электролитом. В температурном диапазоне 1048-1173 °K измерялись ЭДС гальванических элементов:

⊖ U тв.	$NaCl + KCl + UCl_3$ (эквимол.)	U - Me ⊕ тв.р-р

где Me = Mo,Nb. После опытов проводился систематический контроль электролита с помощью химического анализа на присутствие Mo и Nb; переход этих элементов из сплавов в электролит не обнаружен. Особое внимание уделялось обезвоживанию хлоридов перед приготовлением электролита. Влияние коррозии образцов в электролите на величину ЭДС при $T \leq 1173$ °K несущественно. Рассчитанные из значений ЭДС избыточные термодинамические величины в обеих системах имеют сложный, знакопеременный характер в зависимости от состава. ΔS^E в системе U-Mo изменяет знак дважды. Описана модель, на основании которой были получены уравнения для расчета термодинамических функций, исходя из физических свойств твердых растворов. Результаты расчетов для 1100 °K сопоставлены с экспериментальными данными. В системе U-Mo расчетная зависимость ΔS^E практически воспроизвела экспериментальную кривую; расчетная зависимость энтальпии смешения симбатна с экспериментальной. Расчет избыточных термодинамических функций в системе U-Nb позволил предсказать экспериментальную концентрационную зависимость избыточной энтропии смешения (с изменением знака) в хорошем количественном соответствии. Вскрыты физические причины изменений знака ΔS^E в обеих системах и изменения знака ΔH в системе U-Mo. Отмечено, что вибрационная компонента энтальпии смешения, исключая энергию нулевых колебаний, отлична от нуля лишь при учете ангармонизма. Полученный в расчетах отрицательный знак упругой компоненты избыточной энтропии в системе U-Mo и при содержании Nb большем 40 ат% в системе U-Nb свидетельствует о возрастании упругой энергии с увеличением температуры и объясняется превалирующим влиянием увеличения разности периодов решеток чистых компонентов над уменьшением модуля Юнга твердых растворов.

I. Состояние вопроса о термодинамике твердых растворов урана

До недавнего времени в литературе практически отсутствовали сведения о систематическом изучении твердых растворов урана в широком концентрационном диапазоне. Были описаны только термодинамические свойства системы уран-цирконий в диапазоне температур 1030-1184 °K, определенные методом электродвижущих сил (ЭДС) с жидким электролитом, состоявшим из эквимолярной смеси хлоридов натрия и калия с добавкой 0,5 вес.% трихлорида урана. Избыточная энтропия и энтальпия смешения оцк-растворов при 1073 °K изменяют знак с отрицательного на положительный при уменьшении содержания циркония до 20 и ≈10 ат.%, соответственно [1].

Авторами доклада проведены экспериментальные исследования термодинамических свойств оцк-растворов урана с молибденом и ниобием. Избыточные термодинамические функции в обеих системах являются знакопеременными в зависимости от состава [2-3]. Для выяснения причин такого сложного характера концентрационных зависимостей термодинамических свойств растворов, а также для создания термодинамической базы теоретического построения диаграмм состояния урановых систем [4-5] была проведена работа, целью которой являлась разработка физико-химической модели изоморфных неупорядоченных растворов двухкомпонентных систем. Проверка принятой модели осуществлялась путем сопоставления результатов расчетов с экспериментальными термодинамическими данными [2,3,6].

В настоящем сообщении авторами предпринята попытка подвести некоторые итоги термодинамических исследований оцк-растворов урана.

2. Метод экспериментального исследования

Термодинамика растворов γ - урана с молибденом и ниобием изучалась с помощью метода электродвижущих сил (ЭДС), позволяющего определять изменение термодинамического потенциала Гиббса при образовании сплавов в условиях,наиболее близких к равновесным. Помимо общих положений, соблюдение которых необходимо при использовании метода ЭДС, работа с ураном вследствие его значительной химической активности требует особых предосторожностей, чтобы исключить побочные процессы, оказывающие влияние на величину ЭДС элементов

$$\ominus\ \underset{(тв)}{U}\ \Bigg|\ \begin{array}{l}\text{Ионный проводник,}\\ \text{содержащий } U^{n+}\\ \text{(твердый или жидкий)}\end{array}\ \Bigg|\ \underset{\text{(тв.раствор)}}{U\text{-}Me}\ \oplus \qquad (2.I)$$

Здесь $Me = Mo, Nb$, U^{n+}- ионы урана валентности n .

Указанная особенность объектов исследования была учтена при разработке конструкции электрохимической ячейки и оказала влияние на выбор электролита. В работе использовался жидкий электролит, составленный из эквимолярной смеси хлоридов натрия и калия особой чистоты с добавкой 5-6 вес.% UCl_3 (валентность ионов урана n = 3). Выбор состава электролита был сделан на основании анализа литературных данных [7-I5], который показал, что этот состав отвечает требованиям, предъявляемым к жидкому электролиту. В частности, в [7] показано, что в эквимолярной смеси $NaCl + KCl$ в контакте с металлическим ураном трихлорид урана устойчив во всем доступном температурном интервале. Особое внимание уделялось обезвоживанию хлоридов при приготовлении электролита. По данным [I6] после обезвоживания в вакууме 200г. KCl в течение I2 часов при 500°С процент гидролиза равен I,33.I0^{-4}. В настоящей работе обезвоживание производилось при 500° в течение 24 часов с непрерывным вакуумированием. Трихлорид урана добавлялся в электролит с помощью электролиза с растворимым урановым анодом и диафрагмой. Для выяснения влияния коррозии образцов в электролите на величину измеряемой ЭДС было проведено специальное исследование [I7], которое показало, что это влияние при температурах, не превышающих II73°К, несущественно.

Элементы(2.I)работали в атмосфере аргона с содержанием азота не более 0,005%, кислорода-не более 0,00I%, воды-не более 0,02 мг на I л при нормальных условиях. Давление аргона в ячейке P = I,5 кг/см^2.

Чистота применявшихся металлов (вес.%): урана - 99,80; молибдена - 99,988; ниобия - 99,7. Образцы сплавов подвергались гомогенизирующему отжигу в двойных эвакуированных до остаточного давления (5-6).I0^{-3} мм рт.ст. кварцевых ампулах. Между ампулами помещался геттер в виде циркониевой стружки. После закалки образцы подвергались токарной обработке для удаления поверхностного слоя и непосредственно перед сборкой элемента шлифовались и помещались в четыреххлористый углерод.

Измерения ЭДС производились по общепринятому способу. Отклонения температуры от средних значений в рабочем режиме не превышали I-2°. Достижение равновесия в гальванических элементах проверялось воспроизведением температурного хода

ТАБЛИЦА I. КОЭФФИЦИЕНТЫ УРАВНЕНИЯ (2.2) И ОШИБКИ ОПРЕДЕЛЕНИЯ ЭДС В СИСТЕМЕ U-Mo

x	Число экспериментальных точек	$a \cdot 10^6$ в.град$^{-1}$	$b \cdot 10^3$, в	$2\sigma_{\mathcal{E}} \cdot 10^3$, в
0,05	4	-6,4	10,59	±1,16
0,10	18	-34,0	51,74	±1,42
0,15	9	-17,6	41,51	±0,72
0,20	19	-11,7	32,82	±0,45
0,24	10	0,0	18,13	±0,72
0,30	13	15,87	-2,22	±0,15
0,40; 0,60; 0,80	20	42,68	-27,54	±0,90

ЭДС при нагревании и охлаждении, а также воспроизведением значений ЭДС при повторных опытах с образцами из вновь приготовленных сплавов тех же составов. После опытов производился химический анализ электролита на присутствие легирующего элемента. Переход легирующих элементов в электролит не обнаружен. Чувствительность определения молибдена $\approx 2 \cdot 10^{-5}$ г, ниобия $\approx 4 \cdot 10^{-5}$ г в 1 г электролита.

Зависимости ЭДС от температуры во всех случаях хорошо аппроксимировались прямолинейными. Обработка результатов измерений производилась по способу наименьших квадратов. В итоге зависимости ЭДС от температуры представлялись в виде уравнений

$$\mathcal{E} = aT + b \tag{2.2}$$

где T - абсолютная температура, a и b - коэффициенты. За ошибку в определении ЭДС принималось удвоенное среднеквадратичное отклонение ($2\sigma_{\mathcal{E}}$) от величины, найденной по способу наименьших квадратов. Это соответствует доверительной вероятности 0,95.

$$\sigma_{\mathcal{E}} = \sqrt{\frac{\sum_1^n \left| \frac{\sqrt{\sum_1^m |\Delta \mathcal{E}_T|^2}}{m} \right|^2}{n}} \tag{2.3}$$

где $\Delta \mathcal{E}_T = |\mathcal{E}_{\text{измер.}} - \mathcal{E}_{\text{наим.квадр.}}|_{T=const}$; m - число измерений при одной температуре; n - число измерений при разных температурах.

Для вычисления изменения термодинамического потенциала Гиббса при образовании сплавов использовалось графическое интегрирование уравнений Гиббса-Дюгема

$$\Delta G = x \int_0^{1-x} \frac{\Delta \bar{G}_U}{x^2} d(1-x) \tag{2.4}$$

или

$$\lg f_{Me} = -\int_{x=1}^{x} \frac{1-x}{x} d \lg f_U \tag{2.5}$$

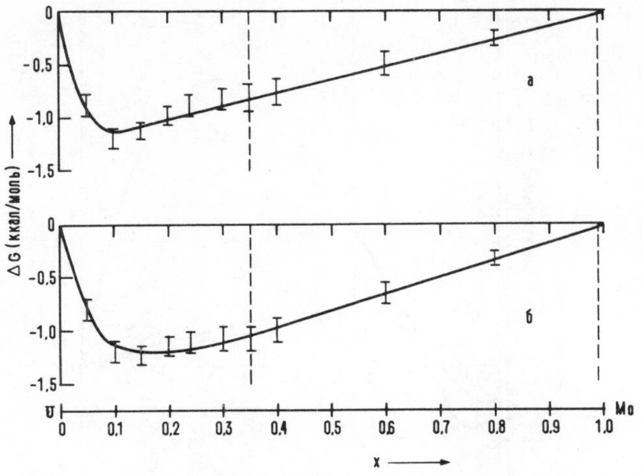

Рис.1. Изменение термодинамического потенциала Гиббса в системе U-Mo а) при 1048 °К; б) при 1173 °К. Пунктиром показаны фазовые границы диаграммы состояния [18].

где ΔG - изменение термодинамического потенциала Гиббса, x - атомная доля легирующего элемента в сплаве, $\Delta \bar{G}_U$ - изменение химического потенциала урана, f_{Me} и f_U - коэффициенты активности легирующего элемента и урана соответственно.

$$\Delta \bar{G}_U = - n\mathcal{F}\mathcal{E} = - 69{,}186\ (aT + b)\ \text{ккал/г.ат}, \qquad (2.6)$$

где $n = 3$; $\mathcal{F} = 23{,}062$ ккал/в.г-экв.; $\mathcal{E}$ - в вольтах.

$$f_U = \frac{a_U}{x} \qquad (2.7)$$

где a_U - активность урана в сплаве.

$$\ln a_U = \frac{\Delta \bar{G}_U}{RT} \qquad (2.8)$$

где R - газовая постоянная. Относительные ошибки изменения термодинамического потенциала Гиббса определялись как относительные ошибки измерения площади по отдельным участкам с учетом их статистического веса.

Энтропия и энтальпия смешения рассчитывались по формулам

$$\Delta S = -\left[\frac{\partial(\Delta G)}{\partial T}\right]_P \quad \text{и} \qquad (2.9)$$

$$\Delta H = \Delta G - T\left[\frac{\partial(\Delta G)}{\partial T}\right]_P \qquad (2.10)$$

причем дифференцирование заменялось отношением конечных приращений, т.е. ΔS и ΔH считались независимыми от температуры в рассматриваемом интервале (1048-1173°К). Избыточные термодинамические функции вычислялись по общеизвестному принципу.

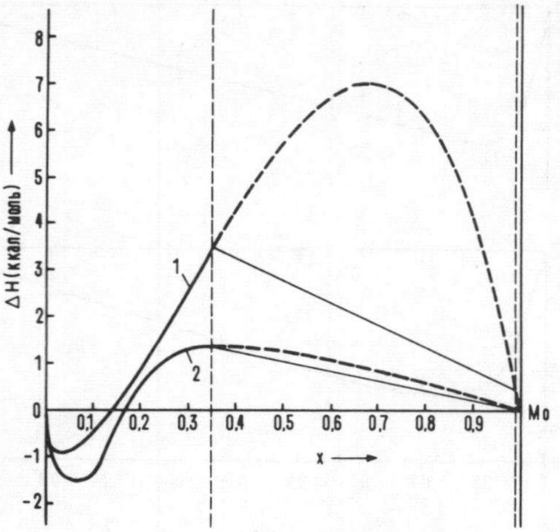

Рис.2. Сопоставление расчетной приведенной энтальпии смешения (1) и экспериментальной энтальпии смешения (2) в системе U-Mo.

3. Результаты экспериментов

3.I. Система $U-Mo$

Характерной особенностью результатов измерений ЭДС является изменение знака ее температурного коэффициента в области существования оцк-растворов на основе урана ($x = 0,00 \div 0,35$). При $x < 0,24$ температурный коэффициент ЭДС меньше нуля, при $x = 0,24$ он равен нулю и при $x > 0,24$ – больше нуля. Коэффициенты уравнения (2.2) представлены в таблице I.

Изменения термодинамического потенциала Гиббса при образовании сплавов показаны на рис.I. Вычисленные по значениям ΔG энтропия смешения и избыточные термодинамические величины имеют отрицательные значения в области малых концентраций и положительные – при более высоком содержании молибдена в сплавах. Изменение знака происходит приблизительно при следующих значениях x : энтропии смешения – 0,II; энтальпии смешения – 0,I7 (рис.2, кривая 2); изотерм избыточного термодинамического потенциала смешения – 0,20. Избыточная энтропия смешения изменяет знак дважды: с отрицательного на положительный при $x \approx 0,15$ и с положительного на отрицательный при $x \approx 0,95$ (рис.3, кривая 2).

3.2. Система $U - Nb$

Особенностью зависимостей ЭДС от концентрации ниобия в сплавах при $T = const$ является довольно резкое возрастание значений ЭДС в интервале $x = 0,4 \div 0,5$ (область равновесия двух твердых растворов согласно диаграмме состояния [19]). Коэффициенты уравнения (2.2) приведены в таблице П. Температурный коэффициент ЭДС для сплавов всех составов, кроме

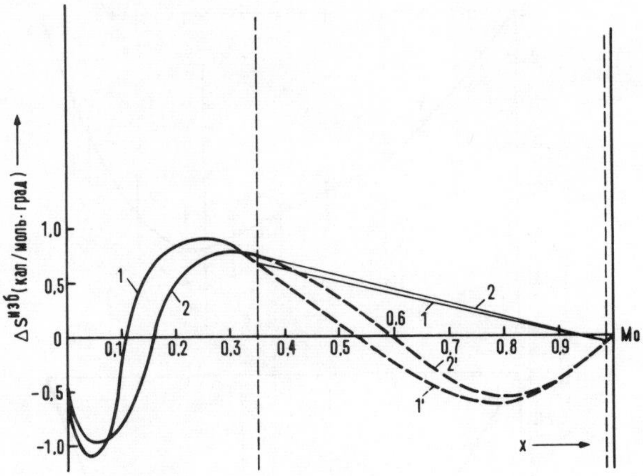

Рис.3. Сопоставление расчетной (1) и экспериментальной (2) избыточной энтропии смешения в системе U-Mo.

ТАБЛИЦА II. КОЭФФИЦИЕНТЫ УРАВНЕНИЯ (2.2) И ОШИБКИ ОПРЕДЕЛЕНИЯ ЭДС В СИСТЕМЕ U-Nb

x	Число экспериментальных точек	$a \cdot 10^6$, в.град$^{-1}$	$b \cdot 10^3$, в	$2\sigma_\varepsilon \cdot 10^3$, в
0,I	4	2,II	-0,99	±0,I32
0,2	I8	I5,50	-I3,59	±0,78
0,3	I9	30,30	-28,42	±0,60
0,4	20	72,40	-70,64	±I,I2
0,5; 0,6	20	I47,50	-I28,49	±I,62
0,7	I0	I78,80	-I47,I95	±2,82
0,8	I5	-I30,00	-26I,I6	±8,92
0,9	4	82,00	58,59	±2,74

x = 0,8, положительный. Изменения термодинамического потенциала Гиббса при образовании сплавов изображены на рис.4. Дальнейшие вычисления показали, что энтропия смешения положительна во всем концентрационном диапазоне. Концентрационные зависимости избыточных термодинамических величин изменяют знак с положительного на отрицательный при увеличении концентрации ниобия. Изменение знака происходит примерно при следующих значениях x : термодинамического потенциала Гиббса (I048°K) - 0,I5; энтальпии - 0,56 (рис.5, кривая 2); энтропии - 0,64 (рис.6, кривая 2).

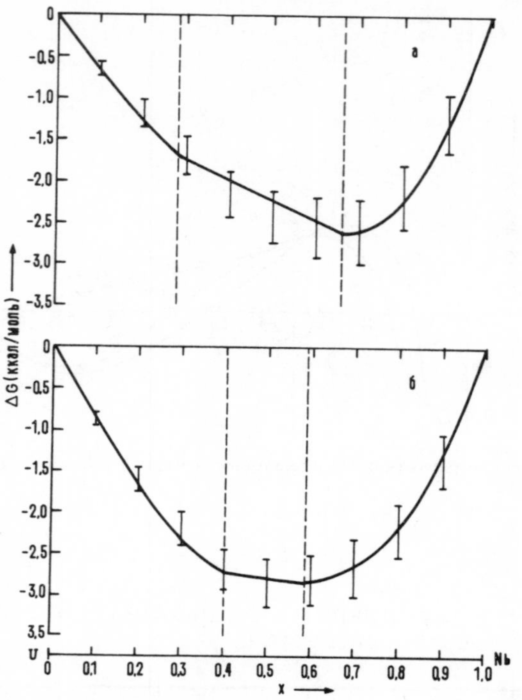

Рис.4. Изменение термодинамического потенциала Гиббса в системе U-Nb а) при 1048 °K; б) при 1173 °K. Пунктиром показаны фазовые границы диаграммы состояния [19].

4. Модельные представления и приближения

4.I. Основные положения

Простейшим вариантом физико-химической модели является установление взаимосвязи между термодинамическими свойствами и достаточно легко измеримыми физическими характеристиками твердых растворов. Неупорядоченный раствор рассматривается как твердое тело, обладающее определенными измеримыми физическими свойствами; с другой стороны, раствор моделируется абстрактной системой в аддитивном приближении – рис.7. Последнее означает, что абстрактная система, введенная в целях теоретического описания раствора, состоит из невзаимодействующих подсистем: I) подсистемы атомных полиэдров "среднего" размера при 0°K, образующих решетку идеального по периодичности кристалла; 2) подсистемы флуктуационных волн статических смещений, описывающих смещения центров "реальных" атомных полиэдров относительно центров атомных полиэдров "среднего" размера; 3) подсистемы фононов, порожденных тепловыми колебаниями атомных остовов, происходящими относительно центров "реальных" атомных полиэдров; 4) подсистемы термически возбужденных электронов; 5) подсистемы перестановок атомных полиэдров "среднего" размера.

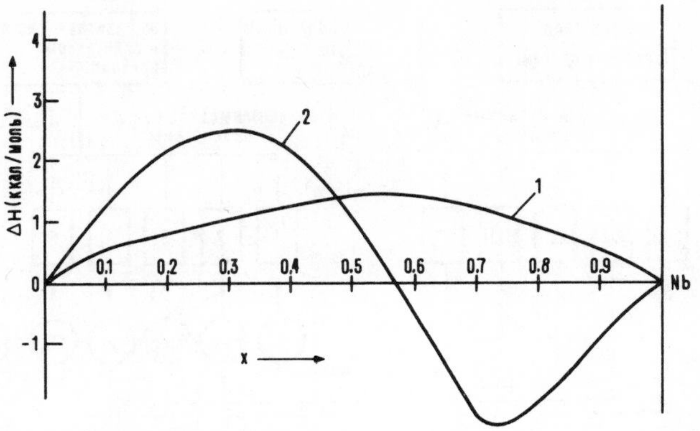

Рис.5. Сопоставление расчетной приведенной энтальпии смешения (1) и экспериментальной энтальпии смешения (2) в системе U-Nb.

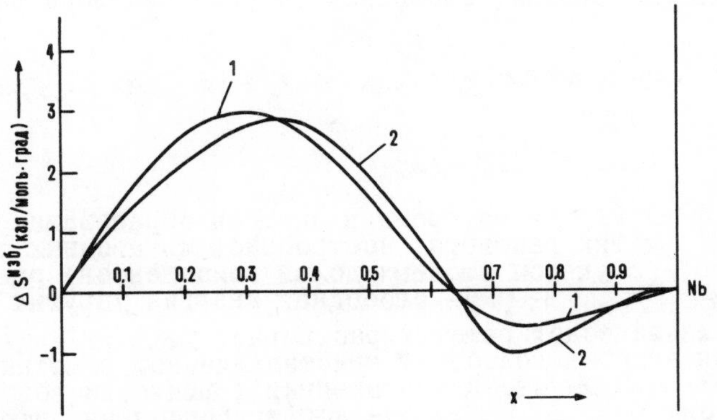

Рис.6. Сопоставление расчетной (1) и экспериментальной (2) избыточной энтропии смешения в системе U-Nb.

Каждая из пяти подсистем вносит соответствующий аддитивный вклад в свободную энергию системы F . Известно, что свободная энергия системы связана со статистической суммой системы Z соотношением $F = -kT\ln Z$. В случае невзаимодействующих подсистем статистическая сумма системы равна произведению статистических сумм подсистем $Z = Z_1 \cdot Z_2 \ldots Z_5$. Тогда свободная энергия системы $F = F_1 + F_2 + \ldots + F_5$, что и показано на рис.7. Именно поэтому в качестве основной термодинамической функции более удобно принять свободную энергию Гельмгольца, несмотря на то, что при обычных условиях проведения эксперимента (температура и давление постоянны) определяются первые (химический потенциал, объем, тепловой эффект) или вторые (теплоемкость, коэффициент термического расширения, модуль упругости) производные термодинамического потенциала Гиббса.

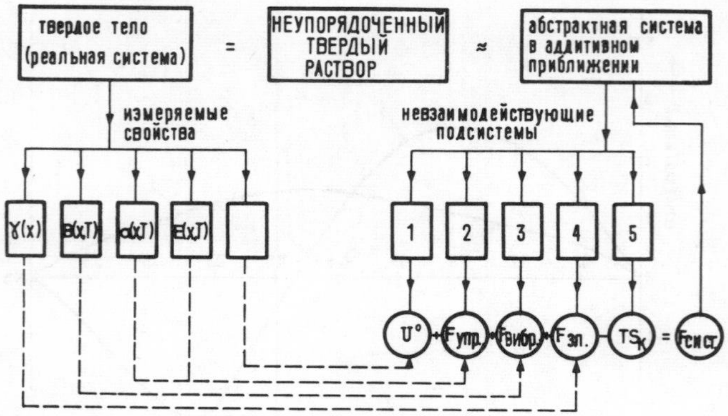

Рис.7. Схематическое изображение модельных представлений.

Таким образом, свободная энергия раствора равна

$$F(x,T,v) = U^{\circ}(x,v) + F_{упр}(x,T,v) + F_{вибр}(x,T,v) + F_{эл}(x,T,v) - TS_{K}(x) \qquad (4.I)$$

Здесь $U^{\circ}(x,v)$ - внутренняя энергия образования кристаллической решетки раствора, построенной из атомных полиэдров "среднего" (с учетом атомных долей компонентов) размера, при 0°К; $F_{упр}(x,T,v)$ - свободная энергия упругих искажений кристаллической решетки раствора; $F_{вибр}(x,T,v)$ - свободная энергия колебаний кристаллической решетки; $F_{эл}(x,T,v)$ - свободная энергия термически возбужденных электронов; $S_{K}(x)$ - конфигурационная энтропия; v - объем.

4.2. Приближения

4.2.I. Приближения модели

При переходе от термодинамического формализма (4.I) к аналитическому описанию физического содержания модельных представлений нами были приняты следующие допущения.

Внутренняя энергия образования кристаллической решетки, построенной из атомных полиэдров „среднего" размера, записывалась в рамках модели ближайших парных соседств. К сожалению, метод псевдопотенциалов для вычисления $\Delta U^{\circ}(x,v)$ [20-25] не может быть использован в связи с тем, что псевдопотенциал урана в настоящее время не определен.

На основании представлений теории псевдопотенциала применительно к сплавам [20] каждый атом сплава можно рассматривать в нулевом приближении как микроскопический кусок соответствующего металла (А или В). Это означает, что для построения решетки из "средних" атомных полиэдров полиэдры А и В нужно деформировать до "средних" размеров. Энергия ис-

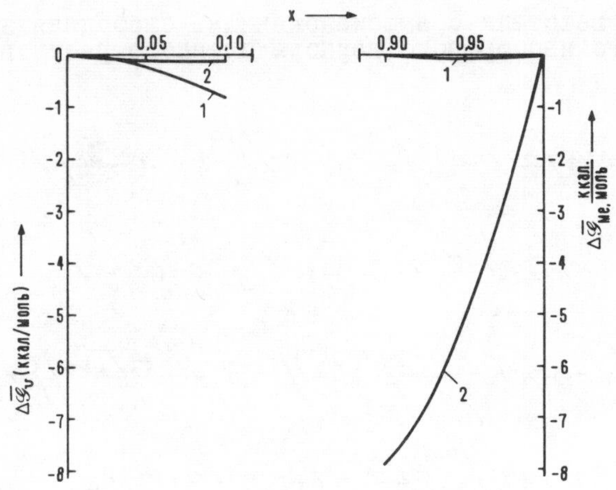

Рис.8. Изменения химического потенциала: слева – γ-урана при растворении в нем молибдена (1) и ниобия (2); справа – молибдена (1) и ниобия (2) при растворении в них γ-урана. T = 1173 °K.

кажения кристаллической решетки после ее релаксации может быть вычислена методами теории упругости. Поскольку рассматривается неупорядоченный раствор, совокупность статических смещений атомов, следуя М.А.Кривоглазу [26], можно представить в виде подсистемы флуктуационных волн (в смысле отклонения от средних размеров) статических смещений. Энергия этой подсистемы в упруго-изотропных кристаллах, как показано в [26], равна упругой энергии, вычисленной в рамках линейной теории упругости. На основании анализа моделей, рассматривающих эффект искажений кристаллической решетки раствора во всей области составов, мы остановились на наиболее последовательной модели И.Л.Аптекаря и Б.Н.Финкельштейна [27], изложенной в приближении ближайших парных соседств. Согласно этой модели зависимость энергий парного межатомного взаимодействия от расстояния между соответствующими атомами записывается в виде их разложения в ряды по степеням смещений. Под смещениями понимаются разности между размерами атомов компонентов раствора и размером "среднего" атома раствора. Конкретное выражение для упругой энергии, полученное на основании упомянутой модели А.Г.Лесником [28] для малолегированных сплавов, было распространено авторами доклада на всю область составов, а также были приняты во внимание температурные зависимости модуля Юнга и периода решетки [2,6].

Свободная энергия колебаний кристаллической решетки рассчитывалась по дебаевской модели с учетом нулевых колебаний. Ангармонизм учитывался в рамках квазигармонической теории.

Свободная энергия термически возбужденных электронов описывалась обычным выражением в приближении электронной подсистемы в виде вырожденного ферми-газа.

В соответствии с вышеизложенным, свободная энергия двухкомпонентного изоморфного неупорядоченного раствора

$$F(x,T,v)=Nz\left[x(1-x)E^{\circ}_{CM}+(1-x)\frac{E^{\circ}_{AA}}{2}+x\frac{E^{\circ}_{BB}}{2}\right]+$$

$$+\frac{3\pi}{16}NzE(x,T,v)\,d(x,T)\left[\Delta d(T)\right]^2 x\cdot(1-x)+$$

$$+RT\left\{3\ln\left[1-\exp\left(-\frac{\theta(x,T,v)}{T}\right)\right]-D\left[\frac{\theta(x,T,v)}{T}\right]\right\}+\frac{9}{8}R\theta(x,0,v)-$$

$$-\frac{\gamma(x,v)}{2}T^2+RT\left[x\ln x+(1-x)\ln(1-x)\right] \qquad (4.2)$$

Здесь I-я, 2-я и 3-я строки соответствуют U°, F упр. и F вибр.; в 4-й строке записаны F эл и TS_K. Соотношение (4.2) устанавливает связь между свободной энергией раствора (а следовательно и ее производными - энтропией и энтальпией) с его физическими свойствами: E - модуль Юнга; a - период решетки (для оцк-решетки $\sqrt{3}a=2d$); θ - температура Дебая; γ - коэффициент электронной теплоемкости. Эта связь показана на рис.7 пунктиром. В (4.2) N - число Авогардо; Z - координационное число; $E^{\circ}_{CM}=E^{\circ}_{AB}-\frac{1}{2}(E^{\circ}_{AA}+E^{\circ}_{BB})$, E°_{ij} - парные энергии связей; $D(\frac{\theta}{T})$ - интеграл Дебая.

4.2.2. Приближения для температурных зависимостей физических характеристик

Нами было принято /2,29/:

$$E(x,T,v)=E[x,v(T)]=E(x,T)=E(x,0)\left[1-\beta(x)t(x)\right]; \qquad (4.3)$$

$$d(x,T)=d(x,0)\left[1+\alpha_{л}(x)\,t(x)\right]; \qquad (4.4)$$

$$\Delta d(T)=d(1;T)-d(0;T); \qquad (4.5)$$

$$\theta(x,T,v)=\theta[x,v(T)]=\theta(x,T)=\theta(x,0)\left[1-\frac{\beta(x)}{2}t(x)\right]; \qquad (4.6)$$

$$\gamma(x,v)=\gamma(x) \qquad (4.7)$$

Здесь β - температурный коэффициент модуля Юнга для $t(x)=T-\theta(x,0)>>0$; $\alpha_{л}$ - линейный коэффициент термического расширения.

Изложенные приближения считаются общими для всех изоморфных растворов, тогда как концентрационные зависимости физических свойств рассматриваются в виде конкретных экспериментальных зависимостей для каждой бинарной системы.

4.3. Выводы

Для расчета термодинамических функций образования растворов изменение внутренней энергии $\Delta U^\circ = Nz E^\circ_{CM} x(1-x)$ в явной форме для урановых систем в настоящее время вычислить не представляется возможным (см.подраздел 4.2.I). Таким образом, для расчета избыточной энтропии (ΔS^E) и приведенной внутренней энергии смешения ($\Delta U^* = \Delta U - \Delta U^\circ$) необходимо и достаточно знать экспериментальные зависимости от состава шести физических характеристик: $E(x,0)$, $a(x,0)$, $\Theta(x,0)$, $\gamma(x)$, $\beta(x)$ и $\alpha_л(x)$.

5. Результаты расчетов

Физические характеристики для оцк-растворов уран-молибден и уран-ниобий были собраны и протабулированы, недостающие оценены авторами [2,3]. Расчетные формулы для избыточных функций были получены традиционным способом: $S^E = -\left(\frac{\partial F}{\partial T}\right)_{x,v} - S_K$;

$$\Delta U^* = \frac{\partial\left(\frac{\Delta F}{T}\right)}{\partial\left(\frac{1}{T}\right)} - \Delta U^\circ = \Delta F + T\Delta S^E - \Delta U^\circ \qquad [2,29].$$

При сопоставлении результатов расчетов с экспериментальными данными было принято во внимание, что $P = const$; объем смешения $\Delta v(x) \approx 0$, так как отклонение параметра решетки от закона Вегарда для растворов U-Mo и U-Nb незначительно [I8,I9]. Поэтому $\Delta U \approx \Delta H$. Расчеты термодинамических функций произведены для T = II00°K.

5.I. Система U - Mo

Результаты расчетов избыточной энтропии и приведенной энтальпии смешения сопоставлены с экспериментальными данными на рис.3 и рис.2, соответственно.

За первое изменение знака $\Delta S^E(x)$ ответственны вибрационная и электронная компоненты избыточной энтропии раствора (названия компонент энтропии и энтальпии соответствуют названиям компонент свободной энергии раствора - подраздел 4.I). Ко второму изменению знака $\Delta S^E(x)$ приводит суммирование электронной и упругой компонент энтропии. Изменение знака $\Delta H^*(x)$ обусловлено в основном вибрационной и упругой компонентами энтальпии смешения.

Упругая компонента энтропии отрицательна для $0<x<1$. Это соответствует возрастанию упругой энергии (упругая компонента свободной энергии) в области ее максимальных значений ($0,6 < x < 0,7$) примерно на 20% при увеличении температуры от 0 до II00°K [30]. Детали расчетов и более подробный анализ результатов приведены в [2,30].

5.2. Система U - Nb

В отличие от расчетов, описанных в подразделе 5.I, расчеты термодинамических функций в данной системе предшествовали обработке экспериментальных данных. Рис.6 показывает, что экспериментальная зависимость ΔS^E практически воспроизвела расчетную кривую. За изменение знака расчетной кривой ответственна вибрационная компонента энтропии. Однако, расчетная зависимость $\Delta H^*(x)$ (рис.5) не изменяет знак, оставаясь положительной для $0 < x < 1$, тогда как экспериментальная зависимость $\Delta H(x)$ изменяет знак с положительного на отрицательный при $x \approx 0{,}56$.

Интересно, что упругая компонента энтропии является знакопеременной функцией концентрации /3,30/.

6. Обсуждение

6.I. Экспериментальное исследование

Из сравнения кривых $\Delta G(x)$ для систем $U-Mo$ и $U-Nb$ - рис.I и рис.4, соответственно - видно, что наиболее термодинамически устойчивы сплавы более богатые ураном (x = 0,I0-0,I6) в системе $U-Mo$ и сплавы, состав которых (x = 0,58-0,68) несколько смещен в сторону ниобия в системе $U-Nb$. Сопоставление концентрационных зависимостей изменения химического потенциала γ - урана и легирующих элементов (Mo, Nb) - рис.8 - показывает, что растворы молибдена в γ - уране термодинамически более устойчивы, чем растворы ниобия и растворы γ - урана в ниобии более устойчивы, чем растворы в молибдене.

Общей чертой термодинамических свойств оцк-растворов в известных нам исследованных системах $U-Zr$ [I], $U-Nb$ и $U-Mo$ (разд.3) является изменение знака избыточных термодинамических характеристик в зависимости от состава; в системе $U-Mo$ $\Delta S^E(x)$ изменяет знак дважды (рис.3) и изменяет знак даже энтропия смешения. Упрощенные теоретические модели (идеальных, регулярных, субрегулярных растворов и др.) не могут описать, не говоря о независимом расчете, такой сложный характер термодинамических свойств оцк-растворов урана.

6.2. Теоретическое исследование

Предложена физико-химическая модель неупорядоченного изоморфного двухкомпонентного раствора. С помощью ее установлена связь между термодинамическими свойствами и достаточно легко измеримыми физическими характеристиками растворов. Оригинальным в этой модели является рассмотрение упругой энергии как компоненты свободной энергии раствора, что привело к возникновению понятия упругой компоненты энтропии. Анализ модели показывает, что вибрационная компонента энтальпии смешения отлична от нуля главным образом лишь при учете ангармонизма. К числу особенностей модели относятся отсутствие формальных параметров и более полный по сравнению с другими моделями учет термических вкладов в свободную энергию.

Расчеты показали значительное влияние ангармонизма на термодинамические свойства твердых растворов урана. В частности, этим влиянием обусловлено возрастание упругой энергии с увеличением температуры в системе $U-Mo$, причина которого заключается в преобладании увеличения размерного фактора с ростом температуры над уменьшением при этом модуля Юнга твердого раствора. Этим же влиянием обусловлены второе изменение знака $\Delta S^E(x)$ и изменение знака $\Delta H^*(x)$ в системе $U-Mo$.

При расчетах энтальпии смешения

$$\Delta H(x,T) = \Delta U^{\circ}(x) + \Delta H^{*}(x,T) = \Delta H^{*}(x,T)[1+\delta]$$

где $\delta = \frac{\Delta U^{\circ}(x)}{\Delta H^{*}(x,T)}$, величина $\Delta U^{\circ}(x)$, как указывалось в подразделе 4.3, не могла быть вычислена. Можно предположить, что в системе $U-Mo$ $\delta \ll$ I (удовлетворительное согласие расчетной и экспериментальной энтальпий смешения), тогда как в системе $U-Nb$ $mod\,\delta \sim$ I (отсутствие изменения знака на расчетной зависимости энтальпии смешения). При этом нужно иметь в виду, что знаменатель δ в системе $U-Mo$ на порядок больше, чем в системе $U-Nb$ для $0,5 < x <$ I.

В целом результаты расчетов показали удовлетворительное согласие с экспериментальными данными. Такое опробование модели является хорошим критерием ее надежности для использования в качестве термодинамической базы теоретического построения диаграмм состояния урановых систем.

ЛИТЕРАТУРА

[1] ФЕДОРОВ, Г.Б., СМИРНОВ, Е.А., Атомная Энергия 21 3(1966) 189.
[2] VAMBERSKIY, YU. V., UDOVSKIY, A.L., IVANOV, O.S. J. Nucl. Mater. 46 2(1973) 192.
[3] ВАМБЕРСКИЙ, Ю.В., УДОВСКИЙ, А.Л., ИВАНОВ, О.С., В сб. Физико-Химический Анализ Сплавов Урана, Тория и Циркония, "Наука", М., 1974, стр. 37.
[4] ИВАНОВ, О.С., УДОВСКИЙ, А.Л., ВАМБЕРСКИЙ, Ю.В., Доклады АН СССР 203 5(1972) 1107.
[5] UDOVSKY, A.L., IVANOV, O.S., J. Nucl. Mater. 49 (1973/1974) 309.
[6] УДОВСКИЙ, А.Л., ВАМБЕРСКИЙ, Ю.В., ИВАНОВ, О.С., Доклады АН СССР 209 6(1973) 1377.
[7] СМИРНОВ, М.В., СКИБА, О.В., Доклады АН СССР 141 4(1961) 904.
[8] ЯРЫМ-АГАЕВ, Н.Л., МЕЛЬНИК, Г.В., Журнал физической химии XXXIX 11(1965) 2650.
[9] СМИРНОВ, М.В., МАКСИМОВ, В.С., Электрохимия 1 6(1965) 727.
[10] СКИБА, О.В., СМИРНОВ, М.В., В сб. Электрохимия Расплавленных Солей и Твердых Электролитов, Вып.2, Свердловск, 1961.
[11] ЛОГИНОВ, Н.А., СМИРНОВ, М.В., В сб. Электрохимия Расплавленных Солей и Твердых Электролитов, Вып.3, Свердловск, 1962.
[12] КУДЯКОВ, В.Я., СМИРНОВ, М.В., Электрохимия 2(1965) 143.
[13] ЛОГИНОВ, Н.А., СМИРНОВ, М.В., РАССОХИН, Б.Г., В сб. Электрохимия Расплавленных Солей и Твердых Электролитов, Свердловск, 7(1965) 9.
[14] СМИРНОВ, М.В., РЫЖИК, О.А., там же, стр.21.
[15] СМИРНОВ, М.В., РЫЖИК, О.А., там же, стр.27.
[16] GARDNER, H.J., BROWN, C.T., and JANS, G.J., J.Phys. Chem. 60 10(1956) 1458.
[17] ВАМБЕРСКИЙ, Ю.В., Диссертация, М., 1973, стр.16.
[18] ИВАНОВ, О.С., СЕМЕНЧЕНКОВ, А.Т., КОЗЛОВА, Н.И., В сб. Строение Сплавов Некоторых Систем с Ураном и Торием, Госатомиздат, М., 1961, стр.68.
[19] ИВАНОВ, О.С., ТЕРЕХОВ, Г.И., там же, стр.20.
[20] ХЕЙНЕ, В., УЭЙР, Д., Теория Псевдопотенциала, "Мир", М., 1973, стр.295.
[21] INGLESFIELD, J.E., Metallurgical Chemistry, Proc. Symp. at Brunel Univ. and NPL, July, 1971, ed. O.Kubaschewski, NPL, London, 1972, p.157.
[22] БЕЛЕНЬКИЙ, А.Я., ГУРСКИЙ, З.А., КРАСКО, Г.Л., Физика Твердого Тела 15 11(1973) 3473.
[23] КРАСКО, Г.Л., ГУРСКИЙ, З.А., Физика Твердого Тела 14 2(1972) 321.
[24] TONIGAWA, S., DOYAMA, M., J. Phys., F. 2 5(1972) L 95.
[25] TONIGAWA, S., DOYAMA, M., Phys. Lett. 43A 1(1973) 17.
[26] КРИВОГЛАЗ, М.А., Теория Рассеяния Рентгеновских Лучей и Тепловых Нейтронов Реальными Кристаллами, "Наука", М., 1967.

[27] АПТЕКАРЬ, И.Л., ФИНКЕЛЬШТЕИН, Б.Н., Журн. Эксп. и Теор. Физ. 21 8(1951) 900.
[28] ЛЕСНИК, А.Г., Модели Межатомного Взаимодействия в Статистической Теории Сплавов, Гос. изд. физ.-мат. лит., М., 1962.
[29] ИВАНОВ, О.С., ВАМБЕРСКИЙ, Ю.В., УДОВСКИЙ, А.Л., В сб. Физико-Химический Анализ Сплавов Урана, Тория и Циркония, "Наука", М., 1974, стр.5.
[30] ВАМБЕРСКИЙ, Ю.В., УДОВСКИЙ, А.Л., ИВАНОВ, О.С., там же, стр.20.

ETUDE PAR TRANSPIRATION DU PROCESSUS DE VAPORISATION DU DIOXYDE D'URANIUM ENTRE 2200 ET 2600 K

G. BENEZECH, J.P. COUTURES, M. FOEX
Laboratoire des ultra-réfractaires,
CNRS,
Odeillo, Font-Romeu,
France

Abstract–Résumé

TRANSPIRATION STUDY OF THE URANIUM DIOXIDE VAPORIZATION PROCESS BETWEEN 2200 K AND 2600 K.

The authors describe a procedure for the transpiration study of sub-stoichiometric uranium oxide vaporization. The experimental set-up involves a radiation concentration system (solar furnace or image furnace). The accent is placed on the practical aspect of the determinations, the technique employed making it possible to work on products subjected to very high thermal gradients.

ETUDE PAR TRANSPIRATION DU PROCESSUS DE VAPORISATION DU DIOXYDE D'URANIUM ENTRE 2200 ET 2600 K.

On décrit un procédé d'étude par transpiration du processus de vaporisation de l'oxyde d'uranium sous-stœchiométrique. L'appareillage est associé à un dispositif à concentration de rayonnement (four solaire ou four à image). L'accent est mis sur l'aspect pratique de ces déterminations, la technique utilisée permettant de travailler sur des produits soumis à de très forts gradients thermiques.

INTRODUCTION

Dans ce travail, nous avons examiné la pression de vapeur au-dessus du dioxyde d'uranium, dans le domaine UO_{2-x}, et ce dans une gamme de températures comprise entre 2200 et 2600K. La méthode expérimentale utilisée est la méthode de transport dans laquelle on s'affranchit des réactions support-échantillon. En effet, afin de pallier ces inconvénients, on a cherché, dans la présente étude, à utiliser les dispositifs de traitement sans contamination et associés à des fours solaires de 2 kW [1], mis au point au Laboratoire des ultra-réfractaires.

1. DISPOSITIF EXPERIMENTAL

L'appareillage utilisé [2] est associé à un four solaire d'axe horizontal comportant un miroir parabolique argenté face arrière (diamètre de la parabole: 2 m, distance focale: 0,85 m, diamètre de la tache focale: 10 mm), desservi par un miroir plan orienteur. Ce dernier, équipé de glaces également argentées face arrière, est asservi de façon à renvoyer les rayons solaires réfléchis parallèles à l'axe de la parabole.

Le dispositif utilisé est représenté schématiquement dans la figure 1. Il comporte un four centrifuge en cuivre, réfrigéré dans sa masse, contenant un creuset de 10 mm de diamètre et de 14 à 16 mm de profondeur.

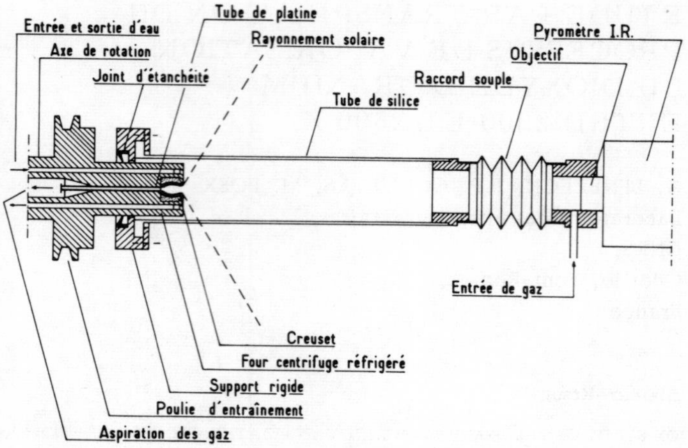

FIG. 1. Dispositif de transpiration associé à un four solaire de 2 kW à axe horizontal.

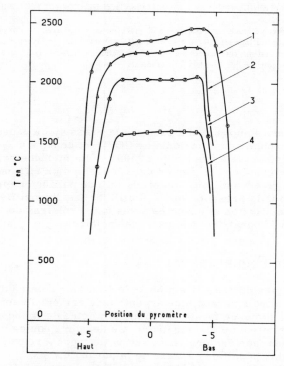

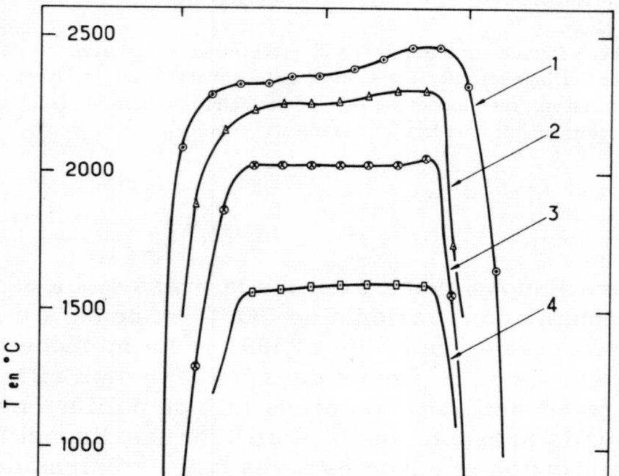

FIG. 2. Répartition de la température à l'intérieur d'un « creuset » Y_2O_3. Entrée du creuset: 1) au foyer du miroir parabolique (f = 850 mm); 2) à 845 mm (courbe décalée de 100°C vers le bas); 3) à 842,5 mm; 4) à 840 mm.

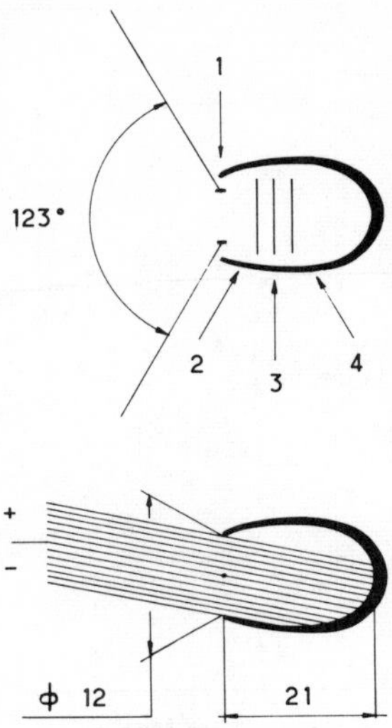

FIG.3. Représentation schématique d'un «creuset» en Y_2O_3 traité au four solaire. Les traits inclinés (11°) indiquent les directions de visée occupées successivement par le pyromètre infrarouge (2,91 μm, surface de visée ~ 1 mm²).

Ce creuset présente vers le fond un orifice de 3 mm de diamètre qui correspond à l'axe creux du four centrifuge. Les gaz chauds, saturés en vapeur dans la cavité au contact de l'oxyde porté à haute température, sont aspirés à travers cet orifice, cependant que les vapeurs se condensent sur un tube de platine disposé dans l'axe du four et indirectement réfrigéré par le courant d'eau parcourant ce dernier. La quantité de vapeur déposée est déterminée par gravimétrie. Afin de travailler en atmosphère contrôlée, on dispose un tube en quartz transparent, s'appuyant sur le support du four centrifuge et sur la tête du pyromètre. La température est mesurée au moyen d'un pyromètre infrarouge (cellule au PbS), équipé d'un filtre interférentiel centré à 2,91 microns, de façon à pouvoir travailler en présence du rayonnement solaire, selon une méthode préalablement décrite [3]. Le débit du gaz transporteur est fixé à l'aide d'une micropompe aspirante et refoulante préétalonnée (débit variable de 0 à 10 l/h).

La possibilité d'effectuer des mesures de température en présence du rayonnement solaire permet d'explorer l'homogénéité en température de la surface du creuset. A titre d'exemple, la figure 2 donne la répartition des températures à l'intérieur d'un «creuset» d'oxyde d'yttrium Y_2O_3 (diamètre d'ouverture: 12 mm, profondeur: 21 mm) mesurées à l'aide d'un pyromètre infrarouge muni d'un filtre interférentiel à 2,91 microns. Afin de décrire l'intérieur de la cavité, le pyromètre est incliné de -12° environ par rapport à l'axe du four. La courbe 1 est obtenue lorsque l'orifice du four est placé au foyer du miroir parabolique, la courbe 2 (décalée vers le bas de 100°C pour plus de commodité de lecture) est relative à un avancement du four de 10 mm dans le cône solaire, et les courbes 3 et

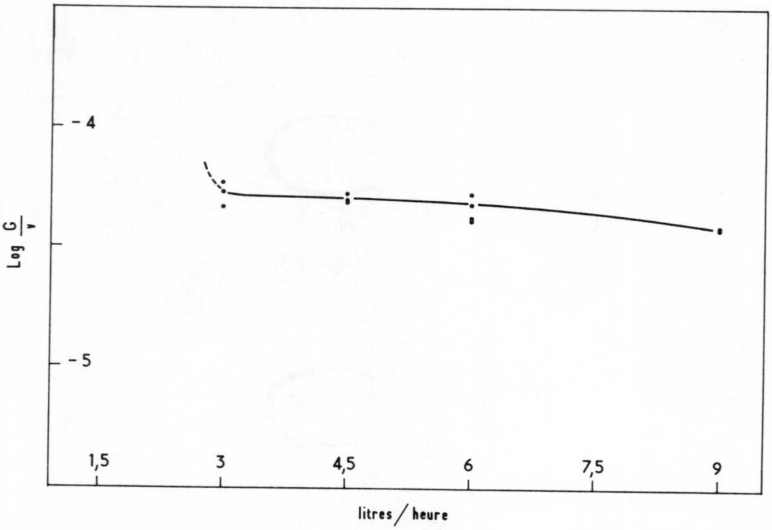

FIG. 4. Evolution du logarithme de G/v (G: quantité de vapeur recueillie en gramme; v: volume total du gaz transporteur en litre) en fonction du débit en l/h.

4 pour des avancements du four de 15 et 20 mm. On a représenté, dans la figure 3, le schéma de principe de ces mesures. Dans le cas de UO_2, afin d'établir un compromis entre homogénéité de température et température maximum d'expérience, l'orifice du four est avancé de 7,5 mm dans le cône solaire, par rapport au foyer théorique.

2. MODE OPERATOIRE

Deux mises en forme différentes des creusets de dioxyde d'uranium ont été utilisées. Dans chaque cas, l'oxyde de départ est UO_2 préalablement calciné pendant 24 heures à 1000°C sous hydrogène, le creuset étant soit fondu à l'air, soit préformé par pressage. Dans le premier cas, on effectue avant toute mesure un balayage d'argon hydrogène à 5% pendant une heure, le creuset étant porté à 2200 K; dans le second cas, ce traitement n'est effectué que pendant quinze minutes.

Avant toute mesure, nous avons déterminé expérimentalement les conditions de vapeur saturante, en examinant l'influence du débit du gaz transporteur sur la quantité de vapeur recueillie. Les résultats obtenus pour une température de 2400 K et exprimés sous la forme de log (G/V), en fonction du débit de gaz transporteur (en l/h), sont représentés dans la figure 4. Le palier de saturation s'étend de 3 l/h à 6 l/h. Nous avons donc choisi de travailler avec un débit de 4,5 l/h, de façon à opérer dans les conditions optimales. On note que pour 9 l/h, la courbe précédente s'infléchit légèrement, semblant indiquer un défaut de saturation du gaz transporteur. En réalité, le condenseur platine est trop petit pour les débits gazeux relativement importants. Le débit de gaz transporteur retenu est légèrement plus faible que celui utilisé par Tetenbaum et Hunt [4] (6 à 9 l/h) et très différent du débit employé par Swarc et Latta [5] (60 l/h). On a utilisé deux mélanges argon-hydrogène (argon à 5% d'hydrogène et argon à 80% d'hydrogène), la durée d'une expérience étant, selon la température choisie, comprise entre une et deux heures.

La composition du produit (rapport O/U) après mesure des pressions de vapeur, est déterminée par microthermogravimétrie dans l'air (température maximale: 900°C) afin d'obtenir U_3O_8 [6]. C'est ainsi que pour des atmosphères d'argon-hydrogène à 5%, le rapport O/U trouvé est constant et égal à 2,000, alors que pour 80% d'hydrogène il est de 1,992.

Il est à noter que la détermination de la composition chimique de la surface du creuset est très difficile, elle ne peut être considérée que comme une valeur par excès. En effet, l'existence d'un gradient radial de température considérable (T surface interne du creuset ~ 2400 K, T surface externe ~ 293 K, épaisseur du creuset: 4 mm) impose un gradient en O/U.

3. RESULTATS

L'équation générale de vaporisation peut s'écrire

$$UO_{2-x} \rightleftarrows \alpha_1 U_{(g)} + \alpha_2 UO_{(g)} + \alpha_3 UO_{2(g)} + \alpha_4 UO_{3(g)} + \alpha_5 O_{(g)} + \alpha_6 O_{2(g)} \quad (1)$$

et la composition de la vapeur peut alors être décrite par les relations indépendantes suivantes:

$$UO_{2(s)} \rightleftarrows U_{(g)} + 2\,O \quad (2)$$

$$U_{(g)} + O_{(g)} \rightleftarrows UO_{(g)} \quad (3)$$

$$UO_{(g)} + O_{(g)} \rightleftarrows UO_{2(g)} \quad (4)$$

$$UO_{2(g)} + O_{(g)} \rightleftarrows UO_{3(g)} \quad (5)$$

$$\tfrac{1}{2}\,O_{2(g)} \rightleftarrows O_{(g)} \quad (6)$$

Les essais étant effectués sous des mélanges argon-hydrogène, il faut également faire intervenir les équations suivantes:

$$\tfrac{1}{2}\,H_{2(g)} \rightleftarrows H_{(g)} \quad (7)$$

$$O_{(g)} + H_{(g)} \rightleftarrows OH_{(g)} \quad (8)$$

$$O_{(g)} + OH_{(g)} \rightleftarrows H_2O_{(g)} \quad (9)$$

Les lois de conservation de la masse (U, O, H), les grandeurs thermodynamiques des réactions indépendantes relatives aux espèces gazeuses étant connues (réactions (3) à (9)), nous permettent, avec les résultats de transpirations, de calculer les pressions partielles des différentes espèces gazeuses pour chaque température.

TABLEAU I. EVOLUTION DES ESPECES GAZEUSES AU-DESSUS DE UO_{2-x} SOUS ARGON A 5% D'HYDROGENE EN FONCTION DE LA TEMPERATURE

T K	P(O)	$P(O_2)$	P(U)	P(UO)	$P(UO_2)$	$P(UO_3)$
2220	$1,54 \cdot 10^{-10}$	$2,13 \cdot 10^{-15}$	$6,46 \cdot 10^{-9}$	$1,60 \cdot 10^{-6}$	$6,90 \cdot 10^{-6}$	$4,46 \cdot 10^{-9}$
2270	$5,96 \cdot 10^{-10}$	$1,75 \cdot 10^{-14}$	$8,94 \cdot 10^{-9}$	$3,41 \cdot 10^{-6}$	$1,15 \cdot 10^{-5}$	$2,87 \cdot 10^{-8}$
2320	$1,61 \cdot 10^{-9}$	$7,10 \cdot 10^{-14}$	$1,21 \cdot 10^{-8}$	$5,16 \cdot 10^{-6}$	$2,03 \cdot 10^{-5}$	$6,65 \cdot 10^{-8}$
2370	$4,75 \cdot 10^{-9}$	$3,55 \cdot 10^{-13}$	$1,64 \cdot 10^{-8}$	$8,82 \cdot 10^{-6}$	$4,63 \cdot 10^{-5}$	$2,28 \cdot 10^{-7}$
2420	$1,23 \cdot 10^{-8}$	$1,40 \cdot 10^{-12}$	$2,19 \cdot 10^{-8}$	$1,36 \cdot 10^{-5}$	$8,59 \cdot 10^{-5}$	$5,64 \cdot 10^{-7}$
2470	$2,82 \cdot 10^{-8}$	$4,76 \cdot 10^{-12}$	$2,85 \cdot 10^{-8}$	$1,88 \cdot 10^{-5}$	$1,32 \cdot 10^{-4}$	$1,03 \cdot 10^{-6}$
2520	$6,10 \cdot 10^{-8}$	$1,27 \cdot 10^{-11}$	$3,74 \cdot 10^{-8}$	$2,50 \cdot 10^{-5}$	$1,89 \cdot 10^{-4}$	$1,74 \cdot 10^{-6}$
2570	$1,41 \cdot 10^{-7}$	$4,23 \cdot 10^{-11}$	$4,92 \cdot 10^{-8}$	$3,71 \cdot 10^{-5}$	$3,28 \cdot 10^{-4}$	$3,73 \cdot 10^{-6}$

TABLEAU II. EVOLUTION DES ESPECES GAZEUSES AU-DESSUS DE UO_{2-x} SOUS ARGON A 80% D'HYDROGENE EN FONCTION DE LA TEMPERATURE

T K	P(O)	P(O_2)	P(U)	P(UO)	P(UO_2)	P(UO_3)
2120	$1,85 \cdot 10^{-12}$	$1,15 \cdot 10^{-18}$	$4,81 \cdot 10^{-8}$	$1,04 \cdot 10^{-6}$	$1,67 \cdot 10^{-7}$	$1,40 \cdot 10^{-11}$
2220	$2,65 \cdot 10^{-11}$	$6,40 \cdot 10^{-17}$	$9,99 \cdot 10^{-8}$	$4,28 \cdot 10^{-6}$	$3,20 \cdot 10^{-6}$	$3,63 \cdot 10^{-10}$
2270	$9,28 \cdot 10^{-11}$	$4,29 \cdot 10^{-16}$	$1,39 \cdot 10^{-7}$	$8,26 \cdot 10^{-6}$	$4,36 \cdot 10^{-6}$	$1,73 \cdot 10^{-9}$
2320	$2,45 \cdot 10^{-10}$	$1,69 \cdot 10^{-15}$	$1,89 \cdot 10^{-7}$	$1,22 \cdot 10^{-5}$	$7,51 \cdot 10^{-6}$	$3,83 \cdot 10^{-9}$
2370	$7,66 \cdot 10^{-10}$	$9,04 \cdot 10^{-15}$	$2,54 \cdot 10^{-7}$	$2,20 \cdot 10^{-5}$	$1,87 \cdot 10^{-5}$	$1,51 \cdot 10^{-8}$
2420	$1,87 \cdot 10^{-9}$	$3,23 \cdot 10^{-14}$	$3,34 \cdot 10^{-7}$	$3,15 \cdot 10^{-5}$	$3,03 \cdot 10^{-5}$	$3,03 \cdot 10^{-8}$
2470	$4,69 \cdot 10^{-9}$	$1,23 \cdot 10^{-13}$	$4,36 \cdot 10^{-7}$	$4,73 \cdot 10^{-5}$	$5,52 \cdot 10^{-5}$	$7,29 \cdot 10^{-8}$

TABLEAU III. ENTHALPIES ET ENTROPIES DE SUBLIMATION DE UO_2 STŒCHIOMETRIQUE

ΔH kcal/mol	ΔS cal/mol K	Méthode expérimentale	Domaine de température (K)	Référence
137,1	36,4	effusion	1600 à 2200	[9]
147,1	42,2	effusion	1920 à 2220	[10]
147,8	42,0	effusion	2200 à 2800	[11]
141,2	39,4	spectrométrie de masse	1890 à 2420	[7]
143,1	39,4	transpiration	2080 à 2705	[4]
134,1	34,5	transpiration	2000 à 2940	[12]

Les résultats obtenus sont rassemblés dans les tableaux I et II. On constate, comme on pouvait s'y attendre, que les pressions partielles des espèces oxydées ($UO_{(g)}$, $UO_{2(g)}$ et $UO_{3(g)}$) sont plus faibles sous argon à 80% d'hydrogène que sous argon à 5% d'hydrogène. On note également que sous fort potentiel d'hydrogène, $P_{U(g)}$ est plus élevé.

4. DISCUSSION

Remarque: Les grandeurs thermodynamiques utilisées dans nos calculs sont tirées de [7].

L'examen des tableaux I et II montre que la répartition des pressions partielles se fait dans l'ordre $P(UO_2) > P(UO) \gg P(UO_3) > P(U) > P(O)$ pour les mélanges d'argon à 5% d'hydrogène et $P(UO) > P(UO_2) \gg P(U) > P(UO_3) \gg P(O)$ pour les mélanges à 80% d'hydrogène. Ces résultats qualitatifs s'avèrent dans notre cas très intéressants car on dispose pour une isothermie (2250 K) de l'évolution des différentes pressions partielles des espèces vaporisant au-dessus de UO_{2-x} pour $O < x < 0,15$. Ces déterminations, effectuées par Drowart et al. [8] par spectrométrie de masse (méthode de Knudsen), nous permettent de mieux situer la composition O/U de la surface interne du creuset d'oxyde d'uranium. C'est ainsi que pour le mélange à 5% d'hydrogène, nous nous trouvons dans un domaine O/U compris entre $O/U = 1,95_3$ et $O/U = 1,97_3$, alors que dans le second cas, $O/U < 1,93_5$. Ces résultats mettent en évidence la nécessité de procéder à une analyse correcte de la composition O/U de la surface du creuset, l'analyse chimique par réoxydation étant en défaut, du fait du fort gradient thermique radial.

Sur la base des résultats expérimentaux précédents, nous avons estimé l'enthalpie et l'entropie moyenne de vaporisation pour l'équation générale de vaporisation (1). C'est ainsi que l'on peut écrire pour les deux atmosphères étudiées

$$\log K_{eq} = -\left(\frac{28\,360}{T}\right) + 7,49 \qquad \text{(A à 5\% } H_2)$$

$$\log K_{eq} = -\left(\frac{25\,340}{T}\right) + 5,99 \qquad \text{(A à 80\% } H_2)$$

ce qui conduit, dans le premier cas, aux valeurs suivantes

$$\Delta H^0_{2400} = 130 \text{ kcal/mol} \qquad \Delta \delta^0_{2400} = 24,2 \text{ cal/mol K}$$

et, dans le second cas,

$$\Delta H^0_{2300} = 116 \text{ kcal/mol} \qquad \Delta \delta^0_{2300} = 27,4 \text{ cal/mol K}$$

Les grandeurs thermodynamiques ainsi déterminées sont plus faibles que les déterminations d'Ackermann [9] et de Pattoret et al. [7], détermination ayant trait à l'enthalpie de vaporisation de UO_2 stœchiométrique. Les différentes données bibliographiques sont rassemblées dans le tableau III.

En ce qui concerne les expérimentations effectuées sous des mélanges argon à 5% d'hydrogène, le logarithme de la pression de vapeur saturante au-dessus du dioxyde d'uranium s'écrit

$$\log_{10} P = -\left(\frac{27\,135}{T}\right) + 7{,}10_2$$

ce qui, dans nos conditions opératoires, conduit à une température d'ébullition de 3820 K.

Les résultats expérimentaux que nous avons obtenus pour la méthode de transport, à l'aide d'un appareillage associé à un dispositif à concentration de rayonnement, s'avèrent, en ce qui concerne les pressions estimées, légèrement plus faibles que les résultats de la littérature. Il faut en chercher la raison dans l'écart à l'équilibre de la phase condensée. Néanmoins, les conditions opératoires employées (gradients radiaux de température très importants) s'apparentent aux conditions pouvant exister dans les gaines à combustibles de réacteurs nucléaires à oxydes. Il faut donc considérer ces valeurs comme des grandeurs pratiques.

REFERENCES

[1] FOEX, M., Rev. Int. Hautes Temp. Réfract. 3 (1966) 281.
[2] BENEZECH, G., FOEX, M., C.R., Ser. C. 268 (1969) 2315.
[3] FOEX, M., COUTURES, J.P., BENEZECH, G., « Mesure des températures des produits traités au four solaire à l'aide de pyromètres infrarouges », Coll. de la coopération méditerranéenne pour l'énergie solaire, Athènes, Oct. 1969.
[4] TETENBAUM, M., HUNT, P.D., J. Nucl. Mater. 34 (1970) 86.
[5] SWARC, R., LATTA, R.E., J. Am. Ceram. Soc. 51 (1968) 264.
[6] SCHAEFER, E.A., MENKE, M.R., HIBBITS, J.O., Determination of O/U ratio in $UO_{2\pm x}$, Tech. Rep. GE-TM-4-9 (Août 1966).
[7] PATTORET, A., DROWART, J.D., SMOES, S., Thermodynamics of Nuclear Materials (Proc. Symp. Vienna, 1967), AIEA, Vienne (1967) 613.
[8] DROWART, J., PATTORET, A., SMOES, S., Mass Spectrometric Studies of the Vaporization of Refractory Compounds (Proc. Br. Ceram. Soc.) 8 (1967) 67.
[9] ACKERMANN, R.S., Argonne National Laboratory, Rep. ANL-5482 (1955).
[10] IVANOV, V.E., KRUGICH, A.A., SAVLOV, U.C., KOVTUN, G.P., AMONENKO, V.M., Thermodynamics of Nuclear Materials (Proc. Symp. Vienna, 1962), AIEA, Vienne (1962) 735.
[11] OHSE, R.W., J. Chem. Phys. 44 (1966) 1375.
[12] ALEXANDER, C.A., OGDEN, J.S., CUNNING, G.W., Battelle Memorial Institute, Rep. BMI-1789 (1967).

DISCUSSION

M. TETENBAUM: Did you attempt to fix the oxygen potential in the carrier gas in order to fix the composition of the condensed phase in the UO_x sample?

J.P. COUTURES: The oxygen potential in the carrier gas was not fixed beforehand since either Ar-5% H_2 or Ar-80% H_2 mixtures were used. On the contrary, the oxygen potential may be measured for each temperature after the platinum condenser and thus correlated with the surface composition of the condensed phase.

DETERMINING THE FREE ENERGY OF ACTINIDE CARBONITRIDES USING AMORPHOUS CARBONS HAVING AN ARBITRARY DEGREE OF GRAPHITIZATION

M. KATSURA, N. SHOHOJI, T. YATO,
T. NOMURA, T. SANO
Osaka University,
Osaka,
Japan

Abstract

DETERMINING THE FREE ENERGY OF ACTINIDE CARBONITRIDES USING AMORPHOUS CARBONS HAVING AN ARBITRARY DEGREE OF GRAPHITIZATION.

It is well known that solid solubility occurs between monocarbide and mononitride in the pseudo-binary systems, ThC-ThN, UC-UN and PuC-PuN. In these systems, there is an equilibrium state where carbonitride exists with free carbon under an atmosphere of nitrogen gas. It is a fundamental requirement that reliable thermodynamic data be obtained for these solid solutions in order to develop these carbonitrides as new nuclear fuels and to consider the changes in carbon and nitrogen potentials caused by changes in composition in these solid solutions due to burn-up. In this paper, a new method using carbon with an arbitrary degree of graphitization as the second phase existing with carbonitride will be proposed for determining the relation between the compositions and carbon and nitrogen potentials in these carbonitrides. A thermodynamical analysis of the systems consisting of carbonitride and free carbon is presented. In order to survey qualitatively the applicability of this method, reactions of $U_{0.75}Ce_{0.25}N$ with graphite and of $U_{0.75}Ce_{0.25}N$ with an amorphous carbon have been studied experimentally.

1. INTRODUCTION

It is well known that solid solubility occurs between monocarbide and mononitride in the pseudo-binary systems, ThC-ThN, UC-UN and PuC-PuN. In order to develop these carbonitrides as new nuclear fuels, it is important to have reliable thermodynamic data for these solid solutions. One of the methods of determining the free energy of these carbonitrides is to study the equilibrium of the system in which the carbonitride coexists with graphite as the second phase at a given nitrogen pressure [1, 2]. The chemical potential of nitrogen in these solid solutions may be controlled by changing the pressure of nitrogen gas to which the carbonitride is subject. The activity of carbon is often fixed by means of a flowing gas technique using a hydrogen-hydrocarbon mixture.

In this paper, a new method for determining the free energy of the carbonitrides using carbon with an arbitrary degree of graphitization as the second phase in place of graphite will be proposed.

Some problems associated with this new method and a thermodynamical analysis of the system consisting of carbonitride and free carbon at a given nitrogen pressure will also be reported. Further, some experiments have been carried out to survey qualitatively the applicability of this method.

2. PRELIMINARY CONSIDERATIONS

In the equilibrium state, where free carbon and carbonitride are present together under an atmosphere of nitrogen gas of given pressure, the chemical potential of the free carbon must be equal to that of the carbon in the carbonitride and the chemical potential of nitrogen in the carbonitride must be equal to one half the chemical potential of nitrogen gas, that is:

$$\mu(C) = RT \ln a_C = \mu(C \text{ in } MC_{1-x}N_x) \tag{1}$$

$$\frac{1}{2}\mu(N_2) = \frac{1}{2} RT \ln p_{N_2} = \mu(N \text{ in } MC_{1-x}N_x) \tag{2}$$

where $\mu(C)$ is the chemical potential of free carbon, $\mu(C \text{ in } MC_{1-x}N_x)$ the chemical potential of carbon dissolved in $MC_{1-x}N_x$, $\mu(N_2)$ the chemical potential of nitrogen gas, and $\mu(N \text{ in } MC_{1-x}N_x)$ the chemical potential of nitrogen dissolved in $MC_{1-x}N_x$; a_C and p_{N_2} are the activity of free carbon and the nitrogen pressure, respectively. Here carbonitrides are assumed to be stoichiometric with respect to non-metallic elements.

The free energy of formation of $MC_{1-x}N_x$ may be given as:

$$\Delta G_f^0(MC_{1-x}N_x) = \mu(M \text{ in } MC_{1-x}N_x) + x\mu(N \text{ in } MC_{1-x}N_x) + (1-x)\mu(C \text{ in } MC_{1-x}N_x) \tag{3}$$

The substitution of Eqs (1) and (2) into (3) leads to:

$$\Delta G_f^0(MC_{1-x}N_x) = \mu(M \text{ in } MC_{1-x}N_x) + RT\left[\frac{x}{2}\ln p_{N_2} + (1-x)\ln a_C\right] \tag{4}$$

Since the activity of the free carbon determines $\mu(C \text{ in } MC_{1-x}N_x)$ and the pressure of the nitrogen gas determines $\mu(N \text{ in } MC_{1-x}N_x)$, the activity of the carbon and the nitrogen pressure combine to determine the composition of $MC_{1-x}N_x$.

If free carbon having an arbitrary degree of graphitization could be obtained and its thermodynamic properties could be known, the reaction of MN (mononitride) with these free carbons under conditions of constant temperature and pressure of nitrogen gas could be used to study the equilibrium of the region of $MC_{1-x}N_x$ and free carbon. According to previous workers [3-5], free carbons having a wide variation in the degree of graphitization can be produced by reacting UC or UC_2 with N_2 or NH_3.

As can be seen from Eq. (4), the free energy of formation of $MC_{1-x}N_x$ can be evaluated from the equilibrium measurement carried out at a constant temperature and nitrogen pressure if the activity of the carbon as the second phase is known.

The thermodynamic properties of $MC_{1-x}N_x$ have usually been determined by changing the nitrogen pressure. In this paper the determination of the free energy of $MC_{1-x}N_x$ will be presented based on a method involving varying the activity of the carbon.

3. SOME PROBLEMS ASSOCIATED WITH THE METHOD

The applicability of thermodynamics to work with amorphous carbon is one of the problems associated with this method. Strictly speaking, thermodynamics can be applied only to stable and metastable states. Amorphous carbon is a solid with some departure from perfect crystallinity, which may, in principle, not be subject to exact thermodynamic treatment.

However, data on the thermodynamic quantities of many kinds of amorphous carbons, including specific heats, exist in the literature [6-8]. The free energy of a solid with some departure from perfect crystallinity may be a function not only of temperature and pressure but also of the degree of departure from perfect crystallinity. If a parameter, z, is introduced to express the degree of departure from perfect crystallinity, the free energy of the solid may be given as a function of T, p and z. The degree of graphitization of amorphous carbons may be considered to correspond to the parameter z. As a result, the free energy of amorphous carbon may vary with the degree of graphitization. As an example, the variation of specific heat of carbon, $C_{p,z}$, with the degree of graphitization, z, will be considered. According to Prigogine and Defay [9], the following relation is obtained:

$$C_{p,z} = C_{p,A=0} + \left(\frac{\partial H}{\partial z}\right)_{T,p}^2 \Big/ T\left(\frac{\partial A}{\partial z}\right)_{T,p}$$

$$= C_{p,A=0} - \left(\frac{\partial H}{\partial z}\right)_{T,p}^2 \Big/ T\left(\frac{\partial^2 G}{\partial z^2}\right)_{T,p} \qquad (5)$$

where H is the enthalpy, $C_{p,A=0}$ corresponds to the specific heat of a liquid carbon, and A is the affinity. Since $(\partial^2 G/\partial z^2)_{T,p}$ must be positive in the stable equilibrium state, $C_{p,A=0}$ is always larger than $C_{p,z}$. This is in good agreement with the measurements of temperature versus specific heat, which show that the specific heat of amorphous carbon decreases with the degree of graphitization over the relevant temperature range. Thus, the amorphous carbons may be subject to thermodynamic treatment if the parameter, z, representing the degree of graphitization is introduced. Here, amorphous carbon is thought of as a solid with all the spatial disorder characteristic of a supercooled liquid carbon. A liquid carbon is treated thermodynamically under certain circumstances where a liquid carbon exists as a stable form of carbon [10].

Another question that may arise is: Up to what temperature can the amorphous carbon exist without changing its degree of graphitization? The answer to this question must be determined by experiment. No reflection lines can be observed in a Debye-Scherrer photograph of an amorphous carbon (E. Merk AG, Darmstadt, Germany) heat-treated to 1400°C for 12 h, so that any change in structure must be of such a nature that it is not revealed in a Debye-Scherrer pattern. The possibility that any structural change is below the detection limit of the X-ray powder diffraction method is not ruled out. However, even soft carbon, which is most likely to be graphitized, can hardly be graphitized at temperatures below 1500°C within a finite time interval. The graphitization of soft carbon begins in vacuum at about

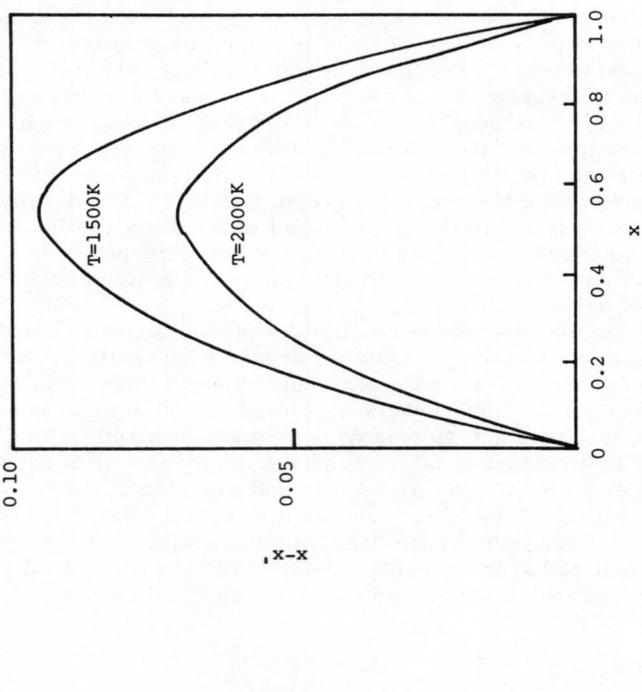

FIG.1. Results of thermodynamic calculations at 1500 and 2000 K.

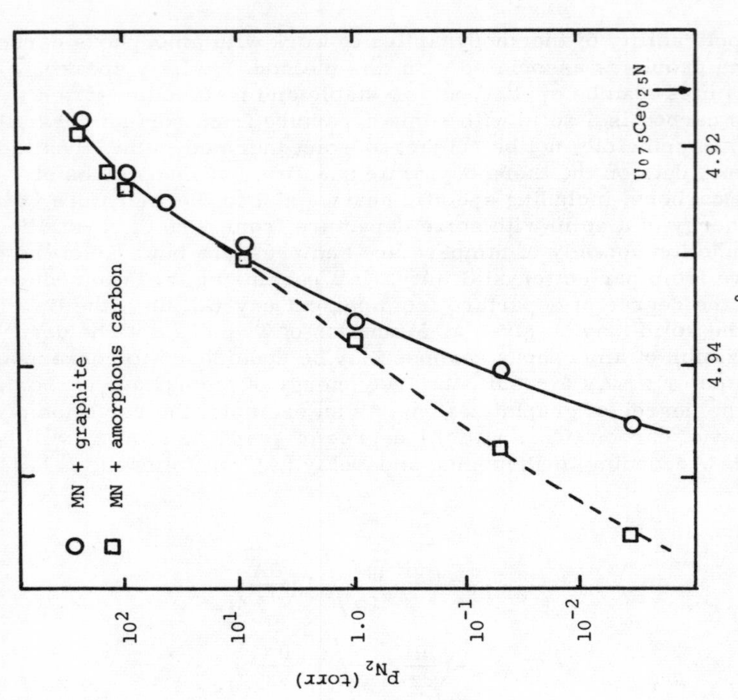

FIG.2. Lattice parameters (Å) of carbonitrides obtained from reactions at 1400°C.

1500°C and proceeds remarkably quickly at temperatures above 2000°C. Actually, structural changes need not be taken into consideration at temperatures below 1500°C. As a result, the temperature range used for this method will be limited to temperatures below 1500°C.

4. THERMODYNAMIC ANALYSIS

The relation between the composition and the carbon and nitrogen potentials of $MC_{1-x}N_x$ can be obtained by the following reaction:

$$MN + C = MC_{1-x}N_x + xC + \frac{1-x}{2} N_2 \qquad (6)$$

In carrying out the reaction, carbon having any desired degree of graphitization was employed and the nitrogen pressure was varied in each experiment.

In what follows, a brief thermodynamic analysis will be given to clarify the influence of the amorphous carbon on the system consisting of $MC_{1-x}N_x$ and carbon.

Since the chemical potential, μ(C in $MC_{1-x}N_x$), varies with x, the chemical potential of free carbon due to its degree of graphitization may be an effective factor for determining the composition of $MC_{1-x}N_x$ co-existing with the free carbon. It is, therefore, important to know quantitatively the effect of the differences in the graphitization of free carbon on the thermodynamic quantities. According to Takahashi and Westrum [8], the entropy value, $S^0_{298.15}$, of the glassy carbon at 298.15 K is 1.406 cal/mol·K, while that of graphite is 1.36 ± 0.02 cal/mol · K [11]. The difference in entropy is about 0.046 cal/mol·K at room temperature. The values -94.047 kcal/mol for graphite and -95.227 kcal/mol for vitreous carbon were obtained from the measurements of the heats of combustion by Lewis et al. [12]. This means that, at room temperature, the enthalpy value of vitreous carbon is higher by about 1.2 kcal/mol than that of graphite. Neglecting the variation with temperature of both enthalpy and entropy differences between graphite and amorphous carbon, the free energy of the amorphous carbon is given by:

$$\Delta G \text{ (cal/mol)} = 1200 - 0.04\, T \qquad (7)$$

It should be noted that the enthalpy of one kind of amorphous carbon and the entropy of another kind of carbon is combined only for convenience. For, to our knowledge, no data are available for both enthalpy and entropy of the same sample of amorphous carbon.

In the case where $MC_{1-x'}N_{x'}$ exists with carbon with an activity a_C at a given nitrogen pressure p_{N_2}, a brief calculation yields the following equation:

$$RT \ln \frac{x'}{1-x'} = \frac{1}{2} RT \ln p_{N_2} - [\Delta G_f^0(MN) - \Delta G_f^0(MC)] - RT \ln a_C \qquad (8)$$

where $\Delta G_f^0(MN)$ and $\Delta G_f^0(MC)$ are the free energies of formation of MN and MC, respectively, and RT ln a_C is the chemical potential of the carbon relative to graphite. If the free carbon as the second phase is graphite, Eq.(8) becomes:

$$RT \ln \frac{x}{1-x} = \frac{1}{2} RT \ln p_{N_2} - [\Delta G_f^0(MN) - \Delta G_f^0(MC)] \quad (9)$$

From Eqs (8) and (9), one obtains:

$$\ln \left[\left(\frac{x}{1-x}\right) / \left(\frac{x'}{1-x'}\right)\right] = \ln a_C \quad (10)$$

Using the relation RT ln a_C = 1200 - 0.04 T (Eq.(7)), Eq.(10) becomes:

$$RT \ln \left[\left(\frac{x}{1-x}\right) / \left(\frac{x'}{1-x'}\right)\right] = 1200 - 0.04\ T \quad (11)$$

Eq.(11) gives the composition difference between the MC-MN solid solution coexisting with graphite and that with amorphous carbon. The results calculated using Eq.(11) are represented graphically in Fig.1. Eq.(11) may also suggest that the carbon content in $MC_{1-x}N_x$ decreases with the degree of graphitization, since the activity of the amorphous carbon decreases with the degree of graphitization.

5. EXPERIMENTAL RESULTS

In order to check the thermodynamic analysis and to survey the applicability of the method, some experiments were carried out. The trend obtained from the thermodynamic analysis had already been verified experimentally for the system consisting of $UC_{1-x}N_x$ and free carbon [13]. In the present work, reactions between $U_{0.75}Ce_{0.25}N$ and excess graphite and those between $U_{0.75}Ce_{0.25}N$ and excess amorphous carbon (E. Merk AG, Darmstadt, Germany) were carried out for various temperatures and nitrogen pressures. The reaction products were analysed by the X-ray diffraction powder method. The reason why $U_{0.75}Ce_{0.25}N$ was chosen as the starting material lay in the fact that the lattice parameters of the carbonitrides $(U_{0.8}Ce_{0.2})C_{1-x}N_x$ increase monotonically with the carbon content [14]. Since the composition of $(U_{0.75}Ce_{0.25})C_{1-x}N_x$ with respect to the metallic component is very close to that of $(U_{0.8}Ce_{0.2})C_{1-x}N_x$, the variation of the lattice parameter of $(U_{0.75}Ce_{0.25})C_{1-x}N_x$ with x may be expected to be similar to that of $(U_{0.8}Ce_{0.2})C_{1-x}N_x$.

Typical experimental results at 1400°C are shown in Fig.2. The results are in good agreement with the thermodynamic analysis.

These results suggest that it may be possible to obtain the relation between the composition and carbon and nitrogen potentials of MC-MN solid solutions through solid-state reaction of metal nitride with amorphous carbon having any desired degree of graphitization under an atmosphere of nitrogen gas of controlled pressure, although some restrictions may exist.

REFERENCES

[1] NAOUMIDIS, A., STÖCKER, H.-J., "Stability of UC-UN solid solutions in the presence of free carbon", Thermodynamics of Nuclear Materials, 1967 (Proc. Symp. Vienna, 1967), IAEA, Vienna (1968) 287.
[2] LEITNAKER, J.M., "The ideality of the UC-UN solid solution", Thermodynamics of Nuclear Materials, 1967 (Proc. Symp. Vienna, 1967), IAEA, Vienna (1968) 317.
[3] KATSURA, M., YUKI, T., SANO, T., SASAKI, Y., J. Nucl. Mater. 39 (1971) 125.
[4] NOMURA, T., KATSURA, M., SANO, T., J. Nucl. Mater. 47 (1973) 58.
[5] SHOHOJI, N., KATSURA, M., SANO, T., to be published.
[6] van der HOEVEN, B.J.C., Jr., KEESOM, P.H., Phys. Rev. 6 (1963) 58.
[7] DELHAES, F., HISHIYAMA, Y., Carbon 8 (1970) 31.
[8] TAKAHASHI, Y., WESTRUM, E.F., Jr., J. Chem. Thermodyn. 2 (1970) 847.
[9] PRIGOGINE, I., DEFAY, R., Thermodynamique Chemique.
[10] BUNDY, F.P., J. Chem. Phys. 38 (1963) 631.
[11] KELLEY, K.K., KING, E.G., US Bureau of Mines Bulletin (1961) 592.
[12] LEWIS, D.C., FRISCH, M.A., MARGRAVE, J.M., Carbon 2 (1965) 431.
[13] KATSURA, M., NOMURA, T., J. Nucl. Mater. 51 (1974) 63.
[14] IHARA, S., TANAKA, K., SUZUKI, M., AKIMOTO, Y., J. Nucl. Mater. 39 (1971) 203.

DISCUSSION

A.S. PANOV: Did I understand you to say that the thermodynamic properties of graphite and amorphous carbon are different?

M. KATSURA: Yes, in the relevant temperature range.

A.S. PANOV: What is the reason for this?

M. KATSURA: Data on the thermodynamic values for many kinds of amorphous carbon, including their specific heats, exist in the literature. The entropy value of amorphous carbon at room temperature is different from that of graphite. Furthermore, the heat of combustion of graphite is different from that of vitreous carbon.

A.S. PANOV: Could it be that the divergence of your experimental data is caused by kinetic rather than by thermodynamic factors?

M. KATSURA: I do not think so. We have observed in our experiments that the equilibrium of UN with carbon can be achieved within 20 hours. As I showed in my oral presentation, the Debye-Scherrer photographs are very sharp and the resolution between $K\alpha_1$ and $K\alpha_2$-doublets is good. I feel that this is evidence for the establishment of equilibrium.

A. NAOUMIDIS: I would first like to congratulate you and your co-authors on your success in demonstrating so clearly that there is a difference in the thermodynamic equilibria in carbonitride systems with various carbon modifications. Like yourself, I do not agree with Mr. Panov's suggestion, since in the system here discussed it is very simple to define the establishment of the equilibrium with the (6 2 0) reflex of U(C,N).

On the basis of your results do your favour UC+N_2 or UN+C (graphite) in your equilibrium measurements in the U(C,N)-C-N_2 system?

M. KATSURA: I prefer the reaction of UN with graphite because the free carbon precipitating from the reaction of UC with N_2 is not graphite but amorphous carbon with some departure from perfect crystallinity. According to our experimental results, equilibrium is established more quickly in the case of the reaction of UN with graphite than in that of UC with N_2.

N. GOLDENBERG: How did you determine the purity of the graphite? We have recently observed that some graphites which we use (Poco and

Thornel 50) release large quantities of gas and moisture at temperatures around 1600°C after brief exposure to air.

M. KATSURA: We have not determined the impurities in graphite or amorphous carbon. The temperatures which prevail in your experiments are somewhat higher than in ours. A Debye-Scherrer photograph of amorphous carbon heat-treated at 1400°C for 12 hours shows no reflection line. Any change in structure must therefore be of such a nature that it is not revealed in a Debye-Scherrer pattern.

THERMODYNAMICS OF FORMATION OF Th-Cu ALLOYS

D.M. BAILEY, J.F. SMITH
Ames Laboratory—USAEC
Iowa State University,
Ames, Iowa,
United States of America

Abstract

THERMODYNAMICS OF FORMATION OF Th-Cu ALLOYS.

Electromotive force cells have been used to determine the Gibbs free energies, enthalpies and entropies of formation for $ThCu_6$, $ThCu_{3.6}$, $ThCu_2$ and Th_2Cu over the temperature range 729-1219 K. Solid CaF_2 was used as the electrolyte. Comparison of the present measurements with earlier measurements on the most copper-rich phase shows that the free energies are reproducible by this technique to ~4%. The magnitudes of the entropies of formation were all measured as less than 1 cal/g-atom·K and are therefore physically reasonable. The values for the entropies of formation of $ThCu_6$ and $ThCu_{3.6}$ were found to be positive, that of $ThCu_2$ to be essentially zero, and that of Th_2Cu to be negative. Comparison with data from other thorium systems shows that positive entropies of formation are atypical for binary intermetallic phases of thorium. The free energies of formation of the Th-Cu phases were found to be roughly one-half as negative as values reported for Zr-Cu phases at comparable temperatures and stoichiometries.

INTRODUCTION

Interest in the thermodynamic properties of thorium alloys is evident by the more than fivefold increase [1] in the available thermodynamic data in the past ten years. The present report of an investigation of Th-Cu alloys by electromotive force techniques with solid CaF_2 electrolytes represents the fourth such investigation in a series at this laboratory; the three previous investigations [2-4] were of Th-Ni, Th-Co, and Th-Fe alloys. Similar investigations have been made of Th-Ru and Th-Rh alloys by Kleykamp and Murabayashi [5,6] at Karlsruhe, and Th-Re alloys by Rezukhina and Pokarev [7] at Moscow, and of the Th-C, Th-S, and Th-B systems by Aronson and coworkers [8-11] at Brookhaven.

The two most recent investigations [12,13] of phase equilibria in the Th-Cu system agree that four intermediate phases exist. They further agree with regard to the stoichiometries of three of these phases: $ThCu_6$, $ThCu_2$, and Th_2Cu. These stoichiometries are further corroborated by crystallographic investigations, that of Buschow and van der Goot [14] which shows $ThCu_6$ to be isostructural with $CeCu_6$ and those of Baenziger et al. [15] and Murray [16] which show that $ThCu_2$ and Th_2Cu are isostructural, respectively, with AlB_2 and $CuAl_2$. The two investigations of phase equilibria differ with regard to the stoichiometry of the fourth intermediate phase with Schiltz et al. [12] reporting $ThCu_{3.6}$ and Berlin [13] reporting $ThCu_3$. However, the more recent crystallographic investigation of Bailey [17] shows that the phase between $ThCu_2$ and $ThCu_6$ is isostructural with $GdAg_{3.6}$ and the stoichiometry is therefore

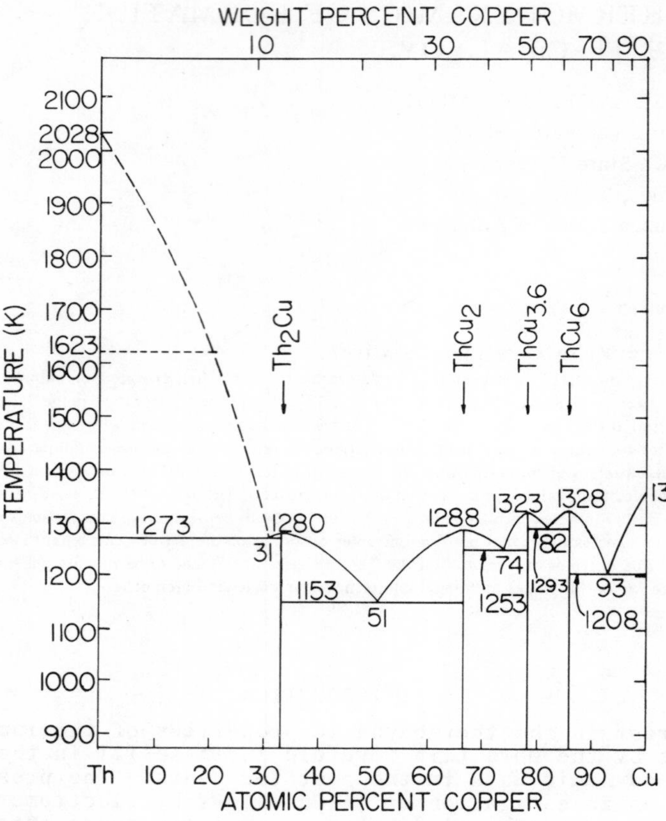

FIG.1. The Th-Cu temperature-composition diagram (after Schiltz et al. [12]).

$ThCu_{3.6}$ ($Th_{14}Cu_{\sim 51}$). The close correspondence between the hexagonal lattice parameters of the phase whose structure was determined by Bailey and the phase $ThCu_3$ reported by Berlin leaves little doubt but that the phases are one and the same. On this basis the phase relationships of Schiltz et al. are corroborated, and their phase diagram is shown in Fig. 1.

EXPERIMENTAL PROCEDURE AND RESULTS

Thorium was obtained from Dr. D. T. Peterson of the Ames Laboratory with chemical analysis indicating purity in excess of 99.97%, and copper was obtained from the American Smelting and Refining Co. with chemical analysis indicating purity in excess of 99.999%. Analyses are shown in Table I. Alloys were prepared by arc melting appropriate amounts of copper and thorium under an atmosphere of purified argon. Homogenization of the alloys was facilitated by successively inverting and remelting the alloy specimens several times. Debye-Scherrer

Table I. Impurity Analysis of Alloying Elements and ThF_4 in ppm

Impurity	Th	Cu	ThF_4
H	<1	-	-
B	-	-	<0.5
C	30	-	-
N	20	-	-
O	85	-	-
Na	<10	-	-
Mg	<20	-	55-120
Al	20	-	<25
Si	<20	<0.1	50-100
S	-	<1	-
Ca	<20	-	150-500
Cr	<20	-	-
Mn	<20	<0.5	<20
Fe	<20	<0.7	-
Ni	<20	<1	-
Se	-	<1	-
Ag	-	<0.2	-
Cd	-	-	0.2
In	-	-	35-45
Sn	-	<1	-
Sb	-	<1	-
Te	-	<2	-
Au	-	<2	-
Pb	-	<1	-
Bi	-	<0.1	-

powder diffraction patterns were taken of all alloys. The patterns were in all cases compatible with the phase diagram of Schiltz et al., though it should be noted that in patterns from compositions lying between $ThCu_{3.6}$ and $ThCu_6$ the number of diffraction lines is very large and some segments of the patterns are unresolvable. Electrode pellets were prepared by mixing pulverized metal with 20 wt% ThF_4 powder and compacting the aggregate to 30 000 lbf/in^2 in a press. Pulverized metal was made from brittle alloys by crushing in a diamond mortar and from non-brittle alloys by filing with a tungsten carbide file. Strain energy in the pulverized metal was relieved by annealing the compacted pellets in situ in the electrochemical cells for an hour or more at elevated temperature before commencing measurements. The electromotive force apparatus was coupled through an access port directly to a dry box, and all operations concerned with fabrication of the electrode pellets and assembly of the cells were performed in the dry box so that the alloys were never exposed to anything but an inert gas environment. Exceptions to this procedure occurred for the initial three alloys in the least reactive portion of the system: 3 at.% Th, 10 at.% Th, and 15.5 at.% Th. In these three cases the material was taken from the dry box in a plastic bag

Table II. Experimental Values for Electromotive Forces at Measured Temperatures for Various Th-Cu Compositions

Comp. (at. %)	Temp. (K)	Electromotive force (mV)	Comp. (at. %)	Temp. (K)	Electromotive force (mV)
Cu-3 Th	729	234.5	Cu-24 Th	1065	155.5
	773	242.9		1021	146.8
	931	232.4		1030	141.8
	1012	248.5		952	150.3
				792	156.3
Cu-10 Th	1052	253.2		952	158.1
	1012	252.4		791	160.3
	972	249.2		1106	145.6
	1092	259.1		1031	154.1
	971	245.3			
	1053	253.7	Cu-30 Th	1120	144.7
	1133	259.7		1046	137.3
	1175	261.1		1101	142.8
	1012	249.8		974	142.8
	1175	260.0		1100	136.1
	1091	255.8		973	142.6
	1134	257.9		1118	135.5
Cu-15.5 Th	1176	254.8	Cu-40.3 Th	1096	70.3
	1010	251.7		1056	68.4
	884	248.5		1015	67.9
	927	250.0		972	73.5
	1089	253.6		992	71.1
	1219	251.7		1035	70.0
	1149	250.9		927	69.4
	1011	248.9		1014	70.6
	1131	246.5		1141	70.5
	926	241.0		886	70.6
	1088	249.8			
Cu-19.4 Th	1120	250.1			
	1024	248.2			
	1035	250.3			
	987	248.6			
	1074	243.4			
	1121	247.5			
	1036	247.5			
	791	242.8			
	988	243.3			
	946	249.0			
	954	247.0			
	1035	245.0			

for a very short interval of time for compaction of the electrode pellets in an external press. Data from a 15.5 at.% Th alloy which was compacted after a press had been placed within the drybox yielded data in good accord with the alloy which had been compacted externally.

The basic details of the experimental arrangement and operation of electromotive force apparatus has previously been described [2]. Data were accumulated over the temperature range 729-1219 K with electrochemical cells of the type

$$Th,ThF_4|CaF_2|ThF_4,Cu,ThCu_6 \quad [1a]$$

$$Th,ThF_4|CaF_2|ThF_4,ThCu_6,ThCu_{3.6} \quad [1b]$$

$$Th,ThF_4|CaF_2|ThF_4,ThCu_{3.6},ThCu_2 \quad [1c]$$

$$Th,ThF_4|CaF_2|ThF_4,ThCu_2,Th_2Cu \quad [1d]$$

with all phases, including the electrolyte, remaining solid throughout the experiments. Cell reactions for these cells are, respectively,

$$Th + 6\ Cu \rightleftarrows ThCu_6 \quad [2a]$$

$$2/5\ Th + 3/5\ ThCu_6 \rightleftarrows ThCu_{3.6} \quad [2b]$$

$$4/9\ Th + 5/9\ ThCu_{3.6} \rightleftarrows ThCu_2 \quad [2c]$$

$$3/2\ Th + 1/2\ ThCu_2 \rightleftarrows Th_2Cu \quad [2d]$$

Reversibilities of cell operations were checked by verifying the absence of significant time variation of voltage at constant temperature, by noting the approach to the same open circuit voltage from temperatures both above and below the temperature of measurement, and by noting recovery to the same voltage after passage of current in either forward or reverse direction. Reversible cell operation at fixed temperature and pressure provides a direct measure of the Gibbs free energy of cell reaction through the relation

$$\Delta G_T = -n\mathcal{F}E$$

where $\mathcal{F}$ is the Faraday constant, E is the open circuit electromotive force, and n is the number of electrons involved in the cell reaction and is 4 per g-atom of thorium.

Experimental datum points are tabulated in Table II. Linear least-squares fits to the experimental EMF's as functions of temperature are shown both graphically and analytically in Fig. 2 with the appropriate reaction also being specified. Also shown in Fig. 2 are the experimental data of Magnani et al. [18] for Cu-5 at.% Th and Cu-7 at.% Th alloys. These latter data were originally interpreted in the belief that the most copper-rich phase was $ThCu_4$, but this interpretation was revised [3] after it had been shown [12,14] that a stoichiometry of $ThCu_6$ was required to explain the x-ray diffraction maxima. The phase studies of Schiltz et al. [12] found negligible solid solubility of thorium in copper and reported no indication of a significant range of homogeneity in any of the four intermediate phases. On this basis the present data were analyzed with pure solid phases as standard states and these were assumed to exist in the alloys without significant deviation from unit activities. With this approach the present electromotive force

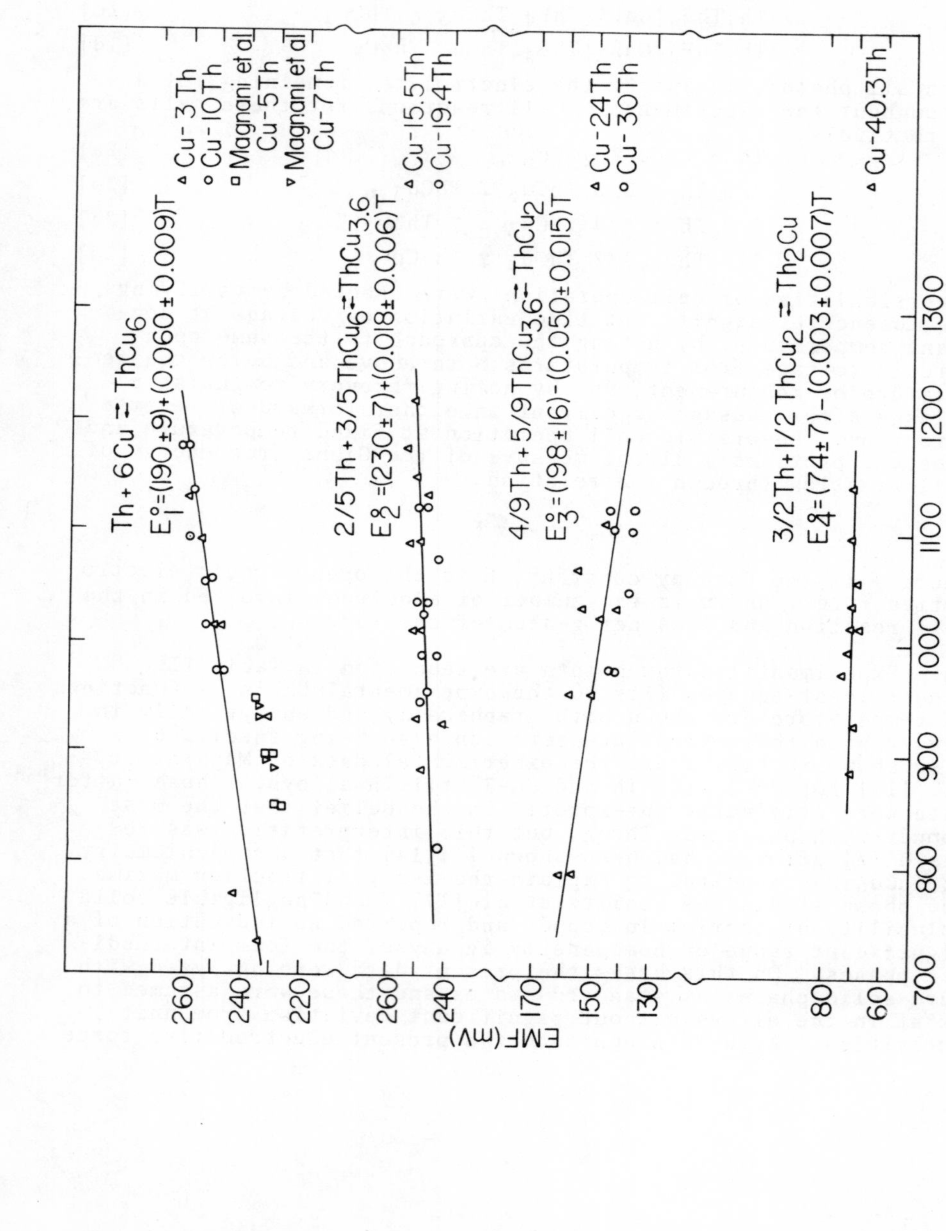

FIG.2. Electromotive force data as functions of temperature for various alloys and reactions.

Table III. Thermodynamic Functions for the Formation of the Four Th-Cu Intermediate Phases

Phase	$-\Delta G^\circ_{973}$ (kcal/g-atom)	$-\Delta H^\circ_{729-1219}$ (kcal/g-atom)	$\Delta S^\circ_{729-1219}$ (cal/g-atom K)	$-\Delta V_{298}$ (%)
$ThCu_6$	3.27±0.01	2.50±0.12	0.79±0.13	1.2
$ThCu_{3.6}$	4.99±0.02	4.12±0.16	0.88±0.20	1.8
$ThCu_2$	6.30±0.04	6.21±0.35	0.07±0.44	4.6
Th_2Cu	6.40±0.05	6.51±0.51	-0.11±0.49	1.9

data yield standard free energies of phase formation as follows with E°_1 through E°_4 belonging to cells [1a] through [1d], respectively:

$$\Delta G^\circ(ThCu_6) = -4\mathcal{F}E^\circ_1$$

$$= -17520 \pm 830 - (5.53\pm 0.92)T \text{ cal/mol}, \quad [3a]$$

$$\Delta G^\circ(ThCu_{3.6}) = -4\mathcal{F}(\tfrac{2}{5} E^\circ_2) + \tfrac{3}{5} \Delta G^\circ(ThCu_6)$$

$$= -4\mathcal{F}[\tfrac{2}{5} E^\circ_2 + \tfrac{3}{5} E^\circ_1]$$

$$= -18990 \pm 740 - (4.06\pm 0.92)T \text{ cal/mol}, \quad [3b]$$

$$\Delta G^\circ(ThCu_2) = -4\mathcal{F}(\tfrac{4}{9} E^\circ_3) + \tfrac{5}{9} \Delta G^\circ(ThCu_{3.6})$$

$$= -4\mathcal{F}[\tfrac{4}{9} E^\circ_3 + \tfrac{2}{9}E^\circ_2 + \tfrac{1}{3} E^\circ_1]$$

$$= -18700 \pm 1000 - (0.2\pm 1.3)T \text{ cal/mol}, \quad [3c]$$

$$\Delta G^\circ(Th_2Cu) = -4\mathcal{F}(\tfrac{3}{2} E^\circ_4) + \tfrac{1}{2}\Delta G^\circ(ThCu_2)$$

$$= -4\mathcal{F}[\tfrac{3}{2} E^\circ_4 + \tfrac{2}{9} E^\circ_3 + \tfrac{1}{9} E^\circ_2 + \tfrac{1}{3} E^\circ_1]$$

$$= -19500 \pm 1500 + (0.3\pm 1.5)T \text{ cal/mol} \quad [3d]$$

Values for the Gibbs free energies of formation at 973 K, which is near the mean of the temperature range of measurement, are listed in Table III for all four intermediate phases in the Th-Cu system. Also shown in Table III are the mean enthalpies of formation and mean entropies of formation as evaluated from the intercepts and slopes of the temperature dependences of the free energies of formation. The chemical reactivity of the phases tends to increase with increasing thorium content. This factor combined with the cumulative nature of the errors causes the experimental uncertainties to increase from $ThCu_6$ to Th_2Cu. Volume contractions accompanying phase formation were computed from crystallographic data, and the resulting values are included as the last column in Table III where it may be noted the trend in values correlates qualitatively with the enthalpies of formation for the three copper-rich phases but not for Th_2Cu.

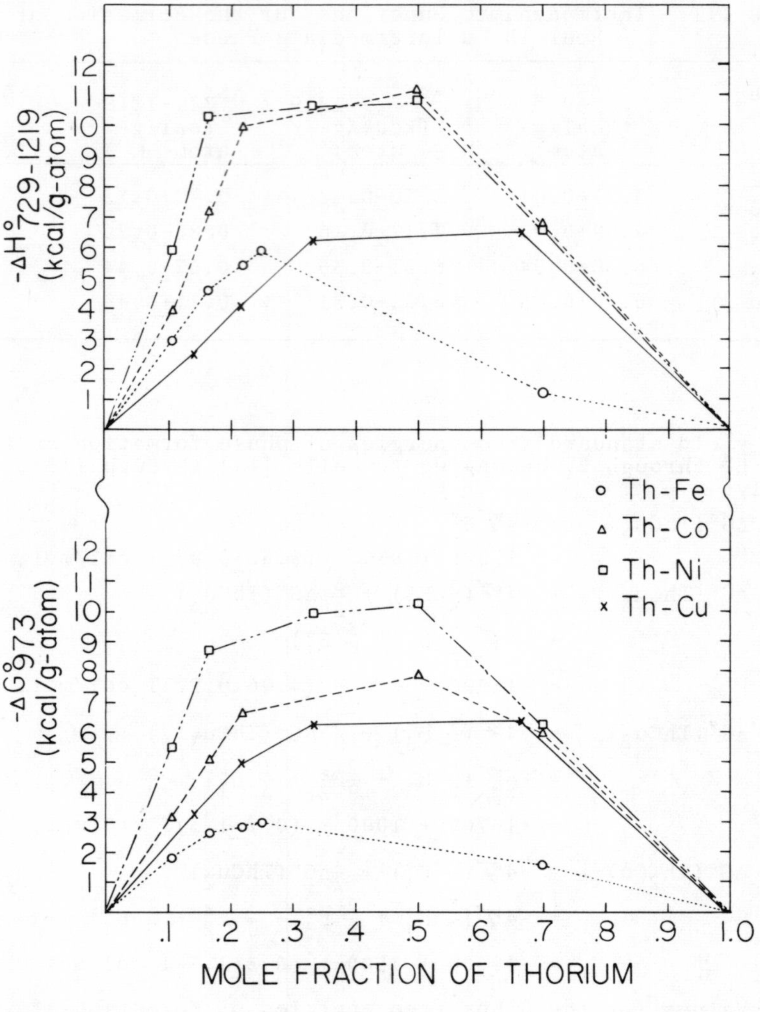

FIG.3. Comparison of enthalpies and free energies of formation as functions of composition for the phases in the Th-Cu, Th-Ni, Th-Co and Th-Fe systems.

DISCUSSION

A check on reproducibility can be made by comparing the present results for $ThCu_6$ with those of Magnani et al. [18] whose data indicate $-\Delta G^{\circ}_{973}$ = 3.13 kcal/g-atom, $-\Delta H^{\circ}_{850-949}$ = 2.09 kcal/g-atom, and $\Delta S^{\circ}_{850-949}$ = 1.06 cal/g-atom K. Corresponding values in Table III show that the free energy of formation is reproducible to ~4% and the enthalpy and entropy of formation are reproducible to ~20%. This agreement supports evaluation of the enthalpies and entropies of phase formation from intercepts and slopes. The greater number of experimental points plus the broader temperature range of measurement favor the present values, and the enthalpy and entropy values for $ThCu_{3.6}$, $ThCu_2$, and Th_2Cu should be of comparable reliability to those for $ThCu_6$.

Fig. 3 shows a comparison of values for $-\Delta G^\circ_{973}$ and $-\Delta H^\circ_{729-1219}$ in kcal/g-atom for all phases in the Th-Cu, Th-Ni, Th-Co, and Th-Fe systems. For the two to three thorium-poor phases in each system where there are few to no thorium-thorium contacts in the crystal structures, both the enthalpies and free energies of phase formation within a given system remain essentially constant when expressed in units of kcal/mol of thorium reactant. In the case of the enthalpies of formation, these values occur in the order -18 for Th-Cu, -28 for Th-Fe, -41 for Th-Co, and -59 for Th-Ni and reflect the order of increasing strength of bonding interactions. In the case of the free energies of formation the positions of Th-Cu and Th-Fe invert in the sequence with values of -16.5 for Th-Fe, -22.7 for Th-Cu, -30.2 for Th-Co, and -52.2 for Th-Ni. This change in sequence is attributable to the small positive entropies of formation of $ThCu_6$ and $ThCu_{3.6}$ which presumably arise from greater vibrational freedom in these two phases.

In Fig. 3 the enthalpies of phase formation for the thorium-rich phases show that for these phases the bonding interactions in the Th-Cu, Th-Co, and Th-Ni systems are quite comparable, and this in combination with the free energies of phase formation implies comparable lattice rigidities and vibrational behavior. Only Th_7Fe_3, which has also been found to have a positive entropy of formation, deviates from this trend. Indeed, among all the phases in the four systems only Th_7Fe_3, $ThCu_6$, and $ThCu_{3.6}$ have positive entropies of formation; the value for $ThCu_2$ is zero within experimental uncertainty and the values for all other phases are small and negative. The data of Kleykamp and Murabayashi [5,6] for Th-Rh and Th-Ru, of Rezukhina and Pokarev [7] for Th-Re, and of Chiotti and Gill [19] for Th-Zn show that small negative entropies of formation are also the rule for those systems, and thus the positive values for Th_7Fe_3, $ThCu_6$, and $ThCu_{3.6}$ are atypical for thorium alloys.

Novokreshchenov et al. [20] have made electromotive force measurements to determine the thermodynamic functions associated with the formation of the five intermediate phases in the Zr-Cu system. Comparison of their results with the present results indicates that at comparable temperatures the free energies of formation of the Zr-Cu phases tend to be more negative than the free energies of formation of the Th-Cu phases by the order of a factor of two. This is in qualitative accord with data [21] indicating that the free energies of formation of phases in the Zr-B, Zr-C, and Zr-Fe systems are considerably more negative than the free energies of formation of phases with comparable stoichiometries in the Th-B, Th-C, and Th-Fe systems.

REFERENCES

1. SMITH, J. F., J. Nucl. Mater. 51 136 (1974).
2. SKELTON, W. H., MAGNANI, N. J., SMITH, J. F., Met. Trans. 1 1833 (1970).
3. SKELTON, W. H., MAGNANI, N. J., SMITH, J. F., Met. Trans. 2 473 (1971).
4. SKELTON, W. H., MAGNANI, N. J., SMITH, J. F., Met. Trans. 4 917 (1973).
5. KLEYKAMP, H., MURABAYASHI, M., J. Less-Common Met. 35 227 (1974).
6. MURABAYASHI, M., KLEYKAMP, H., J. Less-Common Met., in press.
7. REZUKHINA, T. N., POKAREV, B. S., J. Chem. Thermodynamics 3 369 (1971).
8. ARONSON, S., in Nuclear Metallurgy, vol. 10, (WABER, J. T., CHIOTTI, P., MINER, W. N., Eds.) Edwards Bros., Ann Arbor, Mich. (1964), p. 247.
9. ARONSON, S., SADOFSKY, J., J. Inorg. Nucl. Chem. 27 1769 (1965).
10. ARONSON, S., J. Inorg. Nucl. Chem. 29 1611 (1967).
11. ARONSON, S., AUSKERN, A., Proceedings of the Symposium on Thermodynamics, Vienna, 22-27 July 1965, Vol. 1, (International Atomic Energy Agency, Vienna, 1966) p. 165.
12. SCHILTZ, R. J., Jr., STEVENS, E. R., CARLSON, O. N., J. Less-Common Met. 25 175 (1971).
13. BERLIN, B., J. Less-Common Met. 29 337 (1972).
14. BUSCHOW, K. H. J., VAN DER GOOT, A. S., J. Less-Common Met. 20 309 (1970).
15. BAENZIGER, N. C., RUNDLE, R. E., SNOW, A. I., Acta Cryst. 9 93 (1956).
16. MURRAY, J. R., J. Inst. Metals 84 91 (1955-56).
17. BAILEY, D. M., J. Less-Common Met. 30 164 (1973).
18. MAGNANI, N. J., SKELTON, W. H., SMITH, J. F., Nuclear Metallurgy 15 727 (1969): CONF-690801.
19. CHIOTTI, P., GILL, K. J., Trans. Am. Soc. Mining Met. Engrs. 221 573 (1961).
20. NOVOKRESHCHENOV, YU. V., ILYUSHCHENKO, N. G., ROSSOKHIN, B. G., NICHKOV, I. F., Tr. Inst. Elektrokhim., Acad. Nauk SSSR, Ural. Filial No. 12 70 (1969).
21. HULTGREN, R., ORR, R. L., KELLEY, K. K., Supplement to Selected Values of Thermodynamic Properties of Metals and Alloys, University of California, Berkeley, Calif., data sheets through June 1972.

DISCUSSION

G.B. BARBI: I should like to make a general comment on emf measurements with cells employing solid fluorides as electrolytes. With these solid cells it is necessary to work in a highly pre-purified inert gas atmosphere, since very small traces of oxygen during either the preparation or the measurements may cause the formation of spurious compounds — oxides and oxyfluorides — particularly at the triple boundaries between electrode-electrolyte gas phases. Electrochemically speaking this means mixed potentials.

Another point I would like to make is the possibility of extending these measurements to other systems. A further limitation of these cells vis-à-vis those using refractory oxide solid solutions as electrolyte is the volatility of many fluorides, which is generally higher than that of the corresponding oxides. Even if this is not confirmed by your observations — since at your experimental temperatures thorium fluoride shows a low vapour pressure — the use of fluoride cells to determine the thermodynamic functions of other intermetallic compounds, e.g. uranium compounds, appears less promising than that of cells with oxide electrolytes.

J.F. SMITH: I agree entirely. The sensitivity to volatility, impurities and side reactions is very real, and one must be most cautious about extending this technique to other systems. I did not elaborate on the experimental technique because it has become relatively common, with about half the thorium systems using the solid electrolyte emf method. We have taken pains to obtain good analytical data on our starting material and have done all our handling and measurements in a dry box with a purified inert gas atmosphere. I might also note that extending this technique to other materials, such as uranium systems, might introduce complications due to the various competing valence states. This problem does not, however, arise in the case of thorium alloys.

К ТЕРМОДИНАМИКЕ ТРЕХКОМПОНЕНТНЫХ ТУГОПЛАВКИХ СОЕДИНЕНИЙ

Р.А.АНДРИЕВСКИЙ, Э.М.ФЕДОРОВ
Государственный Комитет по использованию атомной энергии СССР,
Москва,
Союз Советских Социалистических Республик

Доклад представлен А.С. Пановым

Abstract–Аннотация

THE THERMODYNAMICS OF HIGH-MELTING THREE-COMPONENT COMPOUNDS.

The authors present and discuss expressions for the partial free energies of three-component carbide interstitial phases with allowance for concentration variation in the heat of formation of vacancies. They analyse the congruent vaporization of the solutions and show that stable congruent vaporization can be expected in the case of solutions with short-range ordering. Expressions for the activities of the components in three-component systems are used to evaluate thermodynamic factors. The values of the latter are discussed in the light of the specific features of diffusion-controlled processes in three-component interstitial phases.

К ТЕРМОДИНАМИКЕ ТРЕХКОМПОНЕНТНЫХ ТУГОПЛАВКИХ СОЕДИНЕНИЙ.

Записаны и обсуждены выражения для парциальных свободных энергий трехкомпонентных карбидных фаз внедрения с учетом концентрационного изменения теплоты образования вакансий. Проанализировано конгруэнтное испарение растворов; показано, что устойчивое конгруэнтное испарение можно ожидать для растворов с ближним упорядочением. Выражения для активностей компонентов в трехкомпонентных системах использованы для оценки термодинамических факторов. Значения последних обсуждены в связи с особенностями диффузионно-контролируемых процессов в трехкомпонентных фазах внедрения.

1. ВСТУПЛЕНИЕ

В продолжение и развитие работ [1-2], которые были доложены на предыдущем симпозиуме по термодинамике атомных материалов, представляло интерес более детально рассмотреть некоторые вопросы термодинамики многокомпонентных тугоплавких фаз. Такое рассмотрение, помимо общего интереса, представляется целесообразным в связи с необходимостью развития расчетных методов, которые бы позволили оценивать такие процессы, как испарение, ползучесть, спекание и диффузию в многокомпонентных топливных и конструкционных тугоплавких соединениях (карбидах, нитридах и т.п.). Влияние облучения на термодинамические свойства в данном случае также может быть оценено в связи с изменением состава в результате выгорания, либо в связи с возможным изменением в расположении легких атомов.

Некоторые из моментов термодинамики трехкомпонентных фаз внедрения обсуждались в работах [3-7].

2. ПАРЦИАЛЬНЫЕ СВОБОДНЫЕ ЭНЕРГИИ В ТРЕХКОМПОНЕНТНЫХ ФАЗАХ ВНЕДРЕНИЯ

Рассматривая обычный термодинамический цикл Борна-Хабера, запишем, следуя [6], выражение для свободной энергии Гиббса фазы $(A_{1-y}\,B_y)_{1-x}\,C_x$

$$F = (1-x)(1-y)F_A^0 + y(1-x)F_B^0 + xF_C^0 \tag{1}$$
$$+\frac{\varphi}{2}\left[(1-y)\,2\Delta F_*^{AC} + y2\Delta F_*^{BC} + \Delta F_{CM}\right] + \left(\frac{\varphi}{2}+x-1\right)H_{M^+}$$
$$+\left(\frac{\varphi}{2}-x\right)H_{C^+} \quad + RT\left[\left(\frac{\varphi}{2}+x-1\right)\ln\left(\frac{\varphi}{2}+x-1\right)\right.$$
$$\left.+\left(\frac{\varphi}{2}-x\right)\ln\left(\frac{\varphi}{2}-x\right) + (1-x)\ln(1-x) + x\ln x - \varphi\ln\frac{\varphi}{2}\right]$$

Здесь $F^0_{A,B,C}$ – свободная энергия образования чистых веществ A,B,C в стандартном состоянии, а $2\Delta F_*^{AC,BC}$ – стехиометрических соединений (типа NaCl) AC и BC соответственно (на 1 г-атом), φ – отношение полного числа мест в решетке к полному числу атомов, H_{M^+} и H_{C^+} – теплоты образования металлических и неметаллических вакансий.

При записи (1) предполагалось, что величины H_{M^+} и H_{C^+} не меняются в области гомогенности, и не учитывался вклад избыточной энтропии в свободную энергию образования вакансий.

Величина φ находится из условия минимума свободной энергии

$$\frac{\delta F}{\delta \varphi} = 0 \tag{2}$$

а парциальные свободные энергии – из соотношений

$$\overline{F}_\alpha = F + (1-x_\alpha)\,\frac{\delta F}{\delta x_\alpha}\left|\,\frac{x_\beta}{x_\gamma} = \text{const}\right. \tag{3}$$

где x_α – молярная концентрация α-компоненты.

Величины H_{M^+} и H_{C^+} для карбидов найдем, воспользовавшись предположениями [8]: карбиды стехиометрического состава равновесны с графитом и энтальпия тепловых вакансий равна теплоте образования карбида. Теперь подстановка (3) в (1) с учетом (2) приводит к следующим выражениям для парциальных свободных энергий

$$\overline{F}_A = F_A^0 + 2\Delta F_*^{AC} + H_{C^+} - y\frac{\varphi-1}{1-x}\,\frac{\delta H_{C^+}}{\delta y} + \epsilon y^2 + RT\ln\frac{(1-x)(1-y)}{\varphi^2/4}\left(\frac{\varphi}{2}-x\right) \tag{4}$$
$$\overline{F}_B = F_B^0 + 2\Delta F_*^{BC} + H_{C^+} + (1-y)\frac{\varphi-1}{1-x}\,\frac{\delta H_{C^+}}{\delta y} + \epsilon(1-y)^2 + RT\ln\frac{(1-x)y}{\varphi^2/4}\left(\frac{\varphi}{2}-x\right)$$
$$\overline{F}_C = F_C^0 - H_{C^+} \quad + RT\ln\frac{x}{\frac{\varphi}{2}-x}$$

где $H_{C^+} = -(1-y)2\Delta H^{AC} \; - y2\Delta H^{BC} \; - RT\ln 2$, (5)

$2\Delta H^{AC}$, $2\Delta H^{BC}$ – теплоты образования соединений AC, BC (на 1 г/атом). Свободная энергия смешения стехиометрических карбидов записана в виде:

$$\Delta F_{CM} = \epsilon y(1-y) + RT\left[y\ln y + (1-y)\ln(1-y)\right] \tag{6}$$

соответствующем приближению регулярных растворов, ϵ – энергия взаимообмена.

Отметим, что при выводе (4) в отличие от нашей работы [6] считалось, что углерод не дает вклада в теплоту смешения. Это соответствует обычному квазихимическому подходу. При отсутствии такого предположения выражение для $\overline{F}_\alpha$ несколько усложняется [6].

Выше предполагалось, что теплота образования вакансий не меняется в области гомогенности. При этом изменение активности компонент в области гомогенности связывается лишь с изменением конфигурационной энтропии для неметаллической подрешетки. Такое приближение не всегда оправдано и может приводить, например для ZrC, к заметному расхождению с экспериментом [9]. Учет зависимости теплоты образования вакансий от состава для простых карбидов выполнен в работе [10]. Учет приближения [10] приводит к удовлетворительному совпадению расчетных и опытных данных активности углерода в области гомогенности ZrC. Ниже рассмотрен случай бинарных растворов, причем зависимость теплоты образования соединения от состава по углероду учитывается с самого начала. Сделаем предварительно одно замечание. В выражении (4) параметр φ может быть с хорошей точностью положен равным $2(1-x)$. Нетрудно показать, что это соответствует пренебрежению вкладом тепловых вакансий в металлической подрешетке в свободную энергию системы, что, очевидно, вполне допустимо ввиду их малой концентрации. Отметим, что это допущение может быть сделано сразу, при этом нет необходимости накладывать ограничение на вид верхней границы области гомогенности.

Теперь выражение для свободной энергии системы может быть записано в виде:

$$\begin{aligned} F &= (1-x)F_A^0(1-y) + (1-x)yF_B^0 + xF_C^0 \\ &+ (1-x)\left[(1-y)\,2\Delta F_*^{AC} + y2\Delta F_*^{BC} + \Delta F_{CM}\right] + (1-x)\int_x^{0,5} h_{C^+}\frac{dx}{(1-x)^2} \\ &+ RT\left[(1-2x)\ln(1-2x) + x\ln x - (1-x)\ln(1-x)\right] \end{aligned} \tag{7}$$

Отсюда следуют выражения для парциальных свободных энергий:

$$\begin{aligned} \overline{F}_A &= F_A^0 + 2\Delta F_*^{AC} + \Delta H_{CM} - y\frac{\delta\Delta H_{CM}}{\delta y} + \int_x^{0,5} h_{C^+}\frac{dx}{(1-x)^2} \\ &+ \frac{x}{1-x}h_{C^+} - y\frac{\delta}{\delta y}\int_x^{0,5} h_{C^+}\frac{dx}{(1-x)^2} + RT\ln\frac{(1-y)(1-2x)}{1-x} \\ \overline{F}_B &= F_B^0 + 2\Delta F_*^{BC} + \Delta H_{CM} + (1-y)\frac{\delta\Delta H_{CM}}{\delta y} + \int_x^{0,5} h_{C^+}\frac{dx}{(1-x)^2} \\ &+ \frac{x}{1-x}h_{C^+} + (1-y)\frac{\delta}{\delta y}\int_x^{0,5} h_{C^+}\frac{dx}{(1-x)^2} + RT\ln\frac{y(1-2x)}{1-x} \\ \overline{F}_C &= F_C^0 - h_{C^+} + RT\ln\frac{x}{1-2x} \end{aligned} \tag{8}$$

Записывая теперь h_{C^+} в соответствии с [10] в форме:

$$h_{C^+} = -\frac{\delta 2\Delta H}{\delta z} \tag{a}$$

где $z = \frac{x}{1-x}$ — формульный показатель, $2\Delta H$ — теплота образования соединения, равная

$$2\Delta H = (1-y)\,2\Delta H^{AC} + y\,2\Delta H^{BC} + \Delta H_{CM} \qquad (10)$$

получим

$$\bar{F}_A = F_A^0 + 2\Delta F_*^{AC} - 2\Delta H_*^{AC} + 2\Delta H^{AC} + \epsilon y^2 - z\left[(1-y)\frac{\delta 2\Delta H^{AC}}{\delta z} + y\frac{\delta 2\Delta H^{BC}}{\delta z}\right] + RT\ln(1-y)(1-z)$$

$$\bar{F}_B = F_B^0 + 2\Delta F_*^{BC} - 2\Delta H_*^{BC} + 2\Delta H^{BC} + \epsilon(1-y)^2 - z\left[(1-y)\frac{\delta 2\Delta H^{AC}}{\delta z} + y\frac{\delta 2\Delta H^{BC}}{\delta z}\right] + RT\ln y(1-z) \qquad (11)$$

$$\bar{F}_C = F_C^0 + y\frac{\delta 2\Delta H^{BC}}{\delta z} + (1-y)\frac{\delta 2\Delta H^{AC}}{\delta z} + RT\ln\frac{z}{1-z}$$

Здесь, как и ранее, знак (*) соответствует стехиометрическому соединению, и теплота смешения вновь для простоты записана в виде $\Delta H_{CM} = \epsilon y(1-y)$ с независящей от состава энергией взаимообмена. Полагая $y=0$, получаем для простого AC — карбида

$$\bar{F}_A = F_A^0 + 2\Delta F_*^{AC} - 2\Delta H_*^{AC} + 2\Delta H^{AC} - z\frac{\delta 2\Delta H^{AC}}{\delta z} + RT\ln(1-z) \qquad (12)$$

$$\bar{F}_C = F_C^0 + \frac{\delta 2\Delta H^{AC}}{\delta z} + RT\ln\frac{z}{1-z}$$

Если в выражениях (4) и (11) выделить члены, принадлежащие соответствующим простым карбидам, то, как легко показать, их можно записать в следующей форме:

$$\bar{F}_A = \bar{F}_A^{AC} + yz\,[\bar{F}_C^{AC} - \bar{F}_C^{BC}] + \epsilon y^2 + RT\ln(1-y)$$

$$\bar{F}_B = \bar{F}_B^{BC} - (1-y)z\left[\bar{F}_C^{AC} - \bar{F}_C^{BC}\right] + \epsilon(1-y)^2 + RT\ln y \qquad (13)$$

$$\bar{F}_C = (1-y)\,\bar{F}_C^{AC} + y\bar{F}_C^{AC}$$

где индексы указывают на парциальную свободную энергию соответствующей компоненты в простом карбиде.

Формулы (13) получаются и непосредственно, если композицию $(AC_z)_{1-y}\ (BC_z)_y$ рассматривать как регулярный раствор с известными термодинамическими свойствами смешиваемых карбидов.

Переход к парциальным давлениям паров компонент осуществляется по известным соотношениям:

$$\bar{F}_A - F_A^0 = RT \ln \bar{P}_A / P_A^0 \qquad (14)$$

где $\bar{P}_A$ и P_A^0 – давление паров А-компоненты над исследуемым соединением и в стандартном состоянии.

Подставляя (13) в (14), получим:

$$\begin{aligned} \bar{P}_A &= \bar{P}_A^{AC} \left[\exp(\epsilon y^2/RT)\right] \left[\frac{a_C^{AC}}{a_C^{BC}}\right]^{yz} (1-y) \\ \bar{P}_B &= \bar{P}_B^{BC} \left[\exp(\epsilon(1-y)^2/RT)\right] \left[\frac{a_C^{AC}}{a_C^{BC}}\right]^{-(1-y)z} y \\ \bar{P}_C &= \left[\bar{P}_C^{AC}\right]^{1-y} \left[\bar{P}_C^{BC}\right]^{y} \end{aligned} \qquad (15)$$

В формулах (15) экспонента учитывает неидеальность раствора в металлической подрешетке, а члены в фигурных скобках соответствуют учету, в соответствии с соотношением Гиббса-Дюгема, влияния изменяющейся при смешении активности углерода.

Отметим, что формулы (15) могут быть использованы и в тех случаях, когда активности компонент в простых карбидах описываются полученными из эксперимента интерполяционными соотношениями.

Аналогичным образом можно получить выражения для парциальных свободных энергий в фазах внедрения других типов – карбонитридах, карбогидридах и т.д. Однако отсутствие достоверных сведений о теплотах образования вакансий в нитридах, моноокислах и гидридах не позволяет нам проводить количественные оценки. Кроме того, для таких фаз, как карбогидриды, вряд ли подходящим окажется регулярное приближение.

3. ИСПАРЕНИЕ РАСТВОРОВ КАРБИДОВ

Применим полученные выражения к анализу испарения растворов карбидов в вакууме. Скорость испарения определяется известной формулой Лэнгмюра и может быть рассчитана по приведенным выше соотношениям (4), (11), (15) как функция состава и температуры поверхности.

Особый интерес представляет конгруэнтное испарение растворов карбидов. Уравнения, определяющие для моноатомного испарения конгруэнтный состав, имеют вид:

$$\frac{\lambda_A \bar{P}_A}{\sqrt{m_A}(1-x)(1-y)} = \frac{\lambda_B \bar{P}_B}{\sqrt{m_B}(1-x)y} = \frac{\lambda_C \bar{P}_C}{\sqrt{m_C}x} \qquad (16)$$

где $m_{A,B,C}$ – атомные веса, $\lambda_{A,B,C}$ – коэффициенты испарения компонент.

Ввиду сложного характера зависимости давления паров от состава система уравнений (16), как правило, может быть решена лишь численно. Рассмотрим уравнение для металлических компонент:

$$\frac{\lambda_A \overline{P}_A}{\sqrt{m_A}(1-x)(1-y)} = \frac{\lambda_B \overline{P}_B}{\sqrt{m_B}(1-x)y} \tag{17}$$

которое с учетом (14) и (3) можно записать в следующей общей форме:

$$\frac{1}{1-x}\frac{\delta F}{\delta y} = RT\ln\frac{\lambda_A}{\lambda_B}\sqrt{\frac{m_B}{m_A}}\frac{y}{1-y} \tag{18}$$

В идеальных растворах карбидов зависимость свободной энергии системы от состава имеет следующий вид:

$$F(x, y, T) = \Phi_1(x, T) + \Phi_2(x, T)y \tag{19}$$

$$+RT(1-x)\Big[y\ln y + (1-y)\ln(1-y)\Big]$$

При подстановке (19) в (18) величина y выпадает из уравнения (18) и, следовательно, оно может удовлетворяться при некоторых значениях x и T только случайным образом. Таким образом, возможность экспериментального обнаружения конгруэнтного испарения в идеальных растворах практически исключена. Отсюда следует, что конгруэнтное испарение связано с неидеальностью системы.

Эти общие соображения непосредственно получаются и в рассмотренных моделях регулярного раствора карбидов (4), (11), (15), в которых уравнение (18) имеет решение:

$$y_C = \frac{1}{2}\left[1 + \frac{M(x, T)}{\epsilon}\right] \tag{20}$$

где для выражения (4):

$$M(x, T) = 2\Delta F_*^{BC} - 2\Delta F_*^{AC} + \frac{\varphi-1}{1-x}\frac{\delta H_C}{\delta y} + RT\ln\frac{\lambda_B}{\lambda_A}\sqrt{\frac{m_A}{m_B}}\frac{P_B^0}{P_A^0}$$

для выражения (11):

$$M(x, T) = 2\Delta F_*^{BC} - 2\Delta F_*^{AC} - 2\Delta H_*^{BC} + 2\Delta H_*^{AC} + 2\Delta H^{BC} - 2\Delta H^{AC}$$

$$+RT\ln\frac{\lambda_B}{\lambda_A}\sqrt{\frac{m_A}{m_B}}\frac{P_B^0}{P_A^0}$$

для выражения (15):

$$M(x,T) = RT \ln \frac{\lambda_B}{\lambda_A} \frac{\overline{P}_B^{BC}}{\overline{P}_A^{AC}} \left[\frac{a_C^{BC}}{a_C^{AC}}\right]^{\frac{x}{1-x}} \sqrt{\frac{m_A}{m_B}}$$

Как видно, значение y_C определяется соотношением величины $M(x,T)$ и энергии взаимообмена ϵ, которая является в рассматриваемой модели единственным параметром, характеризующим неидеальность системы. Имеющие смысл решения ($0 \le y \le 1$) получаются лишь при $|\epsilon| \ge |M|$, т.е. при достаточно сильной неидеальности, причем, что важно отметить, как при положительных, так и при отрицательных ϵ. Для того чтобы выяснить разницу в характере испарения композиций с положительными и отрицательными ϵ, рассмотрим их поведение при малых отклонениях от конгруэнтного состава. В этом случае в образце возникнут диффузионные потоки и необходимо решать соответствующую диффузионную задачу. Граничные условия к ней имеют вид:

$$-N_S D_C \frac{\delta}{\delta r} \frac{x}{1-x} - N_S \frac{x}{1-x} V_\sigma = G_C \qquad (21)$$

$$-N_S D_B \frac{\delta y}{\delta r} - N_S y V_\sigma = G_B$$

и отражают равенство потоков компонент из объема образца к поверхности и от поверхности во внешнюю среду. Здесь r — радиальная координата с положительным направлением от центра образца к поверхности, N_S — суммарная плотность атомов в металлической подрешетке, G_B и G_C — атомные скорости испарения В и С-компонент. В обоих уравнениях первый член — диффузионный поток, второй учитывает конвективный вклад, связанный с перемещением поверхности образца (со скоростью V_σ) при испарении.

Для растворов карбидов естественно считать, что движение поверхности определяется испарением металлических атомов, поэтому:

$$N_S V_\sigma = -(G_A + G_B) \qquad (22)$$

Подставляя (22) в (21), получим:

$$-N_S D_C \frac{\delta}{\delta r} \frac{x}{1-x} = G_C - \frac{x}{1-x}(G_A + G_B)$$

$$(23)$$

$$-N_S D_B \frac{\delta y}{\delta r} = G_B - y(G_A + G_B)$$

Если начальный состав образца таков, что правые части уравнений (23) обращаются в нуль

$$G_C = \frac{x}{1-x}\,(G_A + G_B) \tag{24}$$
$$G_B = y\,(G_A + G_B)$$

то градиенты концентраций не возникают и имеет место конгруэнтное испарение.

Для его устойчивости необходимо, чтобы при малых отклонениях состава образца от конгруэнтного возникли градиенты концентраций, уменьшающие разницу между исходным и конгруэнтным составами. Мы получаем, таким образом, из (23) условия устойчивости:

$$\frac{\delta}{\delta x}\left\{G_C - \frac{x}{1-x}\,(G_A + G_B)\right\} > 0$$
$$\frac{\delta}{\delta y}\left\{G_B - y\,(G_A + G_B)\right\} > 0 \tag{25}$$

Как следует из их вывода, уравнения (24) и (25) справедливы и в более общем случае испарения в газовую среду, а также при наличии гетерогенных химических реакций.

Используя полученные соотношения (4), (11), (15) и (14), теперь нетрудно показать, что первое из условий (25) выполняется во всей области гомогенности, а второе сводится к неравенству $\epsilon < 0$. Таким образом, устойчивое конгруэнтное испарение может происходить лишь в растворах карбидов с отрицательной энтальпией смешения. К этому выводу можно придти и в рамках квазихимического подхода к рассматриваемой фазе внедрения. Не исключено, однако, что конгруэнтное испарение в трехкомпонентных фазах, у которых $\epsilon \nless 0$, может реализоваться кинетически.

Условие $\epsilon < 0$ совпадает с необходимым условием наличия ближнего упорядочения в металлической подрешетке. Можно сказать поэтому, что конгруэнтное испарение может происходить в растворах карбидов, обладающих тенденцией к упорядочению, но не к расслоению. Этот вывод непосредственно получается в модели нерегулярного раствора с ближним порядком в металлической подрешетке. Полученные условия устойчивости не зависят от концентраций и выполняются во всей области существования раствора. Если конгруэнтное испарение имеет место, то в этом случае конгруэнтный состав достижим из любой точки обточки области гомогенности раствора карбидов.

В общем случае, как можно показать, второе из условий устойчивости (25) имеет вид $\delta^2 \Delta H_{CM} / \delta y^2 > 0$. Поэтому поведение системы, в которой функция $\Delta H_{CM}(y)$ имеет несколько экстремумов, может носить более сложный характер.

Наконец, отметим еще два обстоятельства, следующие из рассмотренной модели. Во-первых, как видно из уравнения (20), конгруэнтное испарение может иметь место, вообще говоря, лишь в некотором диапазоне температур и концентраций углерода. Во-вторых, если уравнение (20) имеет физическое решение при положительной энергии взаимообмена, должно происходить антиконгруэнтное испарение — состав (по металлическим компонентам) "разбегается" в разные стороны от этого значения.

4. НЕКОТОРЫЕ ОСОБЕННОСТИ ДИФФУЗИОННО-КОНТРОЛИРУЕМЫХ ПРОЦЕССОВ

В бинарном растворе карбидов имеет место эффект Киркендалла. Используя обычные приближения Даркена о равновесности вакансий и постоянстве плотности узлов решетки, можно получить [11] выражения для полных потоков:

$$\frac{1}{N_S}J_1 = -(c_2 D_{11} + c_1 D_{22})\nabla c_1 - c_1 c_2 (D_{10} - D_{20})\nabla c_0$$
$$\frac{1}{N_S}J_0 = -[D_{00} - c_0(c_1 D_{10} + c_2 D_{20})]\nabla c_0 - c_0 [D_{01} - (D_{11} - D_{22})]\nabla c_1 \qquad (26)$$

которые определяют соответствующие коэффициенты гетеродиффузии.

$$\tilde{D}_{11} = \tilde{D}_{22} = c_2 D_{11} + c_1 D_{22}, \qquad \tilde{D}_{10} = c_1 c_2 (D_{10} - D_{20})$$
$$D_{01} = c_0[D_{01} - (D_{11} - D_{22})], \qquad D_{00} = D_{00} - c_0 (c_1 D_{10} + c_2 D_{20}) \qquad (27)$$

где

$$D_{ik} = D_i^* g_{ik}$$

$$g_{ik} = \frac{\delta \ln a_i}{\delta \ln c_i} \quad \text{при} \quad i = k, \qquad g_{ik} = \frac{\delta \ln a_i}{\delta c_k} \quad \text{при} \quad i \neq k$$

Оценим входящие в коэффициенты диффузии термодинамические факторы, используя полученные ранее выражения (4), (11) и (15) для активностей компонент.

Для модели раствора, описываемой формулами (15) имеем:

$$g_{11} = \frac{\delta \ln a_1}{\delta \ln c_1} = 1 - c_1 c_0 \ln\left(\frac{a_C^{AC}}{a_C^{BC}}\right) - 2\,\frac{\epsilon c_1 c_2}{RT}$$

$$g_{10} = \frac{\delta \ln a_1}{\delta c_0} = \frac{\delta \ln a_A^{AC}}{\delta c_0} + c_2 \ln\left(\frac{a_C^{AC}}{a_C^{BC}}\right) + c_0 c_2 \frac{\delta}{\delta c_0} \ln\left(\frac{a_C^{AC}}{a_C^{BC}}\right)$$

$$g_{22} = \frac{\delta \ln a_2}{\delta \ln c_2} = 1 + c_2 c_0 \ln\left(\frac{a_C^{AC}}{a_C^{BC}}\right) - 2\epsilon\frac{c_1 c_2}{RT} \qquad (28)$$

$$g_{20} = \frac{\delta \ln a_2}{\delta c_0} = \frac{\delta \ln a_B^{BC}}{\delta c_0} - c_1 \ln\left(\frac{a_C^{AC}}{a_C^{BC}}\right) - c_1 c_0 \frac{\delta}{\delta c_0} \ln\left(\frac{a_C^{AC}}{a_C^{BC}}\right)$$

$$g_{00} = \frac{\delta \ln a_0}{\delta \ln c_0} = c_0 \left[c_1 \frac{\delta \ln a_C^{AC}}{\delta c_0} + c_2 \frac{\delta \ln a_C^{BC}}{\delta c_0}\right]$$

$$g_{01} = \frac{\delta \ln a_0}{\delta c_1} = \ln\left(\frac{a_C^{AC}}{a_C^{BC}}\right)$$

Из (28) можно сделать некоторые заключения о величинах g_{ik}. Отрицательная энергия взаимообмена ($\epsilon < 0$) увеличивает коэффициенты диффузии металлических компонент, как это и следует из физических соображений о стремлении к упорядочению в такой системе.

Коэффициент диффузии металлической компоненты того карбида, концентрация которого и активность углерода в достаточное число раз превышает эти величины во втором карбиде, может оказаться отрицательным. Это обстоятельство может иметь место и в идеальном растворе карбидов и связано с наличием углеродной подрешетки. Условием $g_{11} < 0$ служит неравенство

$$c_1 c_0 \ln \frac{a_C^{AC}}{a_C^{BC}} > 1 - 2\frac{\epsilon c_1 c_2}{RT}$$

Ввиду сильного изменения активностей компонент в пределах области гомогенности следует ожидать, что недиагональные коэффициенты диффузии металлических компонент могут быть весьма велики и отрицательны. Как видно из (28), при $c_2 \to 0$ $g_{10} < 0$.

Термодинамический фактор $g_{00} > 0$. Знак g_{01} зависит от отношения активностей углерода в смешиваемых карбидах.

Отмеченные выводы могут быть сопоставлены с расчетом, если использовать модели (4) или (11). Для первой из них имеем:

$$g_{11} = 1 - 2\frac{\epsilon c_1 c_2}{RT} + \frac{c_1 c_0}{RT}(2\Delta H^{BC} - 2\Delta H^{AC})$$

$$g_{10} = -\frac{1}{1-c_0} - \frac{c_2}{RT}(2\Delta H^{BC} - 2\Delta H^{AC})$$

$$g_{22} = 1 - 2\frac{\epsilon c_1 c_2}{RT} - \frac{c_2 c_0}{RT}(2\Delta H^{BC} - 2\Delta H^{AC}) \qquad (29)$$

$$g_{20} = -\frac{1}{1-c_0} + \frac{c_1}{RT}(2\Delta H^{BC} - 2\Delta H^{AC})$$

$$g_{00} = \frac{1}{1-c_0}$$

$$g_{01} = \frac{1}{RT}(2\Delta H^{AC} - 2\Delta H^{BC})$$

и все термодинамические факторы зависят лишь от разности теплот образования карбидов.

Для модели (11):

$$g_{11} = 1 - 2\frac{\epsilon c_1 c_2}{RT} - \frac{c_1 c_0}{RT}\frac{\delta}{\delta c_0}(2\Delta H^{AC} - 2\Delta H^{BC})$$

$$g_{10} = -\frac{1}{1-c_0} + \frac{c^2}{RT}\frac{\delta}{\delta c_0}(2\Delta H^{AC} - 2\Delta H^{BC}) - \frac{c_2 c_0}{RT}\frac{\delta^2 2\Delta H^{BC}}{\delta c_0^2} - \frac{c_1 c_0}{RT}\frac{\delta^2 2\Delta H^{AC}}{\delta c_0^2}$$

$$g_{22} = 1 - 2\frac{\epsilon c_1 c_2}{RT} + \frac{c_2 c_0}{RT}\frac{\delta}{\delta c_0}(2\Delta H^{AC} - 2\Delta H^{BC}) \tag{30}$$

$$g_{20} = -\frac{1}{1-c_0} - \frac{c_1}{RT}\frac{\delta}{\delta c_0}(2\Delta H^{AC} - 2\Delta H^{BC}) - \frac{c_1 c_0}{RT}\frac{\delta^2 2\Delta H^{AC}}{\delta c_0^2} - \frac{c_2 c_0}{RT}\frac{\delta^2 2\Delta H^{BC}}{\delta c_0^2}$$

$$g_{00} = \frac{1}{1-c_0} + \frac{c_0}{RT}\left[c_1\frac{\delta^2 2\Delta H^{AC}}{\delta c_0^2} + c_2\frac{\delta^2 2\Delta H^{BC}}{\delta c_0^2}\right]$$

$$g_{01} = \frac{1}{RT}\frac{\delta}{\delta c_0}(2\Delta H^{AC} - 2\Delta H^{BC})$$

Приведем некоторые численные оценки. По данным [9], для ZrC $2\Delta H = -47{,}8$; для NbC $2\Delta H = -33{,}3$ ккал/г-форм.

Из (29) получаем, что в идеальном растворе ZrC - NbC, $g_{11} > 0$, $g_{10} < 0$, $g_{22} < 0$ при $c_2 \to 1$; $g_{01} < 0$. Абсолютные значения величин g_{10}, g_{20}, g_{00} весьма велики вблизи верхней границы области гомогенности. Качественно аналогичные результаты получаются и по формулам (30), если использовать данные по зависимости теплот образования карбидов ZrC и NbC от состава.

Выражение для g_{ik} типа (28-30) представляются весьма полезными применительно к трехкомпонентным фазам для оценок коэффициентов гетеродиффузии, которые определяют параметры таких диффузионно-контролируемых процессов, как ползучесть, спекание, испарение, насыщение металлических сплавов элементами внедрения и некоторые другие. Доведение оценок до численных результатов не всегда, правда, возможно из-за недостаточности сведений об истинных коэффициентах самодиффузии, но, по крайней мере, качественную информацию такие оценки дают.

ЛИТЕРАТУРА

[1] АНДРИЕВСКИЙ, Р.А., КАЛИНИН, В.П., ПЕПЕКИН, Г.И., Thermodynamics of Nuclear Materials, 1967 (Proc. Symp. Vienna, 1967) IAEA, Vienna(1968)457.

[2] АНДРИЕВСКИЙ, Р.А., ЗАГРЯЗКИН, В.Н., ЛЮТИКОВ, Р.А., Thermodynamics of Nuclear Materials, 1967 (Proc. Symp. Vienna, 1967) IAEA, Vienna(1968)449.

[3] НИКОЛЬСКИЙ, С.С., Порошковая Металлургия, 3(1965)22; 4(1965)61.

[4] KAUFMAN, L., STEPAKOFF, G., in The Performance of High Temperature Systems, Gordon and Breach, New York, 1969.

[5] HOCH, M., YUN, C., YAMAUCHI, S., J. Electrochem. Soc. 118(1971)1498; 119(1972)970.
[6] ФЕДОРОВ, Э.М., АНДРИЕВСКИЙ, Р.А., Теплофизика Высоких Температур 11(1973)419.
[7] ФЕДОРОВ, Г.Б., ГУСЕВ, В.Н., ЗАГРЯЗКИН, В.Н., СМИРНОВ, Е.А., Атомная Энергия 35(1973)267.
[8] KAUFMAN, L., CLOUGHERTY, E., in Metallurgy of High Pressures and Temperatures, AIME, New York (1964)322.
[9] СТОРМС, Э., Тугоплавкие Карбиды, перев. с англ., Атомиздат, М., 1970.
[10] АНДРИЕВСКИЙ, Р.А., ХРОМОНОЖКИН, В.В., ХРОМОВ, Ю.Ф., АЛЕКСЕЕВА, И.С., МИТРОХИН, В.А., Доклады АН СССР 206(1972)896.
[11] ГУРОВ, К.П., АНДРИЕВСКИЙ, Р.А., Физ. Металл. Металловед. 29 (1970)757.

THERMODYNAMIC INTERPRETATION OF SYSTEMS WITH INTERMETALLIC COMPOUNDS

Z. MOSER, K. FITZNER
Polish Academy of Sciences,
Cracow,
Poland

Abstract

THERMODYNAMIC INTERPRETATION OF SYSTEMS WITH INTERMETALLIC COMPOUNDS.

The interpretation of systems with negative deviations from Raoult's law presents a number of difficulties, though such systems are frequently encountered in pyrometallurgical processes connected with nuclear reactor fuels. The present paper gives experimental results for the Mg-Sn system with the mole fraction, X_{Mg}, varying between 0.03 and 0.35. An emf method based on concentration cells with a solid magnesium reference was employed. Results for the Mg-Sn system have been successfully interpreted with Krupkowski's formalism, but this method is not suitable for interpretation of such systems as Mg-Bi or Mg-Sb, which contain more stable intermetallic compounds. For these systems therefore other methods were tested. It was shown that application of the quasi-ideal solution model assuming the formation of MgBi, Mg_2Bi and MgSb, Mg_2Sb results in good agreement with experimental data for magnesium in both the Mg-Bi and Mg-Sb systems and with the respective values of bismuth and antimony calculated by Gibbs-Duhem integration.

Introduction

In the present investigation an emf technique was employed to study the Mg-Sn system. The results were interpreted and compared with previous results from the Mg-Zn, Mg-Bi and Mg-Sb systems. With the exception of the Mg-Zn system which has three, only one solid intermediate phase exists in equilibrium in each of these systems. As measured by the magnitude of the free energy of formation, the stability of the intermediate phases decreases from Mg-Sb, Mg-Bi [1], Mg-Zn [2] to Mg-Sn systems. These systems show negative deviations from Raoult's law and the larger the deviation the more stable the intermediate phases.

The aim of the paper is to report the experimental investigation of the Mg-Sn system and to analyse some methods of interpretation of experimental data relevant to intermetallic compounds.

Experimental results for the Mg-Sn system

The thermodynamic properties of liquid solutions of Mg-Sn were determined by an emf technique with concentration cells of the type:

$$Mg(s) \mid MgCl_2 \text{ in } (LiCl\text{-}KCl)_{eut}\ (l) \mid Mg\text{-}Sn\ (l) \qquad (1)$$

Measurements were made for alloys in the mole fraction range $0.03 < X_{Mg} < 0.35$ and over the temperature range 750-850 K. The experimental apparatus and procedure were similar to those in investigations of the Mg-Zn system [2]. Both Mg and Sn were of 99.99 % pure and a master alloy with $X_{Mg} = 0.35$ was

Table I

EXPERIMENTAL EMF DATA AND CALCULATED VALUES OF $\ln\gamma_{Mg}$, $\Delta\bar{H}_{Mg}$ AND $\Delta\bar{S}_{Mg}$

X_{Mg}	E_{mV}= a + bT (K)	E_{mV} 740	800	860 K	$\ln\gamma_{Mg}$ 740	800	860 K	$\Delta\bar{H}_{Mg}$ $\frac{cal}{mol}$	$\Delta\bar{S}_{Mg}$ $\frac{cal}{mol\cdot K}$
0.03	=174.2+0.1648T	296.2	306.0	316.0	-5.788	-5.378	-5.024	-8038	7.60
0.05	=158.4+0.1532T	271.8	281.0	290.2	-5.533	-5.160	-4.840	-7309	7.07
0.07	=149.5+0.1421T	254.7	263.2	271.7	-5.332	-4.980	-4.677	-6899	6.56
0.10	=151.8+0.1190T	239.8	247.0	254.1	-5.223	-4.866	-4.559	-7002	5.49
0.20	=148.8+0.0790T	207.3	212.0	216.8	-4.895	-4.545	-4.243	-6866	3.64
0.30	=154.5+0.0378T	182.5	184.7	187.0	-4.521	-4.158	-3.845	-7221	1.74
0.35	=155.0+0.0213T	170.8	172.0	173.3	-4.308	-3.943	-3.629	-7150	0.98

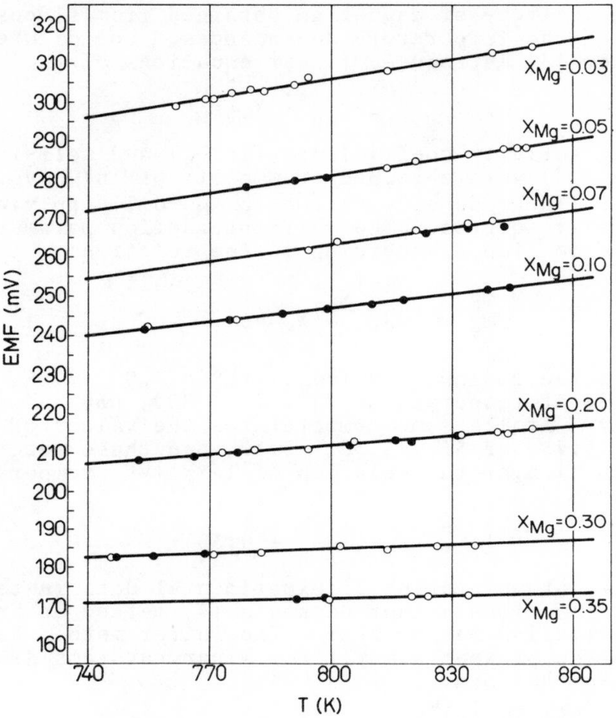

FIG. 1. Temperature dependence of the electromotive force of liquid Mg-Sn alloys relative to pure liquid magnesium.

first made and was subsequently diluted to form other compositions. Measurements were made with a solid magnesium reference electrode, but liquid magnesium was chosen as the standard state and appropriate terms were introduced into the mathematical evaluation of the thermodynamic functions to compensate for this choice. This correction may be represented by the relation,

$$\frac{\Delta G^{o}_{Mg(s\rightarrow\ell)}}{nF} = \frac{1943 - 2.105\ T(K)}{2(23\ 060)} \qquad (2)$$

where $\Delta G^{o}_{Mg(s\rightarrow\ell)}$ is the free energy of fusion of magnesium and the relationship includes the assumption that the heat and entropy of fusion are temperature independent. In Fig. 1 the experimental emf vs. temperature are plotted. Different runs are distinguished by open and filled points. In Fig. 1 the linear dependence of emf on temperature was obtained by the method of least squares. From the linear equations, emf values at arbitrarily selected temperatures of 740, 800 and 860 K were interpolated. Table I shows emf values recalculated into magnesium activity coefficients and also partial molar enthalpy and entropy of magnesium obtained from slopes and intercepts of the temperature dependences. For dilute solutions, Wagner [3] introduced linear equations of the form:

$$\ln\gamma_{Mg} = \ln\gamma^{o}_{Mg} + X_{Mg}\varepsilon^{Mg}_{Mg} \tag{3}$$

The limiting activity coefficients ($\ln\gamma^{o}_{Mg}$) and self-interaction parameters (ε^{Mg}_{Mg}) were obtained from plots of $\ln\gamma_{Mg}$ vs. X_{Mg} in the range of linear dependence (up to X_{Mg}=0.1) applying the method of least squares. The self-interaction parameters were calculated from slopes according to the relation:

$$\varepsilon^{Mg}_{Mg} = \left(\frac{\partial \ln\gamma_{Mg}}{\partial X_{Mg}}\right)_{X_{Mg}=0} \tag{4}$$

resulting in the following values: ε^{Mg}_{Mg} = 7.92, 7.55, 7.20, 6.88 and 6.58 at temperatures 740, 770, 800, 830 and 860 K, respectively. At the same temperatures the values of $\ln\gamma^{o}_{Mg}$ are -5.96, -5.75, -5.55, -5.36, -5.19, and these were used for the determination of the relation of $\ln\gamma^{o}_{Mg}$ vs. temperature in the form:

$$\ln\gamma^{o}_{Mg} = -\left(\frac{4170}{T} + 0.33\right) \tag{5}$$

For the interpretation of experimental data in terms of the partial functions either Darken's [4] method or Krupkowski's [5] formalism may be used. The latter method has been employed in the present study. For binary systems Krupkowski's formulae take the form:

$$\ln\gamma_1 = \omega(T)(1-X_1)^m \tag{6}$$

$$\ln\gamma_2 = \omega(T)\left[(1-X_1)^m - \frac{m}{m-1}(1-X_1)^{m-1} + \frac{1}{m-1}\right] \tag{7}$$

Eq. (7) is derived from Gibbs-Duhem equation assuming that Eq. (6) applied to component 1. The $\omega(T)$ function and m in Eqs. (6) and (7) are evaluated from experimental data. The $\omega(T)$ function usually takes the form $\omega(T) = (\alpha/T) \pm \beta$, which for semiregular solutions may be interpreted to connect α with the partial enthalpy and β with the excess entropy. The parameter m is called the asymmetry coefficient and the closer m to unity, the greater asymmetry in such functions as molar enthalpy and molar excess free energy. In the case of positive deviations from Raoult's law Eq. (6) generally applies to the component with the smaller atom radius [6]. For positive deviations m occurs in the range $1 < m < 2$, but it has been found that for negative deviations from Raoult's law $m > 2$ [7].

In this study Eq. (7) applies to Mg,

$$\ln\gamma^{o}_{Mg} = \omega(T)\,\frac{1}{m-1} \tag{8}$$

The $\omega(T)$ function was calculated equating Eqs. (5) and (8), and it was shown that m = 2.7 gave the best fit to the experimental data. Thus Eqs. (6) and (7) take the form:

$$\ln\gamma_{Sn} = -\left(\frac{7089}{T} + 0.56\right)(1-X_{Sn})^{2.7} \tag{9}$$

$$\ln\gamma_{Mg} = -(\frac{7089}{T})+0.56)[(1-X_{Sn})^{2.7}-1.588(1-X_{Sn})^{1.7}+0.588] \quad (10)$$

The latter equations represent the experimental data better at dilute and moderate Mg concentrations than in the concentrated solution region. From Eqs. (9) and (10) the molar enthalpy $\Delta H'$ may be represented by the relation:

$$\Delta H' = \frac{R\alpha}{m-1}[1-(1-X_{Sn})^{1.7}]X_{Mg} = -8282[1-(1-X_{Sn})^{1.7}]X_{Mg} \quad (11)$$

and, because of the form of $\omega(T)$, the function $\Delta H'$ does not depend on temperature. At X_{Mg} = 0.5 $\Delta H'$ = -2865 cal/mol.

From Eqs. (9) and (10) the Gibbs free energy of formation of solid Mg_2Sn from solid components may be calculated assuming that the solid compound in equilibrium with the liquid is of fixed composition. The calculations have been made for the magnesium-rich side of the Mg-Sn phase diagram at temperatures 500-800 K and from the plot $\Delta G°/T$ vs. $1/T$ the following equation was derived:

$$\Delta G^{o}_{Mg_2Sn} = -22700 + 8.5T \quad (12)$$

thus the standard enthalpy of formation, $\Delta H^{o}_{Mg_2Sn}$, was found to be -22700 cal/mol and the entropy of formation, $\Delta S^{o}_{Mg_2Sn}$ = -8.5 cal/mol·K. At 298 K the value, $\Delta G^{o}_{Mg_2Sn}$ = -20170 cal/mol,

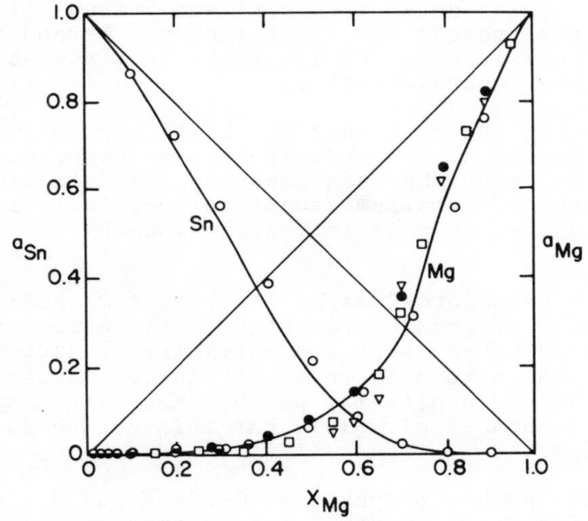

FIG. 2. Activities of magnesium and tin versus X_{Mg} at 1073°K.
◒ experimental present work
— calculated from Eqs (9) and (10)
○ Sharma, Ref. [12] for magnesium from emf data, for tin obtained by graphical integration
● Eremenko, Lukashenko, Ref. [9]
△ Egan, Ref. [10]
□ Eldridge et al. Ref. [11]

was established by linear extrapolation. The Mg-Sn system has been investigated by Ashtakala and Pidgeon [8], Eremenko and Lukashenko [9], Egan [10], Eldridge et al. [11], and Sharma [12]. Sharma made emf measurements similar to the present measurements but with a liquid magnesium reference electrode, and he reported at 298 K, $\Delta G^{o}_{Mg_2Sn} = -18000$ cal/mol, $\Delta H_{Mg_2Sn} = -20000$ cal/mol and $\Delta S^{o}_{Mg_2Sn} = -6.73$ cal/mol·K. Beardmore et al. [13] reported at 298 K $\Delta G^{o}_{Mg_2Sn} = -17700(\pm 900)$ cal/mol $\Delta H^{o}_{Mg_2Sn} = -19300(\pm 180)$ cal/mol and $\Delta S^{o}_{Mg_2Sn} = -5.4(\pm 2.1)$ cal/mol·K. Comparison of the Gibbs free energy of formation at 298 K of Mg_2Sn of the present study with Sharma [12] and Beardmore et al. [13] shows that the deviations are of order of 10%. Higher deviations, however, appear in standard enthalpy and entropy of formation. Comparison of the present experimental data with the other data is shown in Fig. 2 where the solid lines represent the activities calculated from Krupkowski's equations, Eqs. (9) and (10). The activities calculated from Eq. (10) and the present experimental values are similar to those reported in the literature; slightly higher errors appear in dilute magnesium alloys.

Reinterpretation of experimental data for the Mg-Bi and Mg-Sb systems

Krupkowski's method when applied to systems with intermetallic compounds becomes a poorer approximation as the stability of intermetallic compounds increases. The method has been applied to the Mg-Zn system [2] and accord with the direct experimental data of Chiotti and Stevens [14] was obtained. In the present investigation, the method has been used to obtain values for Mg-Sn alloys in reasonable agreement with other experimental results.

However, in the Mg-Bi and Mg-Sb [1] systems with more stable intermetallic compounds it has not been possible to obtain good agreement between experimental data and values from Krupkowski's formalism except for very dilute solutions. Hence different methods of interpretation for those systems were tested.

One such procedures was an analytical method [15] based on the fit of a polynomial function to the experimental data. It was found, however, that the polynomial functions reproduce the experimental data only in a qualitative manner. For both systems promising results may be obtained with a spline method as shown by Ansara et al. [16], but this method is a complicated procedure.

Next, the method proposed by Okajima and Sakao [17] has been tested for Mg-Bi system. This method proceeds from the experimental values of $\ln\gamma_{Mg}$ to an integration of the Gibbs-Duhem equation by introducing an empirical factor k which is dependent on bismuth concentration. The relation then takes the form

$$\ln a_{Bi} = k_{Bi}\left\{-\int_{1}^{X_{Bi}} \frac{X_{Mg}}{X_{Bi}}\, d\ln\gamma_{Mg} + \ln X_{Bi}\right\} \qquad (13)$$

The results are extremely sensitive to the choice of value for k, and the method is not believed to be generally applicable.

Comparatively good agreement with experimental data for the Mg-Bi and Mg-Sb systems has been obtained using Högfeldt's method which is based upon a quasi-ideal solution model [18]. Using this method activity coefficients of magnesium and bismuth may be expressed by

$$\gamma_{Mg} = \frac{1 + 2K_I X'_{Mg} + (3K_{II}-K_I)X'^2_{Mg} - 2K_{II}X'^3_{Mg}}{1 + K_I + 2K_{II}X'_{Mg} - K_{II}X'^2_{Mg}} = \frac{X'_{Mg}}{X_{Mg}} \qquad (14)$$

and

$$\gamma_{Bi} = \frac{1 + 2K_I X'_{Mg} + (3K_{II}-K_I)X'^2_{Mg} - 2K_{II}X'^3_{Mg}}{(1+K_I X'_{Mg} + K_{II}X'^2_{Mg})} = \frac{X'_{Bi}}{X_{Bi}} \qquad (15)$$

This method assumes hypothetical reactions which for the Mg-Bi system are of the type, Mg+Bi = MgBi and 2Mg+Bi = Mg_2Bi. Then two equilibrium constants are considered:

$$K_I = \frac{X'_{MgBi}}{X'_{Mg}X'_{Bi}} \qquad (16)$$

$$K_{II} = \frac{X'_{Mg_2Bi}}{X'^2_{Mg}X'_{Bi}} \qquad (17)$$

Table II

THE VALUES OF $\ln\gamma_{Mg}$ TAKEN FROM REF. [1] AND [19] USED FOR CALCULATIONS OF a_{Mg} and $\ln\gamma_{Bi}$, a_{Bi} AT 973°K BY GRAPHICAL INTEGRATION

X_{Mg}	$\ln\gamma_{Mg}$	$\ln\gamma_{Bi}$	a_{Mg}	a_{Bi}
0.1	-5.721	-0.025	3.0×10^{-4}	8.8×10^{-1}
0.2	-5.629	-0.055	7.0×10^{-4}	7.5×10^{-1}
0.3	-5.537	-0.110	1.2×10^{-3}	6.2×10^{-1}
0.4	-5.200	-0.345	3.1×10^{-3}	4.2×10^{-1}
0.5	-4.500	-0.955	5.5×10^{-3}	1.9×10^{-1}
0.6	-2.320	-3.875	5.9×10^{-2}	8.0×10^{-3}
0.7	-0.540	-6.915	4.1×10^{-1}	2.9×10^{-4}
0.8	-0.260	-7.758	6.2×10^{-1}	8.0×10^{-5}
0.9	-0.070	-8.843	8.4×10^{-1}	1.0×10^{-5}

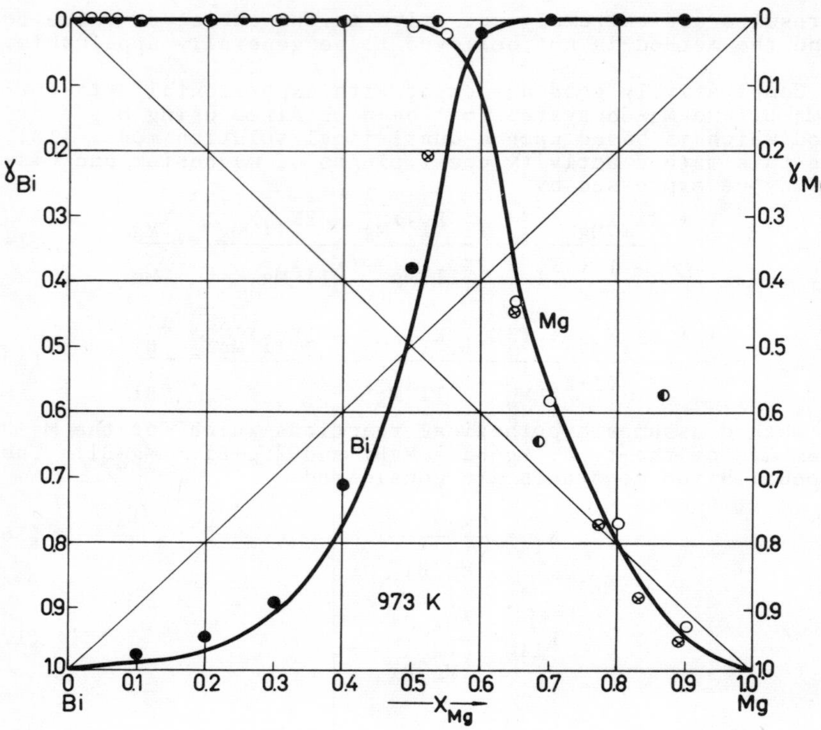

FIG. 3. Activity coefficients γ_{Mg} and γ_{Bi} versus X_{Mg} at 973 K.
○ Egan, Ref. [19]
⊗ Vetter, Kubaschewski, Ref. [20]
● graphical integration, data from Table II
— calculated from Eqs (14) and (15)
◐, ◒ Moser, Krohn, Ref. [1], data from formation and concentration cells, respectively

By definition

$$X'_{Mg} + X'_{Bi} + X'_{MgBi} + X'_{Mg_2Bi} = 1 \qquad (18)$$

$$X_{Mg} + X_{Bi} = 1 \qquad (19)$$

where X_{Mg} and X_{Bi} are the molar fractions when the Mg-Bi system is considered to be formed of Mg and Bi only, and X'_{Mg}, X'_{Bi} are the molar fractions of magnesium and bismuth when the hypothetical compounds are assumed. The values of both equilibrium constants are related to the limiting values of the activity coefficients, namely when $X'_{Mg} \to 0$, then $X'_{Bi} \to 1$ and $\gamma_{Mg} \to \overset{\circ}{\gamma}_{Mg}$ and in this case $\overset{\circ}{\gamma}_{Mg} = 1/(1+K_I)$. Similarly when $X'_{Mg} \to 1$, $X'_{Bi} \to 0$ and $\overset{\circ}{\gamma}_{Bi} = 1/(1+K_I+K_{II})$.

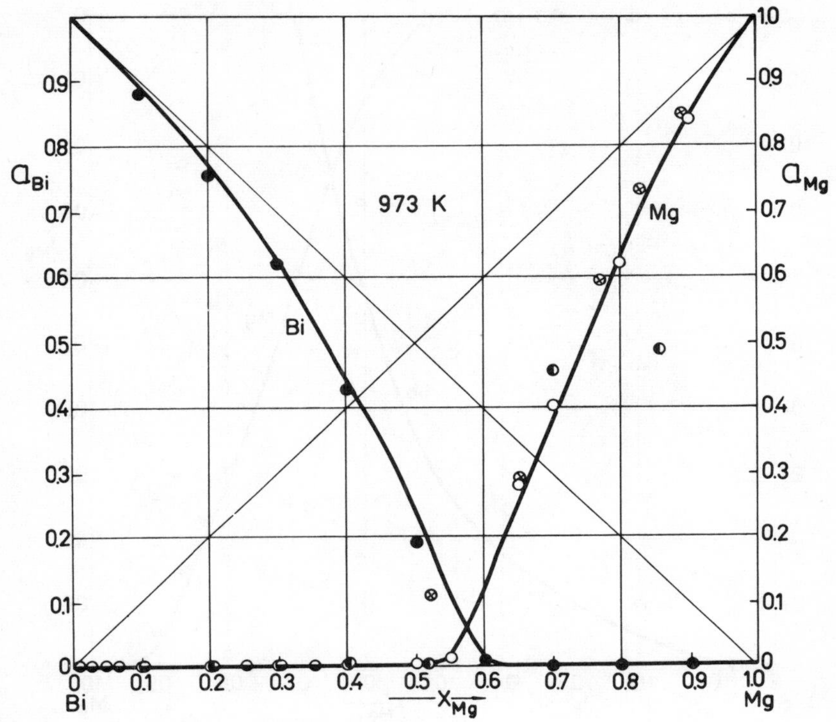

FIG. 4. Activities of magnesium and bismuth versus X_{Mg} at 973 K.
○ Egan, Ref. [19]
⊗ Vetter, Kubaschewski, Ref. [20]
● graphical integration, data from Table II
— calculated from Eqs (14) and (15)
◐, ◒ Moser, Krohn, Ref. [1], data from formation and concentration cells, respectively

It is interesting to note that in Högfeldt's model $X'_{Mg} = a_{Mg}$. Activity of magnesium is determined experimentally and Eqs. (14) and (15) allow evaluation of the thermodynamic functions of alloy solutions on the basis of activities of a single component. These activities are normally obtained by experimental measurements. In the specific case of the Mg-Bi system $K_I = 332$ and $K_{II} = 27444$ at 973 K were obtained for the published value [1,19] of $\ln\gamma_{Mg} = -5.81$ and $\ln\gamma_{Bi} = -10.23$ obtained from graphical integration (see Table II). Next from Eqs. (14) and (15), γ_{Mg}, γ_{Bi} and a_{Mg}, a_{Bi} were calculated as shown in Figures 3 and 4.

Making use of the data of Rao, Patil [21] and Vetter, Kubaschewski [20] similar calculations were carried out for the Mg-Sb system, and the values of $K_I = 2173$ and $K_{II} = 906918$ were obtained at 1073 K. Other results for the Mg-Sb system

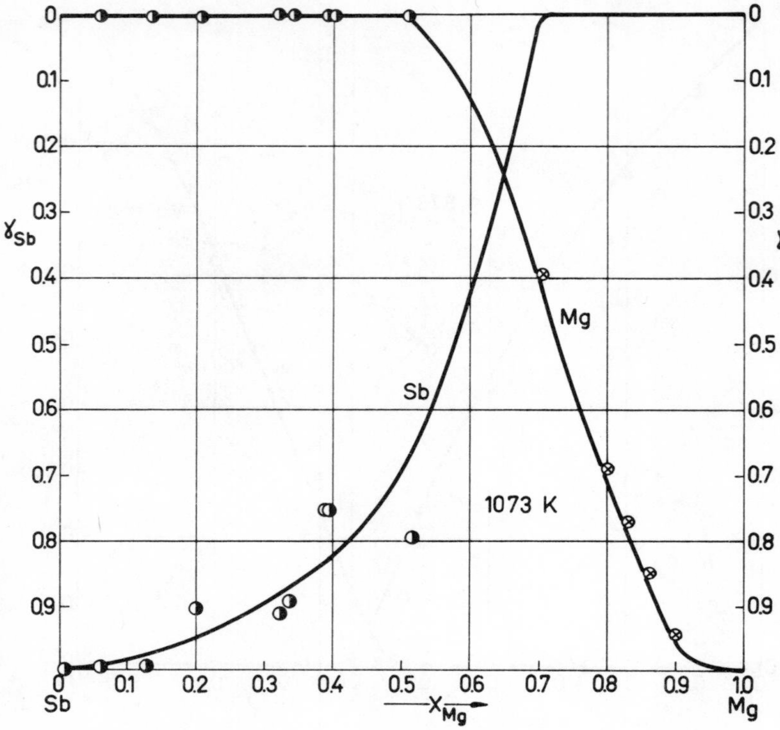

FIG. 5. Activity coefficients γ_{Mg} and γ_{Sb} versus X_{Mg} at 1073 K.
◐ Rao, Patil, Ref. [21] for magnesium from emf data, for antimony obtained by graphical integration
⊗ Vetter, Kubaschewski, Ref. [20]
— Högfeldt's model

are plotted in Figures 5 and 6. Figures 3-6 show that the experimental data for magnesium for both the systems Mg-Bi and Mg-Sb and the graphically integrated data for bismuth and antimony are in good agreement with the Högfeldt model.

Conclusions

With experimental data for the Mg-Sn it has been shown that Krupkowski's method is a useful approximation, and data from dilute solutions may be applied for calculation of thermodynamic functions for the entire Mg-Sn system. As an example the free energy of formation of Mg_2Sn was computed to be $\Delta G^{o}_{298,Mg_2Sn}$ = -20170 cal/mol which is in good agreement with literature data. However, Krupkowski's method shows considerable deviation between experimental and calculated data

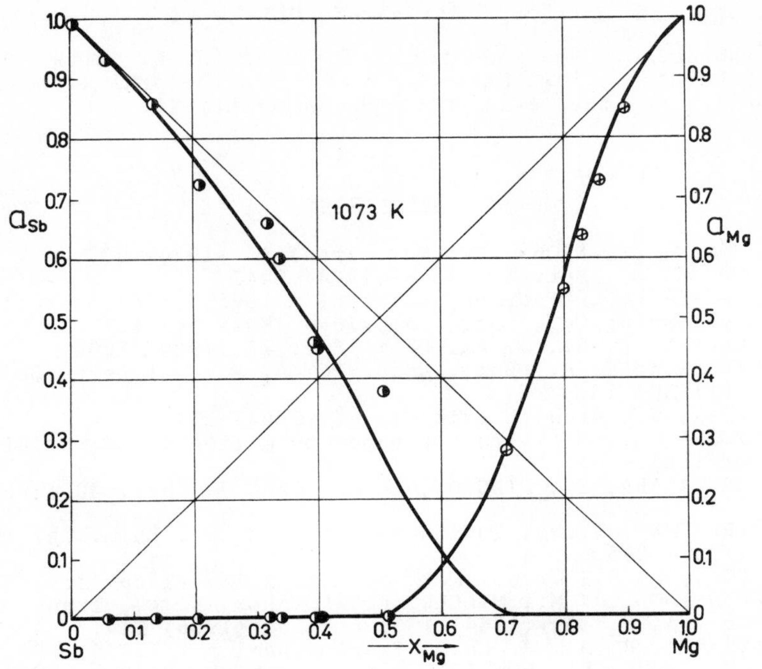

FIG. 6. Activities of magnesium and antimony versus X_{Mg} at 1073 K.
◐ Rao, Patil, Ref.[21] for magnesium from emf data, for antimony obtained by graphical integration
⊗ Vetter, Kubaschewski, Ref.[20]
— Högfeldt's model

in those systems where more stable intermetallic compounds appear, e.g. with extensive negative deviations from Raoult's law. Examples are the Mg-Bi and Mg-Sb systems. For the latter systems other methods of interpretation were tested. First, the polynomial methods were found not to be easily applicable and therefore of limited utility. Second, the method of Okajima and Sakao was shown not to be applicable because of the sensitivity of numerical results to the choice of value of the experimental factor k, which is both temperature and composition dependent.

Promising results were obtained by applying the quasi-ideal solution model of Högfeldt where extrapolation of data from the dilute solution region was shown to be capable of closely approximating direct experimental values in the more concentrated solution region.

Acknowledgements

The authors are indebted to Professor J. F. Smith from Ames Laboratory, Iowa State University for helpful advice, discussion and the revision of the English text.

References

[1] MOSER, Z., KROHN, C., Met. Trans. 5 (1974) 979.
[2] MOSER, Z., Met. Trans. 5 (1974) 1445.
[3] WAGNER, C., "Thermodynamics of Alloys," Addison-Wesley Publishing Co., Inc., Cambridge, Mass. (1952).
[4] DARKEN, L. S., J. Am. Chem. Soc. 72 (1950) 2909.
[5] KRUPKOWSKI, A., Bull. Acad. Polon. Sci. et Lett. Ser. A, 1 (1950) 15.
[6] PTAK, W., Arch. Hutnictwa. 13 (1968) 251.
[7] MOSER, A., Met. Trans. (paper submitted to Editorial Office).
[8] ASHTAKALA, S., PIDGEON, L. M., Can. J. Chem. 40 (1962) 718.
[9] EREMENKO, V. N., LUKASHENKO, G. M., Ukr. Khim. Zh. 29 (1963) 896.
[10] EGAN, J. J., Trans. Met. Soc. AIME 236 (1966) 118.
[11] ELDRIDGE, J. M., MILLER, E., KOMAREK, K. L., Trans. Met. Soc. AIME, 230 (1964) 1361 and 236 (1966) 114.
[12] SHARMA, R. A., J. Chem. Thermodynamics 2 (1970) 373.
[13] BEARDMORE, P., HOWLETT, B. W., LICHTER, B. D., BEVER, B. M., Trans. Met. Soc. AIME 236 (1966) 102.
[14] CHIOTTI, P, STEVENS, E. R., Trans. Met. Soc. AIME 233 (1965) 198.
[15] ANSARA, I., DURAUD, F., DESRE, P., BONNIER, E., Rev. Int. Hautes, Temper. et Refract. 2 (1965) 287.
[16] ANSARA, I., BONNIER, E., Thermodynamics of Nuclear Materials (1967) IAEA Vienna 715.
[17] OKAJIMA, K., SAKAO, H., Trans. JIM 11 (1970) 217.
[18] HÖGFELDT, E., Arkiv Kemi. 7 (1954) 315.
[19] EGAN, J. J., Acta Met. 7 (1959) 560.
[20] VETTER, F. A., KUBASCHEWSKI, O., A. Elektrochem. 57 (1953) 243.
[21] RAO, Y. K., PATIL, B. V., Met. Trans. 2 (1971) 1829.

DISCUSSION

D.D. SOOD: We have recently completed a thermodynamic study of the Mg-Bi system using a transpiration technique, and it will be interesting to see how our results agree with your data, and the model you propose for the thermodynamic analysis. Attempts to apply Krupkowski's model to this system were not successful, as you also found. As for the Mg-Sb system, the data of Rao and Patil may be in error because in their data the logarithm of the activity coefficient does not appear to be a smooth function of composition and it may be better to rely on your measurements for model calculations.

Z. MOSER: We have recently published experimental results for the Mg-Bi system as a comparison of two experimental techniques, namely the use of formation and concentration cells [Metall.Trans. 5 (1974) 979]. Krupkowski's formulae are not applicable to systems with extensive negative deviations from ideality, and therefore when they are applied to the Mg-Bi system (our own results and previous results reported by Egan [Acta Metall. 7 (1959) 560]) and to the Mg-Sb systems (the results of Rao and Patil), no more than promising agreement with experimental data has been obtained for both systems with dilute magnesium concentrations. At the time of writing our paper, only the results of Rao and Patil [Metall.Trans. 2 (1971) 1829] were available. I shall therefore be very interested in examining your values and may introduce some corrections into our Högfeldt model for the Mg-Sb system. However, by applying Krupkowski's method to both systems, Mg-Bi and Mg-Sb, we were able to calculate the standard free energy of formation of Mg_3Bi_2 and Mg_3Sb_2, which turned out to be in reasonable agreement with data in the literature.

A.S. PANOV: Were the samples used in your studies single phase at all temperatures?

Z. MOSER: Yes, our measurements using cells of the type Mg(s)| $MgCl_2$ in LiCl-KCl (l)| Mg-Sb(l) were always performed for a single phase liquid region against a solid reference magnesium electrode. However, the liquid magnesium was chosen as a reference state and therefore our emf data were then corrected to compensate for this choice.

ТЕРМОДИНАМИЧЕСКИЕ СВОЙСТВА ТРЕХКОМПОНЕНТНЫХ γ-ТВЕРДЫХ РАСТВОРОВ УРАНА С ЦИРКОНИЕМ, НИОБИЕМ И МОЛИБДЕНОМ

Г.Б.ФЕДОРОВ, Е.А.СМИРНОВ, В.Н.ГУСЕВ
Московский инженерно-физический институт,
Москва,
Союз Советских Социалистических Республик

Доклад представлен В.В. Ахачинским

Abstract–Аннотация

THERMODYNAMIC PROPERTIES OF THREE-COMPONENT γ-PHASE SOLID SOLUTIONS OF URANIUM WITH ZIRCONIUM, NIOBIUM AND MOLYBDENUM

Using the emf technique, the authors measured the thermodynamic activities of uranium in ternary gamma solid solutions of uranium and zirconium, niobium and molybdenum. The emf from an electrochemical cell of the type

$$U_S \mid U^{3+} + (KCl\text{-}NaCl) \mid U_S\,(\text{alloy})$$

was measured over the temperature range 750-900°C. The alloys were prepared from electrolytic uranium, zirconium iodide and from electron-beam niobium and molybdenum by fusion in an arc furnace in a purified argon atmosphere. The authors studied uranium-rich alloys in the radial cross-sections of a U-Zr-Nb system with Zr/Nb atomic ratios of 3, 1 and 1/3, and a U-Nb-Mo system with an Nb/Mo atomic ratio of 1. To plot lines of constant thermodynamic activity and excess free energy, they used the results of studies on the thermodynamic properties of the binary systems U-Zr and U-Nb, published earlier, and also of the U-Mo system. The thermodynamic activities of the components and the excess thermodynamic free energies of the ternary systems were determined by integrating the Gibbs-Duhem equations, based on Darken's method. In the same way as for ternary systems, negative deviations from the ideal state are observed. In the U-Zr-Nb system the predominant effect on this deviation is exerted by niobium, and in the case of the U-Nb-Mo system by molybdenum.

ТЕРМОДИНАМИЧЕСКИЕ СВОЙСТВА ТРЕХКОМПОНЕНТНЫХ γ-ТВЕРДЫХ РАСТВОРОВ УРАНА С ЦИРКОНИЕМ, НИОБИЕМ И МОЛИБДЕНОМ.

С помощью метода электродвижущих сил измерены термодинамические активности урана в трехкомпонентных γ-твердых растворах урана с цирконием, ниобием и молибденом. Измерения э.д.с. электрохимической ячейки типа $U_{тв} \mid U^{+3} + (KCl - NaCl) \mid U_{тв}$ (сплав) проводились в интервале температур 750 - 900°С. Сплавы приготовлялись из электролитического урана, иодидного циркония и электроннолучевых ниобия и молибдена плавкой в дуговой печи в атмосфере очищенного аргона. Исследовались богатые ураном сплавы, находящиеся на лучевых разрезах системы уран-цирконий-ниобий с соотношением ат% Zr: ат % Nb = 3:1; 1:1; 1:3 и системы уран-ниобий-молибден с соотношением ат% Nb : ат% Mo = 1:1. Для построения изолиний термодинамических активностей и избыточной свободной энергии использовались результаты исследования термодинамических свойств бинарных систем уран-цирконий и уран-ниобий, опубликованные ранее, и системы уран-молибден. Термодинамические активности компонентов и избыточные термодинамические свободные энергии трехкомпонентных систем определялись с помощью интегрирования уравнений Гиббса-Дюгема по методу Даркена. Так же как и для трехкомпонентных систем наблюдаются отрицательные отклонения от идеальности. В системе уран-цирконий-ниобий преимущественное влияние на отклонение от идеальности оказывает ниобий, а в системе уран-ниобий-молибден преобладает влияние молибдена.

В настоящем докладе представлены результаты работ, развивающих исследования, уже опубликованные авторами ранее и посвященные изучению термодинамических свойств бинарных сплавов урана с цирконием [1] и ниобием [2]. В докладе приведены результаты экспериментального исследования термодинамических свойств γ-твердых раство-

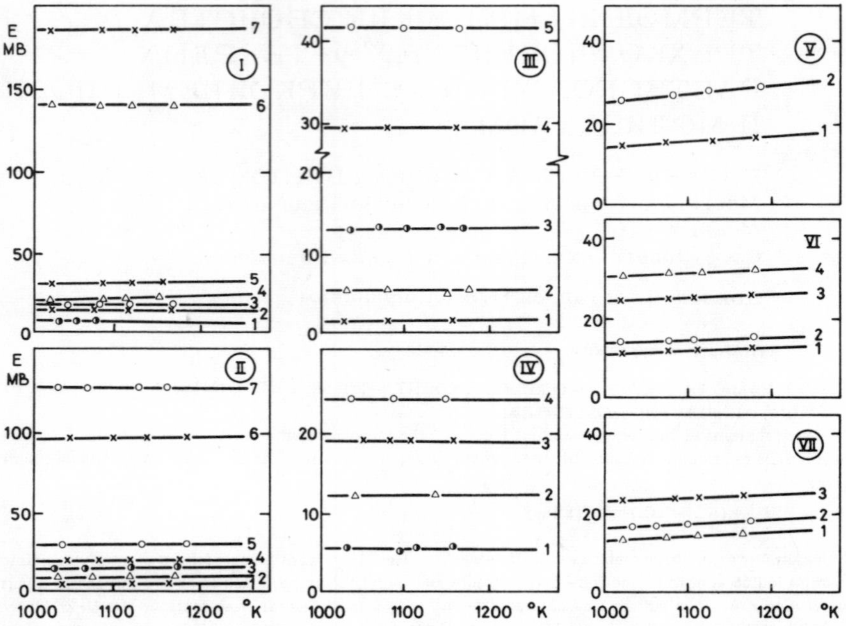

Рис.1. Температурные зависимости э.д.с. сплавов урана (ат%):
I – система уран-ниобий [2]: 1 – 95; 2 – 88; 3 – 80; 4 – 68; 5 – 53; 6 –28; 7 –13.
II – система уран-цирконий [1]: 1— 94; 2 – 89; 3 – 75; 4 – 60; 5 – 41; 6 – 28; 7 – 14:
III – система уран-молибден: 1 – 97,5; 2 – 95; 3 – 90; 4 – 80; 5 – 70.
IV – система уран-ниобий-молибден (ат% Nb : ат% Mo = 1:1): 1 – 95; 2 – 90; 3 – 85; 4 – 80.
V – система уран-цирконий-ниобий (ат% Zr : ат% Nb = 1:3): 1 – 88,7; 2 – 80;
VI – система уран-цирконий-ниобий (ат% Zr : ат% Nb = 1:1): 1 – 92,6; 2 – 89; 3 – 81; 4 – 70,6.
VII – система уран-цирконий-ниобий (ат% Zr : ат% Nb = 3:1): 1 – 89,1; 2 – 84,7; 3 – 80.

ров урана с молибденом, трехкомпонентных систем с цирконием и ниобием, а также с ниобием и молибденом.

Общеизвестно, что термодинамические исследования бинарных и многокомпонентных металлических систем преследуют различные цели. Весьма перспективным в последнее время является использование термодинамических функций и методов для анализа и расчета диаграмм состояния. С этой точки зрения несомненный интерес представляют оценки фазовых равновесий в γ-твердых растворах систем уран-молибден и уран-ниобий, проведенные в работах [3,4]. Кроме того, термодинамические данные необходимы для построения статистических моделей твердых растворов и развития представлений о строении металлических фаз. Наконец, термодинамические свойства металлических систем помогают уточнить представления о физической природе межатомной связи в твердых растворах. С этой целью термодинамические свойства сплавов урана использовались для расчета коэффициентов взаимной диффузии в бинарных и трехкомпонентных системах [1,2,5], температурные и концентрационные зависимости которых позволяют судить об изменении сил межатомного взаимодействия.

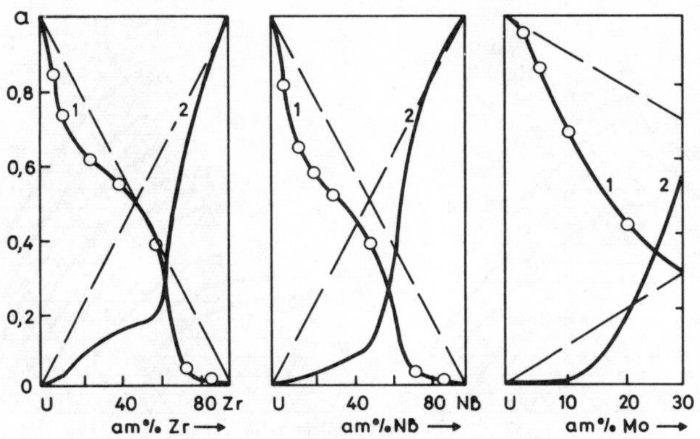

Рис.2. Концентрационные зависимости активностей компонентов в бинарных сплавах урана с цирконием [1], ниобием [2] и молибденом при 1000°С: 1 - активность урана; 2 - активность второго компонента.

Показано, что характеристики взаимной диффузии могут служить диффузионно-термодинамическим критерием жаропрочности [6].

В настоящем докладе представлены результаты исследования методом электродвижущих сил (э.д.с.) термодинамических свойств богатых ураном сплавов с цирконием, ниобием и молибденом. Измерения э.д.с. электрохимической ячейки

$$U_{тв} \left| U^{+3} + (KCl - NaCl) \right| \text{ сплав урана (тв)}$$

проводились в области температур 750 - 900°С. Методика эксперимента описана в работе [1]. Сплавы приготовлялись из иодидного циркония, а также молибдена и ниобия, полученных методом электроннолучевой плавки, и электролитического урана. Плавка образцов проводилась в дуговой печи с нерасходуемым электродом в атмосфере очищенного аргона или гелия. Для лучшей однородности сплава переплавленные 6-7 раз образцы гомогенизировались в вакууме при 900 ÷ 1000°С в течение 100 часов, после чего осуществлялась закалка в воду.

Э.д.с. измеряли компенсационным методом. Температура ячеек поддерживалась с точностью ± 2°С. При каждой температуре э.д.с. измеряли не менее 10 раз с точностью ± 0,05 мВ. Значения э.д.с. при нагреве и охлаждении совпадали в пределах 1 мВ. Для сплавов, богатых ураном, постоянство э.д.с. достигалось быстрее и данные воспроизводились с бóльшей точностью. Усредненные результаты измерения э.д.с. сплавов систем уран-цирконий, уран-ниобий, уран-молибден, уран-цирконий-ниобий (ат % Zr : ат %Nb = 1:3; 1:1; 3:1), уран-ниобий-молибден (ат % Nb : ат % Mo = 1:1) приведены на рис.1. Так как потенциалообразующим процессом в ячейке является перенос урана, термодинамические активности урана могут быть определены по формуле:

$$\lg a_U = -\frac{z_U \cdot E \cdot \overline{F}}{2{,}303 \cdot RT} = -15120\,\frac{E}{T} \qquad (1)$$

где z_U — валентность урана, F — число Фарадея, E — э.д.с. ячейки. Так же как и ранее [1,2], валентность урана принималась равной 3.

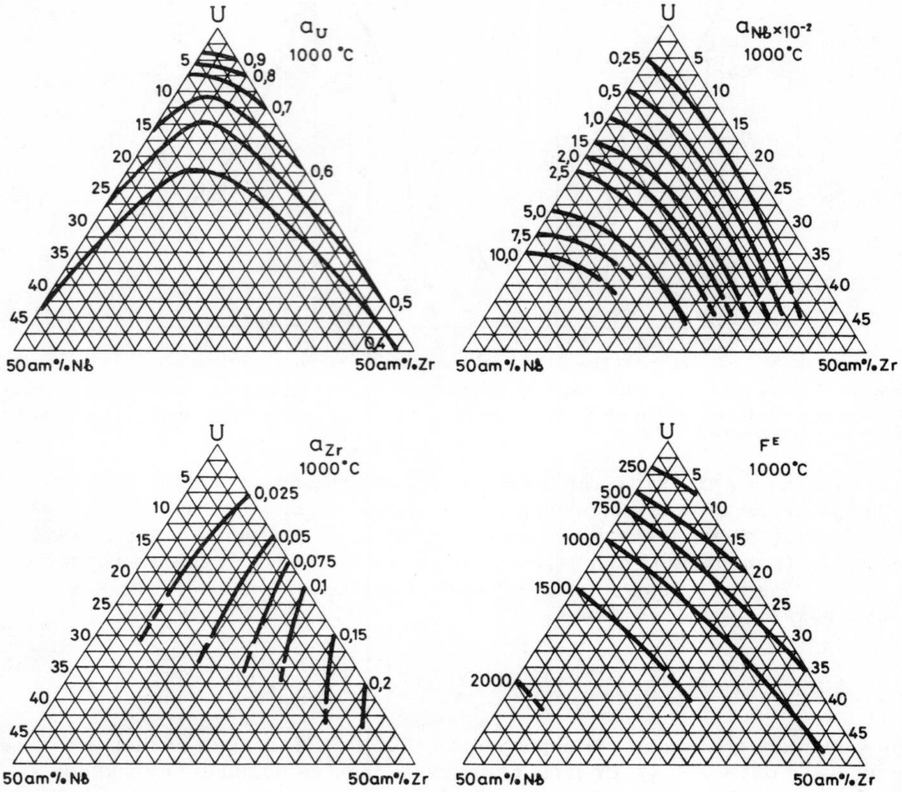

Рис.3. Изолинии термодинамических активностей урана (1), ниобия (2), циркония (3) и избыточной свободной энергии (4) (кал/г-ат) системы уран-ниобий-цирконий при 1000°С.

Активность второго компонента в бинарной системе уран-молибден определялась путем графического интегрирования уравнения Гиббса-Дюгема:

$$\lg a_{Mo} = -\int_0^{x_U} \frac{\lg a_U}{(1-x_U)^2}\, dx_U - \frac{x_U}{1-x_U} \lg a_U \qquad (2)$$

Сравнение ранее полученных результатов по бинарным системам урана с данными настоящей работы показывает, что в исследованном интервале температур и концентраций наблюдаются отрицательные отклонения активностей обоих компонентов от закона Рауля (рис.2). Это свидетельствует об усилении сил связи между атомами урана и легирующих элементов в твердом растворе по сравнению с силами связи между одноименными атомами.

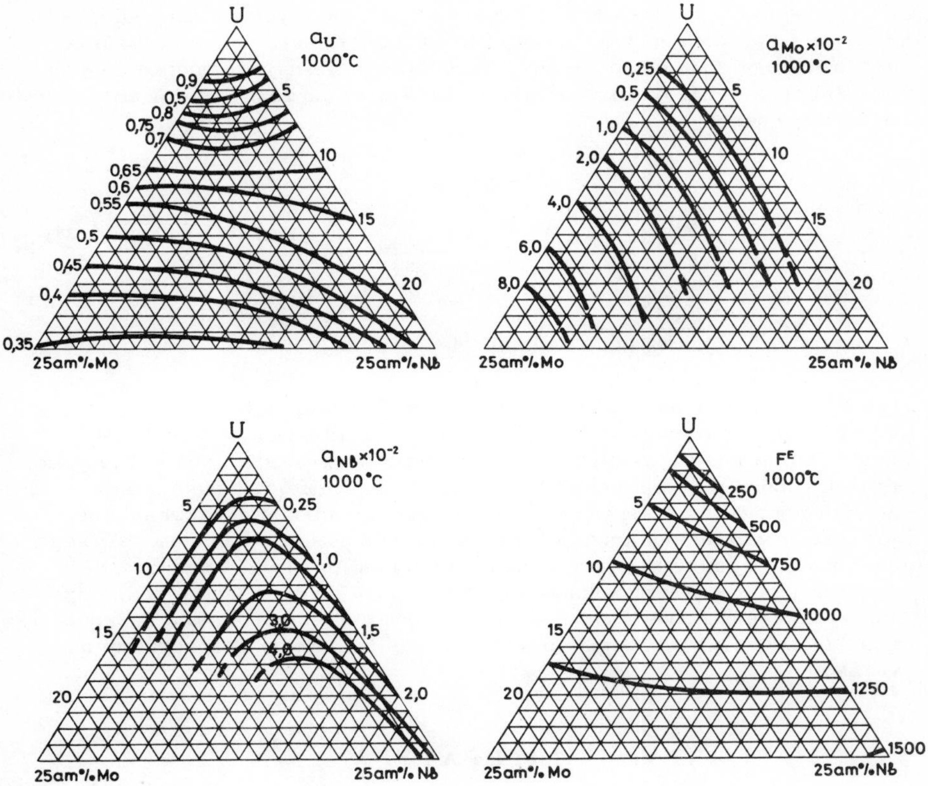

Рис.4. Изолинии термодинамических активностей урана (1), молибдена (2), ниобия (3) и избыточной свободной энергии (4) (кал/г-ат) системы уран-ниобий-молибден при 1000°С.

Использование термодинамических характеристик бинарных систем урана с цирконием, ниобием и молибденом вместе с приведенными в настоящем докладе результатами исследования квазибинарных разрезов систем уран-цирконий-ниобий и уран-ниобий-молибден позволяет построить изолинии термодинамических активностей всех компонентов на концентрационных треугольниках.

Так как метод э.д.с. позволяет определить только парциальные термодинамические характеристики урана, то для нахождения активностей ниобия, а также циркония или молибдена необходимо использовать соответствующие термодинамические соотношения:

$$RT \ln f_i = F^E + (1-x_i)\left(\delta F^E/\delta x_i\right) x_j/x_k \quad (j,k \neq i) \qquad (3)$$

где F^E — избыточная интегральная молярная свободная энергия раствора; f_i — коэффициент термодинамической активности i-го компонента; x_i — мольная доля i-го компонента.

Даркен [7] показал возможность применения уравнений Гиббса-Дюгема к тройным и многокомпонентным системам. Использование этого метода позволило рассчитать избыточные интегральные свободные энергии исследуемых трехкомпонентных систем с помощью следующего выражения:

$$F^E = (1-x_U)\left[\int_1^{x_U} \frac{F_U^E}{(1-x_U)^2}\,dx_U\right]_{\frac{x_{Zr\,(Mo)}}{x_{Nb}}} - x_{Zr\,(Mo)}\left[\int_1^{0} \frac{F_U^E}{(1-x_U)^2}\,dx_U\right]_{x_{Nb}=0}$$

$$- x_{Nb}\left[\int_1^{0} \frac{F_U^E}{(1-x_U)^2}\,dx_U\right]_{x_{Zr\,(Mo)}=0} \qquad (4)$$

где F_U^E — избыточная парциальная свободная энергия урана.

Результаты графических расчетов представлены на рис.3 и 4. Также как и для ограничивающих бинарных систем для всех изученных тройных систем урана наблюдаются отрицательные отклонения активностей компонентов от закона Рауля. Наиболее полное представление о характере изменения прочности межатомной связи в твердом растворе дает рассмотрение концентрационной зависимости F^E. Как следует из рисунков 3 и 4 легирование ниобием более эффективно с точки зрения увеличения прочности межатомной связи по сравнению с цирконием, а легирование молибденом приводит к еще большему возрастанию сил связи.

ЛИТЕРАТУРА

[1] ФЕДОРОВ, Г.Б., СМИРНОВ, Е.А., Атомная Энергия 21 (1966) 189.

[2] ФЕДОРОВ, Г.Б., СМИРНОВ, Е.А., ГУСЕВ, В.Н., Атомная Энергия 32 (1972) 11.

[3] ИВАНОВ, О.С., УДОВСКИЙ, А.Л., ВАМБЕРСКИЙ, Ю.В., В сб. "Химия металлических сплавов", изд-во "Наука", М., (1973) 21.

[4] ИВАНОВ, О.С., УДОВСКИЙ, А.Л., ВАМБЕРСКИЙ, Ю.В., ИСАЙЧЕВ, М.Н., В сб. "Общие закономерности в строении диаграмм состояния металлических систем", изд-во "Наука", М., (1973) 79; см. там же: ВАМБЕРСКИЙ, Ю.В., УДОВСКИЙ, А.Л., ИВАНОВ, О.С., стр.75.

[5] ФЕДОРОВ, Г.Б., СМИРНОВ, Е.А., ГУСЕВ, В.Н., ЖОМОВ, Ф.И., В сб. "Физико-химический анализ сплавов урана, тория и циркония", изд-во "Наука", М., (1974) 104.

[6] ФЕДОРОВ, Г.Б., СМИРНОВ, Е.А., В сб. "Структура и свойства жаропрочных металлических материалов", изд-во "Наука", М., (1973) 171.

[7] DARKEN, L.S., J.Am.Chem.Soc., 72 (1950) 2909.

ETUDE PAR SPECTROMETRIE DE MASSE DES PROPRIETES THERMODYNAMIQUES DE QUELQUES ALLIAGES DE SODIUM

J. TROUVE
CEA, Centre d'études nucléaires de Cadarache,
Saint-Paul-lez-Durance,
France

Abstract–Résumé

MASS-SPECTROMETRIC ANALYSIS OF THE THERMODYNAMIC PROPERTIES OF SOME SODIUM ALLOYS.

The paper describes the use of a quadrupole-filter residual-gas analyser and of an ultra-vacuum pumping unit for obtaining thermodynamic data on liquid-vapour systems such as Na-I and Na-Cs. It describes the apparatus and method used (Knudsen cell) as well as the first experiments performed with NaI, sodium and a Na-5 at.% Cs alloy. These experiments were used in each case to determine: the nature and composition of the vapour phase (monomeric and dimeric types); the sublimation and vaporization heat. The monomeric (NaI) and dimeric (Na_2I_2) types of sodium iodide were investigated and their sublimation heats measured and found to be 43 kcal/mol and 46 kcal/mol, respectively. Assuming that the capture cross-section of the dimer is 1.5, the proportion of the latter is 4%. A vaporization heat of 23.5 kcal/mol and a dimer proportion of 1.5% were recorded for sodium at 330°C. The literature on the subject gives 24.5 kcal/mol and 2%, respectively. A vaporization heat of 17 kcal/mol for caesium and 23 kcal/mol for sodium were found in the case of a Na-5 at.% Cs alloy. In the next stage of the analysis "twin" cells will be used in order to determine directly the activity of each constituent in a metal alloy by making a comparison with the pressure above the pure metal. In practice a cell used with lithium and sodium has shown an interaction between the two holes of approximately 20%, which will make it necessary to align and focus the molecule beam more precisely.

ETUDE PAR SPECTROMETRIE DE MASSE DES PROPRIETES THERMODYNAMIQUES DE QUELQUES ALLIAGES DE SODIUM.

Le présent mémoire rapporte l'utilisation d'un analyseur de gaz résiduels à filtre quadripolaire et d'un groupe de pompage ultra-vide pour la détermination des données thermodynamiques de systèmes liquide-vapeur tels que Na-I, Na-Cs. Il décrit l'appareillage et la méthode utilisée (cellule de Knudsen) ainsi que les premières expériences effectuées avec NaI, Na et un alliage Na-5 at.% Cs. Ces expériences ont permis de déterminer dans chacun des cas: la nature et la composition de la phase vapeur (espèces monomère et dimère); les chaleurs de sublimation ou de vaporisation. Pour l'iodure de sodium (Na-I), les espèces monomère (NaI) et dimère (Na_2I_2) furent détectées et leurs chaleurs de sublimation mesurées, égales à 43 kcal/mol et 46 kcal/mol. En admettant que la section de capture du dimère est de 1,5, la proportion de ce dernier est de 4%. Pour le sodium (Na) à 330°C, une chaleur de vaporisation de 23,5 kcal/mol et une proportion de dimère de 1,5% ont été mesurées. La littérature indique respectivement 24,5 kcal/mol et 2%. Pour un alliage Na-5 at.% Cs, il a été trouvé une chaleur de vaporisation de 17 kcal/mol pour le césium et de 23 kcal/mol pour le sodium. L'étape suivante de cette étude utilisera des cellules «jumelles» afin de déterminer directement les activités de chaque composant dans un alliage métallique, par comparaison avec la pression au-dessus du métal pur. Pratiquement, une cellule utilisée avec du lithium (Li) et du sodium (Na) a montré une interaction entre les deux trous d'environ 20%, ce qui obligera à réaliser un alignement précis et une focalisation plus grande du faisceau moléculaire.

INTRODUCTION

Dans un réacteur rapide, les produits de fission volatils s'évaporent dans le gaz de couverture et il est particulièrement intéressant de déterminer les données thermodynamiques de systèmes liquide-vapeur tels que Na-Cs, Na-I, Na-I-Cs, etc.

Le spectromètre de masse est un outil de choix dans l'étude des phénomènes de vaporisation [1]; il est utilisé pour l'identification des espèces gazeuses, la détermination des pressions partielles [2] et des constantes d'équilibre [3]. La mesure en continu de la pression partielle permet de suivre la cinétique d'une réaction ou l'évolution isotherme de la composition d'un alliage; ceci est obtenu par intégration du flux de chaque espèce gazeuse quittant la cellule de mesure.

Une méthode expérimentale pour la mesure de la pression partielle au-dessus des alliages de sodium, à l'aide d'un spectromètre quadripolaire, a été étudiée.

Il a été nécessaire:

- d'obtenir une teneur en eau d'environ 1 volume par million dans des boîtes à gants, indispensables aux manipulations des alliages de sodium;
- de mettre au point un ensemble «cellule de Knudsen-four» permettant le transport des alliages de sodium sans contamination et leur étude jusqu'à 400°C;
- d'étalonner la cellule.

Le présent mémoire fait le point sur l'utilisation de l'équipement et sur la mise au point de la méthode expérimentale. Il représente un stade intermédiaire qui montre que la méthode est adaptée à la détermination des propriétés thermodynamiques; les premières expériences le prouvent.

1. APPAREILLAGE EXPERIMENTAL

L'utilisation d'une cellule de Knudsen et d'un spectromètre de masse est devenu si courante dans les recherches sur les équilibres de vapeur, qu'il sera sans doute suffisant de mentionner que le montage expérimental est représenté par la figure 1. La cellule est placée à l'intérieur d'une boîte, dont l'étanchéité, durant le transport entre la boîte à gants et le groupe de pompage, est assurée par des joints toriques. Un passage étanche de translation permet le mouvement de la cellule, sous la chambre d'ionisation.

1.1. Cellule de Knudsen

Une cellule de Knudsen en acier inoxydable est employée pour le sodium (Na), l'iodure de sodium (NaI) et les alliages Na-Cs. Le repérage de la température est assuré par deux thermocouples en chromel-alumel, associés à un potentiomètre (Meci) capable de mesurer 0,2°C. Les thermocouples ont été calibrés par la détermination du point de fusion de l'étain. Pour une valeur standard de 232°C, celle obtenue lors de l'étalonnage était de 233°C. Quelques complications sont inhérentes à l'emploi de la cellule de Knudsen, à cause de son orifice non idéal qui a la forme d'un canal de largeur finie. Certaines molécules ne sortiront pas de l'orifice, mais frapperont les parois de ce dernier et seront réfléchies. Ces molécules ont une probabilité finie de retourner dans la cellule après une ou plusieurs réflexions. Le résultat est que le flux sortant de la cellule est diminué et que pour une vitesse d'effusion observée, l'équation théorique donne une pression trop basse.

Un facteur de correction, appelé facteur de Clausing: K_c est inclus dans l'équation théorique. Le facteur de Clausing est évalué numériquement

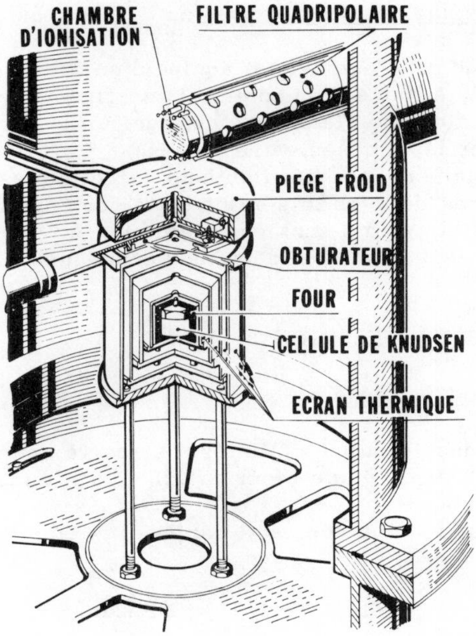

FIG. 1. Montage expérimental de la cellule de Knudsen.

pour différents rapports L/r, où L est la longueur de l'orifice et r son rayon. De Marcus [4], Iczkowski et al. [5] et Freemann et Edwards [6] ont calculé ces valeurs. Pour la cellule utilisée, K_c est égal à 0,25. L'évaporation d'argent (Ag) mesurée en continu dans une thermobalance donne un facteur de Clausing de 0,24.

1.2. Boîtes à gants

Les études sur Na et Cs nécessitent l'emploi d'atmosphère purifiée en oxygène et eau. La boîte à gants utilisée, en acier inoxydable, contient de l'argon purifié sur tamis moléculaire pour l'eau et sur du cuivre pour l'oxygène. Des teneurs inférieures à 1 volume par million d'eau et d'oxygène sont atteintes.

1.3. Système de vide

L'unité de pompage comprend:
- deux pompes préliminaires à tamis moléculaire;
- une pompe ionique;
- une pompe à titane.

Ce système permet d'obtenir une pression d'environ 10^{-9} torr dans l'enceinte.

1.4. Analyseur de gaz résiduel

L'analyseur de gaz résiduel est équipé d'un filtre quadripolaire qui sépare les ions [7] à l'aide d'un potentiel radiofréquence et d'un potentiel continu appliqués à quatre électrodes cylindriques. Le quadrupole utilisé permet de détecter les masses entre 0 et 700.

La mise en place de l'échantillon dans la cellule est faite en boîte à gants. Le transport de la cellule entre cette dernière et le système de vide s'effectue à l'air, mais sans contamination. L'échantillon est chauffé et les différentes espèces d'ions détectées et mesurées.

2. PRINCIPE DE LA METHODE

2.1. Etude thermodynamique

Dans l'étude des équilibres, l'intérêt se porte principalement sur la connaissance des enthalpies de réaction ΔH_0^0 à 0 K ou ΔH_T^0 à la température T.

Cette grandeur est obtenue, soit par la deuxième loi de la thermodynamique:

$$d\ln K_p/d(1/T) = -\Delta H_T^0/R \tag{1}$$

soit par la troisième loi:

$$\Delta G_T^0 = -RT\ln K_p = \Delta H_T^0 - T\Delta S_T^0 \tag{2}$$

$$\Delta G_T^0 = -RT\ln K_p = \Delta H_0^0 + T\Delta[(G_T^0 - H_0^0)/T]$$

$$= \Delta H_0^0 + T\Delta Fef \tag{3}$$

où K_p est la constante d'équilibre à pression constante, p la pression partielle, S_T^0 l'entropie et ΔFef la fonction d'énergie libre.

2.2. Etude cinétique

Dans les études cinétiques, les mesures concernent le flux des espèces s'évaporant comme une fonction de différents paramètres: température, état de surface, pression afin d'établir un mécanisme et de mesurer des constantes de vitesse et des énergies d'activation.

2.3. Relation entre la pression partielle et l'intensité ionique

Un faisceau moléculaire issu de l'orifice d'une cellule de Knudsen est dirigé dans la chambre d'ionisation du spectromètre, du type impact d'électrons. La détection est faite à l'aide d'un multiplicateur d'électrons. Les ions sont identifiés par les moyens classiques de spectrométrie de masse (rapport masse/charge, potentiel d'apparition, etc.). Ces différentes mesures permettent d'identifier sans ambiguïté les molécules neutres à partir desquelles ils sont produits. Une fente d'analyse mobile permet de mesurer les variations d'intensité ionique du faisceau moléculaire issu de la cellule. Ceci permet de connaître d'éventuelles perturbations telles qu'une migration le long de la surface de la cellule.

Pour chaque espèce i, contenue dans la vapeur, la relation entre la pression partielle p_i et le flux Z_i est de la forme

$$p_i = A_i Z_i \sqrt{T} \qquad (4)$$

où A_i est fonction du poids moléculaire de l'espèce effusante et T est la température absolue.

L'équation (4) est appelée équation de Knudsen ou, quelquefois, de Hertz-Knudsen [8, 9].

L'intensité ionique I^+ est proportionnelle au produit du flux Z de particules dans la source de vapeur et au temps t passé dans la région d'ionisation [10] ou inversement proportionnelle à la vitesse v à laquelle les molécules traversent la région d'ionisation:

$$I^+ = \alpha Z t = \alpha \frac{Z}{v} \qquad (5)$$

A partir de la théorie cinétique des gaz, la vitesse moyenne d'une molécule est directement proportionnelle à la racine carrée de la température absolue, d'où

$$I^+ = \alpha \frac{Z}{v} = \alpha \frac{Z}{T^{1/2}}$$

ce qui s'écrit à l'aide de l'équation (4)

$$I^+ = \beta \frac{p}{T} \qquad (6)$$

et finalement

$$p = k I^+ T \qquad (7)$$

où la constante k dépend de la section de capture d'ionisation et de la réponse du multiplicateur d'électrons.

3. APPLICATION DE LA SPECTROMETRIE DE MASSE AUX MESURES THERMODYNAMIQUES

L'application de la deuxième loi (éq. 1) à l'équation (7) conduit à:

$$d \ln I^+ T / d(1/T) = \Delta H_T^0 / R \qquad (8)$$

Différentes mesures furent effectuées sur:

- l'iodure de sodium (NaI);
- le sodium (Na);
- l'alliage Na-5 at.% Cs.

TABLEAU I. INTENSITES RELATIVES DES ESPECES DE NaI (VAPEUR) A 477°C

Masse	Ion	Intensité ionique (unités arbitraires)	Pression partielle
23	^{23}Na	538	
127	^{127}I	820	7×10^{-5} (NaI)
150	$^{23}Na\,^{127}I$	243	
173	$^{26}Na_2^{127}I$	107	$2,8 \times 10^{-7}$ (Na_2I_2)

3.1. Iodure de sodium (NaI)

a) Mesures:

L'iodure NaI, 2 H_2O a été chauffé sous vide (1×10^{-5} torr) à 250°C pendant trois heures, de manière à éliminer l'eau. Après refroidissement, un échantillon (500 mg) a été pesé et introduit dans la cellule de Knudsen en boîte à gants. En chauffant lentement, il a été possible de vérifier la complète déshydratation en notant l'absence du pic de masse 18 (H_2O^+). Le sel anhydre a été chauffé jusqu'à l'apparition du pic 150 (NaI^+). Un balayage du spectre a montré que le pic 127 (I^+) était le pic principal dû à NaI, la pression absolue étant déterminée au moyen du pic 150 et de la vaporisation d'une partie de l'échantillon, par comparaison entre l'intégration du courant ionique NaI^+ et la perte de poids.

Un thermoanalyseur a également été employé pour la détermination de la pression absolue, en utilisant la même cellule de Knudsen. A 750 K (477°C), $p(NaI^+)$ a été mesurée et trouvée égale à 7×10^{-5} atm.

Des mesures ont été effectuées soigneusement à différentes températures; le tableau I rassemble celles faites à 477°C.

Ce spectre de masse a été interprété comme étant dû principalement à l'ionisation du monomère NaI suivant deux processus

$$NaI + e^- \rightarrow NaI^+ + 2e^- \quad (9)$$

$$NaI + e^- \rightarrow Na^+ + I + 2e^- \quad (10)$$

le dimère Na_2I_2 pouvant simultanément être ionisé suivant

$$Na_2I_2 + e^- \rightarrow Na_2I_2^+ + 2e^- \quad (11)$$

et

$$Na_2I_2 + e^- \rightarrow Na_2I^+ + I + 2e^- \quad (12)$$

En utilisant l'intensité de l'ion NaI^+ comme mesure de la concentration des molécules monomères et en représentant Log p (NaI^+) en fonction de 1/T, le changement d'enthalpie par la réaction

$$NaI_{(s)} \rightarrow NaI_{(g)} \quad (13)$$

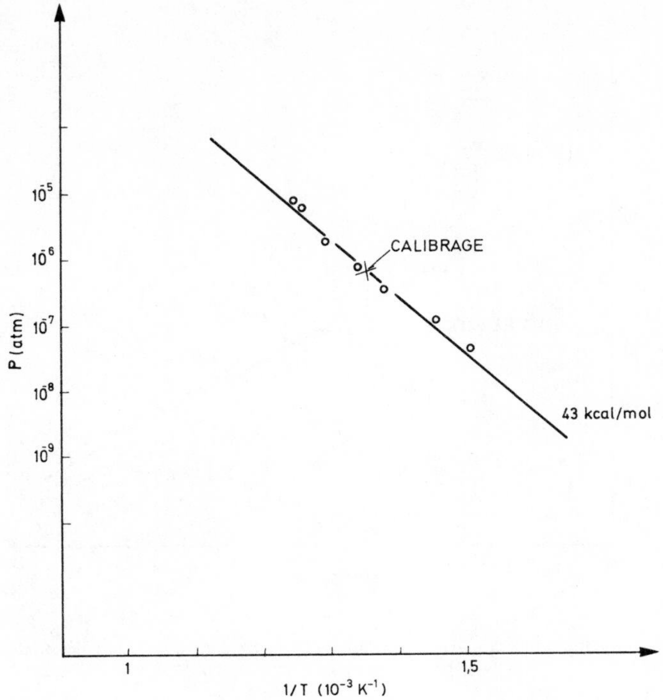

FIG.2. Logarithme de la pression partielle de l'ion monomère NaI^+ en fonction de 1/T.

peut être calculé en mesurant la pente de la droite. La figure 2 représentant cette droite permet de calculer une variation d'enthalpie $\Delta H = 43$ kcal/mol à 470°C pour la réaction représentée par l'équation (13). Si Na_2I^+ est considéré comme indiquant la concentration des molécules dimères, la pente de la droite de

$$\mathrm{Log}[(Na_2I^+)/NaI^+] \quad \text{en fonction de } 1/T$$

doit être proportionnelle au changement d'enthalpie de la réaction

$$NaI_{(s)} + NaI_{(g)} \rightarrow Na_2I_{2(g)} \qquad (14)$$

Nous trouvons ainsi 3 kcal/mol pour l'équation (14). En combinant les résultats des équations (13) et (14), nous trouvons 40 kcal/mol pour la chaleur de dimérisation de NaI à 470°C et 46 kcal/mol pour la réaction:

$$2\,NaI_{(s)} \rightarrow Na_2I_{2(g)} \qquad (15)$$

(b) Discussion des résultats:

Il est utile de comparer la chaleur de sublimation déterminée à l'aide de la figure 2 avec la valeur obtenue avec les pressions mesurées par

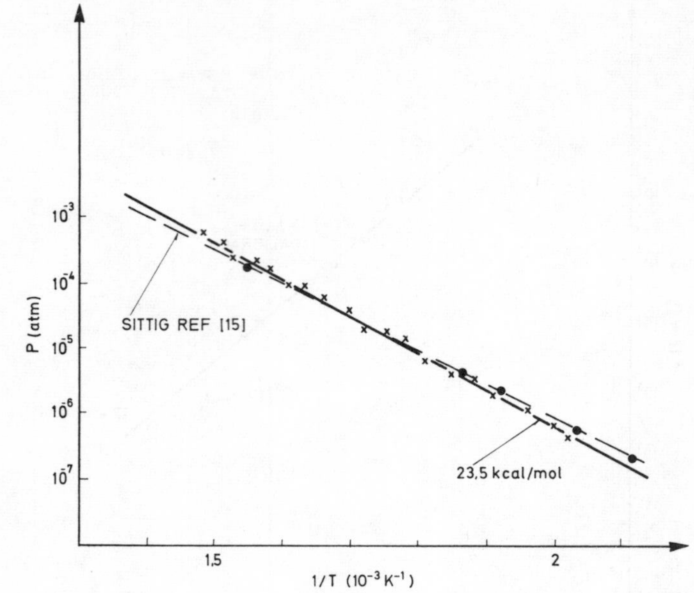

FIG.3. Logarithme de la pression partielle de l'ion monomère Na^+ en fonction de 1/T.

l'intermédiaire de l'entropie ou de la fraction d'énergie libre de ces espèces (ΔFef). Malheureusement, le manque d'information sur l'espèce NaI vapeur conduit à supposer que ΔFef (NaI) est sensiblement la même que celle de KCl (36 cal/deg. mol), NaCl (37 cal/deg. mol), KI (34 cal/deg. mol). Avec ΔFef NaI = 35 cal/deg. mol, nous obtenons une chaleur de sublimation de 47 kcal/mol, soit 4 kcal/mol de plus que le précédent calcul (pente de la droite). Si nous supposons, comme Berkowitz et al. [11, 12] que la section de capture relative du dimère est 1, 5, alors la vapeur contient 4% de l'espèce dimère (Na_2I_2). Pour comparaison, signalons que la bibliographie pour la chaleur de sublimation donne 39 ± 5 kcal/mol [13] ou 49 kcal/mol [14].

3.2. Sodium (Na)

L'étude du sodium et des alliages sodium-césium se complique à cause de leur contamination possible durant leur manipulation, particulièrement celle due à l'eau.

En conséquence, l'assemblage de la cellule de Knudsen a été étuvé à 250°C sous vide (10^{-5} torr) avant d'y placer l'échantillon.

La même étude que celle effectuée pour NaI montre (fig.3):

- une chaleur de vaporisation à 330°C (603 K) égale à 23, 5 kcal/mol (deuxième loi) ou 26 kcal/mol (troisième loi); à cette température, Sittig [15] donne 24,5 kcal/mol;
- une proportion de dimère de 1,5%. La littérature [15, 16] en signale 2%.

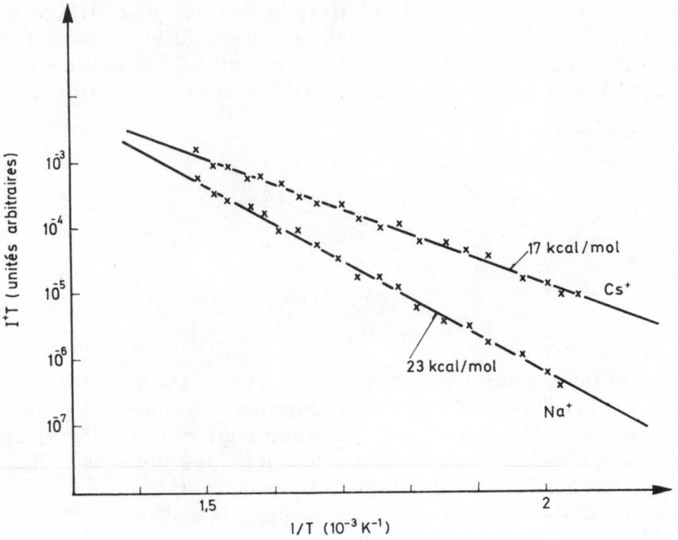

FIG.4. Alliage Na-5 at.% Cs.

3.3. Alliage sodium 5 at.% Cs

Un alliage Na-5 at.% Cs a été préparé et examiné; les résultats (fig. 4) montrent, à 600 K, une chaleur de vaporisation de 23 kcal/mol pour le sodium et 17 kcal/mol pour le césium; Janaf [16] donne pour ce dernier 18,1 kcal/mol.

CONCLUSION

Les premières expériences faites sur l'iodure de sodium, les alliages Na-Cs, le sodium, montrent qu'avec un spectromètre quadripolaire, il est possible de déterminer:
– la nature et la composition de la phase vapeur;
– la chaleur de vaporisation des différents éléments.

La cinétique d'évaporation, étape suivante de cette étude, est envisagée, particulièrement l'emploi de cellules de Knudsen jumelles pour la détermination directe des activités des composants dans les alliages métalliques. En effet, un tel montage expérimental permet la comparaison immédiate de la pression partielle d'un métal sur le corps pur et sur un alliage. Théoriquement, le rapport des pics donne l'activité directement. Pratiquement, les difficultés expérimentales proviennent du contrôle de l'alignement. Ainsi, des cellules jumelles utilisées avec du lithium (Li) et du sodium (Na) ont montré une action réciproque entre les deux trous d'environ 20%, d'où la nécessité de revoir la géométrie du faisceau moléculaire.

Nous envisageons d'employer une méthode indirecte suggérée par Berkowitz et Chupka [17]. Ils montrent que la comparaison des rapports du monomère au dimère présents pour le corps pur et l'alliage donne l'activité sans référence aux pressions absolues ou aux concentrations de la phase liquide. Soient $I_0(A)$ et $I_0(A_2)$ les intensités des ions du monomère et du dimère à l'état standard, $I_1(A)$ et $I_1(A_2)$ ceux de l'alliage, on peut écrire:

$$K = \frac{I_0(A)^2}{I_0(A_2)} = \frac{I_1(A)^2}{I_1(A_2)} \tag{16}$$

De même:

$$a_A = \frac{I_1(A)}{I_0(A)} = \frac{I_1(A)/I_1(A_2)}{I_0(A)/I_0(A_2)} \tag{17}$$

où K est la constante pour l'équilibre $A - A_2$ et a_A l'activité de A dans l'alliage. Il paraît donc suffisant de déterminer le rapport monomère/dimère à l'état de référence et sur l'alliage pour obtenir l'activité. Dans le cas de l'alliage Na-5 at.% Cs, les intensités des pics Na^+, Na_2^+ ont été mesurées sur Na pur et sur l'alliage. L'activité a_{Na} a été trouvée égale à 0,94 au début de l'expérience. Nous avons l'intention d'utiliser cette méthode dans les futures expériences.

REMERCIEMENTS

Nous tenons à remercier G.J. Laplanche pour son travail expérimental ainsi que le CEA qui a donné l'autorisation de publier ce mémoire.

REFERENCES

[1] BUCHLER, A., BERKOWITZ, J.B., Vapor Pressure Methods, Techniques of Metals Research 2 Part 1, Interscience Publishers (1970) 161.

[2] DROWART, J., GOLDFINGER, P., «The role of mass spectrometry in high temperature chemistry», Advances in Mass Spectrometry 3, Institute of Petroleum (1966) 923.

[3] PATTORET, A., Etudes thermodynamiques par spectrométrie de masse sur les systèmes uranium-oxygène et uranium-carbones, Thèse, Université de Bruxelles (1969).

[4] DE MARCUS, W.C., The Problem of Knudsen fluw, Report N° K 1302, Parts 1-6, Oak Ridge Gazeous Diffusion Plant (1957).

[5] ICZKOWSKI, R.P., et al., J. Phys. Chem. 38 (1963) 2064.

[6] FREEMANN, R.D., EDWARDS, J.G., The Characterization of High Temperature Vapors, Wiley, New York (1967) App.C.

[7] PAUL, W., REINHARD, H.P., Von ZHAN, U., Z. Phys. (1958) 143-52.

[8] KNUDSEN, M., Ann. Phys. (1909) 29-179.

[9] HERTZ, H., Ann. Phys. (1982) 17-177.

[10] SCHLISSEL, P.O., J. Chem. Phys. 26 (1957) 1276.

[11] BERKOWITZ, J., MARQUART, J.R., Mass Spectrometric Study of the Magnesium Halides, J. Chem. Phys. 37 (1963) 1853.

[12] BERKOWITZ, J., TASMAN, H.A., CHUPKA, W.A., J. Chem. Phys. 36 (1962) 2170.

[13] PASCAL, P., Nouveau Traité de Chimie Minérale 2, MASSON, Paris (1966) 383.

[14] KUBASCHEWSKI, O., EVANS, E.LL., Metallurgical Chemistry, Pergamon Press, Londres (1968) 299.

[15] SITTIG, M., Sodium – Its Manufacture Properties and Uses, Reinhold Publishing Corporation, New York (1956) 432.

[16] Thermodynamical Tables JANAF, The Dow Chemical Company, Midland (1962, 1968).

[17] BERKOWITZ, J., CHUPKA, W.A., Trans. N.Y. Acad. Sci. 79 (1960) 1073.

CRYOGENIC HEAT CAPACITIES AND NEEL TRANSITIONS OF TWO URANIUM PNICTIDES, UAs_2 AND USb_2*

E.F. WESTRUM, Jr., J.A. SOMMERS, D.B. DOWNIE**
University of Michigan,
Ann Arbor, Mich.,
United States of America

F. GRØNVOLD
University of Oslo,
Blindern,
Norway

Abstract

CRYOGENIC HEAT CAPACITIES AND NEEL TRANSITIONS OF TWO URANIUM PNICTIDES, UAs_2 AND USb_2.

The heat capacities of uranium diantimonide (USb_2) and uranium diarsenide (UAs_2) — with anti-Cu_2Sb structures — have been measured by adiabatic calorimetry from 5 to above 700 K. Both heat capacity curves show sharp, lambda anomalies indicative of cooperative phenomena. The corresponding heat capacity maxima at 272.2 and 202.5 K for UAs_2 and USb_2, respectively, are related to maxima in respective magnetic susceptibility curves at 283 and 206 K, occasioned by transitions between antiferro- and paramagnetic states (Néel points). The heat capacities at 300 and 350 K (in $cal \cdot K^{-1} \cdot mol^{-1}$) are 19.11 and 19.18 for UAs_2, 19.16 and 19.33 for USb_2 (in comparison with 19.2 and 20.3 for (isostructural) UP_2 for which the Néel transition was found by Stalinski, et al. at 203 K). The differences in the temperature dependency and excess heat capacity between UP_2 and the others beyond the Dulong and Petit limit are especially pronounced even at 300 K. Lack of monotonic progression of Néel temperatures with anionic size may be due to the relatively small radius ratio in UP_2. Thermodynamic properties for both pnictides as well as the thermodynamics of the transitions have been evaluated.

INTRODUCTION

Compounds of uranium with the Group V elements (the pnictides) exhibit a variety of solid state phenomena--consequences of the outer electron behavior of uranium--which are interesting from both experimental and theoretical points of view. The compounds which are of interest currently are those of stoichiometries UX, U_3X_4, and UX_2. For a given stoichiometry there is a striking similarity of properties despite the large variation in pnigogen size. All the UX compounds possess the NaCl structure and those of the U_3X_4 stoichiometry have the Th_3P_4 structure. In the dipnictide series, UN_2 is an exception and is stable only under high N_2 pressure and has the fluorite structure [1]. The remaining members UP_2, UAs_2, USb_2, and UBi_2 are isostructural with tetragonal Cu_2Sb [2-4]. Lattice parameters are shown in Table I; the unit cell and magnetic structures are shown in Figure 1. Magmetic susceptibility [5-7] and neutron diffraction measurements

* This work was supported at the University of Michigan by the National Science Foundation, USA.

** Present address: Department of Metallurgy, University of Strathclyde, Glasgow, Scotland.

TABLE I. Lattice constants of the uranium dipnictides* (Cu_2Sb structure)

Compound	a (nm)	b (nm)	a/b	Reference
UP_2	0.3808	0.7778	2.043	[2]
UAs_2	0.3962	0.8132	2.053	[2]
USb_2	0.4281	0.8759	2.044	[3]
UBi_2	0.4454	0.8926	2.004	[4]

* The values have been transformed from kX to nm by the factor 0.100202.

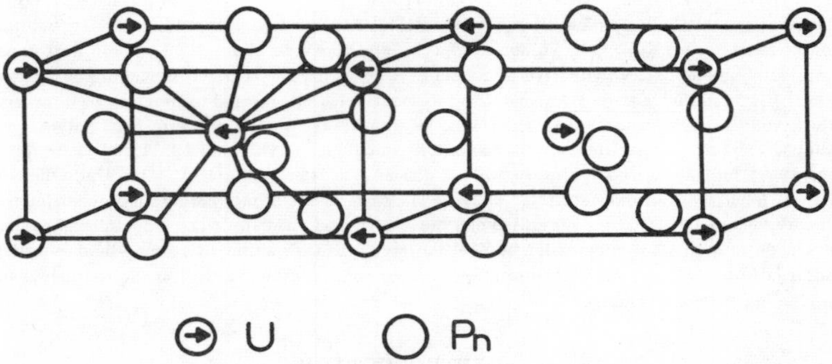

FIG. 1. Unit cell for the uranium dipnictides and magnetic unit cell for UP_2, UAs_2, and USb_2 [11].

[8-10] have been made on the dipnictides UP_2 through UBi_2. The dependence of inverse magnetic susceptibility on temperature indicates that the compounds all order antiferromagnetically and that Curie-Weiss behavior obtains above the ordering (Néel) temperature. The derived parameters are shown in Table II. The paramagnetic moment increases in monotonic progression with increasing pnigogen size, but Néel temperatures do not. The magnetic coupling may either be anomalously weak in UP_2 or anomalously strong in UAs_2.

The antiferromagnetic structures of UP_2, UAs_2, and USb_2 determined by neutron diffraction (in the ordered state) consist of ferromagnetic sheets of uranium ions stacked perpendicularly to the c-axis and in the sequence +--+ as shown in Figure 1. The magnetic unit cell is twice as large as the chemical one due to

TABLE II. Magnetic properties of the uranium dipnictides

Compound	Magnetic moment, μ_B Para-magnetic	Antiferro-magnetic	T_N (K)	Weiss constant	Ref.
UP_2	2.50	1.0 ± 0.1	203	+30	[6,8]
UAs_2	2.94	1.61 ± 0.01	283	+34	[5,8]
USb_2	3.04	0.94 ± 0.03	206	+18	[5,8]
UBi_2	3.40	2.1 ± 0.1	183	-54	[7,8]

its doubled c-axis. UBi_2 contrasts in that the chemical and magnetic unit cells are identical and the ferromagnetic sheets have the order +-+-. The compounds display metallic conduction [with ρ(293 K) = 1.83×10^{-4} and 1.95×10^{-4} ohm cm for single crystal samples of UP_2 and UAs_2, respectively][11].

Enthalpy of formation (ΔH_f) measurements on lower arsenides and phosphides together with the observation that ΔH_f°(298 K) seems to be a linear function of the pnigogen to uranium ratio give an estimate [12] of ΔH_f°(298 K) for UP_2 and UAs_2 of -69 kcal mol^{-1} and -60 kcal mol^{-1}, respectively. ΔH_f°(298 K) for USb_2 has been found to be -(41.5 ± 2.6) kcal mol^{-1}, and -26.3 for UBi_2 [13]. It is noted that ΔH_f° increases with decreasing electronegativity (cf. Table IV).

Only the heat capacity of UP_2 (22.5 to 350 K) has been reported previously [14]. Here a prominent lambda anomaly at 203.2 K, accords well with the transition to the antiferromagnetic state as observed in magnetic susceptibility measurements.

EXPERIMENTAL

The dipnictides were prepared by direct union of stoichiometric amounts of the elements in sealed evacuated quartz tubes at elevated temperatures at Oslo [15]. The resulting grey powders were handled in an inert atmosphere. The heat capacities of UAs_2 and USb_2 have been measured by adiabatic calorimetry on samples of 147 g and 169 g in the temperature range 5 K to 350 K at Ann Arbor and from 300 to above 700 K at Oslo.

RESULTS

The most interesting aspects of the heat capacities shown in Figures 2 and 3 and Table III are the sharp maxima observed at

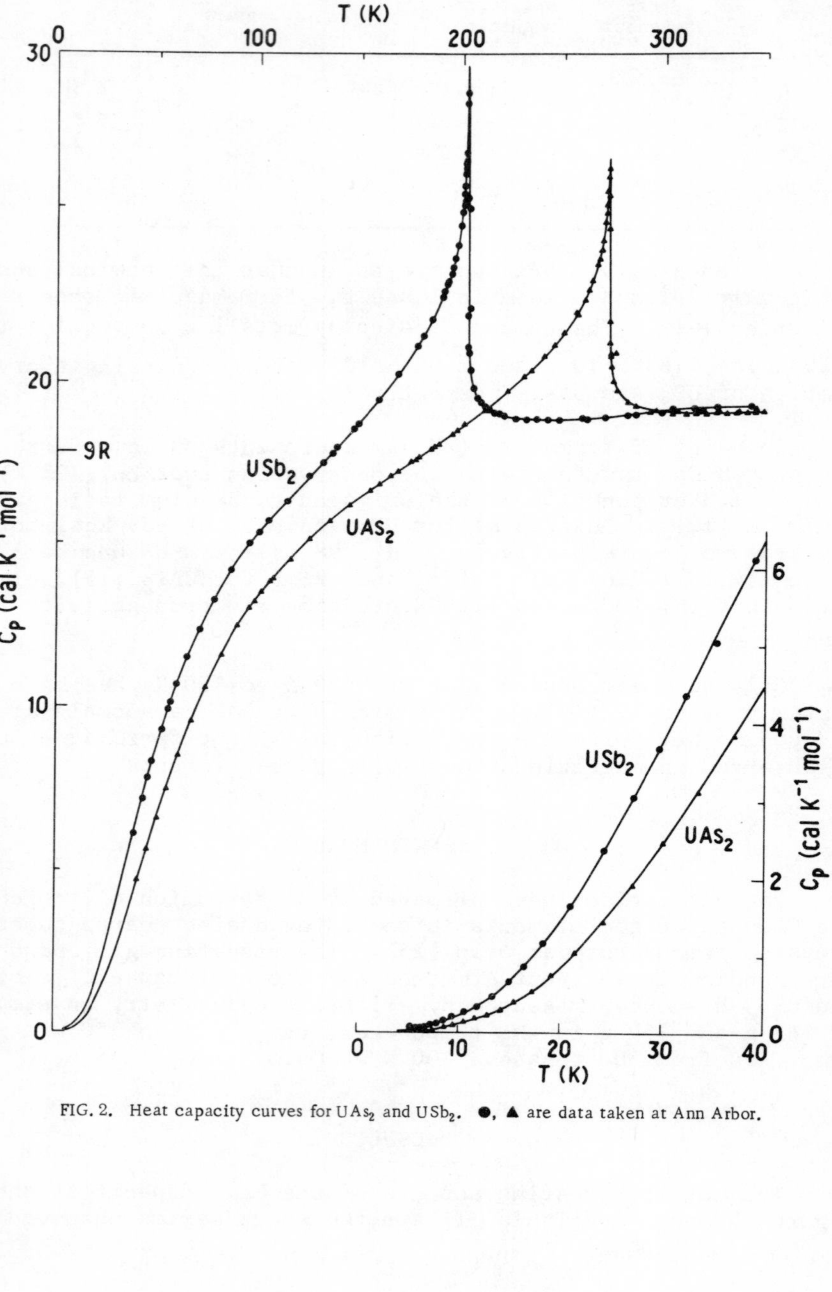

FIG. 2. Heat capacity curves for UAs_2 and USb_2. ●, ▲ are data taken at Ann Arbor.

TABLE III. Selected heat capacity values for the uranium dipnictides*

T (K)	C_p (cal $K^{-1}mol^{-1}$) UP_2 †	UAs_2	USb_2
10	----	0.110	0.207
25	----	1.559	2.504
50	4.30	6.541	8.741
100	10.47	13.43	15.38
150	15.00	16.49	18.86
200	20.00	18.68	25.16
250	17.84	21.49	18.91
298.15	19.12	19.11	19.16
300	19.16	19.11	19.17
350	20.31	19.18	19.33
400	----	19.47	19.84
500	----	19.79	20.29
600	----	19.97	20.63
700	----	20.24	20.80

* cal = 4.184 J.

† From Stalinski et al.[14]

272.2 K and 202.5 K for UAs_2 and USb_2, respectively. The peak for UAs_2, however, is at a temperature rather lower than that indicated by the susceptibility. Inspection of the latter work shows that the maximum in the inverse susceptibility versus temperature is somewhat rounded and that the slope is steepest at a temperature somewhat lower than that of the maximum (283 K). In the present work, the ΔH's through the transitions were reproducible to well within 0.1 per cent regardless of rates of heating or cooling through the transition region and the same result was obtained after the samples were taken to 4 K.

The thermodynamic contributions--the enthalpy and entropy increments (ΔH_{mag} and ΔS_{mag}) due to the magnetic ordering transitions--were determined by plotting effective Debye θ's as f(T) for the total heat capacities and interpolating over the gap caused by the enhanced heat capacity in the transition region. The θ values thus obtained over this interval were used to generate "lattice" heat capacity points which were then integrated. The results are

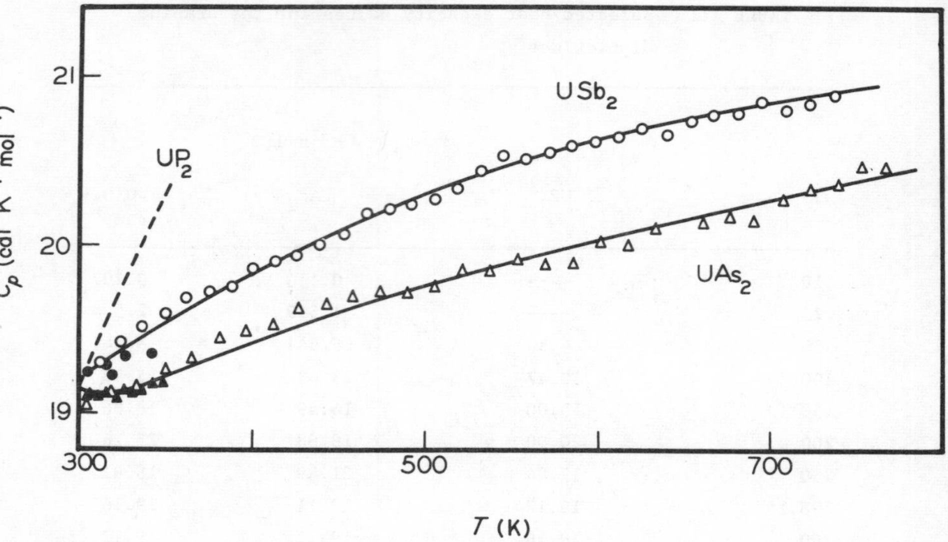

FIG. 3. Higher temperature heat capacity curves for UP_2 [14], UAs_2, and USb_2. Broken line represents the heat capacity of UP_2; ●, ▲ are taken at Ann Arbor; and ○, △ are data taken at Oslo.

shown in Table IV, together with those for UP_2. Magnetic quantities for the latter compound are from the analysis of Stalinski et al. [14] obtained by subtraction of a lattice heat capacity approximated by a trial and error sum of Debye functions. The values for γ, the temperature coefficient of electronic heat capacity, were obtained from a C_p/T versus T^2 plot of heat capacity points below about 20 K for UAs_2 and USb_2.

DISCUSSION

Difficulties associated with determination of magnetic heat capacity contributions are well known and are encountered with uranium compounds involving Group IV, V, and VI elements [16,17]. In the absence of heat capacity data for the isostructural diamagnetic analogs (i.e., the corresponding thorium compounds) such resolutions must be regarded as tentative at best. It is interesting to note, however, that the heat capacity of USe_2 [19] (with nearly the same molecular mass as UAs_2 although the crystal structures differ) agrees rather well with the heat capacity of the latter compound for some ranges both above and below the transition and with our independently selected lattice contribution.

The intermediate temperature data in Figure 3 are also noteworthy in that the magnitudes of the UP_2 heat capacity are already well above the Dulong and Petit limit and the trend of these data contrast so markedly with the UAs_2 and USb_2 data.

TABLE IV. Some thermodynamic quantities for the uranium pnictides[a]

Quantity	UP_2	UAs_2	USb_2	UBi_2
		At 298.15 K		
C_p° (cal K^{-1} mol^{-1})	19.12[b]	19.11	19.16	(19.2)[c]
S° (cal K^{-1} mol^{-1})	24.34[b]	29.41	33.81	(29.1)[d] (37.5)[e]
$H^\circ - H_0^\circ$ (cal mol^{-1})	3679[b]	4282	4596	----
ΔC_{pf}° (cal K^{-1} mol^{-1})	1.12	0.73	0.60	(0.2)
ΔH_f° (kcal mol^{-1})	(-69)[c]	(-60)[c]	-41.5±2.6	(-26.3)[d]
ΔS_f° (cal K^{-1} mol^{-1})	-7.33	0.58	0.02	(-1.64)
ΔG_f° (kcal mol^{-1})	(-71.2)[c]	(-60.2)[c]	-41.5±3	(-25.8)[c]
ΔV_f° (cm^3 mol^{-1})	-1.61[f]	-0.13	-0.82	-1.78
		At other temperatures		
$H_{700}^\circ - H_0^\circ$ (cal mol^{-1})	----	12203	12711	----
$S_{700}^\circ - S_0^\circ$ (cal K^{-1} mol^{-1})	----	46.16	50.94	----
$T_{Néel}$ (K)	203.2[b]	272.2	202.5	----
ΔH_{mag} (cal mol^{-1})	204.5	234.4	258.5	----
ΔS_{mag} (cal K^{-1} mol^{-1})	1.31[b]	0.99	1.70	----
$\gamma \times 10^3$ (cal K^{-2} mol^{-1})	5[b]	1.36	2.98	----

[a] cal = 4.184 J
[b] From Stalinski et al.[14]
[c] Parentheses indicate quantities in which estimates are involved.
[d] Rand and Kubaschewski [13] based on Cosgarea et al. [18] and others.
[e] Estimate from USb_2 and magnetic contribution.
[f] From black phosphorus.

Relevant to the interpretation of the ΔS_{mag} is the amount of entropy to be expected--despite reservations about well defined oxidation states for conducting compounds involving the pnigogens. Naively, one may at least initially adopt a model with a specified oxidation state for uranium (often U^{+4}) and proceed from the free-ion, ground term ($^3H_4$) under the assumption of Russell-Saunders coupling to the expected magnetic entropy of R ln(2J + 1) = 4.37 cal K^{-1} mol^{-1}. Failure of the experimental results to account for this amount of entropy is usually attributed to crystal field quenching of the orbital moment. Occasionally in these systems the paramagnetic moment and the excess entropy can be identified more closely with spin-only values, which in the present case would be R ln(2S + 1) = R ln3 = 2.18 cal K^{-1} mol^{-1}. It is seen in Table II that paramagnetic moments are all less than the free-ion values (3.58 μ_B) and vary from below the spin-only value (2.83 μ_B) for UP_2 to near the free-ion value for UBi_2. However, it is shown in Table IV that the ΔS_{mag} values are well below even the spin-only values.

These low values are evidence of ligand field effects. Other evidence of the importance of this phenomenon is derived from the magnetic susceptibility measurements. In the case of UP_2 [6], the paramagnetic moment increases with temperature. This is expected from the gradual population of higher-lying levels of the ground state multiplet with increasing temperature. Evidence of a similar effect was not noted by Trzebiatowski et al. [5] for UAs_2 or USb_2. Chechernikov et al. [20] in reporting the results of susceptibility measurements on a series of USb_2-$ThSb_2$ solid solutions claimed that the increase of the magnetic moment per uranium ion on substitution of Th for U confirms the $^3H_4$ ground term (though for each of the prepared solutions this value is less than that reported [5] for pure USb_2 and the trend--if real--is within the experimental uncertainties. However, their claim that the Weiss constant decreases with such substitution is noted also in the comments of Hutchison and Candela [21] on the susceptibility of UO_2-ThO_2 solid solutions.

Implication of the importance of crystal field splitting is mentioned by Counsell et al. [16]. A similar argument on a more quantitative basis has been made by Grunzweig-Genossar [22]. He advocates that the full expected entropy will not be developed in the cooperative transition from the magnetically ordered state but that the cooperative transition will be followed at higher temperatures by an anomaly of the Schottky type and that these together account for the full degeneracy of the ground term. He then calculates, based on the reported entropy of transition increments, the splitting of the ground term (i.e., the distance from the assumed ground singlet to the next higher component, assuming that still higher-lying components are too far removed to be populated) for several uranium compounds. It should be pointed out that it is likely that these resolutions do not account for the full cooperative entropy and thus would cause the splitting to be overestimated. His argument precludes analyses of heat capacity data which treat only the cooperative excess and underscores the necessity for heat capacity data on thorium compounds. In contrast to the prominent cooperative contribution, Schottky peaks--especially at higher temperatures--are much more subtle and easily overlooked.

A potentially more serious deficiency of such attempts at interpretation is related to the metallic nature of the uranium pnictides as noted by Counsell et al. [16]. Robinson and Erdös [23] recently proposed a model which takes into account the closeness of the 5f (localized) and 6d-7s (band) levels in the uranium monopnictides and have applied it to measurements on two members of that series (UP and UAs) and to NpC. They are able to attribute the sharp heat capacity anomaly at 22.5 K in UP to a cooperative transfer of electrons from the conduction band to the actinide ion. The spins of the actinide ion disorder cooperatively at a higher temperature. The former transition is accompanied by a 10 per cent drop in magnetic moment, and by a sudden change in resistivity. This model is a departure from those admittedly flawed ones previously employed to interpret heat capacity measurements in that it does not require assumptions about a definite ionic state and allows for the essential interdependence of magnetic and conduction phenomena. It is not clear at present how this model could be applied to the heat capacity measurements presented here since only a single transition is observed and other studies such as resistivity, crystallography, and susceptibility do not cover a sufficient range of temperature. Generalization of this model (able to accommodate the complication of metallic conductivity) to the dipnictides and those of the U_3X_4 stoichiometry and more extensive investigation of the properties of the pnictides are some promising lines of future investigation.

REFERENCES

[1] HANSEN, M., "Constitution of Binary Alloys", 2nd Ed., McGraw-Hill, New York (1958).

[2] IANDELLI, A., Atti acad. nazl. Lincei, Rend. Classe sci. fis., mat. e nat. 13 (1952) 138, 144.

[3] FERRO, R., Ibid. 14 (1953) 89.

[4] FERRO, R., Ibid. 13 (1952) 53.

[5] TRZEBIATOWSKI, W., SEPICHOWSKA, A., ZYGMUNT, A., Bull. Acad. polon. Sci. Sér. sci. chim. 12 (1964) 687.

[6] TRZEBIATOWSKI, W., TROC, R., Ibid. 11 (1963) 661.

[7] TRZEBIATOWSKI, W., ZYGMUNT, A., Ibid. 14 (1966) 495.

[8] LECIEJEWICZ, J., TROC, R., MURASIK, A., ZYGMUNT, A., Phys. stat. sol. 22 (1967) 517.

[9] TROC, R., LECIEJEWICZ, J., CISZEWSKI, R., Phys. stat. sol. 15 (1966) 515.

[10] OLES, A., J. Phys. (Paris) 26 (1965) 561.

[11] TRZEBIATOWSKI, W., HENKIE, Z., Personal communication cited by Leciejewicz, et al. in reference [5].

[12] BASKIN, Y., SMITH, D. S., J. Nucl. Mater. 37 2 (1972) 209.

[13] RAND, M. H., KUBASCHEWSKI, O., "The Thermochemical Properties of Uranium Compounds, Oliver and Boyd, London (1963).

[14] STALIŃSKI, B., BIEGAŃSKI, Z., TROĆ, R., Bull. Acad. polon. Sci., Sér. sci. chim. 15 (1967) 238.

[15] GRØNVOLD, F., ZAKI, M. R. Unpublished X-ray and magnetic susceptibility results.

[16] COUNSELL, J. F., DELL, R. M., MARTIN, J. F., Trans. Faraday Soc. 62 (1966) 1736.

[17] WESTRUM, E. F., JR., LYON, W. G. "Thermodynamics of Nuclear Materials", 'Thermodynamics of Semimetallic Compounds', Proc Conf. Vienna, IAEA, Vienna (1968).
[18] COSGAREA, A., HUCKE, E. E., RAGONE, D. V., Acta Metall. 9 (1961) 225.
[19] WESTRUM, E. F., JR., GRØNVOLD, F., J. inorg. nucl. chem. 32 (1970) 2169.
[20] CHECHERNIKOV, V. I., KUZ'MAN, R. N., CHACHKHIANI, L. G., GOLOVIN, V. A., IRKAEV, S. M., SLOVYANSKIKH, V. K., SHAVISHVILI, T. M., Sov. Phys. JETP 30 (1970) 73.
[21] HUTCHISON, C. A., Jr., CANDELA, G. A., J. Chem. Phys. 27 (1957) 707.
[22] GRUNZWEIG-GENOSSAR, J., Solid State Commun. 8 (1970) 1673.
[23] ROBINSON, J. M., ERDÖS, P., Phys. Rev. B. 8 (1973) 4333; Ibid. 9 (1974) 2187.

DISCUSSION

H.E. FLOTOW: I noted that your values of γ, the electronic heat capacity coefficient, seemed to vary quite a bit from compound to compound. Lower temperature heat capacity data (below 4 K) would be needed in order to obtain more accurate values for γ. To resolve the total heat capacity, above the Néel temperature, the electronic contribution would have to be taken into account. It is possible that the low temperature γ's will not be the correct ones to use above the Néel temperatures.

E.F. WESTRUM, Jr.: I agree that lower temperature data are desirable — for the reasons which you have mentioned as well as on other grounds. Nevertheless, γ's extrapolated for comparable compounds from 5 K have by and large proven surprisingly reliable. We consider that with data on thorium homologues we can resolve Schottky, electronic and magnetic (tail) contributions adequately. Schottky contributions are significant in many transition element compounds, and recent work in our laboratory on $EuBn_3$ has shown that we can resolve them even at 300 K.

THE HEAT CAPACITY OF URANIUM MONOPHOSPHIDE FROM 80-1080 K, AND THE ELECTRONIC CONTRIBUTION

H. YOKOKAWA, Y. TAKAHASHI, T. MUKAIBO
University of Tokyo,
Tokyo,
Japan

Abstract

THE HEAT CAPACITY OF URANIUM MONOPHOSPHIDE FROM 80 - 1080 K, AND THE ELECTRONIC CONTRIBUTION.

The heat capacity of uranium monophosphide from 80 to 1080 K was determined by an improved laser flash method. Below room temperature, the results agreed very well with those of Counsell et al., including the region of antiferroparamagnetic transition near 121 K. At higher temperatures, no sharp anomalous peak was observed up to 1080 K; a slight anomalous temperature dependence in heat capacity was found above room temperature. Thermodynamic functions of UP from 298 to 1050 K were calculated from the observed heat capacity and are tabulated. Attempts to resolve the observed heat capacity into various contributions are also reported. The lattice component of the heat capacity was estimated by the corresponding states method. The excess heat capacity, which was found to be high at all temperatures investigated, was analysed by following the localized $5f^2$ and $5f^3$ configuration models.

1. INTRODUCTION

Among the semi-metallic uranium compounds which have an NaCl-type structure, uranium monophosphide is one that has not been investigated extensively to determine its thermal properties such as high-temperature heat capacity. Recently, Men'shikova et al. [1] studied the high-temperature properties of UP and observed anomalous behaviour in its heat capacity, thermal expansion, thermal diffusivity and electrical resistivity at temperatures from 800 to 1000 K. From this anomalous behaviour, they suggested that UP might have a second-order phase transition at about 1000 K, which had not been reported anywhere.

In view of the accumulated knowledge about the heat capacity and electronic structure of NaCl-type uranium compounds, more precise measurements of UP over a wide range of temperature seemed to be desirable. We endeavoured to establish a method of precise measurement of heat capacity by using laser flash heating and have reported [2] its feasibility for a temperature range from 80 to 1100 K. This paper deals with the experimental determination of the heat capacity of UP by the laser flash method from 80 to 1080 K. Attempts to resolve the heat capacity into the various contributions are also reported.

2. EXPERIMENTAL

The sample used was a sintered pellet of UP, which was supplied by the Central Research Laboratory, Mitsubishi Metal Corporation. The results of the chemical analysis done by the Laboratory showed that the

TABLE I. HEAT CAPACITY OF URANIUM MONOPHOSPHIDE
1 mol = 269.00 g

T (K)	Cp ($cal \cdot K^{-1} \cdot mol^{-1}$)	T (K)	Cp ($cal \cdot K^{-1} \cdot mol^{-1}$)	T (K)	Cp ($cal \cdot K^{-1} \cdot mol^{-1}$)	T (K)	Cp ($cal \cdot K^{-1} \cdot mol^{-1}$)
Series I		Series III		Series IV		Series V	
295.00	11.98	131.8	9.80	294.3	12.00	286.7	11.88
Series II		134.3	9.79	327.3	12.25	518.8	12.99
86.7	6.98	136.6	9.76	350.1	12.41	617.7	13.33
90.4	7.33	138.8	9.79	368.4	12.52	620.7	13.28
93.9	7.70	141.1	9.86	385.8	12.67	674.3	13.40
100.2	8.32	145.5	9.96	404.5	12.68	708.6	13.44
102.8	8.64	149.9	10.03	423.7	12.78	716.3	13.56
105.5	8.95	154.6	10.12	441.2	12.84	763.7	13.59
108.1	9.35	159.4	10.21	467.7	12.95	793.7	13.73
110.5	9.76	164.4	10.32	487.0	12.98	823.8	13.81
112.7	10.48	169.2	10.35	509.1	13.07	851.9	13.89
114.8	10.82	174.1	10.44	527.8	13.11	872.7	13.94
116.7	11.49	179.0	10.49	546.3	13.20	887.5	14.04
118.6	12.46	183.5	10.66	567.1	13.22	903.3	14.09
119.7	15.28	193.2	10.72	587.4	13.39	923.0	14.13
120.2	20.47	198.0	10.82	608.0	13.31	941.7	14.17
120.5	26.07	202.7	10.84	627.9	13.38	960.4	14.17
120.8	32.34	207.3	10.91	647.9	13.37	977.9	14.22
121.0	48.24	211.8	10.99	666.7	13.39	980.5	14.31
121.1	60.83	216.3	11.06			985.5	14.25
121.7	24.20	220.6	11.09			985.9	14.32
121.7	19.25	227.3	11.20			992.1	14.24
122.0	14.23	233.9	11.25			999.5	14.32
122.4	12.40	240.2	11.34			1011.5	14.28
123.1	9.88	246.5	11.45			1028.3	14.42
123.7	9.89	252.7	11.54			1042.4	14.36
126.6	9.80	259.9	11.59			1062.8	14.43
129.4	9.71	264.9	11.67			1084.2	14.41
132.1	9.75	271.1	11.69				
134.3	9.84	278.1	11.73				
137.3	9.76	284.4	11.90				
139.4	9.84	290.7	11.94				
		297.9	12.02				
		307.4	12.15				
		314.5	12.19				

uranium, phosphorus and oxygen contents were 88.10, 11.79 and 0.05 wt%, respectively, giving a nominal composition of $UP_{1.027}$. Spectroscopic analysis showed the presence of the following impurities: boron – 4 ppm; copper – 11 ppm, silicon – 16 ppm. X-ray diffraction analysis showed the material to be in a single phase, and its lattice parameter to be 5.589 Å ± 0.002 Å. The size of the sample was about 10 mm in diameter and 5 mm in thickness.

Measurements of the heat capacity were performed by the improved laser flash method, details of which were described elsewhere [2]. In this method, the experimental procedure for the heat capacity determination is divided into two steps. The first step is to determine the absolute heat capacity of the sample at room temperature by utilizing a sample of pure Al_2O_3 single crystal as a standard material. A series of preliminary measurements at room temperature on several Al_2O_3 specimens having different weights confirmed that the experimental results were reproducible to about ± 0.5% on different runs. As the second step, the temperature dependence of the heat capacity is determined relative to the absolute value determined at room temperature. The precision of the determination of the heat capacity was confirmed to be better than ± 0.5% below 700 K, and better than ± 2% above this temperature.

In the present experiments, an 'absorbing disc' made of glassy carbon, details of which were described in our previous paper [2], was used at temperatures below 300 K. At higher temperatures, instead of using the absorbing disc, a thin layer of colloidal graphite was coated onto the sample surface.

The weights of the UP sample, the absorbing disc and the silver paste (Du Pont, 4817) used were 3.5501, 0.0376 and 0.0010 g, respectively. The heat capacity of the UP sample was determined by subtracting the contribution of the absorbing disc from the gross heat capacity, using the literature value for the heat capacity of the glassy carbon [3]. No correction was made for the silver paste and the colloidal graphite, as their heat capacities were confirmed to be negligible as compared with the experimental error.

The temperature increment obtained from the laser flash was usually 1.5 to 5 K, and typically 2 K. In the temperature range of a phase transition, the input energy of the laser flash was controlled so as to be almost constant and the minimum temperature increment, which appeared at the maximum peak of the heat capacity, was about 0.5 K.

3. EXPERIMENTAL RESULTS AND THERMAL PROPERTIES OF UP

The heat capacity of UP that was determined is presented in Table I in a chronological sequence. These results are presented in terms of the defined thermochemical calorie of 4.1840 J. The sample was regarded as stoichiometric and no correction for impurities was made.

In Fig. 1, the present results are shown together with those of other investigators [1, 4-7]. Below room temperature the agreement between the present results and those of Counsell et al. [4], obtained by the adiabatic calorimetry, is very good. The heat capacity starts to increase very sharply from 119 K, and reaches its maximum value, 60.8 $cal \cdot K^{-1} \cdot mol^{-1}$, at 121 K. This sharp increase in the heat capacity is consistent with the observation of the sharp drop of the ordered magnetic moment obtained by

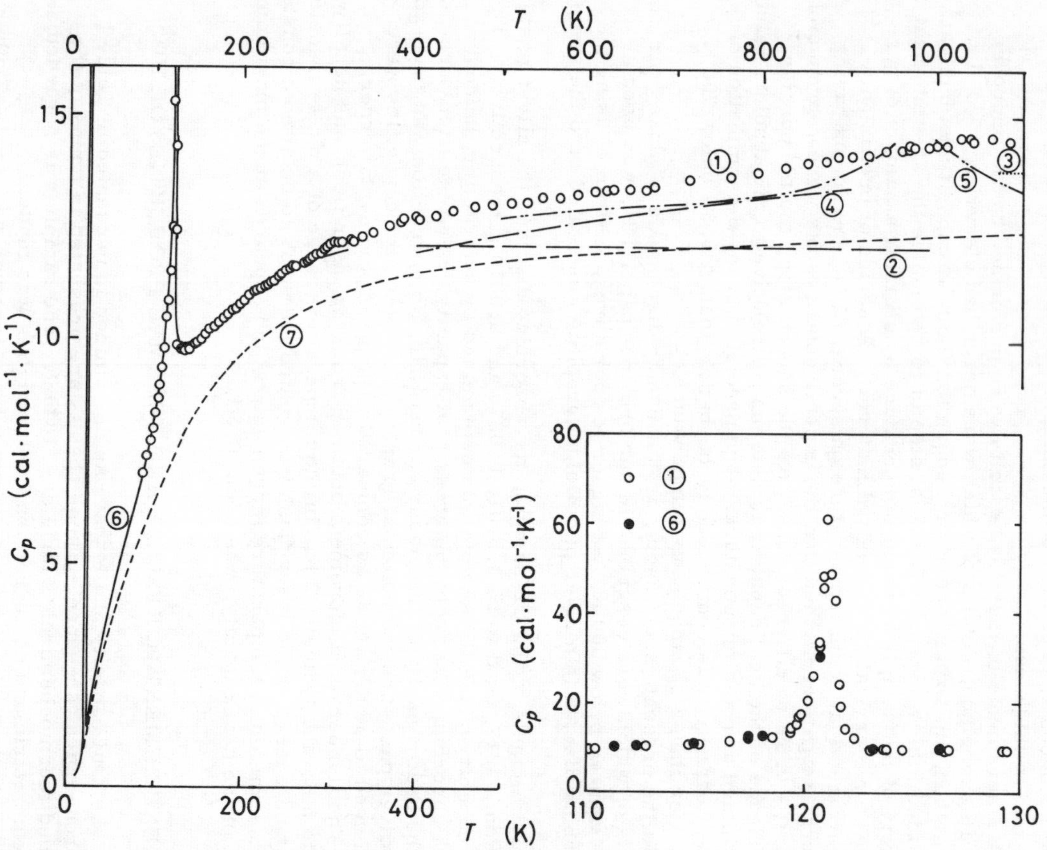

FIG. 1. Heat capacity of UP. ① present results (o); ② Moser and Kruger [5]; ③ Brugger [6]; ④ Ono et al. [7]; ⑤ Men' Shikova et al. [1]; ⑥ Counsell et al. [4]; ⑦ Estimated lattice heat capacity in the present study.

NMR measurements [8]. The observed peak temperature, 121 K, also agreed well with the determinations made using magnetic susceptibility measurements [9, 10]. At higher temperatures, no anomalous peak in the heat capacity was observed in the present study, whereas Men'shikova et al. [1] had reported an anomalous peak at about 1000 K, although the method and accuracy of the measurement had not been reported in their paper. In the present study, however, the heat capacity of UP above 700 K was found to have a slight upward deviation from the usual temperature dependence. Such behaviour, as seen in the temperature dependence of the heat capacity at high temperatures, was also observed in the thermal diffusivity [11] and electrical resistivity [12] of UP.

The experimental results on heat capacity were well fitted by a function of the form:

$$C_p = a + bT + cT^2 + dT^{-2}$$

A least-squares method gave the following equation for the heat capacity of UP from 298 to 1080 K, with a standard deviation of 0.3%:

$$C_p/\mathrm{cal\cdot K^{-1}\cdot mol^{-1}} = 13.714 - 1.379 \times 10^{-3}\,T/K + 2.095 \times 10^{-6}\,(T/K)^2 - 1.302 \times 10^{5}\,(T/K)^{-2} \qquad (1)$$

The smoothed heat capacity and derived thermodynamic functions at selected temperatures, which were calculated by the above equation, are shown in Table II, and the smoothed heat capacity below room temperature calculated by the least-squares method is shown in Table III.

Compared with other NaCl-type uranium compounds, the heat capacity of UP in the present study is almost equal to that of US [13] from 300 to 700 K, and above that temperature becomes close to that of UN [14]. A deviation from the usual temperature dependence, similar to that of UP in the present study, has been observed in the heat capacity of UC [15]. However, the deviation in UC starts at a higher temperature, 1600 K, than in UP, and is considered to be caused by the formation of lattice defects [15].

4. RESOLUTION OF THE HEAT CAPACITY

4.1. Lattice contribution

For lack of information on the lattice dynamics from neutron inelastic diffraction studies on the NaCl-type uranium compounds (except for UC), the method of corresponding states has been used to evaluate the lattice heat capacity. Thus, Flotow et al. [16] and Danan et al. [17] estimated the lattice contribution to the heat capacity in US by utilizing the data available on the heat capacity of ThS. In the present study, the lattice heat capacity at constant pressure, C_L, of UP was estimated by using this method in connection with Lindemann's formula, in which the available heat capacity data of ThS [16] – instead of ThP – were used, no data being available on ThP. From Lindemann's formula and the literature values of the melting

TABLE II. THERMODYNAMIC FUNCTIONS OF URANIUM MONOPHOSPHIDE I
1 mol = 269.00 g

T (K)	C_p $(cal \cdot K^{-1} \cdot mol^{-1})$	S $(cal \cdot K^{-1} \cdot mol^{-1})$	$H - H_{298}$ $(cal \cdot mol^{-1})$	$-(G - H_{298})/T$ $(cal \cdot K^{-1} \cdot mol^{-1})$
298.15	12.02	18.71	0	18.71
300	12.04	18.78	22	18.71
350	12.43	20.67	635	18.86
400	12.68	22.35	1263	19.19
450	12.87	23.85	1902	19.63
500	13.03	25.22	2550	20.12
550	13.16	26.47	3204	20.64
600	13.28	27.62	3865	21.17
650	13.39	18.68	4532	21.71
700	13.51	29.68	5205	22.25
750	13.63	30.62	5883	22.77
800	13.75	31.50	6567	23.29
850	13.88	32.34	7258	23.80
900	14.01	33.13	7955	24.30
950	14.15	33.90	8659	24.78
1000	14.30	34.63	9370	25.25
1050	14.46	35.33	10090	25.72

TABLE III. THERMODYNAMIC FUNCTIONS OF URANIUM MONOPHOSPHIDE II
1 mol = 269.00 g

T (K)	C_p $(cal \cdot K^{-1} \cdot mol^{-1})$	T (K)	C_p $(cal \cdot K^{-1} \cdot mol^{-1})$
90	7.29	200	10.82
100	8.28	210	10.96
110	9.70	220	11.09
120	18.30	230	11.22
130	9.77	240	11.35
140	9.84	250	11.47
150	10.03	260	11.59
160	10.21	270	11.71
170	10.38	280	11.83
180	10.53	290	11.94
190	10.68		

point, the molar volume and the atomic mass of UP and ThS, the ratio of the characteristic temperature of UP to that of ThS was calculated as:

$$\Theta(UP)/\Theta(ThS) = 1/0.902 \quad (2)$$

The lattice heat capacity of UP was then derived from this relation by the following equation:

$$C_L(UP, T) = C_L(ThS, 0.902T) \quad (3)$$

which is almost equal to that of US estimated by Flotow et al. [16]. It should be noted that the agreement between the heat capacity calculated using Eq. (3) and that observed by Counsell et al. [4] at low temperatures was very good. Above room temperature, the most likely values were assumed for Debye temperature (430 K) and the dilation term, $C_p - C_v = AC_v^2T$, (where $A = 4.1 \times 10^{-6}$ $cal^{-1} \cdot mol$), and the evaluation of the lattice contribution was extended up to 1100 K. The excess heat capacity, $C_E = C_p - C_L$, was derived by subtracting the evaluated lattice contribution from the observed heat capacity, and is shown in Fig. 2.

4.2. Electronic contribution

It has been recognized [18] that the electronic state of 5f elements is more complicated than that for 3d, 4f elements. The band calculation for the NaCl-type uranium compounds [19] shows hybridized f-d bands contributing to the Fermi surface. However, as was indicated by Kuznietz [18], many features of the NaCl-type uranium compounds have been satisfactorily explained by following the localized 5f electron models: $5f^2$ (Grunzweig-Genossar et al. [20-23]); $5f^3$ (Chan and Lam [24, 25], Long and Wang [26]). In the present study, we estimated the excess heat capacity above the Néel temperature by the simple assumption of a localized 5f electron model.

As can be seen from Fig. 2, the observed excess heat capacity was high at all temperatures from the Néel temperature up to 1100 K. It seems reasonable to consider that this excess heat capacity (C_E) in the paramagnetic region is to be attributed to the conduction electron heat capacity (C_e) and the Schottky heat capacity of localized 5f electrons (C_s). The contribution of the tail of the magnetic heat capacity should also be considered.

The coefficient of electronic heat capacity, γ, of UP was estimated as $(0.8 \text{ to } 1.2) \times 10^{-3}$ $cal \cdot mol^{-1} \cdot K^{-2}$, by assuming that it is temperature independent and comparable with that of thorium compounds having NaCl-type structure: ($\gamma = (0.7 \text{ to } 1.0) \times 10^{-3}$ $cal \cdot mol^{-1} \cdot K^{-2}$ [16, 17, 27, 28]).

In an attempt to explain the rest of the excess heat capacity, we assumed crystalline field splitting, so as to get the best fit to the observed heat capacity in the case of the $5f^2$ (Γ_1 ground-state level) and $5f^3$ (Γ_8^1 or Γ_6 ground-state level) configurations. At first, the large C_E in the temperature range from Néel temperature to 250 K was considered. For the $5f^3$ (Γ_8^1) configuration, the analysis of crystalline field splitting was performed by subtracting the magnetic heat capacity, which was assumed as follows;

$$C_{mag}/cal \cdot mol^{-1} \cdot K^{-1} = 3.18 \times 10^4 \, (T/K)^{-2} \quad (4)$$

since C_E in this temperature range seems to be a magnetic rather than a

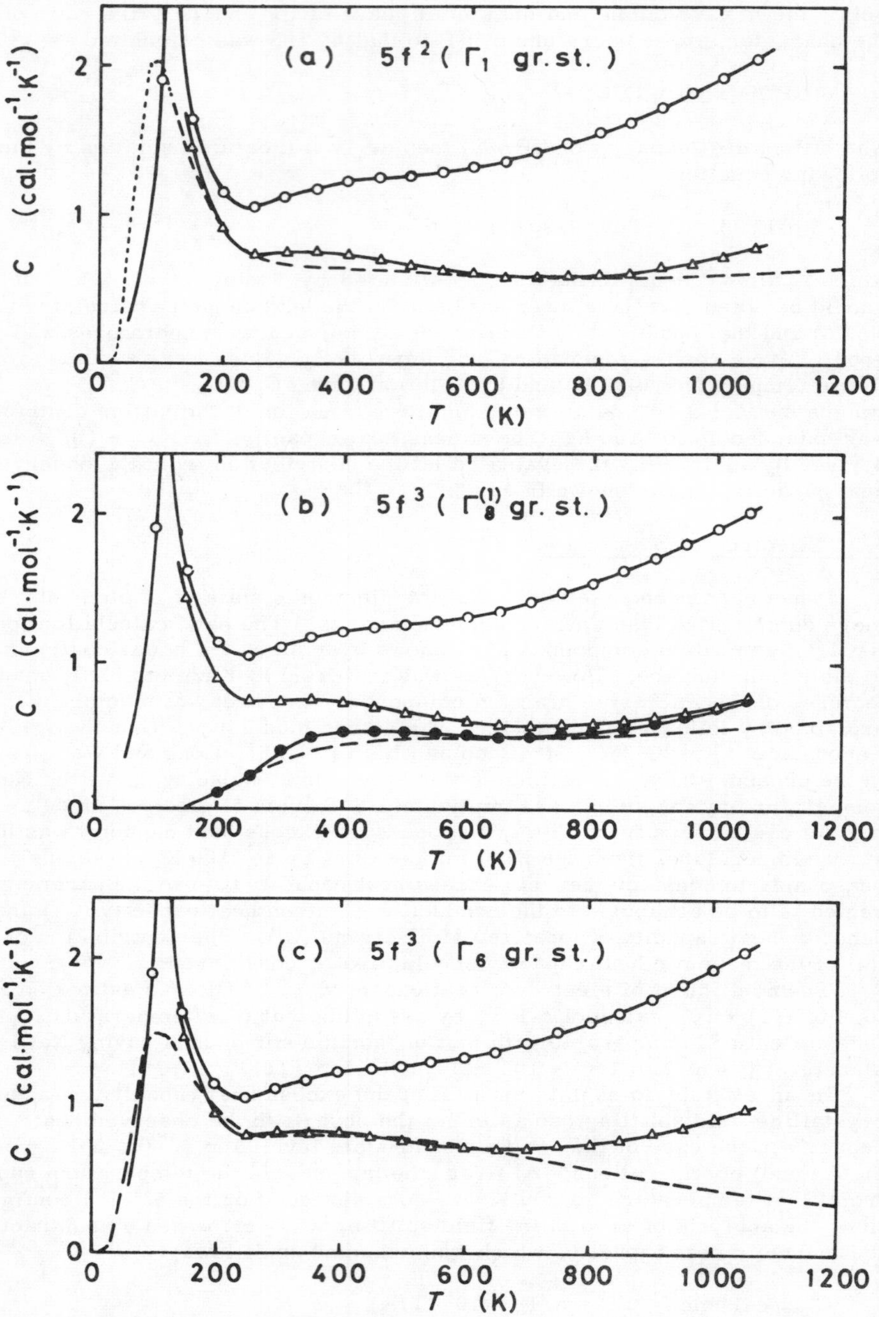

FIG.2. Resolution of excess heat capacity. —○—, gross excess heat capacity (C_E); —△—, C_E minus conduction electron heat capacity (C_e); —●—, C_E - C_e minus magnetic heat capacity (C_{mag}); ---, calculated Schottky heat capacity; (a) for the case of the $5f^2$ configuration (Γ_1 ground state level); (b) for $5f^3(\Gamma_8^1)$; (c) for $5f^3(\Gamma_6)$.

TABLE IV. RESULTS OF RESOLUTION OF EXCESS HEAT CAPACITY

configuration (ground state level)	(a) $5f^2(\Gamma_1)$		(b) $5f^3(\Gamma_8^1)$		(c) $5f^3(\Gamma_6)$	
γ ($cal \cdot mol^{-1} \cdot K^{-2}$)	1.2×10^{-3}		1.2×10^{-3}		1.0×10^{-3}	
E (footnote a)	Γ_1	0	Γ_8^1	0	Γ_6	0
	Γ_4	250	Γ_6	1100	Γ_8^1	280
	Γ_3	1300	Γ_8^2	4500	Γ_8^2	1400
	Γ_5	4500				
S_{mag} ($cal \cdot mol^{-1} \cdot K^{-1}$)	1.86		2.78		0.74	

[a] E is energy of crystalline field levels of ground multiplet presented in units of K.

Schottky heat capacity. In the cases of the $5f^2$ (Γ_1) and the $5f^3$ (Γ_6) configurations, on the other hand, C_E in this temperature range was explicable as consisting mainly of Schottky contributions. The results of the calculation in these three cases are shown in Table IV, and graphically in Fig. 2. At higher temperatures, the discrepancy between the experimental C_E and the calculated one is rather large, since only the crystalline field levels of the ground multiplet were considered in all cases. Unfortunately, we have not enough information about higher excited states, and if they were taken into account, the discrepancy might be decreased.

By using the values of C_e and C_s obtained with these models, we also estimated the entropy change associated with the magnetic transition at 121 K, for which the data available from Counsell et al. [4][1] below 90 K were also used. The results are also given in Table IV. In this estimation, C_e was not considered in the antiferromagnetic region, since its contribution in this region is still uncertain, as pointed out by Danan et al. [17]. These values were strongly dependent on the models of electronic levels. By following $5f^2$ model [20-23], in the paramagnetic region the ground state level is non-magnetic Γ_1, and the next higher level is magnetic Γ_4. In the magnetically ordered region, the Schottky term becomes the magnetic heat capacity in the presence of magnetic fields. Thus, the energy difference, $E(\Gamma_4) - E(\Gamma_1)$, can be calculated from the value of the entropy change [21]. This resulted in 235 K, when the entropy change of the transition at 22.5 K was not included. This value is consistent with the analytical value (250 K) shown in Table IV and also with the value (270 K) obtained from the magnetic properties by Grunzweig-Genossar [21]. This energy difference of UP is the same order as that of UN (210 K) [29] and US (315 K) [17], obtained by following the $5f^2$ model. However, it should be noted that in this model the total crystalline field splitting of the ground multiplet, J = 4, is considerably larger than was expected by Grunzweig-Genossar et al. [20].

[1] We should like to thank Dr. Martin for sending us the tables of their experimental values.

Recently, Lam and Fradin [30] studied magnetic and NMR properties of actinide monophosphides by applying the non-perturbative crystalline field theory [25, 26], and concluded that the electron configuration of UP would be $5f^3$. Their calculated value of the energy difference between the levels of the ground state Γ_8^1 and the first excited Γ_6 was 1380 K, and the value of the density of states was 0.83 states/eV spin (i.e. $\gamma = 0.95 \times 10^{-3}$ cal · mol^{-1} · K^{-2}). Their values shown above are in good agreement with values in the case of $5f^3(\Gamma_8^1)$ in the present study, $E(\Gamma_6) - E(\Gamma_8^1) = 1100$ K, $\gamma = 1.2 \times 10^{-3}$ cal · mol^{-1} · K^{-2}. In addition, the calculated value of the entropy change of the transition at 121 K became 2.78 cal · mol^{-1} · K^{-1} in this model, which agreed fairly well with R log4 = 2.76 cal · mol^{-1} · K^{-1}.

We also considered the case of $5f^3$ (Γ_6 ground state). In this case, the crystalline field splitting was analysed on the assumption that the peak of the Schottky term would be close to that of C_{mag}, and the total crystalline field splitting of ground multiplet obtained became 1400 K. At present, we have too little information to proceed further with a discussion of this model.

In conclusion, the results of the resolution showed that the cause of the excess heat capacity of UP from 80 to 1100 K could be explained by both the simple localized $5f^2$ and $5f^3$ configuration models. The slightly anomalous behaviour in the heat capacity of UP at high temperatures might be attributed to the Schottky heat capacity. Further discussion on the specific electronic configuration of a uranium ion in UP from heat capacity data are limited because of the inevitable uncertainty in the estimation of lattice heat capacity. It would be necessary to examine the other properties of UP, such as spin-spin interaction above the Néel temperature and its relationship with the heat capacity, and also to observe the crystalline field levels of the localized state directly using inelastic neutron diffraction, as has been done for the rare-earth compounds [31]. The developments of the theoretical treatments, such as the band approach [19, 32], and the new theory proposed recently by Robinson and Erdös [33], will also contribute to the further understanding of the NaCl-type uranium compounds.

ACKNOWLEDGEMENTS

We are grateful to Dr. Y. Akomoto, Central Research Laboratory of Mitsubishi Metal Corporation for providing us with the sample. The partial financial support of the Toray Science Foundation is also gratefully acknowledged.

REFERENCES

[1] MEN'SHIKOVA, T.S., et al., Proc. 4th Int. Conf. Peaceful Uses At. Energy (Proc. Conf. Geneva, 1971) 10, UN, New York & IAEA, Vienna (1972) 217.
[2] TAKAHASHI, Y., J. Nucl. Mater. 51 (1974) 17.
[3] TAKAHASHI, Y., WESTRUM, E.F., Jr., J. Chem. Thermodyn. 2 (1970) 847.
[4] COUNSELL, J.F., DELL, R.M., JUNKINSON, A.R., MARTIN, J.F., Trans. Faraday Soc. 63 (1967) 72.
[5] MOSER, J.B., KRUGER, O.L., J. Appl. Phys. 38 (1967) 3215.
[6] BRUGGER, J.E., USAEC Rep. ANL-7175 (1967) 142.
[7] ONO, F., KANNO, M., MUKAIBO, T., J. Nucl. Sci. Technol. 10 (1973) 764.
[8] CARR, S.L., LONG, C., MOULTON, W.G., KUZNIETZ, M., Phys. Rev. Lett. 23 (1969) 786.
[9] GULICK, J.M., MOULTON, W.G., Phys. Lett., A35 (1971) 429.
[10] ADACHI, H., IMOTO, S., KUKI, T., Phys. Lett., A44 (1973) 491.

[11] KAMIMOTO, M., TAKAHASHI, Y., MUKAIBO, T., J. Nucl. Sci. Technol. 11 (1974) 158.
[12] WARREN, I.H., PRICE, C.E., Can. Metall. Q. 3 (1964) 183.
[13] MACLEOD, A.C., J. Inorg. Nucl. Chem. 33 (1971) 2419.
[14] OETTING, F.L., LEITNAKER, J.M., J. Chem. Thermodyn. 4 (1972) 199.
[15] OETTING, F.L., NAVRATIL, J.D., STORMS, E.K., J. Nucl. Mater. 45 (1972/73) 271.
[16] FLOTOW, H.F., OSBORNE, D.W., WALTERS, R.R., J. Chem. Phys. 55 (1971) 880.
[17] DANAN, J., GRIVEAUN, B., MARCON, J.P., GATESONPE, J.P., DE NOVION, C.H., Rare Earths and Actinides (Conf. Digest, Durham, 1971) 3, The Institute of Physics, London and Bristol (1971) 176.
[18] KUZNIETZ, M., Rare Earths and Actinides (Conf. Digest, Durham, 1971) 3, The Institute of Physics, London and Bristol (1971) 162.
[19] DAVIS, H.L., Plutonium and Other Actinides (Pro. Conf. Santa Fe, 1970) 1 1970, Nucl. Metall. 17, The Metallurgical Society of AIME, New York (1970) 209.
[20] GRUNZWEIG-GENOSSAR, J., KUZNIETZ, M., FRIEDMAN, F., Phys. Rev. 173 (1968) 562.
[21] GRUNZWEIG-GENOSSAR, J., Solid State Commun. 8 (1970) 1673.
[22] GRUNZWEIG-GENOSSAR, J., FIBICH, M., Rare Earths and Actinides (Conf. Digest, Durham, 1971) 3, The Institute of Physics, London and Bristol (1971) 101.
[23] GRUNZWEIG-GENOSSAR, J., CAHN, J.W., Int. J. Magnetism 4 (1973) 193.
[24] CHAN, S.K., LAM, D.J., Plutonium and other Actinides (Proc. Conf. Santa Fe, 1970) 1, Nucl. Metall. 17, The Metallurgical Society of AIME, New York (1970) 219.
[25] CHAN, S.K., LAM, D.J., Rare Earths and Actinides (Conf. Digest, Durham, 1971) 3, The Institute of Physics, London and Bristol (1971) 92.
[26] LONG, C., WANG, Y.L., Phys. Rev., B 3 (1971) 1656.
[27] HARNESS, J.B., MATTHEWS, J.C., MORTON, N., Brit. J. Appl. Phys. 15 (1964) 963.
[28] DANAN, J., DE NOVION, C.H., DALLAPORTA, H., Solid State Commun. 173 (1972) 775.
[29] DE NOVION, C.H., C.R., Ser. B 273 (1971) 26.
[30] LAM, D.J., FRADIN, F.Y., Phys. Rev., B 9 (1974) 238.
[31] BIRGENEAU, R.J., BUCHER, E., MAITA, J.P., PASSELL, L., TURBERFIELD, K.C., Phys. Rev., B 8 (1973) 15.
[32] ADACHI, H., IMOTO, S., J. Nucl. Sci. Technol. 6 (1969) 371.
[33] ROBINSON, J.M., ERDÖS, P., Phys. Rev., B 8 (1973) 4333; B 9 (1974) 2187.

DISCUSSION

A.S. PANOV: You said that the samples chosen for investigation were very pure. But could you state precisely how much oxygen there was in the UP and what was the initial stoichiometry?

Y. TAKAHASHI: The oxygen content was 500 ppm, as determined by chemical analysis, and the initial composition of the sample was $UP_{1.027}$.

A.S. PANOV: Did you register a change in the composition of the UP during the experiment?

Y. TAKAHASHI: No, we did not.

A.S. PANOV: Since your composition corresponded to the formula mentioned, did you not observe a second phase, for example UP_2, in the uranium monophosphide?

Y. TAKAHASHI: No, the sample was single phase and no second phase was detected.

E.F. WESTRUM, Jr.: In further defence of your sample purity, I should like to comment that the presence of a separate-phase UP_2 contaminant could readily have been detected by the Néel anomaly in the heat capacity. No such phase was observed. I might add that I was very much impressed by the ingenuity of your interpretation.

Since laser-flash calorimetry appears moreover to be exceedingly useful for investigations of thermophysical properties of nuclear materials, I should like to ask you for details of the sample requirements, the duration of the measurements and the lower as well as the upper temperature limits associated with the technique.

Y. TAKAHASHI: The optimum dimensions of the samples are a diameter of approximately 10 mm and a thickness of 1 to 5 mm. Thus the mass of the sample required is 0.3 to 5 grams. The time required for a single determination of C_p at a given temperature is less than 5 minutes, and therefore it takes approximately two days to go from 80 to 1100 K.

As for the temperature limits, I do not anticipate any difficulty in extending the measurements to the liquid helium temperature. At temperatures above 1100 K, however, there may still be some difficulties in obtaining reliable results, for example, the effects of radiation heat loss.

ТЕРМОДИНАМИЧЕСКИЕ СВОЙСТВА МОНОКАРБИДОВ ПЕРЕХОДНЫХ МЕТАЛЛОВ

В.Н.ЗАГРЯЗКИН, А.С.ПАНОВ, Е.В.ФИВЕЙСКИЙ
Физико-энергетический институт,
Обнинск,
Союз Советских Социалистических Республик

Abstract–Аннотация

THERMODYNAMIC PROPERTIES OF MONOCARBIDES OF TRANSITION METALS.

On the basis of published data and using the methods of statistical thermodynamics, the authors demonstrate the possibility of calculating partial thermal functions of transition metal monocarbide components. They calculate the temperature dependences of the congruent vaporization of zirconium, hafnium and niobium carbide. Questions of calculating the thermodynamic properties of complex carbides are considered in the regular approximation. The results are used in the thermodynamic analysis of a fuel composition based on uranium monocarbide with tungsten.

ТЕРМОДИНАМИЧЕСКИЕ СВОЙСТВА МОНОКАРБИДОВ ПЕРЕХОДНЫХ МЕТАЛЛОВ.

На основании литературных данных с использованием метода статистической термодинамики показана возможность расчета парциальных термодинамических функций компонент монокарбидов переходных металлов. Проведен расчет температурной зависимости состава конгруэнтного испарения карбидов циркония, гафния и ниобия. Рассмотрены в регулярном приближении вопросы расчета термодинамических свойств сложных карбидов. Полученные результаты использованы для термодинамического анализа взаимодействия топливной композиции на основе монокарбида урана с вольфрамом.

Монокарбиды переходных металлов имеют широкие области гомогенности. Отклонения от стехиометрических составов обусловлены наличием структурных вакансий в углеродной подрешетке. В металлической подрешетке структурных вакансий не обнаружено [1,2]. Наличие структурных дефектов оказывает существенное влияние на термодинамические свойства монокарбидов. Для описания термодинамических свойств нестехиометрических монокарбидов существует несколько модельных подходов. Нами с использованием известных приемов статистической термодинамики [3] записаны выражения концентрационной и температурной зависимостей активности компонент нестехиометрических карбидов MeC_z в следующем виде [4]:

$$RT \ln a_1 = F_1 + 0{,}5F_{11} - 0{,}5\,F_{22} \cdot z^2 + RT \ln(1-z),\ \frac{\text{ккал}}{\text{г·атом}} \qquad (1)$$

$$RT \ln a_2 = F_2 + F_{12} + F_{22} \cdot z + RT \ln\left(\frac{z}{1-z}\right),\ \frac{\text{ккал}}{\text{г·атом}} \qquad (2)$$

где индекс "1" относится к металлу;
индекс "2" – к углероду;
a_1 и a_2 – активности металла и углерода;

F_1 и F_2 – свободные энергии перевода 1 г/атома вещества из стандартного состояния в решетку монокарбида;
F_{11} и F_{22} – свободные энергии парного взаимодействия атомов;
z – стехиометрический коэффициент;
R – газовая постоянная;
T – абсолютная температура.

В качестве стандартных состояний выбраны чистые металлы и графит.

Разделяя F_{ij} на энтальпийную и энтропийную части и определяя численные значения указанных параметров на основании данных [1], была показана возможность использования фазовых диаграмм металл-углерод для вычисления концентрационной и температурной зависимостей термодинамических характеристик в области гомогенности монокарбидов переходных металлов [4]. Такие расчеты были выполнены для карбидных систем TiC, ZrC, HfC. Точность расчетов в данном случае определяется в первую очередь точностью используемой фазовой диаграммы Me-C.

В настоящее время экспериментально наиболее полно изучены термодинамические свойства монокарбида циркония [1,5-7]. Сравнение экспериментальных результатов с оцененными нами [4] значениями активностей компонент в области гомогенности ZrC показывает, что результаты расчета и экспериментов удовлетворительно согласуются.

Расчет концентрационной зависимости парциальных теплот испарения циркония и углерода в области гомогенности ZrC_z, выполненный на основании данных [7], показал, что расчетные и экспериментальные результаты согласуются во всем интервале составов. На верхней и нижней фазовых границах ZrC_z наблюдаются скачкообразные изменения величин $\Delta\overline{H}_i$, причина появления которых заключается в температурной зависимости фазовых границ.

Согласно работе [4], выражения для теплоты испарения металлической и углеродной компонент в зависимости от состава карбидной фазы и температуры имеют следующий вид:

$$\Delta\overline{H}^{\text{исп}}_{\text{Me}[x,T]} = -\Delta H^{o}_{\text{Me}} + E_1 + 0{,}5E_{11} - 0{,}5E_{22}\left(\frac{x}{1-x}\right)^2 + \\ + [RT^2(1-x)^{-1}(1-2x)^{-1} + F_{22}\cdot T\cdot x(1-x)^{-3}]\cdot\frac{dx}{dT}, \frac{\text{ккал}}{\text{г·атом}} \quad (3)$$

$$\Delta\overline{H}^{\text{исп}}_{\text{C}[x,T]} = -\Delta H^{o}_{\text{C}} + E_1 + E_{12} + E_{22}\left(\frac{x}{1-x}\right) + \\ + [RT^2\cdot x^{-1}(1-2x)^{-1} + F_{22}\cdot T\cdot(1-x)^2]\cdot\frac{dx}{dT}, \frac{\text{ккал}}{\text{г·атом}} \quad (4)$$

где E_{ij} - энергии парного взаимодействия;
ΔH^{o}_{i} - теплоты испарения чистых металлов и углерода;
x - мольная доля углерода в карбиде.

Появление скачков на фазовых границах обязано членам уравнений (3-4), содержащим сомножителем величину $\frac{dx}{dT}$. Для составов внутри области гомогенности $\frac{dx}{dT}=0$ и уравнения (3-4) существенно упрощаются.

Выражения (3-4) были использованы нами ранее [4] при анализе экспериментальных данных [8] для монокарбида ниобия. С помощью выражений (3-4) экспериментальные данные [8] по концентрационной зависимости парциальных теплот испарения ниобия и углерода получили убедительное объяснение.

Факт наличия скачкообразных изменений в значениях парциальных теплот испарения, полученных расчетным путем и подтвержденных экспериментально для системы ZrC_z, является, на наш взгляд, еще одним доказательством правомерности предложенного метода расчета термодинамических характеристик на основании анализа диаграмм состояния металл-углерод [1].

Подход, развитый в работе [4] и подтвержденный экспериментально для систем цирконий-углерод, ванадий-углерод [7] и ниобий-углерод [8], позволяет проводить вычисления активностей еще целого ряда термодинамических характеристик, как например, концентрационную и температурную зависимости свободных энергий и теплот образования монокарбидов и т.п.

Представляет интерес использовать данные [4] для вычисления температурной зависимости составов конгруэнтного испарения монокарбидов.

Согласно работе [1] монокарбиды титана и ванадия любого состава при нагреве теряют преимущественно металлическую составляющую и не имеют конгруэнтно испаряющихся составов. Конгруэнтно испаряющийся состав в системе тантал-углерод не установлен, но предполагается, что он слабо зависит от температуры и находится несколько ниже $TaC_{0,5}$ [1]. Поэтому расчет температурной зависимости составов конгруэнтного испарения проведен для монокарбидов циркония, гафния и ниобия.

Приравнивая отношение скоростей испарения отношению мольных долей компонент, для монокарбида MeC_z можно записать следующее условие конгруэнтного испарения

$$\frac{1}{z} = \frac{a_1 \cdot a_2}{P_1^o \cdot P_2^o} \cdot \left(\frac{A_1}{A_2}\right)^{-1/2} \tag{5}$$

где a_1 и a_2 - активности металла и углерода в карбиде;
P_1^o и P_2^o - упругости паров чистых металла и углерода при данной температуре;
A_1 и A_2 - атомные веса компонент паровой фазы.

Логарифмируя правую и левую часть выражения (5), умножая на RT и подставляя (1) и (2) в (5) получим

$$2RT\ \ln(1-z) - F_{22}(0{,}5z^2+z) = (F_2+F_{12}) - (F_1+0{,}5F_{11}) +$$
$$+RT\left[(\ln P_2^o - \ln P_1^o) - 0{,}5(\ln A_2 - \ln A_1)\right] \tag{6}$$

Здесь z имеет смысл состава конгруэнтного испарения для заданнной температуры. Учитывая, что для заданной температуры правая часть соотношения (6) есть константа, и решая уравнение (6) относительно z, можно получить температурную зависимость составов конгруэнтного испарения.

Решение трансцендентного уравнения (6) проводилось методом итерации для трех произвольно выбранных температур: 2000; 2500:и 3000°K. Необходимые для расчетов значения F_{ij} и P_i^o заимствованы из работ [4,9].

Определенные таким образом значения составов конгруэнтного испарения $z_{[T]}^{конгр}$ в области гомогенности монокарбидов циркония, гафния и ниобия представлены в табл.I. Возможная ошибка в значениях величин $z_{[T]}^{конгр}$ оценена равной ±0,02.

ТАБЛИЦА I. ТЕМПЕРАТУРНАЯ ЗАВИСИМОСТЬ СОСТАВОВ КОНГРУЭНТНОГО ИСПАРЕНИЯ КАРБИДНЫХ ФАЗ

	$Z_{[T]}^{конгр}$		
	2000°K	2500°K	3000°K
ZrC	0,91	0,89	0,86
HfC	0,87	0,86	0,83
NbC	0,86	0,84	0,81

Данные настоящего расчета значений $z_{[T]}^{конгр}$ для карбида циркония согласуются с экспериментальными данными [7]. Расхождения наблюдаются при сравнении данных настоящего расчета для монокарбидов циркония, гафния и ниобия с результатами расчета [10], что связано с рядом недостатков модели Л.Кауфмана [10], в которой парциальная теплота испарения компонент не зависит от состава.

Таким образом, на примере карбида циркония подтверждена правомерность рассмотрения монокарбидов переходных металлов как объединенного ансамбля частиц [3] и возможность использования диаграмм состояния металл-углерод для расчета термодинамических свойств карбидных фаз в зависимости от состава и температуры [4].

Представляет интерес использование термодинамических данных для анализа процессов взаимодействия в системе карбид-тугоплавкий металл. Например, применительно к проблемам ядерного металловедения, практически важным является рассмотрение взаимодействия топливной композиции на основе монокарбида урана с материалом оболочки.

Рассмотрим реакцию взаимодействия монокарбида урана, легированного карбидом циркония, с вольфрамом

$$[U_{1-y}\,Zr_y]C_z + 2\Delta z \cdot W \rightarrow \Delta z \cdot W_2\,C + [U_{1-y}\,Zr_y]C_{z-\Delta z} \qquad (7)$$

Здесь

z - исходный состав топливной композиции;
Δz - изменение состава в результате взаимодействия;
y - мольная доля легирующей добавки.

В результате взаимодействия в зоне контакта топливная композиция – оболочка может образоваться карбид вольфрама W_2C. Как обычно, вероятность протекания реакции (7) определяется знаком и величиной изменения свободной энергии реакции. Отсутствие взаимодействия в системе $[U_{1-y}Zr_y]C_z - W$ или невозможность протекания реакции (7) определяется соотношениями

$$\Delta z \cdot \Delta G_{W_2C} + \Delta G_{[U_{1-y}Zr_y]C_z} - \Delta G_{[U_{1-y}Zr_y]C_{z-\Delta z}} \geq 0 \quad (8)$$

или

$$\Delta G_{W_2C} \geq \frac{\Delta G_{[U_{1-y}Zr_y]C_z} - \Delta G_{[U_{1-y}Zr_y]C_{z-\Delta z}}}{\Delta z} \quad (9)$$

где ΔG - свободная энергия образования соответствующих соединений.

При $\Delta z \to 0$ в правой части неравенства (9) получаем значение производной свободной энергии образования по составу при постоянной температуре. Дифференцируя выражение для свободной энергии сложного карбида $[U_{1-y}Zr_y]\cdot C_z$ по составу и сравнивая величину производной со значением ΔG_{W_2C}, можно сделать априорное заключение о возможности взаимодействия и определить температуру его начала.

Выражение для свободной энергии образования $[U_{1-y}Zr_y]\cdot C_z$ записано следующим образом

$$\Delta G_{[U_{1-y}Zr_y]C_z} = RT\,[(1-y)\ln a_U + y\cdot \ln a_{Zr} + z\ln a_C] \quad (10)$$

где a_i - активности компонент в сложном карбиде. Связь между активностями компонент в сложном карбиде $[U_{1-y}Zr_y]C_z$ и монокарбидах UC_z и ZrC_z задавалась в форме, представленной в работе [11]. Выражения для активностей компонент в монокарбиде урана представлены в форме уравнений (1-2), определение значений констант в которых проведено с использованием экспериментальных данных [1].

Учитывая, что $z = x(1-x)^{-1}$, где x - мольная доля углерода в карбиде, дифференцирование следует проводить по мольной доле.

Окончательное выражение скорости изменения свободной энергии образования $[U_{1-y}Zr_y]C_z$ с составом записывается следующим образом

$$\frac{\partial(\Delta G_{[U_{1-y}Zr_y]\cdot C_z})}{\partial x} = RT\,[(1-y)(\ln a_C - \ln a_U) + y(\ln a_C - \ln a_z) - \frac{y(1-y)}{RT}\cdot L - y\cdot \ln y - (1-y)\ln(1-y)],\ \frac{\text{ккал}}{\text{г}\cdot\text{атом}} \quad (11)$$

где L - энергия взаимообмена в псевдобинарной системе UC - ZrC, численное значение которой определено в работе [12].

Учитывая, что активность углерода в сложном карбиде $U_{1-y}Zr_yC_z$ подчиняется закону аддитивности [11] и является функцией активностей углерода в UC и ZrC, легирование монокарбида урана карбидам циркония предельной стехиометрии следует считать нецелесообразным, поэтому сопоставление свободных энергий проведено нами для составов, отличающихся от стехиометрического.

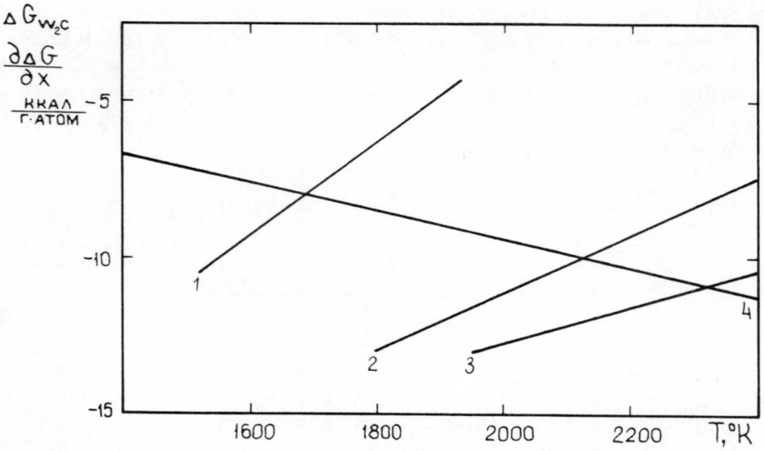

Рис.1. Сопоставление свободных энергий образования и определение температур начала взаимодействия в системе топливная композиция-вольфрам.
1 - $\frac{\partial \Delta G}{\partial x}$ для монокарбида урана $UC_{0,99}$; 2 - $\frac{\partial \Delta G}{\partial x}$ для твердого раствора $[U_{0,9}\ Zr_{0,1}]C_{0,90}$; 3 - $\frac{\partial \Delta G}{\partial x}$ для твердого раствора $[U_{0,9}\ Zr_{0,1}]C_{0,86}$; 4 - свободная энергия образования W_2C.

На рис.1 представлена в графической форме схема определения температуры начала взаимодействия в системе $U_{0,9}Zr_{0,1}C_z$ - W. Расчет температурной зависимости производной свободной энергии образования по составу выполнен нами для двух составов сложного карбида z = 0,9 (прямая 2) и z = 0,86 (прямая 3). Сведения о температурной зависимости свободной энергии образования W_2C цитируются по данным [9]. Для сравнения приведена температурная зависимость производной свободной энергии образования по составу для монокарбида урана состава $UC_{0,99}$.

Как видно из рис.1, точки пересечения прямых (1), (2) и (3) с прямой температурной зависимости свободной энергии образования W_2C (4) характеризуют температуры начала взаимодействия в системах топливная композиция - вольфрам, согласно выражению (9). Легирование монокарбида урана 10 % мольными ZrC и снижение стехиометрии по углероду повышают температуру начала взаимодействия примерно на 500° по сравнению с чистым монокарбидом урана. Так, температура начала взаимодействия в системе $U_{0,9}Zr_{0,1}C_{0,9}$ - W составляет 2150°K, а в системе $U_{0,9}Zr_{0,1}C_{0,86}$ -W — 2300°K. Точность определения соответствующих температур оценена равной ± 50°.

Проведенные здесь величины температур начала взаимодействия не являются окончательными из-за неопределенностей в используемых в расчетах термодинамических характеристиках. Однако представляется, что априорные оценки температур начала взаимодействия с помощью критерия (9) достаточно надежны и могут быть использованы не только для системы карбид-вольфрам, но и для других систем.

ЛИТЕРАТУРА

[1] СТОРМС, Э., Тугоплавкие Карбиды, Атомиздат, М., 1970.

[2] ГОРБУНОВ, Н.С. и др., Известия АН СССР 11 (1961) 2093.

[3] ФАУЛЕР, Р., ГУГГЕНГЕЙМ, Е.А., Статистическая Термодинамика, ИЛ, М., 1949.

[4] ЗАГРЯЗКИН, В.Н. и др., Журнал Физической Химии 47 8 (1973) 1946-1951.

[5] BOWMAN, M., Nuclear Application of Nonfissionable Ceramics (Proc. Conf. Illinois, 1964) ANS 267.

[6] АНДРИЕВСКИЙ, Р.А. и др., Доклады АН СССР 206 4 (1972) 896.

[7] STROMS, E.K., GRIFFIN, J., High Temp. Sci. 5 (1973) 291.

[8] STROMS, E.K., et.al., High Temp. Sci. 1 (1969) 430.

[9] SCHICK, H., Thermodynamics of Certain Refractory Compounds, Acad. Press. New-York (1966).

[10] KAUFMAN, L., SARNEY, A., Compounds of Interest in Nuclear Reactor Technology, AIME, Michigan (1964) 267.

[11] De POORTER, G., WALLACE, T., "Fundamentals of Refractory Compounds", (Hausner H.H., Bowman M.G., Eds), Plenum Press, New York (1968) 1.

[12] АНДРИЕВСКИЙ, Р.А. и др., Thermodynamics of Nuclear Materials (Proc. Symp. Vienna, 1967) IAEA (1968) 449.

CALORIMETRIC STUDIES ON ACTINIDE COMPOUNDS AND MATERIALS OF INTEREST IN REACTOR TECHNOLOGY*

P.A.G. O'HARE, M. ADER, W.N. HUBBARD,
G.K. JOHNSON, J.L. SETTLE
Argonne National Laboratory,
Argonne, Ill.,
United States of America

Abstract

CALORIMETRIC STUDIES ON ACTINIDE COMPOUNDS AND MATERIALS OF INTEREST IN REACTOR TECHNOLOGY.

Standard enthalpies of formation, ΔH_f^0, at 298.15 K, based on solution-reaction or fluorine bomb calorimetric measurements are reported for the following compounds: α-Li_2UO_4, K_2UO_4, Rb_2UO_4, $Cs_2U_2O_7$, $Cs_2Mo_2O_7$, Cs_2CrO_4, Li_3N, Li_2O, $ThS_{1.029}$, $MoC_{0.502}$, $MoC_{0.651}$, U_3Si, and β-US_2. The results are compared with earlier determinations and the significance of some of the data vis-à-vis the chemistry of processes in irradiated fuel pins is discussed.

1. INTRODUCTION

Recent calorimetric research in the Chemical Engineering Division at Argonne National Laboratory has been concerned with materials whose chemistry bears on a variety of nuclear reactor processes. This research yielded standard enthalpies of formation at 298.15 K, $\Delta H^\circ_{f\,298.15}$, and high-temperature (to 1500 K) thermodynamic properties for a number of compounds. Parallel studies by D. W. Osborne and H. E. Flotow of the Chemistry Division at Argonne gave thermodynamic properties from 5 to 350 K. Examples of previously reported investigations include those on (a) Na_3UO_4[1], α-Na_2UO_4[2], and $NaUO_3$[3], of interest in connection with the interaction of oxide fuel and liquid sodium coolant; (b) Cs_2MoO_4[4], Rb_2MoO_4[5], $BaMoO_4$[6], and Cs_2O[7], all of which apparently form in irradiated fuel pins; (c) Cs_2UO_4[8], which evidently plays a significant role in the swelling of fuel rods; and (d) Pu_2C_3[9], which may be an important material in the realm of advanced fuels.

The objective of this paper is to present in succinct form some of our hitherto unpublished enthalpies of formation for α-Li_2UO_4, K_2UO_4, Rb_2UO_4, $Cs_2U_2O_7$, $Cs_2Mo_2O_7$, Cs_2CrO_4, Li_3N, Li_2O, $ThS_{1.029}$, $MoC_{0.502}$, $MoC_{0.651}$, U_3Si, and β-US_2. Implications of the results are also briefly discussed. The calorimetric data given are in condensed form; however, detailed information on the experiments and calculations can be obtained from the authors upon request.

2. GENERAL EXPERIMENTAL PROCEDURES

Enthalpies of formation are based on either solution-reaction calorimetry in aqueous media or fluorine bomb calorimetry. The solution calorimeter was a commercial unit (LKB-8700 Precision Calorimetric System),

* Work performed under the auspices of the USAEC.

TABLE I. REACTION SCHEME AND ENTHALPY VALUES[a] FOR CALCULATIONS OF $\Delta H_f^0(M_2UO_4)$

	Reaction			ΔH(kcal) M=Li	ΔH(kcal) M=K	ΔH(kcal) M=Rb
1.	$(2M^+ \cdot UO_2^{2+} \cdot 4Cl^-) \cdot 183.80\ (HCl \cdot 55.572\ H_2O)$	=	$M_2UO_4(c) + 187.80\ (HCl \cdot 54.378\ H_2O)$	41.77 ± 0.02[b]	42.07 ± 0.05[b]	41.30 ± 0.05[b]
2.	$UO_2Cl_2(c) + 183.80\ (HCl \cdot 55.572\ H_2O)$	=	$(UO_2^{2+} \cdot 2Cl^-) \cdot 183.80\ (HCl \cdot 55.572\ H_2O)$	-22.49 ± 0.13[b]	-22.49 ± 0.13[b]	-22.49 ± 0.13[b]
3.	$2MCl(c) + (UO_2^{2+} \cdot 2Cl) \cdot 183.80\ (HCl \cdot 55.572\ H_2O)$	=	$(2M^+ \cdot UO_2^{2+} \cdot 4Cl^-) \cdot 183.80\ (HCl \cdot 55.572\ H_2O)$	-16.38 ± 0.03[b]	8.61 ± 0.03[b]	8.30 ± 0.02[b]
4.	$4\ (HCl \cdot 55.572\ H_2O)$	=	$2\ H_2(g) + 2Cl_2(g) + 222.288\ H_2O(l)$	158.11 ± 0.09[c]	158.11 ± 0.09[c]	158.11 ± 0.09[c]
5.	$187.80\ (HCl \cdot 54.378\ H_2O) + 224.233\ H_2O(l)$	=	$187.80\ (HCl \cdot 55.572\ H_2O)$	-1.05 ± 0.05[d]	-1.05 ± 0.05[d]	-1.05 ± 0.05[d]
6.	$U(c) + O_2(g) + Cl_2(g)$	=	$UO_2Cl_2(c)$	-291.37 ± 0.81[e]	-291.37 ± 0.81[e]	-291.37 ± 0.81[e]
7.	$2H_2(g) + O_2(g)$	=	$2H_2O(l)$	-136.63 ± 0.02[f]	-136.63 ± 0.02[f]	-136.63 ± 0.02[f]
8.	$2M^+(aq, m{=}1) + 2Cl^-(aq, m{=}1)$	=	$2MCl(c)$	17.70 ± 0.10[g]	-8.23 ± 0.02[g]	-8.26 ± 0.15[g]
9.	$2M(c)$	=	$2M^+(aq, m{=}1)$	-133.10 ± 0.04[h]	-120.54 ± 0.05[h]	$-120.04 \pm -.06$[h]
10.	$Cl_2(g)$	=	$2Cl^-(aq, m{=}1)$	$-79.87 + 0.04$[h]	$-79.87 + 0.04$[h]	$-79.87 + 0.04$[h]
11.	$2M(c) + U(c) + 2O_2(g) = M_2UO_4(c)$			-463.31 ± 0.84	-451.39 ± 0.83	-452.00 ± 0.85

[a] Unless designated otherwise, all species are aqueous.

[b] Measured in present study.

[c] Calculated from $\Delta H_f^\circ(Cl^-, aq, m{=}1)$ [13] and dilution data for HCl [14].

[d] Dilution data from [14].

[e] $\Delta H_f^\circ(UO_2Cl_2, c)$ from [1].

[f] $\Delta H_f^\circ(H_2O, l)$ from [13].

[g] Enthalpies of solution from [14].

[h] ΔH_f° values for $(M^+, aq, m{=}1)$ and $(Cl^-, aq, m{=}1)$ from [15] and [13], respectively.

and the fluorine bomb equipment has been described in other publications from this Laboratory [10]. In both systems, temperatures were measured with an accuracy of 1×10^{-4} K by means of a quartz-crystal thermometer (Hewlett-Packard Model 2801A). All air- or moisture-sensitive materials were handled exclusively in a helium-atmosphere glovebox in which the H_2O and O_2 levels were customarily below 2 ppm by volume.

3. SOLUTION-REACTION CALORIMETRIC RESULTS

3.1. Monouranates of Li, K, and Rb

Ippolitova, Faustova, and Spitsyn [11] described a thermochemical study of the five alkali metal monouranates, M_2UO_4. They deduced $\Delta H_f^\circ(M_2UO_4)$ from calorimetric measurements of the enthalpies of reaction with aqueous 1M HCl. Recently, however, results obtained by Cordfunke and Loopstra [12] for α-Na_2UO_4 and by the present authors for both α-Na_2UO_4 [2] and Cs_2UO_4 [8] have disagreed with those reported by Ippolitova et al. Consequently, it was decided to redetermine ΔH_f° for Rb_2UO_4, α-Li_2UO_4 and K_2UO_4.

Our immediate interest was Rb_2UO_4 which, by analogy with Cs_2UO_4 [8], could form from fission-product rubidium and the urania-plutonia in reactor fuel. Hence, a reliable determination of $\Delta H_f^\circ(Rb_2UO_4)$ could help to unravel the complex chemistry of irradiated fuels. Although neither Li_2UO_4 nor K_2UO_4 is involved in reactor chemistry, we decided to measure their enthalpies of formation in order to complete a careful, systematic thermochemical study on pure, well-characterized samples of the alkali metal monouranates.

The monouranates were synthesized by heating Li_2CO_3, K_2CO_3 or $RbNO_3$ with U_3O_8 in air. Chemical analyses for the alkali metals and uranium, and emission spectrochemical analyses for trace metals, gave the following M/U ratios: 1.997 ± 0.005, 1.999 ± 0.005, and 1.999 ± 0.005 for the lithium, potassium, and rubidium salts, respectively. The maximum amount of trace metal impurities in any of the preparations was 0.048 mass per cent.

Enthalpies of reaction of the monouranates with ~1M HCl(aq), and enthalpies of solution of UO_2Cl_2 and the alkali metal chlorides are given in Table I. This table also includes the auxiliary thermochemical data used to derive the values for $\Delta H_f^\circ(M_2UO_4)$.

3.2. Cesium Diuranate

Swelling and the resultant failure of irradiated fuel pins have been attributed [16] to cesium monouranate formed by the reaction between fission-product cesium and oxide fuel. It is known [17] that at elevated temperatures Cs_2UO_4 converts to the diuranate, $Cs_2U_2O_7$, according to the reaction

$$2Cs_2UO_4 = Cs_2U_2O_7 + 2Cs(g) + 1/2O_2(g) \quad (1)$$

Because a typical fuel pin is subjected to elevated temperatures, the formation of diuranate as well as monouranate is possible. Since reliable thermodynamic data are available for all the species in (1) except the diuranate, we have determined $\Delta H_f^\circ(Cs_2U_2O_7)$ based on dissolution in 1M HCl.

The calorimetric specimen was synthesized from $CsNO_3$ and U_3O_8. Chemical analyses for Cs and U, and emission spectrochemical analyses for trace metals indicated a Cs/U ratio of 1.003 ± 0.005.

The following thermochemical cycle was used to determine $\Delta H_f^\circ(Cs_2U_2O_7)$:

(i) $2CsCl(aq) + 2UO_2Cl_2(aq) + 3H_2O(l) = Cs_2U_2O_7(c) + 6HCl(aq)$,
$\Delta H = 46.09 \pm 0.10$ kcal;

(ii) $2UO_2Cl_2(c) = 2UO_2Cl_2(aq)$, $\Delta H = -44.98 \pm 0.26$ kcal;

(iii) $2CsCl(c) = 2CsCl(aq)$, $\Delta H = 8.04 \pm 0.02$ kcal;

(iv) $6HCl(aq) = 3H_2(g) + 3Cl_2(g) + nH_2O$, $\Delta H = 236.06 \pm 0.08$ kcal [13];

(v) $2U(c) + 2O_2(g) + 2Cl_2(g) = 2UO_2Cl_2(c)$, $\Delta H = -582.74 \pm 1.62$ kcal [1];

(vi) $2Cs(c) + Cl_2(g) = 2CsCl(c)$, $\Delta H = -211.78 \pm 0.20$ kcal [8];

(vii) $3H_2(g) + 3/2O_2(g) = 3H_2O(l)$, $\Delta H = -204.94 \pm 0.03$ kcal [13].

Enthalpies of reactions (i), (ii), and (iii) were measured in the present study. Addition of the ΔH terms for reactions (i) through (vii) gave $\Delta H_f^\circ(Cs_2U_2O_7$, c, 298.15 K$) = -754.3 \pm 1.7$ kcal mol^{-1}. The uncertainty, and those attached to all other ΔH_f° values in this paper, is the uncertainty interval equal to twice the standard deviation of the mean.

3.3. Cesium Dimolybdate

Substantial quantities of fission-product cesium and molybdenum have been found in the region between the oxide fuel and the cladding of irradiated fuel pins [18]. These elements are believed to be combined in part as cesium orthomolybdate, Cs_2MoO_4. For this reason, the thermodynamic properties of Cs_2MoO_4 have been determined over a wide temperature range [4].

Our interest in cesium dimolybdate, $Cs_2Mo_2O_7$, stems from the observation [19] that at temperatures in the vicinity of 1200 K, Cs_2MoO_4 appears to disproportionate as follows:

$$2Cs_2MoO_4 = Cs_2Mo_2O_7 + 2Cs(g) + 1/2O_2(g) \qquad (2)$$

A knowledge of ΔG as a function of temperature for this process would enable the equilibrium constant to be deduced and, therefore, the partial pressures of the gaseous species.

It is well known that isopolyanions of molybdenum form readily from alkali metal molybdates in acidic solution [20]. These isopolyanions revert to $MoO_4^{2-}(aq)$ in alkaline solutions. Accordingly, our determination of ΔH_f° was based on measurements of the enthalpy of the reaction

$$Cs_2Mo_2O_7(c) + 2CsOH(aq) = 2Cs_2MoO_4(aq) + H_2O(l) \qquad (3)$$

Since ΔH_f° values are already available for $Cs_2MoO_4(aq)$, $CsOH(aq)$, and $H_2O(l)$, $\Delta H_f^\circ(Cs_2Mo_2O_7)$ can be readily derived.

Cesium dimolybdate was prepared by the reaction of equimolar amounts of high-purity Cs_2MoO_4 and MoO_3. Analytical results for Cs, Mo, and trace metals indicated a Cs/Mo ratio of 1.007 ± 0.008.

The thermochemical cycle for determining $\Delta H_f^\circ(Cs_2Mo_2O_7)$ follows:

(i) $2Cs_2MoO_4(aq) + H_2O(l) = Cs_2Mo_2O_7(c) + 2CsOH(aq)$, $\Delta H = 7.78 \pm 0.09$ kcal;

(ii) $2MoO_3(c) + 4CsOH(aq) = 2Cs_2MoO_4(aq) + 2H_2O(l)$, $\Delta H = -37.29 \pm 0.06$ kcal [4];

(iii) $2Cs(c) + 2H_2O(l) = 2CsOH(aq) + H_2(g)$, $\Delta H = -96.33 \pm 0.02$ kcal [21];

(iv) $H_2(g) + 1/2O_2(g) = H_2O(l)$, $\Delta H = -68.32 \pm 0.01$ kcal [13];

(v) $2Mo(c) + 3O_2(g) = 2MoO_3(c)$, $\Delta H = -356.16 \pm 0.16$ kcal [22,23].

The enthalpy of reaction (i) was measured by us. Summation of items (i) through (v) gave $\Delta H_f^\circ(Cs_2Mo_2O_7$, c, 298.15 K) $= -550.3 \pm 0.2$ kcal mol^{-1}.

3.4. Cesium Chromate

Thermodynamic properties of Cs+Cr+O compounds are of interest in connection with intergranular attack of the stainless steel cladding of fuel pins [24]. Accordingly, we have embarked on a program to determine the standard enthalpies of formation and high- and low-temperature thermodynamic properties of Cs_nCrO_4 (n = 2,3,4) and $Cs_2Cr_2O_7$.

To date, final results have been obtained for only Cs_2CrO_4. The enthalpy of solution in water according to

$$Cs_2CrO_4(c) + 21111H_2O(l) = (2Cs^+\cdot CrO_4^{-2})\cdot 21111H_2O \qquad (4)$$

was found to be 7622 ± 28 cal mol^{-1}. For the dilution process

$$(2Cs^+\cdot CrO_4^{-2})\cdot 21111\,H_2O + \infty H_2O(l) = 2Cs^+(aq) + CrO_4^{2-}(aq) \qquad (5)$$

we estimated $\Delta H = -110 \pm 20$ cal mol^{-1}, somewhat less than the limiting Debye-Hückel value, -130 cal mol^{-1}, and greater than that for the same concentration of Cs_2SO_4, -89 cal mol^{-1} [25]. Addition of the ΔH terms for reactions (4) and (5) gives $\Delta H^\circ = 7.51 \pm 0.03$ kcal mol^{-1} for

$$Cs_2CrO_4(c) + \infty H_2O(l) = 2Cs^+(aq) + CrO_4^{-2}(aq). \qquad (6)$$

We have taken ΔH_f° values of -61.67 ± 0.03 [15] and -210.60 ± 0.20 kcal mol^{-1} [22] for $Cs^+(aq)$ and $CrO_4^{-2}(aq)$, respectively, and deduce $\Delta H_f^\circ(Cs_2CrO_4$, c, 298.15 K) $= -341.45 \pm 0.21$ kcal mol^{-1}.

3.5. Nitride and Oxide of Lithium

Since liquid lithium is a potential blanket material for fusion reactors [26], information concerning the effects of dissolved impurities on its compatibility with likely materials of construction is important. Lithium nitride, Li_3N, and oxide, Li_2O, which form by spontaneous combination of the elements, are very common contaminants in lithium and have a significant deleterious effect on alloys and ceramics [27,28] that are being considered in the design of fusion reactors. Accordingly, reliable and precise thermodynamic data for both compounds would be helpful for the prediction and interpretation of their interactions with other materials.

Although two values for $\Delta H_f^\circ(Li_3N)$ ostensibly in agreement with each other have been reported, we decided to redetermine this property. It is likely that one of the previous studies [29] had employed a specimen of questionable purity; and that the other study [30], which yielded incorrect ΔH_f° values for other metallic nitrides, might likewise have given an erroneous result for Li_3N.

Several values for $\Delta H_f^\circ(Li_2O)$ have been reported; however, the JANAF selection [31] assigned a ± 0.5 kcal mol^{-1} uncertainty due, apparently, to reservations concerning the purity of the Li_2O specimens. An uncertainty of this magnitude in a basic datum is, we believe, unacceptable, and thus $\Delta H_f^\circ(Li_2O)$ has been redetermined.

Lithium nitride was prepared by direct combination of high-purity lithium and nitrogen; lithium oxide, by thermal decomposition under vacuum of high-purity Li_2CO_3. Both preparations were thoroughly analyzed. The only significant contaminant found in the Li_3N was oxygen (0.59 ± 0.05 mass per cent) which was assumed to be present as Li_2O. In the Li_2O, Na (0.1 mass per cent) and C (0.037 mass per cent) were the major impurities, and they were assumed to be present as Na_2CO_3 and Li_2CO_3.

Two essentially independent determinations of $\Delta H_f^\circ(Li_3N)$ were made: one was based on reaction in an acidic medium, the other on reaction in an alkaline medium.

The net reaction with acid was as follows:

$$Li_3N(c) + 4HCl(aq) = 3LiCl(aq) + NH_4Cl(aq). \tag{7}$$

Combination of the enthalpy of reaction (7), -192.04 ± 0.30 kcal mol^{-1}, with our auxiliary measurements of the enthalpies of solution of LiCl and NH_4Cl and with literature data for HCl [13] yielded $\Delta H_f^\circ(Li_3N) = -39.38 \pm 0.36$ kcal mol^{-1}.

Lithium nitride reacts with aqueous alkali according to

$$Li_3N(c) + 3H_2O(l) = 3LiOH(aq) + NH_3(aq). \tag{8}$$

For the above reaction, we found $\Delta H = -139.01 \pm 0.34$ kcal mol^{-1}. Combination of this result with literature data for H_2O [13], LiOH(aq) [21, 15], and NH_4Cl [32,22] yielded $\Delta H_f^\circ(Li_3N) = -39.47 \pm 0.37$ kcal mol^{-1}. Our selected result for $\Delta H_f^\circ(Li_3N$, c, 298.15 K), -39.42 ± 0.26 kcal mol^{-1}, is a weighted mean of the two determinations.

The Li_2O study consisted primarily of measurements of the enthalpy of the reaction:

$$Li_2O(c) + nH_2O(l) = 2[LiOH\cdot(\tfrac{n-1}{2})H_2O] \tag{9}$$

for which we found $\Delta H = -31.488 \pm 0.024$ kcal mol^{-1} with n = 1670. This result, in conjunction with auxiliary thermochemical data for H_2O [13], and LiOH [21, 15], gave $\Delta H_f^\circ(Li_2O$, c, 298.15 K) $= -142.902 \pm 0.066$ kcal mol^{-1}.

3.6. Thorium Monosulfide

The investigation of ThS was undertaken as part of our ongoing program to obtain quality data for binary actinide compounds. Our calorimetric specimen, which was part of the batch used by Flotow et al [33] for low-temperature calorimetry, contained 3.77 mass per cent Th_2S_3, 1.47 mass per cent ThOS, 88 ppm C, and 15 ppm N. The S/Th ratio of the monosulfide phase was calculated to be 1.029 ± 0.019.

The measured enthalpy of reaction of the thorium sulfide with hydrochloric acid was combined with auxiliary thermochemical data from the literature, as illustrated in the cycle below, to yield $\Delta H_f^\circ(ThS_{1.029 \pm 0.019}, c, 298.15\ K) = -96.9 \pm 0.9\ kcal\ mol^{-1}$.

(i) $(ThCl_4 \cdot 1196HCl \cdot 9776H_2O \cdot 1.029H_2S) + 0.971H_2(g) = ThS_{1.029}(c) + (1200HCl \cdot 9776\ H_2O)$, $\Delta H = 94.18 \pm 0.44$ kcal;

(ii) $Th(c) + (2400HCl \cdot 19452H_2O) = (ThCl_4 \cdot 2396HCl \cdot 19452H_2O) + 2H_2(g)$, $\Delta H = -181.4 \pm 0.5$ kcal [34];

(iii) $(1200HCl \cdot 9776H_2O) + (ThCl_4 \cdot 2396HCl \cdot 19452H_2O) = (2400HCl \cdot 19452H_2O) + (ThCl_4 \cdot 1196HCl \cdot 9776H_2O)$, $\Delta H = 0 \pm 0$ kcal;

(iv) $1.029H_2(g) + 1.029S(rh) = 1.029H_2S(g)$, $\Delta H = -5.02 \pm 0.15$ kcal [35];

(v) $1.029H_2S(g) + (ThCl_4 \cdot 1196HCl \cdot 9776H_2O) = (ThCl_4 \cdot 1196HCl \cdot 9776H_2O \cdot 1.029H_2S)$, $\Delta H = -4.63 \pm 0.62$ kcal [13].

4. FLUORINE BOMB CALORIMETRIC RESULTS

4.1. Molybdenum Carbides ($MoC_{0.502}$ and $MoC_{0.651}$)

It is anticipated that reliable thermodynamic data will be required to help elucidate the reactions of fission-product molybdenum in carbide fuels. Carefully prepared and characterized specimens of $MoC_{0.502}$ (usually called α-Mo_2C) and $MoC_{0.651}$ ("Mo_3C_2") were reacted with high-purity fluorine in a bomb calorimeter. The standard energies of combustion, after correction for production of minor amounts of C_2F_6 and C_3F_8, were as follows:

$$MoC_{0.502}(c) + 4.004F_2(g) = MoF_6(g) + 0.502CF_4(g),\ \Delta U_c^\circ = -476.3 \pm 0.3\ kcal\ mol^{-1}; \qquad (10)$$

$$MoC_{0.651}(c) + 4.302F_2(g) = MoF_6(g) + 0.651CF_4(g),\ \Delta U_c^\circ = -510.8 \pm 0.4\ kcal\ mol^{-1}. \qquad (11)$$

These results, when corrected to constant-pressure conditions and combined with ΔH_f° values for $MoF_6(g)$ [36] and $CF_4(g)$ [37], yielded $\Delta H_f^\circ(MoC_{0.502}) = -6.5 \pm 0.7$ and $\Delta H_f^\circ(MoC_{0.651}) = -5.1 \pm 0.8\ kcal\ mol^{-1}$.

4.2. Triuranium Silicide (U_3Si)

Triuranium silicide, because of its potential as a fuel in water-cooled nuclear reactors, has been extensively investigated at the Chalk River, Ontario, Laboratory of Atomic Energy of Canada, Ltd. At Argonne, high-temperature enthalpy increments have been determined [38], and a low temperature calorimetric study is planned.

The U_3Si was synthesized at Chalk River via the peritectoid reaction between U and U_3Si_2 at 1200 K, followed by vacuum annealing for 120 h at 1070°K to ensure completion of the reaction. Calorimetric experiments were performed in a two-chamber reaction vessel [10b] by means of which fluorine could be isolated from the U_3Si until ignition was desired. A small quantity of sulfur was used to bring about rapid combustion. Reaction to UF_6 and SiF_4 was complete except for 20-40 mg residues of solid uranium fluoride of approximate composition U_2F_9.

Calorimetric data for the combustion experiments are shown in Table II. Most of the symbols are either self-explanatory or have been discussed elsewhere [10a, 39]. The correction for the energy of combustion of the sulfur to SF_6, ΔU_{sulfur}, is based on the results from [40]; $\Delta U_{residue}$ is the correction for the (hypothetical) conversion of the U_2F_9 residue to UF_6(c) and is based on $\Delta H^\circ_f(U_2F_9)$ from [41]; and, $\Delta U_{sub(UF_6)}$ is the correction for the condensation of any UF_6(g) formed to UF_6(c). The impurity correction is due largely to zirconium and oxygen contaminants, both of which are present to the extent of 500 ppm.

The value for $\Delta U^\circ_c/M(U_3Si)$, which refers to the reaction

$$U_3Si(c) + 11F_2(g) = 3UF_6(c) + SiF_4(g) \tag{12}$$

was corrected to constant pressure conditions and combined with $\Delta H^\circ_f(UF_6, c)$ [42] and $\Delta H^\circ_f(SiF_4, g)$ [43] to yield $\Delta H^\circ_f(U_3Si, c, 298.15\ K) = -24.9 \pm 4.6$ kcal mol^{-1}.

TABLE II. RESULTS OF U_3Si COMBUSTION EXPERIMENTS

Combustion No.	2	3	4	5	6	8
m'(sample) (g)	1.55574	1.56530	1.50180	1.51029	1.51273	1.53061
m"(sulfur) (g)	0.00300	0.00989	0.00351	0.00993	0.00830	0.00718
$\Delta\Theta_c$ (K)	1.24268	1.26902	1.20099	1.22248	1.22374	1.23625
ε(calor)x($-\Delta\Theta_c$) (cal)	-4032.51	-4117.98	-3897.22	-3966.96	-3971.05	-4011.64
$\Delta U_{contents}$ (cal)	-8.36	-8.38	-8.23	-8.11	-8.14	-8.34
$\Delta U_{sub(UF_6)}$ (cal)	-11.08	-11.09	-11.08	-11.08	-11.07	-11.07
ΔU_{sulfur} (cal)	+27.19	+89.63	+31.81	+89.99	+75.22	+65.07
$\Delta U_{residue}$ (cal)	-5.87	-3.15	-3.15	-4.95	-4.26	-5.44
ΔU_{blank} (cal)	+4.69	+4.69	+4.69	+4.69	+4.69	+4.69
ΔU_{gas} (cal)	-0.19	-0.19	-0.18	-0.18	-0.18	-0.18
$\Delta U^\circ_c/M$(sample) (cal g^{-1})	-2587.92	-2585.11	-2585.80	-2580.03	-2587.90	-2591.72

Average $\Delta U^\circ_c/M$(sample) = -2586.41 ± 1.59 cal g^{-1} (std. dev.)

Impurity correction = -4.86 ± 5.00 cal g^{-1}

$\Delta U^\circ_c/M(U_3Si)$ = -2591.27 ± 5.93 cal g^{-1}

[a] ε(calor), the energy equivalent of the calorimetric system, was 3245.01 cal K^{-1}.

4.3. β-Uranium Disulfide (US_2)

Two stable uranium sulfides of nominal composition US_2 are known. We have measured the energy of combustion in fluorine of the β-modification whose composition is close to US_2, and a similar study is in progress on α-US_2, whose composition is close to $US_{1.9}$.

Our specimen of β-US_2 was synthesized at Stanford Research Institute, Menlo Park, California, by heating a mixture of uranium metal and powdered sulfur in a quartz tube. Metallography showed the specimen to be a single-phase material. X-ray diffraction analyses gave the following cell parameters (Å): a = 8.480, b = 7.115, and c = 4.122, in good agreement with published measurements [44]. From chemical and spark-source mass spectrometric analyses, an anion/cation ratio of 1.992 ± 0.002 was deduced.

The combustion technique was similar to that used for U_3Si. Corrections had to be made for small amounts of combustion residues which were shown by chemical analyses to consist of US_2 and U_4F_{17}. For the reaction

$$US_{1.992} + 8.976F_2(g) = UF_6(c) + 1.992SF_6(g) \tag{13}$$

the enthalpy of combustion was -981.18 ± 0.56 kcal mol^{-1}. This result, when combined with $\Delta H_f^\circ(UF_6,c)$ [42] and $\Delta H_f^\circ(SF_6,g)$ [40] gave $\Delta H_f^\circ(US_{1.992}$, c, β, 298.15 K) = -122.7 ± 2.1 kcal mol^{-1} from which we estimate a value of -123.0 ± 2.1 kcal mol^{-1} for $\Delta H_f^\circ(US_2$, c, β, 298.15 K).

5. DISCUSSION

Ippolitova, Faustova, and Spitsyn [11] reported the enthalpies of reaction of the alkali metal monouranates with ~1M HCl; for Li_2UO_4 they obtained -47.27 ± 0.12, for K_2UO_4, -33.72 ± 0.10, and for Rb_2UO_4, -33.42 ± 0.05 kcal mol^{-1}. These values deviate significantly from those given in Table I. Similar discrepancies were found recently for α-Na_2UO_4 [2a] and Cs_2UO_4 [8]. One would anticipate that the enthalpies of reaction of all five uranates would be within a few kcal mol^{-1} of each other, as we have found. On the other hand, the results of Ippolitova et al show a spread of about 24 kcal mol^{-1}. Their analyses of the Na, K, and Cs salts, for which $M/U < 2$, suggest the presence of diuranate, and the Li/U ratio is disturbingly high at 2.05. We are unable to give an objective assessment of this work because of the paucity of experimental data and other details. However, the analytical results suggest that the samples may have been inadequate for reliable calorimetric studies.

We have carried out thermodynamic calculations for the purpose of predicting whether $Cs_2U_2O_7$ is likely to form from Cs_2UO_4 in a fuel pin. The following processes have to be considered:

$$2Cs(g) + UO_2(c) + O_2(g) = Cs_2UO_4 \tag{14}$$

$$2Cs_2UO_4(c) = Cs_2U_2O_7(c) + 2Cs(g) + 1/2O_2(g) \tag{15}$$

Based on ΔH_f° and S° values from the literature and $\Delta H_f^\circ(Cs_2U_2O_7)$, the following approximate expressions were derived:

$$\Delta G(14)_T = -230{,}800 + 98.9T \quad \text{cal mol}^{-1} \tag{16}$$

$$\Delta G(15)_T = 189{,}300 - 117.4T \quad \text{cal mol}^{-1} \tag{17}$$

From equation (16), and taking the activity of UO_2 to be 0.8, we deduce

$$\ln P_{Cs} = [\Delta G(14) - \Delta\overline{G}(O_2)]/2RT + 0.1116 \quad (18)$$

Similarly, from equation (17) we obtain

$$\ln P_{Cs} = [-\Delta G(15) - 1/2\ \Delta\overline{G}(O_2)]/2RT \quad (19)$$

In the above equations, P denotes partial pressure (atm), and $\Delta\overline{G}(O_2)$, the oxygen potential, equals RT ln P_{O_2}.

The partial pressure of Cs(g) as a function of $\Delta\overline{G}(O_2)$ is given in Table III for T = 1000 K. The values of the oxygen potential embrace those typical of oxide-fuel systems in the region contiguous with the fuel-cladding interface.

TABLE III. CESIUM PARTIAL PRESSURES FOR THE Cs+U+O SYSTEM AT 1000 K

$\Delta\overline{G}(O_2)$ (cal mol^{-1})	$P_{Cs}(14)$ (atm)	$P_{Cs}(15)$ (atm)
-90 000	2.96×10^{-5}	7.91×10^{-8}
-95 000	1.04×10^{-4}	1.48×10^{-7}
-100 000	3.66×10^{-4}	2.78×10^{-7}
-105 000	1.29×10^{-3}	5.22×10^{-7}
-110 000	4.53×10^{-3}	9.79×10^{-7}

The data in Table III suggest that $Cs_2U_2O_7$ should form at lower Cs pressures than Cs_2UO_4. However, our calculations show that the activity of $Cs_2U_2O_7$ will be low and Cs_2UO_4 will be the predominant Cs+U+O compound.

Similar calculations for the Cs+Mo+O system indicate that the activity of $Cs_2Mo_2O_7$ in a fuel pin is likely to be very low. The partial pressures of Cs(g) in the Cs+Mo+O system are calculated to be 4 to 5 orders of magnitude lower than those in the Cs+U+O system. On this basis, it appears that in a fuel pin Cs_2MoO_4 will form far more readily than Cs_2UO_4.

For Cs_2CrO_4, the enthalpy of solution, 7.51 ± 0.03 kcal mol^{-1}, is in agreement with the 7.4 ± 0.1 kcal mol^{-1} reported by Shidlovskii, Balakireva, and Voskresenskii [45].

There have been two previous determinations of $\Delta H_f^\circ(Li_3N)$, one based on solution calorimetry, the other on direct combination of the elements. Guntz [29] reported the enthalpy of reaction (8) to be -131.1 kcal mol^{-1} at 291.15 K. The final concentration of the LiOH was not specified. However, even after making reasonable assumptions about the LiOH concentration and correcting the enthalpy of reaction to 298.15 K, the result of Guntz still differs by about 8 kcal mol^{-1} from our determination, -139.01 ± 0.34 kcal mol^{-1}. A discrepancy of this magnitude could be caused by unaccounted for LiOH or Li_2O in the Li_3N used by Guntz.

Neumann, Kröger, and Haebler [30] measured the enthalpy of reaction of lithium with nitrogen at 5 atm pressure by heating at 893 K for one minute. They obtained $\Delta H_f^\circ(Li_3N) = -47.2 \pm 1$ kcal mol^{-1}. As part of the same investigation, they reported ΔH_f° values for AlN, Mg_3N_2, and Be_3N_2. Each of the latter results has been superseded by more modern determinations which differ from the data of Neumann et al by as much as 10 kcal mol^{-1}. The discrepancy of $\sim$7 kcal mol^{-1} between our value for $\Delta H_f^\circ(Li_3N)$ and that of Neumann et al is therefore in harmony with the findings of several other investigators.

The agreement between the present $\Delta H_f^\circ(Li_3N)$ values, obtained by two essentially independent routes, as well as the careful synthesis, characterization, and manipulation of the Li_3N, are substantive reasons for preferring our determination over those of Guntz and Neumann et al. In addition, our result indicates that decomposition of the nitride to the elements at 1000 K leads to a nitrogen pressure several orders of magnitude greater than that based on the previous determinations of $\Delta H_f^\circ(Li_3N)$. This observation is supported by the results of preliminary tensimetric measurements on the Li+N system [46].

Our determination of $\Delta H_f^\circ(Li_2O)$ is in excellent agreement with an earlier study [47], and indicates that JANAF [31] was much too conservative in the assignment of an uncertainty to this value.

The value of $\Delta H_f^\circ(ThS_{1.029})$, -96.9 ± 0.9 kcal mol^{-1}, may be compared with Aronson's estimate [48] of -104 kcal mol^{-1} for $\Delta H_f^\circ(ThS)$, based on solid-state emf measurements between 1100 and 1200 K. Our result may also be expressed as $\Delta H_f^\circ(Th_{0.493}S_{0.507}) = -47.7$ kcal g-atom^{-1}, which may be compared with the similarly expressed value for Th_2S_3, $\Delta H_f^\circ(Th_{0.4}S_{0.6}) = -51.7$ kcal g-atom^{-1} [49].

From $\Delta H_f^\circ(MoC_{0.502}) = -6.5 \pm 0.7$ kcal mol^{-1}, we deduce $\Delta H_f^\circ(Mo_2C) = -13.0 \pm 1.5$ kcal mol^{-1}, in agreement with Mah's determination [50] of -11.0 ± 0.7 kcal mol^{-1} by oxygen bomb calorimetry.

Neither $\Delta H_f^\circ(U_3Si)$ nor $\Delta H_f^\circ(\beta\text{-}US_2)$ has been reported previously. The results for U_3Si and $\beta\text{-}US_2$, -6.2 and -41 kcal g-atom^{-1}, respectively, are consistent with similarly expressed values for other uranium silicides [51] and uranium monosulfide [52].

ACKNOWLEDGEMENTS

We wish to thank H. R. Hoekstra, E. K. Storms, I. J. Hastings, V. A. Maroni, and R. M. Yonco for synthesizing many of the compounds used in this study. Analyses were performed by K. J. Jensen, I. Fox, M. I. Homa, J. P. Faris, R. V. Schablaske, M. Adams, J. Williams and K. E. Anderson. J. Boerio, R. T. Grow and C. Zahradnik assisted with several of the measurements. We have had the benefit of frequent discussions with I. Johnson.

REFERENCES

[1] (a) O'HARE, P. A. G., SHINN, W. A., MRAZEK, F. C., MARTIN, A. E., J. Chem. Thermodyn. 4, (1972) 401; (b) OSBORNE, D. W., FLOTOW, H. E., J. Chem. Thermodyn. 4, (1972) 411; (c) FREDRICKSON, D. R., CHASANOV, M. G., J. Chem. Thermodyn. 4 (1972) 419; (d) BATTLES, J. E., SHINN, W. A., BLACKBURN, P. E., J. Chem. Thermodyn. 4 (1972) 425.

[2] (a) O'HARE, P. A. G., HOEKSTRA, H. R., J. Chem. Thermodyn. 5 (1973) 769; (b) OSBORNE, D. W., FLOTOW, H. E., DALLINGER, R. P., HOEKSTRA, H. R., J. Chem. Thermodyn. 6 (1974) 751.

[3] O'HARE, P. A. G., HOEKSTRA, H. R., J. Chem. Thermodyn. 6 (1974) 965.

[4] (a) O'HARE, P. A. G., HOEKSTRA, H. R., J. Chem. Thermodyn. 5 (1973) 851; (b) OSBORNE, D. W., FLOTOW, H. E., HOEKSTRA, H. R., J. Chem. Thermodyn. 6 (1974) 179; (c) FREDRICKSON, D. R., CHASANOV, M. G., "The enthalpy and heat of transition of Cs_2MoO_4 by drop calorimetry", Analytical Calorimetry Vol. III (JOHNSON, J. F., Ed.), American Chemical Society, in press.

[5] O'HARE, P. A. G., HOEKSTRA, H. R., J. Chem. Thermodyn. 6 (1974) 117.

[6] O'HARE, P. A. G., J. Chem. Thermodyn. 6 (1974) 425.

[7] SETTLE, J. L., JOHNSON, G. K., HUBBARD, W. N., J. Chem. Thermodyn. 6 (1974) 263.

[8] O'HARE, P. A. G., HOEKSTRA, H. R., J. Chem. Thermodyn. 6 (1974) 251.

[9] JOHNSON, G. K., VAN DEVENTER, E. H., HUBBARD, W. N., J. Chem. Thermodyn. 6 (1974) 219.

[10] (a) HUBBARD, W. N., "Fluorine bomb calorimetry", Ch. 6, Experimental Thermochemistry, Vol. 2 (SKINNER, H. A., Ed.), Interscience London (1962); (b) SETTLE, J. L., GREENBERG, E., HUBBARD, W. N., Rev. Sci. Instr. 38 (1967) 1805.

[11] IPPOLITOVA, E. A., FAUSTOVA, D. G., SPITSYN, V. I., "Investigations in the field of Uranium Chemistry," Argonne National Laboratory translation, USAEC Rep. ANL-Trans-33 (1961), 170.

[12] CORDFUNKE, E. H. P., LOOPSTRA, B. O., J. Inorg. nucl. Chem. 33 (1971) 2427.

[13] WAGMAN, D. D., EVANS, W. H., PARKER, V. B., HALOW, I., BAILEY, S. M. SCHUMM, R. H., N.B.S. Tech. Note 270-3, U. S. National Bureau of Standards, Washington, D. C., 1968.

[14] PARKER, V. B., Thermal properties of aqueous uni-univalent electrolytes, NSRDS-NBS-2, U. S. National Bureau of Standards, Washington, D. C., 1965.

[15] WAGMAN, D. D., U. S. National Bureau of Standards, Washington, D. C., private communication.

[16] NEIMARK, L. A., LAMBERT, J. D. B., Reactor development program progress report, USAEC Rep. ANL-RDP-11 (1972) p. 6.18.

[17] EFREMOVA, K. M., IPPOLITOVA, E. A., SIMANOV, YU.P., SPITSYN, V. I., Dokl. Akad. Nauk SSSR 124 (1959) 1057.

[18] JOHNSON, C.E., STALICA, N. R., SEILS, C. A., ANDERSON, K. E., USAEC Rep. ANL-7675 (1969), p. 102.

[19] JOHNSON, I., Argonne National Laboratory, private communication.

[20] AVESTON, J., ANACKER, E. W., JOHNSON, J. S., Inorg. Chem. 3 (1964) 735.

[21] GUNN, S. R., J. phys. Chem. 71 (1967) 1386.

[22] WAGMAN, D. D., EVANS, W. H., PARKER, V. B., HALOW, I., BAILEY, S. M., SCHUMM, R. H., N.B.S. Tech. Note 270-4, U. S. National Bureau of Standards, Washington, D. C., 1969.

[23] NUTTALL, R. L., CHURNEY, K. L., KILDAY, M. V., U. S. National Bureau of Standards Rept. No. 10481 (1971).

[24] STALICA, N. R., SEILS, C. A., CROUTHAMEL, C. E., USAEC Rep. ANL-7575 (1968), p. 102.

[25] ROSSINI, F. D., WAGMAN, D. D., EVANS, W. H., LEVINE, S., JAFFE, I., "Selected values of chemical thermodynamic properties," U. S. National Bureau of Standards Circular 500, Washington, D. C. (1952).

[26] GOUGH, W. C., EASTLUND, B. J., Scientific American 224 2 (1971) 50.

[27] MARONI, V. A., CAIRNS, E. J., CAFASSO, F. A., USAEC Rep. ANL-8001 (1973).

[28] SESSIONS, C. E., DEVAN, J. H., Nucl. Appl. 9 (1970) 250.

[29] GUNTZ, A., Compt. Rend. 123 (1896) 995.

[30] NEUMANN, B., KRÖGER, C., HAEBLER, H., Z. Anorg. Allg. Chem. 204 (1932) 81.

[31] JANAF Thermochemical Tables, The Dow Chemical Co., Midland, Mich., March (1964).

[32] (a) VANDERZEE, C. E., MÅNSSON, M., WADSÖ, I., SUNNER, S., J. Chem. Thermodyn. 4 (1972) 541; (b) VANDERZEE, C. E., KING, D. L., WADSÖ, I., J. Chem. Thermodyn. 4 (1972) 685.

[33] FLOTOW, H. E., OSBORNE, D. W., WALTERS, R. R., J. chem. Phys. 55 (1971) 880.

[34] EYRING, L., WESTRUM, E. F., JR., J. Amer. Chem. Soc. 72 (1950) 5555.

[35] JANAF Thermochemical Tables, The Dow Chemical Co., Midland, Mich., December (1965).

[36] SETTLE, J. L., FEDER, H. M., HUBBARD, W. N., J. phys. Chem. 65 (1961) 1337.

[37] GREENBERG, E., HUBBARD, W. N., J. phys. Chem. 72 (1968) 222.

[38] FREDRICKSON, D. R., Argonne National Laboratory, private communication.

[39] GREENBERG, E., SETTLE, J. L., HUBBARD, W. N., J. phys. Chem. 66 (1962) 1345.

[40] O'HARE, P. A. G., SETTLE, J. L., HUBBARD, W. N., Trans. Faraday Soc. 62 (1966) 558.

[41] RAND, M. H., KUBASCHEWSKI, O., The Thermochemical Properties of Uranium Compounds, Interscience, New York (1963).

[42] SETTLE, J. L., FEDER, H. M., HUBBARD, W. N., J. phys. Chem. 67 (1963) 1892.

[43] WISE, S. S., MARGRAVE, J. L., FEDER, H. M., HUBBARD, W. N., J. phys. Chem. 67 (1963) 815.

[44] SUSKI, W., GIBIŃSKI, T., WOJAKOWSKI, A., CZOPNIK, A., Phys. stat. solidi, 9 (1972) 653.

[45] SHIDLOVSKII, A. A., BALAKIREVA, T. N., VOSKRESENSKII, A. A., Zh. Fiz. Khim. 45 (1971) 1857.

[46] VELECKIS, E., Argonne National Laboratory, private communication.

[47] KOLESOV, V. P., SKURATOV, S. M., ZAĬKIN, I. D., Zh. Neorg. Khim. 4 (1959) 1233.

[48] ARONSON, S., J. Inorg. nucl. Chem. 29 (1967) 1611.

[49] EYRING, L., WESTRUM, E. F., JR., J. Amer. Chem. Soc. 75 (1953) 4802.

[50] MAH, A. D., "Heats of combustion and formation of carbides of tungsten and molybdenum," U. S. Bureau of Mines Rept. of Investigation No. 6337 (1963).

[51] GROSS, P., HAYMAN, C., CLAYTON, H., "Heats of formation of uranium silicides and nitrides," Thermodynamics of Nuclear Materials, Proc. Conf. Vienna, 1962, IAEA, Vienna (1962) 653.

[52] O'HARE, P. A. G., SETTLE, J. L., FEDER, H. M., HUBBARD, W. N., "The thermochemistry of some uranium compounds," Thermodynamics of Nuclear Materials, Proc. Conf. Vienna, 1967, IAEA, Vienna (1968) 265.

DISCUSSION

F. SCHMITZ: Your results, if I understand correctly, appear to indicate that caesium molybdate exists in typical fast breeder reactor fuel. Do you think that there is still sufficient oxygen for appreciable clad corrosion to occur, as the oxygen produced by fission is completely bound up by the formation of the Cs-Mo compound in addition to the other oxide forms?

Also, do you consider that the formation of molybdate-type caesium compounds is favourable or unfavourable to fuel element operation?

P.A.G. O'HARE: I feel that Mr. Johnson is better qualified to answer this question.

I. JOHNSON: At normal clad temperatures pure Cs_2MoO_4 does not lead to intergranular attack of the stainless steel clad. At temperatures of upwards of about 825°C intergranular attack is observed. However, we do not at the present time believe that Cs_2MoO_4 as a pure substance is an essential compound in the intergranular attack of the clad. It probably does play a role in the complex chemical reactions which occur at the fuel/clad interface, particularly in reactions between caesium and the chromium oxides at the clad surface. We are not yet certain whether the formation of Cs_2MoO_4 at the fuel/clad interface favours or inhibits attack on the cladding.

V.A. MARONI: I should like to make an additional comment on the work on Li_3N reported in this paper. If one were to extrapolate the results of the two prior studies of ΔG_f^0 for Li_3N (which were fortuitously in agreement), one would find that, at temperatures between 800 and 900°C, the predicted equilibrium nitrogen pressure would be very small ($\ll$ 1 torr). Use of Mr. O'Hare's data together with the heat capacity data of Flotow et al. gives equilibrium nitrogen pressures of several tens of torr (up to 60 torr) at these temperatures. Recent tensimetric studies on solid Li_3N carried out in our laboratory by Veleckis and Yanco have in fact confirmed the nitrogen activities predicted by Mr. O'Hare's work.

THERMODYNAMIC PROPERTIES OF GASEOUS URANIUM HYDROXIDE

S.R. DHARWADKAR, S.N. TRIPATHI,
M.D. KARKHANAVALA, M.S. CHANDRASEKHARAIAH
Bhabha Atomic Research Centre,
Trombay, Bombay,
India

Presented by D.D. Sood

Abstract

THERMODYNAMIC PROPERTIES OF GASEOUS URANIUM HYDROXIDE.

Thermodynamic data for the homogeneous reaction of $UO_3(g)$ with $H_2O(g)$, in the temperature range of 1325 to 1675 K, were evaluated by studying the partial pressure of uranium vapour over U_3O_8 in the presence of dry and moist oxygen. The standard free energy change (ΔG_1^0) for the reaction of U_3O_8 with dry oxygen could be represented as:

$$\Delta G_1^0 \text{ (kJ)} = (369.9 \pm 7.9) - (145.2 \pm 4.9) \times 10^{-3}\, T$$

and the free energy change (ΔG_2^0) for the reaction of U_3O_8 with wet oxygen (water vapour of 23 torr partial pressure) could be represented as:

$$\Delta G_2^0 \text{ (kJ)} = (183.8 \pm 7.7) - (52.2 \pm 5.6) \times 10^{-3}\, T$$

From ΔG_1^0 and ΔG_2^0 the free energy change (ΔG_3^0) for the reaction of $UO_3(g)$ with $H_2O(g)$ was evaluated as:

$$\Delta G_3^0 \text{ (kJ)} = -(186.1 \pm 15.6) + (93.0 \pm 10.5) \times 10^{-3}\, T$$

From the large enthalpy change for this gas phase reaction it was deduced that a species of the type $UO_2(OH)_2$ is formed in the vapour phase, rather than a structure with fewer than four U-O bonds.

1. INTRODUCTION

In the use of a ceramic body at high temperatures, the material loss through vapour transport is an important consideration. In a vacuum or in an inert gaseous environment, the vapour pressure of a ceramic oxide determines the maximum vapour phase transport. But the situation alters when the gaseous environment contains a reactive gas. If there is any significant homogeneous gas phase reaction between any of the vapour species of the oxide and the surrounding gas, then the material transport rate will be increased. Hence, the net material transport of an oxide ceramic at a definite temperature depends on its vapour pressure and any reaction equilibrium between the vapour and the gaseous environment.

Moisture is an ubiquitous component of common gaseous environments and, as such, a knowledge of its influence on the volatility of oxides is of practical significance. The available quantitative information is meagre and has been summarized by Glemser and Wendlandt [1]. The increase in the volatility of an oxide on the introduction of moisture may be as large as a few orders of magnitude, as demonstrated in the case of BeO [2] and B_2O_3 [3]. Hence a programme was initiated to investigate the thermodynamic equilibria involving U_3O_8 ceramic oxide and moisture at elevated temperatures.

The U_3O_{8-z} phase was selected for a thorough investigation because of its role in the fabrication of the oxide fuels for power reactors. In addition, it provides a convenient source of $UO_3(g)$, free from the interference of other oxides, for a gas phase interaction of the following kind:

$$UO_3(g) + nH_2O(g) \rightleftharpoons UO_3 \cdot nH_2O(g) \qquad (I)$$

where n is an integer. Dependable thermodynamic data for this reaction has a practical application in reactor safety calculations. The vapour composition over urania is highly O/U dependent and it has been reported [4] that UO_3 will be the predominant vapour species when the composition becomes $O/U > 2.02$. Under this condition, total uranium transported via the vapour phase will be determined by the $UO_3(g)$ species concentration and any other gas phase reaction in which this is a reactant, e.g. reaction (I). In an accidental clad failure of a fuel rod with ingress of steam into the fuel chamber, there will be a reaction between the hot surface of the urania and steam — resulting in surface oxidation. This would make the surface layer of the fuel hyperstoichiometric, even if it was not already hyper-stoichiometric as a result of the temperature gradient causing a redistribution of O/U in the fuel pellet [5]. Hence the predominant vapour species will be $UO_3(g)$. This $UO_3(g)$ may be released into the cooling water. But if the reaction (I) becomes significant under these conditions, the transport of uranium species from the fuel surface will be increased. Reliable free energy information for reaction (I) would facilitate the estimation of the maximum possible uranium transport in such an accident.

The free energy change for the reaction (I) was determined by measuring the free energy changes of the following heterogeneous reactions, utilizing a modified transpiration method.

$$\tfrac{1}{3}U_3O_{8-z}(s) + \tfrac{1}{6}(1+z)\,O_2(g) + nH_2O(g) \rightleftharpoons UO_3 \cdot nH_2O(g) \qquad (II)$$

$$\tfrac{1}{3}U_3O_{8-z}(s) + \tfrac{1}{6}(1+z)\,O_2(g) \rightleftharpoons UO_3(g) \qquad (III)$$

Pure oxygen at one atmosphere pressure or oxygen containing a definite amount of moisture was employed as the reactive carrier gas. The results are presented here.

2. EXPERIMENTAL

A nuclear grade urania sample (total metallic impurities < 200 ppm; supplied by Atomic Fuels Division of BARC) of average size - 100 to + 300 mesh was employed in all measurements. All other chemicals were of reagent grade; where necessary, they were further purified by standard techniques.

Platinum-rhodium resistance wire wound on an alumina tube served as the furnace. Another ancillary heater was provided to improve the temperature uniformity. The temperature of the furnace was maintained within + 0.5 K of the set temperature using a SCR regulated proportional

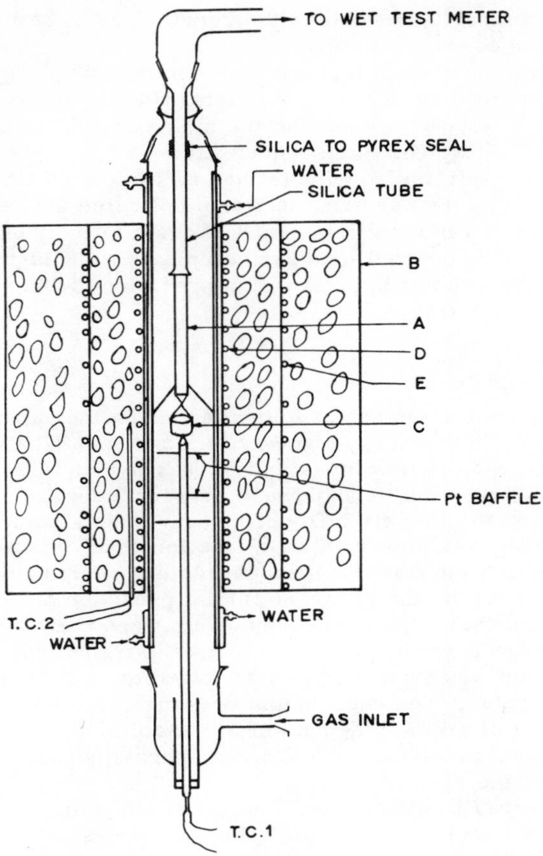

FIG.1. Transpiration apparatus (A: platinum collector; B: furnace; C: Pt-Rh gauge crucible; D: Pt-Rh wire primary heater; E: Pt-Ir wire secondary heater).

temperature controller (Eurotherm Model PID). The temperature was measured by another calibrated S-type thermocouple. A platinum collector was shaped in such a way as to minimize the diffusional contribution while at the same time providing the maximum facility to condense and collect all the vapours as soon as the vapours left the reaction zone [9]. This design also prevented the vapour-saturated carrier gas from making multiple passes on the outside of the collector. A schematic diagram of the transpiration set-up is presented in Fig.1.

Pure oxygen gas (Indian Oxygen Ltd.) was dried by passing it sequentially over Drierite and anhydrous magnesium perchlorate columns before it was used in the measurement of ΔG^0 of reaction (III). In a subsequent series of runs, the gas was saturated with moisture by bubbling it through a tube containing distilled water kept in a thermostatically controlled water bath whose temperature was maintained to $\pm 0.1°C$ of the desired temperature.

Changing this temperature enabled the carrier gas to be mixed with a known partial pressure of water vapour.

The flow rate of the carrier gas was controlled by a glass capillary flow meter described earlier [6]. The total volume of the gas passed was measured with a wet-test meter and the moles of moisture by condensing H_2O at the exit and weighing the condensate.

The uranium-bearing species (either as UO_3 or as $UO_3 \cdot nH_2O$) which were swept away by the carrier gas were collected inside the platinum collector, dissolved in hot nitric acid and quantitatively estimated colorimetrically as the di-benzoyl methane complex [7]. This procedure yielded the uranium content up to 1 μg/ml with a precision of 1%.

3. RESULTS

To calculate the influence of water vapour on the vapour transport of U_3O_{8-z} phase, the free energy change of the reaction (III) was required. Hence, the first step in this investigation was to redetermine the equilibrium pressure, p_{UO_3}, over the U_3O_{8-z} phase between 1500 and 1675 K. This also provided the test of reliability of measurements made with this set-up.

The volatility was measured as a function of oxygen flow rates at 1573 K to establish the flow-independent plateau region. With flow rates of between 1.2 to 2.7 l/h, the apparent partial pressure of UO_3 was found to exhibit a plateau. Subsequent measurements were therefore made with flows in the range 1.5 to 2.0 l/h. Prior to the transpiration experiment, the entire system was evacuated and degassed to remove any moisture inside the apparatus. The equilibrium pressure of $UO_3(g)$ was calculated from the amount of uranium deposited in the collector for the passage of 1 mole of oxygen, and using the ideal gas approximation. Table I presents the summary of the results.

The data were fitted to obtain a linear relationship by means of the least-squares method, yielding the following expression:

$$- \log p_{UO_3} = \frac{(19318 \pm 409)}{T} - (12.60 \pm 0.26) \tag{1}$$

where p is in pascals and T in kelvin. From expression (1), the change in the free energy of the reaction (III) was calculated:

$$\Delta G_1^0(\text{kJ}) = (369.9 \pm 7.9) - (145.2 \pm 4.9) \times 10^{-3}\ T \tag{2}$$

The agreement with the Second Law heats ($\Delta H^0 = 348.5 \pm 14.2$ kJ) and entropies ($\Delta S^0 = 130.9 \pm 10.9$) reported by Ackermann et al. [8] was good. Non-availability of reliable molecular parameters for UO_3 precluded us from evaluating the Third Law heats.

The heterogeneous reaction (II) was investigated in two steps. The vertical transpiration assembly employed in the measurement of volatility in dry oxygen was used for volatility-rate measurements with moisture contents of up to 23 torr. The carrier gas was first bubbled through distilled water kept at a predetermined temperature and then passed over the sample located in the isothermal zone of the furnace. The total uranium thus transported was collected in the platinum collector and analysed. The geo-

TABLE I. EQUILIBRIUM PRESSURE OF UO_3 (g) OVER U_3O_{8-z}(s) IN PURE OXYGEN

Temperature, T (K ± 0.5 K)	1/T (10^{-4} K^{-1})	No. of moles of uranium collected (× 10^{-6})	No. of moles of oxygen passed	log p_{UO_3} (footnote a)
1525	6.557	3.85	0.440	-0.0532
1525	6.557	4.25	0.440	- 0.0092
1573	6.359	5.15	0.265	0.2938
1573	6.359	5.15	0.264	0.2948
1572	6.361	5.00	0.264	0.2828
1572	6.361	5.15	0.264	0.2948
1572	6.361	5.05	0.264	0.2868
1571	6.365	5.05	0.264	0.2868
1605	6.231	5.05	0.132	0.5878
1626	6.150	6.85	0.133	0.7168
1627	6.148	6.85	0.133	0.7168
1675	5.997	6.50	0.0523	1.0988
1675	5.997	6.50	0.0526	1.0988
1572	6.364	5.05	0.265	0.2848

[a] p in Pa.

metry of the apparatus limited the highest moisture content to p_{H_2O} = 23 torr. The total uranium transported by 1 mole of oxygen containing 23 torr of water vapour was measured as a function of temperature and is presented in Table II. From the total uranium collected, the fraction removed as UO_3(g) owing to the equilibrium (III) was calculated from the expression (1). The difference was attributed to the $UO_3 \cdot nH_2O$ species formed as a result of the reaction (II). The partial pressure data of $UO_3 \cdot nH_2O$ thus evaluated were fitted to a linear expression using the least-squares method to calculate the enthalpy and the entropy change for the reaction (II). The least-squares expression for the hydrate was:

$$- \log p_{hyd} = \frac{(9612 \pm 416)}{T} - (6.13 \pm 0.28) \quad \text{for} \quad 1273 \leq T \leq 1625 \text{ K} \quad (3)$$

and

$$\Delta G_2^0 \text{(kJ)} = (183.8 \pm 7.7) - (52.2 \pm 5.6) \times 10^{-3}\ T \quad (4)$$

TABLE II. THE URANIUM TRANSPORT FROM U_3O_{8-z} AS $UO_3 \cdot nH_2O$ BY MOIST OXYGEN AT 23, 55, 92 AND 149 torr PARTIAL PRESSURE OF OXYGEN

Temperature T (K ± 0.5 K)	No. of moles of oxygen passed	No. of moles of water passed	No. of moles of uranium transported ($\times 10^{-6}$)	$P_{U, total}$	$P_{UO_3 \cdot nH_2O}$
A. Partial pressure of water vapour = 149 torr					
1323	0.209	0.036	0.693	0.29	0.28
	0.202	0.035	0.662	0.28	0.27
1345	0.164	0.028	0.95	0.50	0.48
	0.195	0.042	1.10	0.47	0.46
	0.130	0.026	1.58	1.03	0.98
1391	0.100	0.020	1.25	1.06	1.01
	0.132	0.029	1.41	0.88	0.83
	0.082	0.014	1.35	1.43	1.31
	0.082	0.015	1.09	1.14	1.02
1427	0.095	0.016	1.02	0.94	0.82
	0.083	0.015	1.09	1.12	1.00
	0.087	0.018	1.01	0.98	0.86
	0.064	0.010	1.20	1.64	1.41
1460	0.043	0.006	0.97	1.99	1.76
	0.041	0.006	0.66	1.46	1.23
B. Partial pressure of water vapour = 92 torr					
	0.188	0.007	0.36	0.19	0.18
1323	0.193	0.007	0.36	0.18	0.17
	0.217	0.008	0.36	0.17	0.16
	0.240	0.008	0.49	0.20	0.19
	0.251	0.033	2.04	0.72	0.67
1391	0.192	0.026	1.67	0.77	0.73
	0.203	0.027	1.66	0.73	0.68
	0.208	0.027	1.59	0.68	0.63
C. Partial pressure of water vapour = 55 torr					
	0.216	0.016	0.97	0.42	0.37
1391	0.241	0.018	1.17	0.46	0.41
	0.206	0.015	0.90	0.41	0.36

TABLE II (cont.)

Temperature T (K ± 0.5 K)	No. of moles of oxygen passed	No. of moles of water passed	No. of moles of uranium transported ($\times 10^{-6}$)	$p_{U,total}$	$p_{UO_3 \cdot nH_2O}$
	D. Partial pressure of water vapour = 23 torr				
1373	0.436	0.013	0.63	0.15	0.11
	0.429	0.013	0.71	0.17	0.13
1423	0.341	0.010	1.23	0.36	0.26
	0.332	0.010	1.15	0.35	0.25
1473	0.334	0.010	2.30	0.70	0.39
	0.332	0.010	2.24	0.68	0.38
	0.333	0.010	2.23	0.68	0.38
1523	0.265	0.008	3.56	1.36	0.55
	0.266	0.008	3.70	1.41	0.60
	0.266	0.008	3.64	1.39	0.58
1573	0.266	0.008	8.52	3.25	1.22
	0.266	0.008	8.50	3.24	1.20
	0.266	0.008	8.60	3.28	1.25
1623	0.110	0.003_3	6.70	6.15	1.32

These data (Tables I and II) unambiguously establish that the introduction of moisture enhances the vapour transport of uranium species significantly. From this enhancement the existence of a volatile uranium oxide hydrate ($UO_3 \cdot nH_2O$) was inferred.

The next step was to establish the molecular formula of this species. At a given temperature and fixed oxygen pressure, the partial pressure of $UO_3(g)$ will be constant. However, the total uranium transported in the presence of moisture is determined by the heterogeneous equilibrium (reaction II). Thus, the total uranium transported in the presence of moisture is given by:

$$p_{U,total} = p_{UO_3} + K_2\, p^n_{H_2O} \qquad (5)$$

where K_2 is the equilibrium constant of reaction (II). Hence, a measurement of total uranium [$p_{U,total}$] transported by oxygen containing different amounts of moisture at a definite temperature would give the value of n according to the reaction (II) (assuming that there was no other reaction involving these species). A plot of $p_{U,total}$ versus p_{H_2O} at 1573 K between 0 - 23 torr (Fig.2) gave linear relationship, indicating that n = 1.0. Thus the hydrate could be assigned the formula, $UO_3 \cdot H_2O$.

To confirm the influence of water on the volatility of U_3O_{8-z} and to confirm the correctness of n = 1, further measurements were made. Three series of measurements with p_{H_2O} = 55 torr, 92 torr and 149 torr,

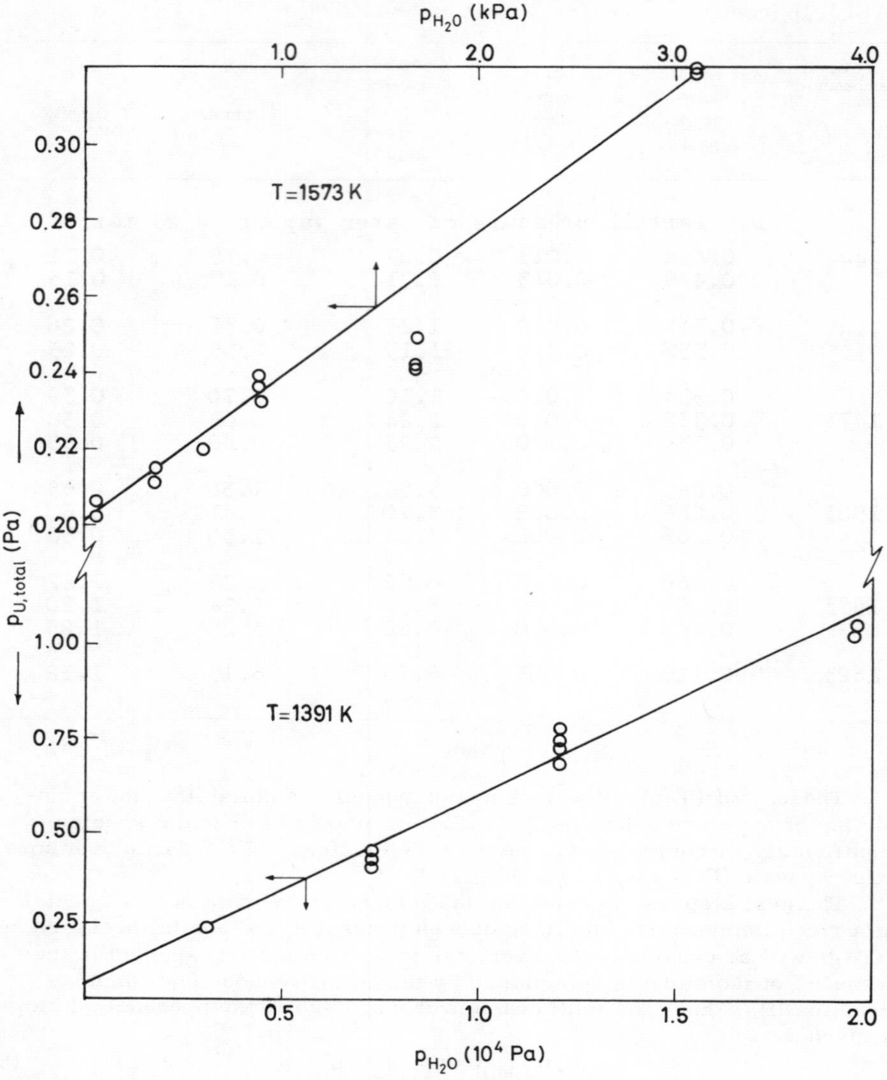

FIG.2. Plot of $p_{U,total}$ against p_{H_2O} at 1573 and 1391 K.

respectively, were completed between 1325 K and 1474 K. For this purpose, the apparatus was modified and a horizontal arrangement with heating tapes on all exposed glass parts on the inlet side was adopted. Oxygen gas was saturated with water vapour by bubbling it through distilled water whose temperature was thermostatically controlled and it was then passed over the sample. On the exit side, the moisture was first condensed in a weighed, cooled U-tube before the oxygen was admitted to the wet-test meter. This enabled us to determine the amount of water that had passed over the sample.

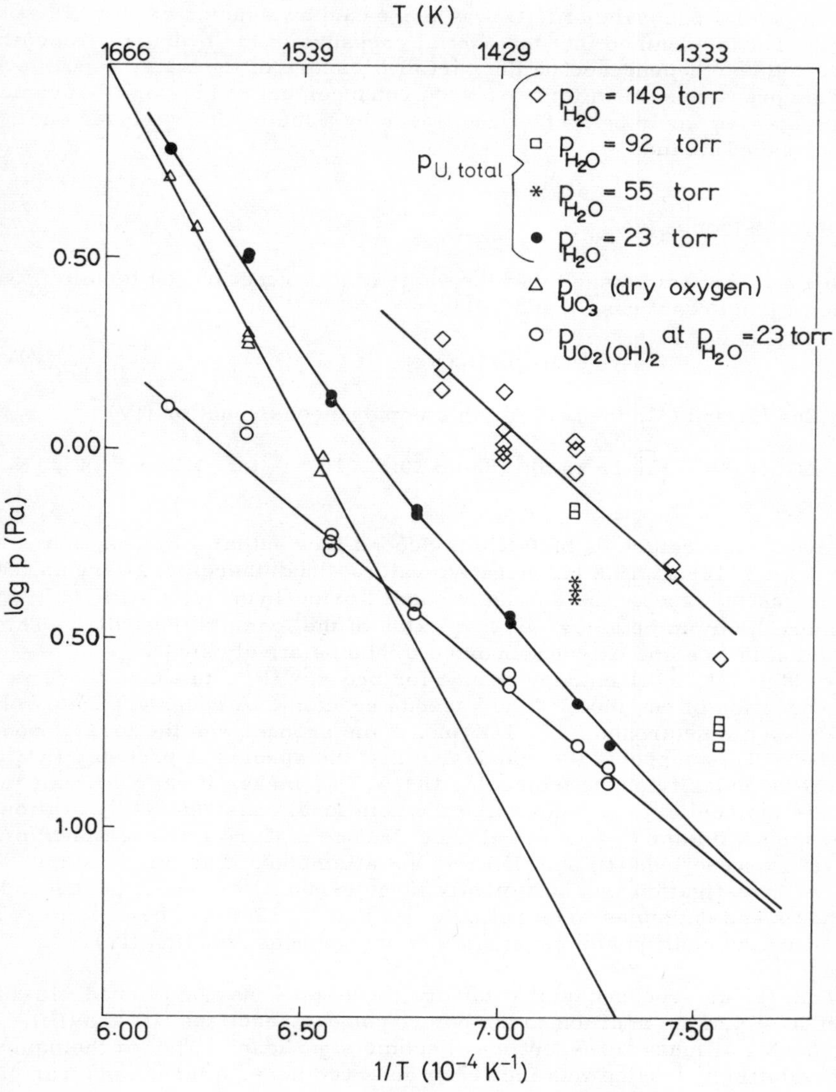

FIG.3. Volatility of the U_3O_{8-z} phase in moist oxygen.

The data are presented in Table II. As before, the total uranium collected inside the collector was analysed and the moles of uranium transported as the hydrate were evaluated with the help of equation (1). The plot of $\log p_{U,total}$ versus $\log p_{H_2O}$ at T = 1391 K had a slope of 1.08 ± 0.10, thus confirming the value of n = 1.0 in this pressure region.

These observations, together with the results using dry oxygen as the carrier gas, are summarized in Fig.3. It is seen that the total uranium-bearing species transported in the vapour phase by wet oxygen containing

149 torr partial pressure of H_2O was increased by a factor of 30 to 40 at 1325 K. If it is assumed that the partial pressure of the hydrate will continue to have a linear dependence on the partial pressure of the water vapour even at higher pressures, then this observed enhancement of the volatile uranium-bearing species would be further increased by a factor of five under one atmosphere of steam.

4. DISCUSSION

Since it was established that the value of n in reaction (II) is unity, it follows from reactions (II) and (III) that

$$UO_3(g) + H_2O(g) = UO_3 \cdot H_2O(g) \qquad \text{(IV)}$$

From Eqs (2) and (4), the ΔG^0 for this homogeneous reaction (IV) is:

$$\Delta G_3^0\,(\text{kJ}) = -(186.1 \pm 15.6) + (93.0 \pm 10.5) \times 10^{-3}\,T \quad \text{for} \quad 1325 \leq T \leq 1575\,K \qquad (6)$$

At a mean temperature of 1450 K this Second Law enthalpy of reaction has a large (- 186.1 ± 15.6 kJ) negative value. It is, therefore, very unlikely that the gaseous species is a loosely bonded oxide-hydrate, $UO_3 \cdot H_2O$: it is more likely to be a gaseous hydroxide of the formula $UO_2(OH)_2$. This conclusion is to some extent supported by the relatively small (ΔS^0 = - 93.0 J/K·mol) entropy change for process (IV), in which there is a net reduction of one mole of the gaseous reactant. A rough estimate for this ΔS^0 would be around - 167 J/K·mol if the product was the loosely bonded $UO_3 \cdot H_2O$. This supports the conclusion that the species is probably $UO_2(OH)_2$.

The complexity of the molecule, $UO_2(OH)_2$, makes it very difficult to estimate molecular parameters like the rotational constants and vibrational frequencies with any degree of reliance. Hence a Third Law treatment of the data for reactions (II) and (III) was not attempted. The temperature range of investigation was sufficiently large (~300 K) to expect the Second Law heats and entropies to be reliable, so that Eq.(6) could be used for estimating the equilibrium constants of homogeneous reaction (IV) with confidence.

This linear dependence of volatility on the p_{H_2O} may be altered with high pressure steam: in addition to the heterogeneous reactions (II) and (III), other modes of mass transport may become significant. Though the nature of the volatile hydroxide was indirectly inferred here, a mass spectrometric measurement would unambiguously establish this fact.

ACKNOWLEDGEMENTS

The authors thank Shri O.M. Sreedharan, Chemistry Division, Bhabha Atomic Research Centre, for his assistance in carrying out preliminary experiments and Shri A.S. Kerkar, also of Chemistry Division, in the fabrication of the apparatus. Our indebtedness to the personnel of the glass workshop is acknowledged.

REFERENCES

[1] GLEMSER, O., WENDLANDT, H.G., Advances in Inorganic and Nuclear Chemistry (EMELEUS, H.J., SHARP, A.G., Eds) 5, Academic Press (1961) 215-58.

[2] CROSSWEINER, L.I., SEIFERT, R.L., J. Am. Chem. Soc. 74 (1952) 2701.

[3] RANDALL, S.P., MARGRAVE, J.L., J. Inorg. Nucl. Chem. 16 (1960) 29.

[4] DANIELSON, P., CHANDRASEKHARAIAH, M.S., EDWARDS, R.K., High Temp. Sci. 1 (1969) 98.

[5] FRYXELL, R.E., AIETKEN, E.A., J. Nucl. Mater. 30 (1969) 50.

[6] SREEDHARAN, O.M., DHARWADKAR, S.R., CHANDRASEKHARAIAH, M.S., Bhabha Atomic Research Centre Rep. BARC/I-239 (1973).

[7] ATHAVALE, V.T., PATKAR, H.J., RAO, B.L., J. Sci. Ind. Res., B 21 (1962) 231.

[8] ACKERMANN, R.J., THORN, R.J., TETENBAUM, M., ALEXANDER, C., J. Phys. Chem. 64 (1960) 350.

[9] MERTEN, U., BELL, W.E., Chapter 4, The Characterization of High-Temperature Vapors (MARGRAVE, J.L., Ed.), Wiley, New York (1967).

ЭНТАЛЬПИЯ ОБРАЗОВАНИЯ ДИБОРИДА ТИТАНА

В.В.АХАЧИНСКИЙ, Н.А.ЧИРИН
Государственный комитет по использованию атомной энергии СССР,
Москва,
Союз Советских Социалистических Республик

Abstract–Аннотация

ENTHALPY OF FORMATION OF TITANIUM DIBORIDE.

The values given in the literature for the enthalpy of the formation of titanium diboride, as obtained experimentally and by theoretical estimation, range between -32 and -74.4 kcal/mol. In this paper the authors use the method of direct synthesis from elements in a Calvet calorimeter to determine the enthalpy of formation, $\Delta H^0_{f,298}$, of titanium diboride with the composition $Ti_{1.000\pm0.002}B_{2.056\pm0.006}C_{0.009}N_{0.003}$, which was found to be -76.78 ± 0.83 kcal/mol. They calculate that $\Delta H_{f,298}(TiB_{2.056})$ = -76.14 ± 0.85 kcal/mol. The procedure employed makes it possible to carry out the titanium diboride synthesis reaction with the calorimeter at room temperature.

ЭНТАЛЬПИЯ ОБРАЗОВАНИЯ ДИБОРИДА ТИТАНА.

Имеющиеся в литературе значения энтальпии образования диборида титана, полученные как экспериментально, так и путем оценки, находятся в интервале от -32 до -74,4 ккал/моль. В данной работе методом прямого синтеза из элементов в калориметре Кальве определена энтальпия образования $\Delta H^0_{f,298}$, диборида титана состава $Ti_{1,000\pm0,002}B_{2,056\pm0,006}C_{0,009}N_{0,003}$, оказавшаяся равной -76,78±0,83 ккал/моль. Вычислено, что $\Delta H^0_{f,298}$ $TiB_{2,056}$ = = -76,14±0,85 ккал/моль. Использованная методика позволила осуществить реакцию синтеза диборида титана при комнатной температуре калориметра.

1. ВВЕДЕНИЕ

Диборид титана – хороший поглотитель нейтронов, что обеспечило ему применение в ядерной технике. Высокая температура плавления (~2900°C) и достаточная коррозионная стойкость позволяют считать, что сфера применения TiB_2 будет расти. В связи с этим необходимо всестороннее знание его свойств и, в частности, одной из важнейших характеристик – энтальпии образования.

К сожалению, имеющиеся в литературе данные об энтальпии образования TiB_2 сильно различаются (см. табл. I). Многие из них носят оценочный характер [3,9], или получены косвенными способами [1,2,5,6,8]. Калориметрические измерения методом сожжения в кислороде [4,10] и прямого синтеза из элементов [7] также не дали согласующихся результатов. Несмотря на высокую точность результата работы [10] и хорошую репутацию ее автора, нельзя безоговорочно принять найденную им величину окончательной, т.к. она существенно выше всех ранее предложенных. Поэтому было целесообразно провести новое определение энтальпии образования диборида титана, по возможности не менее точное, чем в работе [10], и, желательно, другим способом.

ТАБЛИЦА I. ЛИТЕРАТУРНЫЕ ДАННЫЕ ОБ ЭНТАЛЬПИИ ОБРАЗОВАНИЯ TiB_2

Автор	Год	Метод	$-\Delta H^0_{f,298}$ (ккал/моль)	Литература
Г.В.Самсонов	1955	По константе равновесия реакции $2TiO+B_4C=2TiB_2+CO_2$	70,04	[1]
L.Brewer, H.Haraldsen	1955	Оценка по реакциям с участием боридов	71,4	[2]
Г.В.Самсонов	1956	Оценка по Кубашевскому	73,0	[3]
В.А.Эпельбаум М.И.Старостина	1958	Калориметрия, сжигание в кислороде	65,8±3	[4]
P.O.Schissel, W.S.Williams	1959	Масс-спектрометрия и метод Кнудсена	32	[5]
W.S.Williams	1961	Критический анализ и экспериментальная проверка работ [1,2,5]	50÷60	[6]
C.E.Lowell, W.S.Williams	1961	Прямой синтез при 1500°C	50±5	[7]
P.O.Schissel, O.C.Trulson	1962	Масс-спектрометрия испарения в системе Ti-B	52,1±6	[8]
Л.А.Резницкий	1966	Оценка по Кирееву и Карапетьянцу	66±3	[9]
E.J.Huber	1966	Калориметрия. Сжигание в кислороде.	77,4±0,9[a]	[10]

[a] Для соединения $TiB_{2,022}$

2. ВЫБОР МЕТОДИКИ

Анализ методики эксперимента в работе [7] привел нас к выводу, что столь низкое значение теплоты образования диборида титана получено в результате частичного протекания реакции во время прогрева смеси для ее обезгаживания.

Несмотря на явную неудачу авторов работы [7] при определении ΔH^0_f TiB_2 методом прямого синтеза, а также заявление авторов работы [4] о том, что в случае боридов определение ΔH^0_f путем синтеза из элементов вообще невозможно, мы выбрали именно метод прямого синтеза диборида титана в калориметре непосредственно из элементов титана и бора.

Выбор основывался на литературных данных о том, что реакция между мелкодисперсными порошками титана и бора протекает быстро и нацело с образованием TiB_2, если элементы взяты в стехиометрическом соотношении [11]. Наши исследования подтвердили эти данные. В таком случае метод прямого синтеза имеет преимущество перед всеми остальными.

ТАБЛИЦА II. РЕЗУЛЬТАТЫ АНАЛИЗА ТИТАНА И БОРА

Определяемый элемент	Анализируемый элемент		Метод анализа
	Титан вес %	Бор вес %	
Ti	97,45±0,20	-	Сплавл. с персульфатом и калориметрирование
B	-	98,6±0,8	Растворение в HNO_3 и титрование щелочью
O	1,90±0,10	-	Нейтронно-активационный
-"-	2,20±0,10	0,50±0,03	Спекание с углеродом при 3000°C и хроматография CO
-"-	2,10±0,30	-	По разности 100 % – –%(Ti+O+N+C+H+Fe+H_2O)
N	0,15	0,05	Спекание с углеродом при 3000°C и хроматография
-"-	0,21±0,01	0,004±0,001	По Кьельдалю
C	0,05±0,01	1,10±0,05	Сжигание в кислороде
H	0,11±0,01	0,001	Спектральный
H_2O	0,010±0,005	0,05±0,01	Кулонометрический с нагревом пробы до 500°C
Fe	0,035±0,003	0,20±0,01	Растворение в кислоте и колориметрирование
Σ	99,96±0,22	100,45±0,90	

3. ЭКСПЕРИМЕНТАЛЬНАЯ ЧАСТЬ

3.1. Разработка способа инициирования реакции

Способ инициирования реакции – один из важнейших элементов методики, от которого зависит конструкция калориметра и точность результатов.

Температура вспышки спрессованной смеси порошков состава Ti+2B составляет, по нашим измерениям, около 1000°C. Выше этой температуры в условиях теплоизоляции смесь за доли секунды саморазогревается до температуры около 2900°C и реакция протекает полностью.

Мы разработали простой способ, позволяющий не нагревать всю смесь до 1000°C для начала саморазвивающейся реакции. Реакцию инициировали вольфрамовой проволокой, разогреваемой импульсом разрядного тока конденсатора до переплавления вольфрама. Использовали проволоку длиной 8 мм и диаметром 0,1 мм, которая касалась таблетки спрессованной смеси на участке длиной не более 2 мм.

При таком способе инициирования реакция начиналась и заканчивалась мгновенно при комнатной температуре калориметра. В результате исключались источники ошибок, имевшие место в работе [7], и отпадала необходимость пересчета теплоты реакции к стандартной температуре 298°K.

Энергия, вводимая в калориметр для инициирования реакции, была невелика и составляла всего 0,8 кал. Возможно, она может быть значительно уменьшена.

Способ инициирования некоторых реакций с помощью переплавляемой импульсом тока вольфрамовой проволоки может в ряде случаев оказаться более выгодным и удобным, чем рекомендуемый в работе [12] импульсный нагрев с помощью оптического квантового генератора.

3.2. Исходные вещества

В работе использовали порошок титана крупностью не более 40 мкм (основная фракция 3-10 мкм), полученный гидрированием иодидного титана с последующим измельчением гидрида и термическим разложением его в вакууме.

Работа с порошком и его хранение осуществлялись на воздухе, поэтому он содержал значительное количество кислорода. Однако длительное наблюдение за изменением веса порошка титана показало, что он "стабилизировался", т.к. вес оставался постоянным с точностью до 0,05 % в течение 10 дней. Порошок был использован в течение месяца после его анализа.

Поскольку основной и наиболее важной примесью в титане был кислород, его определению было уделено большое внимание. Кислород был определен тремя независимыми методами.

Из анализов на содержание азота мы отдали в дальнейшем предпочтение результатам, полученным по методу Кьельдаля, т.к. при высокотемпературном нагреве возможно неполное выделение азота.

Использованный в работе аморфный бор предварительно прокаливали в вакууме при 1450°С в течение 1 часа, после чего бор терял высокую адсорбционную способность и не окислялся на воздухе. Его вес в течение 10 дней увеличился всего на 0,1 %, а затем не изменялся в течение 3 месяцев наблюдений.

Результаты анализа титана и бора приведены в табл. II.

Для приготовления смеси взяли 0,6993 г титана и 0,3153 г бора. После тщательного перемешивания порошков из смеси прессовали таблетки диаметром 3,5 мм и высотой около 1 мм при давлении 7300 кг/см^2.

3.3. Калориметрическая установка

3.3.1. Калориметр и измерительные приборы

Определение теплоты реакции проводили на калориметрической установке французской фирмы "Setaram" с калориметром Кальве, фотокомпенсационным усилителем "Amplispot" и гальванометрическим самописцем "Graphispot". Объем ячеек калориметра – 15 см^3 (диаметр – 17 мм, высота – 80 мм). Порог чувствительности установки при мгновенном источнике энергии – около $3 \cdot 10^{-3}$ калорий, постоянная времени калориметра – 200 с.

3.3.2. Реакционный контейнер

С целью увеличения точности измерений и повышения производительности установки нами был разработан специальный реакционный контейнер, состоящий из 3-х состыкованных элементов, каждый из ко-

ТАБЛИЦА III. ЗАВИСИМОСТЬ КОНСТАНТЫ КАЛОРИМЕТРА ОТ МОЩНОСТИ И ПОЛОЖЕНИЯ ИСТОЧНИКА ЭНЕРГИИ

Введено тепла (кал)	Мощность нагревателя (кал/с)	Константа калориметра (кал/дел)		
		Элемент контейнера		
		Верхн.	Средн.	Нижн.
47,0	0,425	0,1537	0,1506	0,1505
44,0	1,703	0,1532	0,1498	0,1491

торых представлял собой медный цилиндрик длиной 22 мм и диаметром 12 мм с внутренней полостью диаметром 6 мм и глубиной 20 мм. Внутрь каждой полости введены два электрода, к которым крепилась вольфрамовая проволочка. На проволочку клали таблетку из исследуемой смеси. Для предохранения таблетки от контакта с медной поверхностью внутрь полости вставляли фарфоровую трубочку, которая одновременно центрировала таблетку и удерживала ее на проволочке.

Контейнер опускали на дно длинной стеклянной ампулы, которую сверху закрывали пришлифованной крышкой с вваренными в нее 4 электродами и вакуумным краном.

К электродам присоединялись провода, идущие к реакционному контейнеру и к источнику энергии для инициирования реакции или калибровки калориметра.

Ампулу опускали в ячейку калоримтра, эвакуировали и заполняли аргоном до атмосферного давления. Точно такую же ампулу с контейнером опускали во вторую ячейку калориметра.

Таким образом, конструкция реакционного контейнера позволяла без извлечения его из калориметра произвести последовательно 6 или, при несколько большей точности, 4 опыта, т.е. закончить собственно измерительную калориметрическую часть работы за 2-3 дня.

3.3.3. Калибровка калориметра

При калибровке калориметра кроме константы калориметра, т.е. площади под кривой, записываемой самописцем, приходящейся на единицу выделившейся в ячейке энергии, определяли также зависимость константы от положения источника энергии в реакционном контейнере и от скорости тепловыделения (мощности источника). Было найдено, что без разборки контейнера константа воспроизводится с точностью ± 0,2 % и почти не зависит от мощности вводимой энергии (табл. III).

Из табл. III видно, что константа калориметра практически одинакова для среднего и нижнего элементов реакционного контейнера и на 1,5 % больше для верхнего, что обусловлено влиянием подводных проводов.

В одном из элементов ячейки всегда находился нагреватель, поэтому калибровка калориметра проводилась в каждом опыте без разборки калориметра.

ТАБЛИЦА IV. РЕЗУЛЬТАТЫ ИЗМЕРЕНИЯ ТЕПЛОТЫ РЕАКЦИИ СМЕСИ Ti + 2B

№№ п/п	Навеска (мг)	Константа калориметра (кал/дел)	Выделилось тепла (кал)	Q (кал/г)
1	39,66	0,1491	39,38	992,9
2	38,55	0,1498	38,40	996,1
3	42,52	0,1497	42,16	991,5
4	34,115	0,1497	34,07	998,7
5	39,72	0,1469	39,68	998,9
6	36,80	0,1468	36,49	991,6
			Среднее	995,0±3,5[a]

[a] Погрешность с надежностью 0,95

При увеличении мощности в 7 раз константа калориметра уменьшалась всего на 0,5 %. Учитывая тепловую инерцию ячейки, можно считать, что при калибровке со скоростью ввода тепла около 2 кал/с достаточно хорошо воспроизводился исследуемый процесс, т.к. термокинетические кривые процесса и калибровки практически совпадали.

3.4. Калориметрические измерения

Снаряженные исследуемыми образцами и нагревателями реакционные контейнеры в ампулах опускали в ячейки калориметра, находящегося при комнатной температуре 22-23°C, эвакуировали и заполняли аргоном. После установления в системе теплового равновесия поджигали первую таблетку, затем проводили калибровку калориметра, поджигали вторую таблетку и снова калибровали калориметр. Всего было проведено 6 опытов, результаты которых приведены в табл. IV.

Специальными опытами было установлено, что для инициирования реакции вводится 0,77 ± 0,06 кал. Эта величина учтена при вычислении значений q и Q, приведенных в табл. IV.

3.5. Анализ продуктов реакции

Для вычисления теплоты образования знание состава продуктов реакции так же важно, как и состава исходных компонентов. Ниже приводятся сведения о проделанных анализах продуктов реакции и их результатах.

а) Установлено (с точностью метода рентгенофазового анализа), что продукт реакции содержит только одну фазу — TiB_2 с параметрами решетки a=3,033 ± 0,001 Å, c =3,226 ± 0,001 Å.

б) Методом металлографического анализа обнаружено наличие второй фазы по границам зерен TiB_2. Микротвердость фазы TiB_2 равна 3480 кг/мм^2, размер зерна 2-10 мкм. Микротвердость второй фазы определить не удалось из-за отсутствия участка нужных размеров. В плоскости шлифа площадь второй фазы составляла $<3\%$ от площади основной фазы.

в) Найдено, что основной продукт реакции содержит $0,50 \pm 0,03$ вес% кислорода, $0,06 \pm 0,006$ вес% азота и $0,16 \pm 0,02$ вес% углерода.

г) Установлено, что TiB_2 не содержит свободного титана с точностью до 0,01 вес% (путем дейтерирования продукта при давлении 7 атм и температуре 450°C в течение 5 часов и последующего определения в нем дейтерия).

д) В атмосфере аргона в газоволюмометрической установке были "сожжены" таблетки того же состава, что и при калориметрических измерениях, но весом 1 г. Установлено, что в результате реакции

- выделилось 14,0 норм.см^3 газа
- вес выделившегося газа равен 9 ± 2 мг
- газ содержит 57 ± 10 об% ($CO + N_2$) и 43 ± 10 об% H_2
- из таблетки испарилось 26 ± 3 мг вещества, сконденсировавшегося в виде темного налета на стенках реакционного сосуда
- налет содержит (в вес%) $40 \pm 0,4$ титана, $8,6 \pm 0,8$ бора и $2,1 \pm 0,4$ железа.

е) Найдено, что при реакции в калориметре в газовую фазу переходит 0,3 вес% кислорода (в виде CO) и 0,05 вес% азота.

4. ОБРАБОТКА ЭКСПЕРИМЕНТАЛЬНЫХ ДАННЫХ

4.1. Обработка результатов анализа и установление уравнения реакции, протекающей в калориметре

На основании результатов анализа и сведений о фазовых диаграммах систем титана и бора с примесными элементами [13,14,15,16], а также данных о термодинамических свойствах этих систем [15,16,17, 18,19] были рассчитаны содержание примесей в исходных компонентах и теплоты их образования в расчете на 1 г смеси. Результаты расчета приведены в табл.V. Распределение кислорода в титане на кислород, содержащийся в виде твердого раствора и связанный в TiO_2, произведено на основании измерения параметров решетки использованного нами порошка. Постоянная гексагональной решетки "c" оказалась равной 4,686 Å, что соответствует содержанию в титане 0,2 вес% растворенного кислорода. Правда, такому же параметру решетки соответствует примерно такое же содержание в виде твердого раствора азота, однако ошибка в распределении кислорода не внесет существенной ошибки в теплоту образования TiB_2, т.к. количество кислорода в виде твердого раствора невелико, а разница в энергии связи кислорода в TiO_2 и твердом растворе также мала.

Для сведения материального равенства исходных и конечных продуктов считали, что в возогнавшемся веществе кроме бора, титана и железа содержится весь кислород, который не вошел ни в основное вещество, ни в состав выделившихся газов (в виде CO). Кроме того, полагали, что титан и бор к моменту анализа окислялись до TiO_2 и B_2O_3. При этих допущениях оказалось, что из 1 г смеси в процессе

ТАБЛИЦА V. СОДЕРЖАНИЕ И ТЕПЛОТА ОБРАЗОВАНИЯ ПРИМЕСЕЙ В ТИТАНЕ И БОРЕ

Титан				Бор			
Примесь	Кол-во молей на 1 г смеси	$-\Delta H^0_{f,298}$ (ккал/моль)	$-\Delta H^0_{f,298}$, примеси в 1 г смеси (кал)	Примесь	Кол-во молей на 1 г смеси	$-\Delta H^0_{f,298}$ (ккал/моль)	$-\Delta H^0_{f,298}$, примеси в 1 г смеси (кал)
O тв.р-р	$0,86 \cdot 10^{-4}$	134	11,5±0,6	B_2O_3	$0,30 \cdot 10^{-4}$	300	9,00±0,05
TiO_2	$4,1 \cdot 10^{-4}$	225	92,2±4,0	B_4C	$2,8 \cdot 10^{-4}$	14	3,9±0,4
C тв.р-р	$0,29 \cdot 10^{-4}$	45	1,3±0,2	H_2O	$0,08 \cdot 10^{-4}$	68	0,5±0,1
N тв.р-р	$1,0 \cdot 10^{-4}$	80	8,0±0,04	Fe тв.р-р	$0,12 \cdot 10^{-4}$	16	0,19±0,01
TiH_2	$4,0 \cdot 10^{-4}$	30	12,0±1,2	N тв.р-р	$3 \cdot 10^{-6}$	60	0,18
M тв.р-р	$4,1 \cdot 10^{-6}$	5	0,02	M тв.р-р	$0,15 \cdot 10^{-4}$	10	0,15

реакции возгоняется $1{,}52 \cdot 10^{-4}$ молей B_2O_2, $1{,}96 \cdot 10^{-4}$ молей TiO и $1{,}29 \cdot 10^{-4}$ молей Ti. Железа в возгоне содержалось $1{,}1 \cdot 10^{-4}$ моля. Оставшемуся веществу (~2 вес%) мы приписали молекулярный вес 40 и энергию связи с бором 10 ккал/г-атом.

Уменьшение содержания углерода в продукте по сравнению с содержанием в исходных веществах, равное $1{,}9 \cdot 10^{-4}$ молей, полностью соответствует количеству CO, найденному в газе.

На основании объема и веса выделившегося газа можно рассчитать, что он содержит 47 об% $(CO+N_2)$ и 53 об% H_2, что хорошо согласуется с данными масс-спектрального анализа газа.

Поскольку объем выделяющегося при реакции газа и содержание в нем окиси углерода определены довольно точно, мы приписали газовой фазе следующий состав:

$1{,}9 \cdot 10^{-4}$ молей CO, $0{,}3 \cdot 10^{-4}$ молей N_2 и $4{,}1 \cdot 10^{-4}$ молей H_2.

Учитывая все вышеизложенное, можно написать следующее уравнение протекающей в калориметре реакции (в расчете на 1 г смеси):

$$132{,}1 \cdot 10^{-4}\, Ti\,(C_{0,002}\; N_{0,008}\; O_{0,0065}\; M_{0,0003})^1 + 4{,}09 \cdot 10^{-4}\, TiO_2 +$$

$$+ 4{,}0 \cdot 10^{-4}\, TiH_2 + 269{,}4 \cdot 10^{-4}\, B\,(Fe_{0,0005}\; N_{0,0001}\; M_{0,0006}) +$$

$$+ 2{,}8 \cdot 10^{-4}\, B_4C + 0{,}30 \cdot 10^{-4}\, B_2O_3 \cdot 0{,}08 \cdot 10^{-4}\, H_2O \longrightarrow$$

$$135{,}3 \cdot 10^{-4}\, TiB_{2,056}\,(N_{0,003}\; C_{0,009}) + 1{,}57 \cdot 10^{-4}\, TiO_2 +$$

$$+ 1{,}9 \cdot 10^{-4}\, CO + 4{,}1 \cdot 10^{-4}\, H_2 + 0{,}3 \cdot 10^{-4}\, N_2 + 1{,}50 \cdot 10^{-4}\, B_2O_2 +$$

$$+ 2{,}0 \cdot 10^{-4}\, TiO + 1{,}3 \cdot 10^{-4}\, Ti + 1{,}1 \cdot 10^{-4}\, Fe + 1{,}1 \cdot 10^{-4}\, M$$

Существенным подтверждением этого уравнения, основанного только на анализе исходных веществ и побочных продуктов реакции, явился анализ основного продукта реакции, в результате которого оказалось, что он содержит $68{,}04 \pm 0{,}10$ вес% титана. Эта величина полностью совпадает с содержащейся в уравнении.

4.2. Вычисление энтальпии образования TiB_2

При вычислении энтальпии образования TiB_2 кроме приведенных в табл. V значений использовали следующие величины: $\Delta H^0_{f,298}$: $TiO_{(тв)}$ – 124 ккал/моль, $CO_{(г)}$ – 26,4 ккал/моль, $B_2O_{2(тв)}$ – 200 ккал/моль (наша оценка). В результате расчета оказалось, что

$$\Delta H^0_{f,298}\; Ti_{1,000 \pm 0,002}\; B_{2,056 \pm 0,006}\,(N_{0,003}\; C_{0,009}) =$$

$$= -76{,}78 \pm 0{,}83 \text{ ккал/моль}.$$

Приняв теплоту растворения азота и углерода в дибориде титана равной теплоте растворения в титане, можно рассчитать, что

$$\Delta H^0_{f,298}\; Ti_{1,000 \pm 0,002}\; B_{2,056 \pm 0,006} = -76{,}14 \pm 0{,}85 \text{ ккал/моль}.$$

Таким образом, найденная нами величина отличается от полученной Хьюбером [10] на 1,3 ккал, что находится в пределах ошибок опыта.

---

1 M = сумме всех определенных примесей, кроме O, N, C, H и Fe в случае бора.

Если пересчитать данные Хьюбера, приняв для теплоты образования B_2O_3 значение − 299,71 ± 0,41 [20] вместо − 300 ± 0,7, принятого Хьюбером, то оказывается, что $\Delta H^0_{f,298}$ $TiB_{2,022}$ = -77,1 ккал/моль, что отличается от полученной нами величины уже всего на 1 ккал.

ВЫВОДЫ

На основании проделанной нами работы и данных Хьюбера [10] можно для величины энтальпии образования диборида титана рекомендовать значение -76,6 ± 0,6 ккал/моль. Результаты всех других определений имеют лишь историческое значение.

ЛИТЕРАТУРА

[1] САМСОНОВ, Г.В., Ж.Прикл.Химии 28 (1955) 1018.
[2] BREWER, L., HARALDSEN, H., J.Electrochem.Soc. 102 (1955) 399.
[3] САМСОНОВ, Г.В., Ж.Физич.Химии 30 (1956) 2057.
[4] ЭПЕЛЬБАУМ, В.А., СТАРОСТИНА, М.И., В сб. Труды Конференции по Химии Бора и Его Соединений, Госхимиздат, М., (1958) 97.
[5] SCHISSEL, P.O., WILLIAMS, W.S., Bull.Am.Phys.Soc.Ser. 11 4 3 (1959).
[6] WILLIAMS, W.S., J.Phys.Chem. 65 (1961) 2213.
[7] LOWELL, C.E., WILLIAMS, W.S., Rev.Sci.Instrum. 32 (1961) 1121.
[8] SCHISSEL, P.O., TRULSON, O.C., J.Phys.Chem. 66 (1962) 1492.
[9] РЕЗНИЦКИЙ, Л.А., Ж.Физич.Химии 40 (1966) 134
[10] HUBER, E.J., J.Chem.Eng.Data 11 3 (1966) 430.
[11] МЕРЖАНОВ, А.Г., БОРОВИНСКАЯ, И.П., Докл. АН СССР 204 2 (1972) 366.
[12] ВИДАВСКИЙ, Л.М., Шестая Всесоюзная конференция по калориметрии, Расширенные тезисы докладов, Тбилиси, 1973, стр. 534.
[13] ХАНСЕН, М., АНДЕРКО, К., Структуры Двойных Сплавов, т.т.I, II, Металлургиздат, М., 1962 г.
[14] МОРОЗ, Л.С., ЧЕЧУЛИН, Б.Б. и др. Титан и Его Сплавы, т. I, Судпромгиз, Л., 1960 г.
[15] СТОРМС, Э., Тугоплавкие Карбиды, Атомиздат, М., 1970 г.
[16] Гидриды Металлов, Под ред. В.Мюллера и др. Атомиздат, М., 1973 г.
[17] Термодинамические Свойства Индивидуальных Веществ, Под ред. В.П.Глушко и др., т. II, изд. АН СССР, 1962 г.
[18] КРЕСТОВНИКОВ, А.Н.,и др., Справочник по Расчетам Равновесий Металлургических Реакций, Металлургиздат, М., 1963 г.
[19] KOMAREK, K.L., SILVER, M., Thermodynamics of Nuclear Materials (Proc.Symp. Vienna, 1962) IAEA, Vienna (1962) 749.
[20] Термические Константы Веществ. Под ред. В.П.Глушко и др., вып. V, М., 1971 г.

EXPERIMENTAL HEAT CAPACITIES OF $^{242}PuF_3$ AND $^{242}PuF_4$ FROM 10 TO 350 K, AND OF $^{244}PuO_2$ FROM 4 TO 25 K

Derived entropies and other thermodynamic properties at 298.15 K*

H.E. FLOTOW, D.W. OSBORNE,
S.M. FRIED, J.G. MALM
Argonne National Laboratory,
Argonne, Ill.,
United States of America

Abstract

EXPERIMENTAL HEAT CAPACITIES OF $^{242}PuF_3$ AND $^{242}PuF_4$ FROM 10 TO 350 K, AND OF $^{244}PuO_2$ FROM 4 TO 25 K: DERIVED ENTROPIES AND OTHER THERMODYNAMIC PROPERTIES AT 298.15 K.

The heat capacities of $^{242}PuF_3$ and $^{242}PuF_4$ have been determined from 20 to 350 K by the quasi-adiabatic method with intermittent electrical heating. From 18 to 58 K the measured temperature drifts with no electrical heating were compared with the measured heat capacities to evaluate the radioactive self-heating rate, $(1.210 \pm 0.004) \times 10^{-4}$ W/g Pu. Heat capacities between 10 and 20 K were determined from the radioactive self-heating rate and temperature drifts with no electrical heating. The data for $^{242}PuF_3$ showed a transition, believed to be non-magnetic in origin, in the region of 119 K. Magnetic susceptibility measurements indicate that $^{242}PuF_3$ becomes antiferromagnetic below about 9 K. The results for $^{242}PuF_4$ showed no anomalous effects. The heat capacity of $^{244}PuO_2$ has been obtained between 10 and 25K by the intermittent electrical heating technique. No thermal anomaly was detected from 4 to 25 K. Evidence for a slight amount of radiation crystal damage was detected for $^{242}PuF_3$ after holding its temperature below 9 K overnight. $^{242}PuF_4$ and $^{244}PuO_2$ showed no radiation damage effects. Heat capacities and derived entropies, enthalpies and Gibbs energies divided by temperature are presented in tabular form at selected temperatures to 350 K. Comparisons of these data with other data and estimates in the literature are made.

1. INTRODUCTION

Low temperature heat capacity measurements of ^{239}Pu (half-life 2.44×10^4 a) compounds are difficult and uncertain because of the high rate of self-heating (19.3×10^{-4} W/g) and serious crystal damage effects. This rate of self-heating limits the lowest practical working temperatures for accurate heat capacity measurements to 15 to 20 K. Radiation damage from alpha decay can result in energy storage at low temperatures which is partially or completely released at higher temperatures; if this occurs spuriously low heat capacity values are obtained. The U. S. Atomic Energy Commission, Division of Research, has made available to us gram quantities of the longer lived, less radioactive isotopes ^{242}Pu (half-life, 3.87×10^5 a) and ^{244}Pu (half-life, 8.26×10^7 a). A program has been initiated to

* Work performed under the auspices of the USAEC.

measure the low-temperature heat capacities of selected plutonium compounds of these isotopes to obtain reliable experimental entropies and other thermodynamic quantities at 298.15 K.

In this paper we present a summary of our results for $^{242}PuF_3$ and $^{242}PuF_4$ from 10 to 350 K and for $^{244}PuO_2$ from 4 to 25°K. No previous heat capacities for plutonium trifluoride or plutonium tetrafluoride have been reported. The heat capacity of $^{239}PuO_2$ from 11.9 to 325 K was reported by Sandenaw [1]; however, since he experienced adverse experimental difficulties of uncertain origins it seemed advisable to remeasure the low-temperature heat capacity of plutonium dioxide and to use $^{244}PuO_2$ at the lowest temperatures and $^{242}PuO_2$ above 20 K. The $^{242}PuO_2$ sample is being prepared at Argonne National Laboratory.

2. CALORIMETRIC SAMPLES

2.1. Plutonium Trifluoride

The $^{242}PuF_3$ was prepared at Argonne National Laboratory by the dissolving pure ^{242}Pu metal in hydrobromic acid, centrifuging this solution, and then precipitating the $^{242}PuF_3$ from the plutonium solution by bubbling an excess of pure hydrogen fluoride gas through the solution. The $^{242}PuF_3$ was dried at 500-600°C in a stream of hydrogen fluoride and hydrogen. Throughout the preparation reducing conditions were present so that the only possible oxidation state for the plutonium after dissolving the metal was Pu^{+3}. An x-ray powder pattern of this $^{242}PuF_3$ gave lattice constants in agreement with those previously reported by Zachariasen [2]. The sample was analyzed chemically for plutonium and fluorine by conversion of a weighed portion of $^{242}PuF_3$ to $^{242}PuO_2$ at 1120 K in a stream of oxygen and steam in a platinum tube; the evolved hydrogen fluoride was collected in dilute aqueous sodium hydroxide. The amount of fluorine was determined by titration with a thorium nitrate solution and with Alizarin Red S as the end-point indicator; the plutonium content was calculated from the weight of the $^{242}PuO_2$. A F/Pu ratio of 3.002 ± 0.008 was obtained. The mass of the calorimetric $^{242}PuF_3$ sample was 17.508 g.

2.2. Plutonium tetrafluoride

The $^{242}PuF_4$ sample was prepared by the reaction of the $^{242}PuF_3$ sample with pure fluorine at 450 to 500°C over a period about 6 d. After heating for 1 d it was found by weight gain and also by the volume of fluorine used that the reaction was 92% completed to form $^{242}PuF_4$. The sample was then heated for the remaining 5 days to complete the reaction. An analysis for fluorine and plutonium using the same method given above for plutonium trifluoride gave F/Pu ratio of 3.97 ± 0.02. The mass of the calorimetric $^{242}PuF_4$ sample was 20.8717 g.

2.3. Plutonium dioxide

The $^{244}PuO_2$ was prepared at Oak Ridge National Laboratory by heating precipitated plutonium oxalate in a porcelain crucible to 650°C in air for 4 h and then to 900°C for 4 h. Next, the sample contained in a tantalum crucible was heated in a vacuum furnace ($P = 1 \times 10^{-5}$ torr) to 1250°C for a period of 4.5 h and then allowed to cool to room temperature under vacuum conditions. The final heat treatment was to assure the correct $^{244}PuO_{2.000}$ stoichiometry and to improve the crystallinity of the particles. An x-ray analysis gave the lattice constant 5.395 ± 0.001 Å at 298 K, in good agreement with the value 5.396 ± 0.001 Å reported in the literature [3,4]. The mass of the sample was 3.6429 g.

3. CALORIMETRIC TECHNIQUES

3.1. Plutonium fluorides

The $^{242}PuF_3$ and $^{242}PuF_4$ samples were manipulated in the following manner. Each was encapsulated in a weighed 0.006 cm-wall high purity, oxygen-free, high conductivity copper container which fitted snugly into a gold-plated copper calorimeter [5] (laboratory designation: 6-GS-II). The calorimeter was sealed with a gold gasket [6] and had an internal volume of 5.83 cm^3. To assure rapid heat transfer within the calorimeter a small amount of helium gas (10^{-5} mol) was added, and also a small amount of Apiezon T grease was used between the copper capsule and the calorimeter.

The heat capacities were determined in an adiabatic calorimetric apparatus as described in the literature [7,8]. Recent modifications to this equipment to permit digital read-out and recording of the data, and a description of the calibration of the platinum resistance thermometer have also been published [9].

The loaded calorimeter was cooled to the lowest temperature (about 8 K) by raising the calorimeter and adiabatic shield into contact with a container of liquid helium (or liquid nitrogen) by means of a nylon cord and a windlass. Shortly after reaching the minimum temperature the calorimeter and shields were lowered into the operating position, and as soon as the adiabatic shield was under control (at about 10 K) a series of time-temperature readings was made. These observations were continued without any electrical heating of the calorimeter up to 20 K, at which point the temperature drift became low enough for heat capacity measurements to be made by our usual quasi-adiabatic method involving discontinuous electrical heating followed by a series of temperature drift measurements. For $^{242}PuF_3$ at 10 K the drift was 1.5 K/min, at 20 K it was 0.18 K/min, at 40 K it was 0.032 K/min, and at 350 K it was 0.0045 K/min. In a separate series of experiments on the empty calorimeter the drift was 0.013 K/min at 10 K, 0.0009 K/min at 20 K, 0.0005 K/min at 40 K, and -0.0026 K/min at 350 K.

For the measurements on the loaded calorimeter between 20 and 40 K the usual method of extrapolating the temperature-time curves to the middle of the heating period had to be modified. The temperatures T observed in the periods before and after the electrical heating period (after equilibrium was attained) were each fitted with an exponential equation of the form

$$T = a + b \exp(-\lambda t) \qquad (1)$$

where a, b, and λ are constants and t is the time. Corrected temperatures T_1' and T_2' were calculated from the relations

$$T_1' = T_1 + D_1 t_h/2 \quad (2)$$

and

$$T_2' = T_2 - D_2 t_h/2 \quad (3)$$

where t_h is the heating time. T_1 and D_1 are the temperature and drift (dT/dt), respectively, at the start of the heating period. Both of these were calculated from the equation (1) fitting the temperature-time observations before the heating. Similarly, T_2 and D_2 are the temperature and drift, respectively, at the end of the heating period, calculated from the equation (1) fitting the temperature-time observations made after equilibrium had been attained following the heating. The mean heat capacity $\overline{C}_p$ at the mean temperature $\overline{T} = (T_1' + T_2')/2$ was calculated from Equation (4),

$$\overline{C}_p = Q/(T_2' - T_1') \quad (4)$$

where Q is the electrical heat. This method corrects for the radioactive self-heating, for the heating due to the thermometer current, and for the heat interchange due to the slight deviation from the adiabatic conditions.

Above 40 K equation (1) could be replaced by

$$T = a + bt \quad (5)$$

and equations (2) and (3) reduced to the usual linear extrapolation of the temperature to the middle of the heating period.

3.2. Plutonium dioxide

The calorimeter was a copper (ASARCO) tube with a 0.476-cm inner diameter, a 0.038-cm wall thickness and a sample length of 6.2 cm. After loading the sample into the calorimeter the sides of the tube were flattened by positioning the calorimeter tube between steel plates and applying pressure to the steel plates with a hydraulic press. The result was that the sample had a average thickness of 0.076 cm and the sample was in good contact with the copper walls. The calorimeter was evacuated and sealed vacuum-tight.

A description of the calorimetric apparatus and cryostat that was used to measure the heat capacity of $^{244}PuO_2$ has not been published. Briefly, the refrigeration is supplied by a ^3He-^4He dilution refrigerator and contact between the refrigerator and the calorimeter is made by means of mechanically operated jaws and a gold wire connected to the calorimeter. The temperature of the calorimeter was measured with a calibrated germanium resistance thermometer (laboratory designation Ge No. 3); its calibration and testing by measuring the 1965 Calorimetry Conference Copper Standard sample has been reported [10].

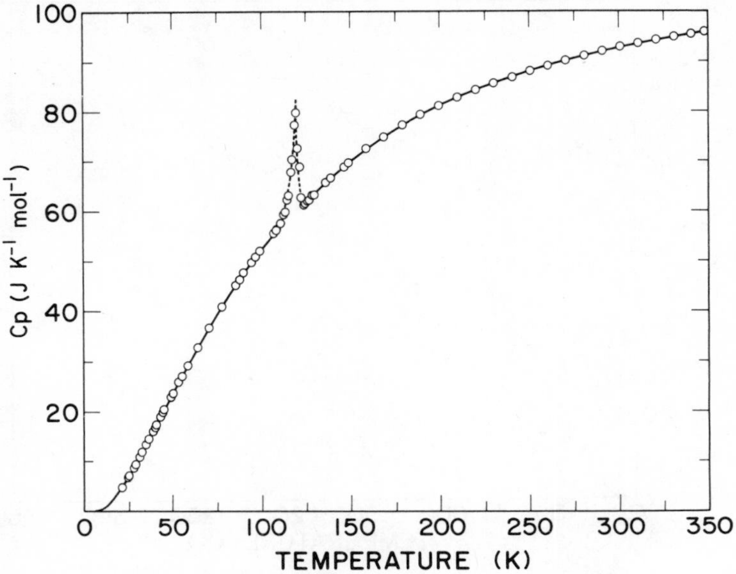

FIG.1. Heat capacity of $^{242}PuF_3$ from 20 to 350 K. The circles represent the experimental data obtained as described in the text.

The method of taking and evaluating the heat capacity data was very similar to the procedure used for the plutonium fluorides. Between 4 and 10 K temperature drift data were used to evaluate heat capacities and between 10 and 25 K both drift data and the conventional intermittent electric heating technique were used. The heat capacity of the empty calorimeter amounted to about 95% of the loaded calorimeter. The heat capacity of an identical empty calorimeter was evaluated in a separate series of measurements.

4. HEAT CAPACITY RESULTS

4.1. C_p° of $^{242}PuF_3$ and $^{242}PuF_4$ from 20 to 350 K

The experimental heat capacity data for $^{242}PuF_3$ and from $^{242}PuF_4$ from 20 to 350 K, obtained by intermittent electrical heating, are given in Figs. 1 and 2, respectively. The molecular weight for the $^{242}PuF_3$ sample is 299.051 and for the $^{242}PuF_4$ sample is 318.050. A correction equal to $-(d^2C_p^\circ)/dT^2)(\Delta T)^2/24$ has been applied for curvature, i.e., for the difference between C_p° and $\bar{C}_p$, except for runs for $^{242}PuF_3$ extending into the anomaly near 119 K.

The heat capacity of $^{242}PuF_3$ is anomalously high from 110-122 K, with the maximum occurring near 119°K as can be seen in Fig. 1. The $^{242}PuF_3$ sample was examined by G. H. Lander and

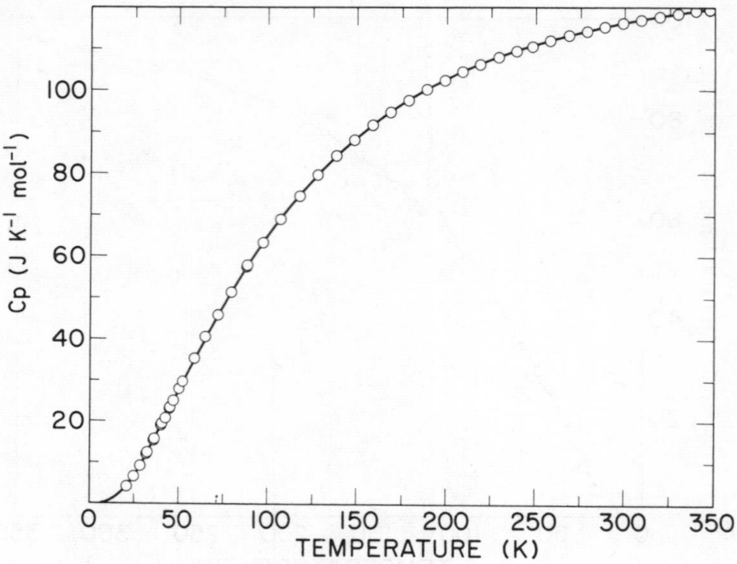

FIG.2. Heat capacity of $^{242}PuF_4$ from 20 to 350 K. The circles represent the experimental data obtained as described in the text.

M. H. Mueller of Argonne National Laboratory by neutron diffraction [11] in an attempt to determine the cause of this transition. They reported the following: "On cooling to 80 K no additional peaks were observed, nor any change in the two low order nuclear lines (at d = 3.67 and 3.22 Å). We conclude that $^{242}PuF_3$ is not magnetic at 80 K, unless the ordered magnetic moment per Pu atom is less than 0.5 μ_B. One small additional line was, however, observed at $2\theta = 27.5°$ (d = 2.64 Å). We think this line may be due to a structural transition. The temperature dependence of this line was such that it disappeared at ∿125 K."

4.2. C_p° of $^{242}PuF_3$ and $^{242}PuF_4$ from 10 to 20 K

The radioactive self-heating rate was needed for the calculation of heat capacities from temperature drift measurements below 20 K. It was evaluated from a comparison of the drift rates with the measured heat capacities from 20 to 58 K. The self-heating rate, $\dot{q}_{rad}$, was calculated with the equation

$$\dot{q}_{rad} = C_p^\circ \, (dT/dt) - \dot{q}_{leak} \quad (6)$$

where C_p° is the total heat capacity obtained from equation (4) less the curvature correction, dT/dt is the drift with no electrical heating and $\dot{q}_{leak}$ is the small leak due to the deviation from adiabatic conditions and to the effect of the

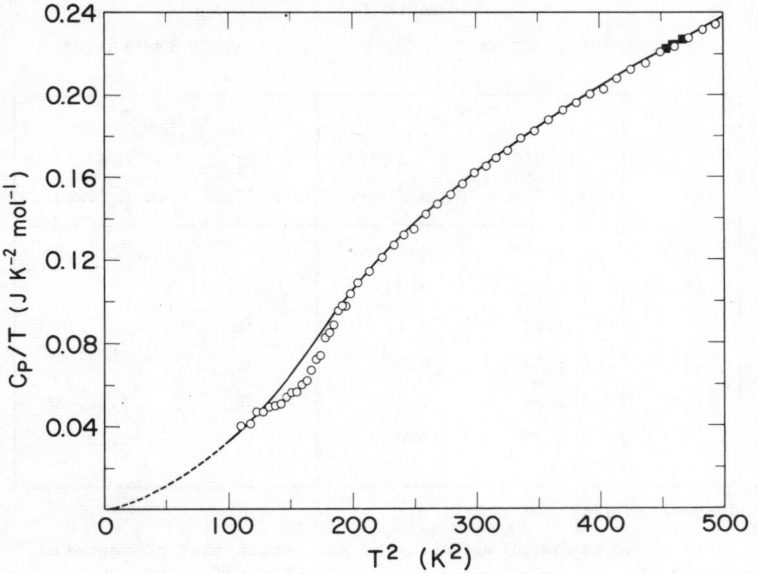

FIG.3. Heat capacity of $^{242}PuF_3$ at the lowest temperatures. The solid line above 10 K represents the smoothed heat capacities. The circles show the abnormally low heat capacities observed between 10 and 14 K after storage for 12 h at 8 K. The squares are heat capacity data obtained by the normal intermittent heating method. The dotted line below 10 K is the extrapolation to 0 K.

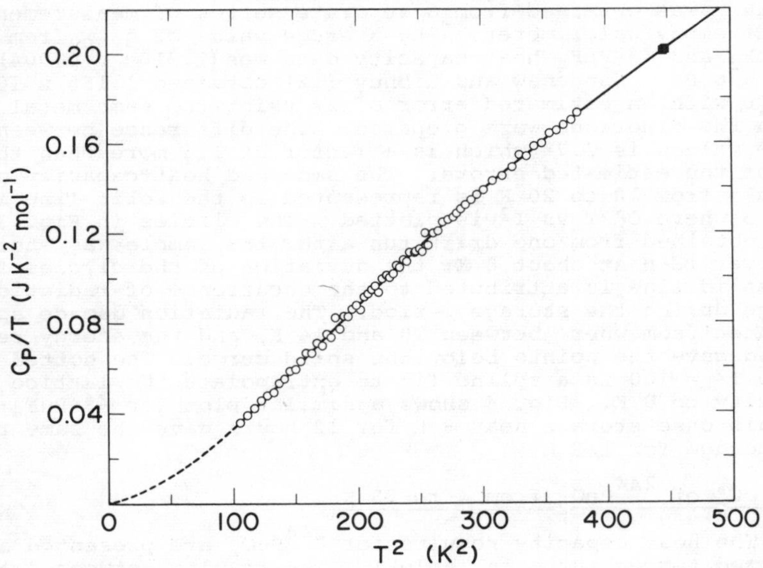

FIG.4. Heat capacity of $^{242}PuF_4$ at the lowest temperatures. The solid line above 10 K represents the smoothed heat capacities. The circles are values obtained from the drifts. The squares are data obtained by the normal intermittent heating method. The dotted line below 10 K is the extrapolation to 0 K.

TABLE I

HEAT CAPACITY AND ENTROPY OF PuO_2 AT SELECTED TEMPERATURES

T (K)	$^{244}PuO_2$ C_p° ($J\ K^{-1}\ mol^{-1}$)	$^{244}PuO_2$ S° ($J\ K^{-1}\ mol^{-1}$)	$^{239}PuO_2$ [a] C_p° ($J\ K^{-1}\ mol^{-1}$)	$^{239}PuO_2$ [a] S° ($J\ K^{-1}\ mol^{-1}$)
4	0.0090	(0.0032) [b]		
10	0.111	0.039		
15	0.43	0.13	1.21	
20	1.12	0.34	2.01	
25	2.36	0.71	3.35	1.80
	±0.24	±0.07		±0.13

[a] Reference [1].

[b] This value was obtained with the assumption that no magnetic ordering or other anomalous effect occurs below 4 K.

thermometer current. The value of $\dot{q}_{leak}$ as a function of temperature was obtained from a separate series of measurements on the empty calorimeter. The average value of $\dot{q}_{rad}$ from the $^{242}PuF_3$ and $^{242}PuF_4$ heat capacity data was $(1.210 \pm 0.004) \times 10^{-4}$ W/g Pu. Sandenaw and Gibney [12] obtained 1.165×10^{-4} W/g Pu with an estimated error of 2% using the same metal from which the fluorides were prepared. The difference between these values is 3.7%, which is a factor of 1.6 more than the sum of the estimated errors. The smoothed heat capacity of $^{242}PuF_3$ from 10 to 20 K is represented as the solid line in Fig. 3, where C_p°/T vs T^2 is plotted. The circles in Fig. 3 were obtained from one drift run after the sample was stored for over 12 h at about 8 K; the deviation of the circles from the solid line is attributed to the occurrence of radiation damage during the storage period. The radiation damage appears to anneal somewhere between 10 and 14 K, and the energy released gave the points below the solid curve. The dotted line below $T^2 = 100$ is a spline fit to extrapolate the lattice heat capacity to 0 K. Fig. 4 shows a similar plot for $^{242}PuF_4$; in this case storage near 8 K for 12 hours gave the same result as storage for 1/2 h.

4.3. C_p° of $^{244}PuO_2$ from 4 to 25 K

The heat capacity results for $^{244}PuO_2$ are presented at selected temperatures in Table I. The results between 4 and 25 K lie on a smooth curve and no radiation damage effects were observed even after storage near 4 K for 48 h. The molecular weight for the $^{244}PuO_2$ sample is 276.028.

TABLE II

THERMODYNAMIC FUNCTIONS OF $^{242}PuF_3$

T (K)	C_p° ($J\ K^{-1}\ mol^{-1}$)	S° ($J\ K^{-1}\ mol^{-1}$)	$H^\circ - H_0^\circ$ ($J\ mol^{-1}$)	$-(G^\circ - H_0^\circ)/T$ ($J\ K^{-1}\ mol^{-1}$)
10	0.331	(5.846) [a]	(52.53) [a]	(0.593) [a]
25	6.861	8.240	99.10	4.276
50	23.59	18.041	476.7	8.506
100	52.43	43.818	2419.7	19.621
150	70.06	69.43	5604	32.07
200	81.25	91.27	9412	44.21
250	88.10	110.18	13657	55.55
300	92.78	126.69	18187	66.06
350	96.09	141.24	22910	75.78
298.15	92.64	126.11	18015	65.69
	±0.28	±0.36	±54	±0.20

[a] Extrapolated as discussed in text.

5. THERMODYNAMIC FUNCTIONS

5.1. Plutonium trifluoride

The heat capacity, C_p°, entropy, S°, enthalpy increment, $H^\circ - H_0^\circ$, and Gibbs energy divided by temperature $(G^\circ - H_0^\circ)/T$, are given at various temperature in Table II. Magnetic susceptibility measurements on our sample of $^{242}PuF_3$ were made by D. J. Lam of Argonne National Laboratory from 3 to 300 K [13]. His results gave a sharp maximum in the susceptibility, indicating antiferromagnetic ordering, at 9.0 ± 0.5 K. From a linear plot of $1/\chi$ versus T from 20 to 100 K, the paramagnetic moment was calculated to be 1.15 μ_B. The values of the entropy and enthalpy changes associated with the antiferromagnetic transition were taken as $R\ \ell n\ 2$ and $9.0\ R\ \ell n\ 2$, respectively. The thermodynamic functions at 10°K, shown in parentheses in Table II, were obtained as the sum of the lattice contribution and the magnetic contribution. The lattice contribution was evaluated assuming the dotted line extrapolation of C_p°/T to 0 K as shown in Fig. 3.

The entropy of plutonium trifluoride at 298.15 K has been estimated by Rand to be 128 ± 13 $J\ K^{-1}\ mol^{-1}$, which agrees with the experimental value of 126.11 ± 0.36 $J\ K^{-1}\ mol^{-1}$ given in Table II.

TABLE III

THERMODYNAMIC FUNCTIONS OF $^{242}PuF_4$

T (K)	C_p° ($J\ K^{-1}\ mol^{-1}$)	S° ($J\ K^{-1}\ mol^{-1}$)	$H^\circ - H_0^\circ$ ($J\ mol^{-1}$)	$-(G^\circ - H_0^\circ)/T$ ($J\ K^{-1}\ mol^{-1}$)
10	0.330	(0.082)[a]	(0.66)[a]	(0.016)[a]
25	6.657	2.236	42.85	0.522
50	27.61	13.003	461.4	3.774
100	65.10	44.499	2842.0	16.078
150	88.66	75.73	6734	30.84
200	102.69	103.35	11550	45.60
250	110.97	127.22	16907	59.59
300	116.35	147.97	22601	72.63
350	119.80	166.19	28512	84.72
298.15	116.19 ±0.35	147.25 ±0.44	22385 ±67	72.17 ±0.22

[a] Extrapolated as discussed in the text.

5.2. Plutonium tetrafluoride

Table III gives the thermodynamic functions of $^{242}PuF_4$. The heat capacity was extrapolated from 10 to 0 K as shown by the dotted line in Fig. 4. The functions at 10 K shown in parentheses in Table III were calculated assuming this extrapolation. The magnetic susceptibility of $^{239}PuF_4$ was measured from about 4 to 300 K by Hendricks [15], and he reported that below 50 K the paramagnetic susceptibility was independent of temperature. Therefore no magnetic ordering below 10 K is expected and we assume our extrapolated heat capacity curve is appropriate.

Rand [14] has estimated the entropy of plutonium tetrafluoride at 298.15 K to be 162 ± 2 $J\ K^{-1}\ mol^{-1}$. This estimate does not agree with our experimental value of 147.25 ± 0.44 $J\ K^{-1}\ mol^{-1}$.

5.3. Plutonium dioxide

The entropies of $^{244}PuO_2$ as evaluated from our experimental heat capacities at selected temperature are given in Table I. We obtained the entropy at 4 K, shown in parenthesis, assuming that C_p° varies as T^3 between 4 and 0 K. Since plutonium dioxide has a temperature independent magnetic susceptibility from 4 to 1000 K [16], we assume only a lattice contribution to the heat capacity below 4 K.

Some results of Sandenaw [1] for $^{239}PuO_2$ are included in Table I for comparison. We note that his heat capacity and entropy at 25 K are somewhat higher than our values. We have prepared a 20 g sample of $^{242}PuO_2$ and plan to determine its heat capacity from 10 to 350 K in the near future. The lesser amount of radioactivity in our $^{242}PuO_2$ sample compared to the $^{239}PuO_2$ sample used by Sandenaw [1] should permit us to obtain a more reliable value of the entropy of plutonium dioxide at 298.15 K.

ACKNOWLEDGEMENTS

The authors acknowledge the assistance of a number of colleagues at Argonne National Laboratory who contributed their time and talents to this research: J. Faris, I. Fox, D. Lam, G. Lander, M. Mueller, D. Rokop, E. Sherry, and S. Siegel. There are also several individuals at Oak Ridge, Tennessee who assisted in the preparation of the $^{244}PuO_2$ sample; we thank the following for their contribution: E. Bomar, J. Burns, C. Hamby, E. Kobisk, and F. Scheitlin. We also are indebted to the many people associated with the U. S. Atomic Energy Commission who prepared and made available the separated plutonium isotopes. Special thanks are due to W. Ramsey of Lawrence Radiation Laboratory who exchanged a sample of ^{242}Pu metal that was isotopically less pure for some of the metal studied by Sandenaw and Gibney [1].

REFERENCES

[1] SANDENAW, T. A., J. Nucl. Mater. 10 (1963) 165.
[2] ZACHARIASEN, W. H., Acta Cryst. 2 (1949) 388.
[3] GARDNER, E. R., MARKIN, T. L., STREET, R. S., J. Inorg. Nucl. Chem. 27 (1965) 541.
[4] International Atomic Energy Agency, The Plutonium-Oxygen and Uranium-Plutonium-Oxygen Systems: A Thermochemical Assessment, Technical Reports Series No. 79, IAEA, Vienna (1967) 9.
[5] FLOTOW, H. E., OSBORNE, D. W., Rev. Sci. Instru. 37 (1966) 1414.
[6] FLOTOW, H. E., KLOCEK, E. E., Rev. Sci. Instru. 39 (1968) 1578.
[7] WESTRUM, E. F., Jr., HATCHER, J. B., OSBORNE, D. W., J. Chem. Phys. 21 (1953) 419.
[8] OSBORNE, D. W., WESTRUM, E. F., Jr., J. Chem. Phys. 21 (1953) 1884.
[9] OSBORNE, D. W., SCHREINER, F., FLOTOW, H. E., MALM, J. G., J. Chem. Phys. 57 (1972) 3401.
[10] OSBORNE, D. W., FLOTOW, H. E., SCHREINER, F., Rev. Sci. Instru. 38 (1967) 159.
[11] LANDER, G. H., MUELLER, M. H., Private communication.
[12] SANDENAW, T. A., GIBNEY, R. B., J. Chem. Thermodynamics 3 (1971) 85.
[13] LAM, D. J., Private communication.
[14] RAND, M. H., "Thermochemical properties", Plutonium: Physico-Chemical Properties of its Compounds and Alloys, IAEA, Vienna (1966) 43.
[15] HENDRICKS, M. E., Ph.D. Thesis, University of South Carolina (1971).
[16] RAPHAEL, G., LALLEMENT, R., Solid State Commun. 6 (1968)383.

DISCUSSION

R.J. ACKERMANN: May I clarify one statement you made in your oral presentation. The high value of 19.7 cal/K·mol for the absolute entropy of PuO_2(s) at 298 K should not be attributed to Mr. Rand. Actually, it was the value adopted by the 1967 Vienna Panel, which chose it from the information available in the literature at that time.

H.E. FLOTOW: Thank you for this clarification. I realize that the assessment of 19.7 cal/K·mol was made before the magnetic susceptibility data were published.

M.H. RAND: I should also like to thank Mr. Ackermann for his comment. May I add that I am rather surprised that my estimated value for PuF_4(s) was so much in error, since it was based on measurements on the $PuF_4(s) + F_2(g) = PuF_6(g)$ equilibrium. Since the entropies of F_2(g) and PuF_6(g) are reasonably well-known, the value for PuF_4(s) at 600 K should be fairly accurate, if the 'Second Law' slope is not in error. Perhaps the ΔC_p values are much larger than we estimated. It would be interesting to have high temperature heat capacity data for this compound.

H.E. FLOTOW: I agree.

SOLUBILITY BEHAVIOUR OF PuF_3 IN FLUORIDE SALT MIXTURES OF INTEREST IN MOLTEN SALT REACTOR TECHNOLOGY

D.D. SOOD, P.N. IYER
Bhabha Atomic Research Centre,
Trombay, Bombay,
India

Abstract

SOLUBILITY BEHAVIOUR OF PuF_3 IN FLUORIDE SALT MIXTURES OF INTEREST IN MOLTEN SALT REACTOR TECHNOLOGY.

The solubility of plutonium trifluoride in molten salts containing fluorides of lithium, beryllium and thorium was determined in the temperature range of 550-800°C to assess the chemical feasibility of a plutonium-fuelled molten salt reactor. The minimum solubility was found to be 0.85 mol% at 550°C, which is higher than the PuF_3 concentration calculated as the minimum requirement for a plutonium start-up molten salt reactor. Empirical correlations which can predict the solubility of PuF_3 in these ternary fluoride mixtures within 5-10% were derived from the solubility data. The solubility of PuF_3 was also determined in binary melts containing fluorides of lithium and thorium over a similar temperature range to obtain an understanding of the solubility behaviour of PuF_3. The data on the solubility of PuF_3 in binary and ternary salt mixtures containing LiF, BeF_2 and ThF_4 were used to calculate the activity coefficients of PuF_3 in these molten salt media. The partial molal heat of mixing and excess partial molal entropy of mixing of liquid PuF_3 with molten fluoride media were also calculated. For all the salt compositions, the heat and entropy values are negative, ranging from -1.5 to -7.5 kcal/mol for H_{PuF_3} and -2.0 to -8.9 entropy units for $S^E_{PuF_3}$. In the binary salt melts, the heat and entropy effects are low for compositions close to $2LiF\text{-}BeF_2$ and $2\,LiF\text{-}ThF_4$, indicating that these melts are more ordered. In the ternary salt melts, the heat and entropy effects have minima for salt compositions having a small value for parameter $D_t = ThF_4/(ThF_4 + 1.7\ BeF_2)$, and approach a plateau beyond $D_t = 0.5$.

1. INTRODUCTION

The molten salt reactor is one of the advanced breeder concepts being developed to utilize thorium for the production of nuclear power. The concept is under development at Oak Ridge National Laboratory, and visualizes the use of fluorides of fissile materials (UF_4, PuF_3) dissolved in a mixture of lithium fluoride, beryllium fluoride and thorium fluoride as fuel [1]. The fuel salt also acts as a heat exchanger medium and the suggested minimum and maximum temperatures in the fuel-salt circuit are 565°C and 704°C, respectively. The chemical feasibility of the use of plutonium in such a reactor has not been investigated so far, and one of the first requirements for that purpose was the demonstration of adequate solubility of PuF_3 in molten salts containing LiF, BeF_2 and ThF_4 over the temperature range of 500-700°C. The solubility of PuF_3 in LiF-BeF_2 melts has been reported by Barton et al. [2] and Mailen et al. [3], and the solubility of PuF_3 in one LiF-BeF_2-ThF_4 composition has been reported by Bamberger et al. [4]. Additional data were, however, required to predict the feasibility of a plutonium-fuelled molten salt reactor. To this end, this paper presents the results of the measurements of the solubility of PuF_3 in a number of LiF-BeF_2-ThF_4 compositions of interest.

The solubility of PuF_3 in a number of $LiF-ThF_4$ melts was also determined to assist in understanding the solubility behaviour of PuF_3 in these fluoride melts. The solubility data obtained for binary and ternary salt compositions were used to derive empirical correlations for the solubility of PuF_3 as a function of composition. Values of activity coefficients, partial molal heats of mixing and excess partial molal entropies of mixing of liquid PuF_3 were also derived from the solubility data.

2. EXPERIMENTAL

The solubility studies were carried out in an argon-atmosphere dry box to guard against contamination of the various salt mixtures by moisture or oxygen. A mixture of $LiF-BeF_2-ThF_4$, purified by HF and H_2 treatment, was used for the first experiment. For the subsequent experiments the composition of the salt was adjusted by the addition of either purified $LiF-BeF_2$ eutectic, LiF or ThF_4. Plutonium trifluoride prepared by hydrofluorination of plutonium dioxide was used for the studies. For each solubility experiment, PuF_3 was equilibrated with the salt mixture in a nickel crucible by bubbling high-purity argon through the melt. The temperature of the melt was maintained constant to within ± 1°C. The salt was sampled by sucking the melt into a nickel sampling stick having a 10-15 micron porosity filter. Samples were withdrawn both during the cooling and heating cycles to check the reproducibility of the data. The plutonium content in the salt was estimated by potentiometric and radiometric methods. The composition of the base salt was also analysed from melt samples. Apparatus and procedure have been reported in more detail earlier [5-7].

3. RESULTS AND DISCUSSION

The solubility of PuF_3 in eight ternary mixtures containing LiF, BeF_2 and ThF_4 and four binary mixtures containing LiF and ThF_4 was determined over the temperature range of 550-800°C. The solubility data for all the mixtures could be represented by an equation of the form:

$$\log \left\{ \text{solubility (mol\% } PuF_3) \right\} = C_1 + C_2/T \qquad (1)$$

where C_1 and C_2 are composition-dependent constants and T is the absolute temperature. The values of the constants C_1 and C_2 obtained in the present investigation and those reported in the literature are presented in Table I, together with the values of the solubility of PuF_3 at 600 and 700°C. From the solubility behaviour of PuF_3 in binary-salt compositions of $LiF-BeF_2$ and $LiF-ThF_4$, an attempt was made to obtain an empirical correlation for the solubility of PuF_3 in these fluoride melts as a function of composition. The solubilities of PuF_3 in $LiF-BeF_2$ and $LiF-ThF_4$ melts are plotted as a function of mol% BeF_2 and mol% ThF_4 in Figs 1 and 2, respectively. The solubility of PuF_3 in pure LiF was obtained by extrapolation from the phase diagram of $LiF-PuF_3$ system [8]. From Figs 1 and 2 it is seen that:

(a) The solubility of PuF_3 is a maximum in pure LiF and decreases with the addition of BeF_2 or ThF_4. The minima in solubility are observed near the salt compositions $2LiF-BeF_2$ and $2LiF-ThF_4$. This behaviour may be attributed to the availability of free fluoride ions in pure LiF, the concentration of which decreases by the formation of ions of the type BeF_4^{2-}, ThF_5^- and ThF_7^{3-} as BeF_2 and ThF_4 are added to lithium fluoride.

(b) For similar additions of BeF_2 and ThF_4 to LiF, BeF_2 tends to decrease the solubility of PuF_3 much more than does ThF_4. This probably occurs because the Pu^{3+} ion with a charge-to-size ratio (z/r) of 3.0 competes much more effectively with Th^{4+} (z/r = 4.0) than with Be^{2+} (z/r = 6.7) for interaction with fluoride ions [9].

(c) The curves of the solubility of PuF_3 in these binary mixtures do not vary symmetrically around the minima, the change being much sharper for the high-LiF composition side.

From these observations two composition parameters were defined which could account for the variation of solubility of PuF_3 in fluoride mixtures containing LiF, BeF_2 and ThF_4. Parameters of this type were previously used by Barton et al. to account for the solubility behaviour of CeF_3 in similar fluoride mixtures [10]. For the present study parameters D_t and E_t as defined by Eqs (2) and (3) were chosen.

TABLE I. SOLUBILITY OF PuF_3 IN $LiF-BeF_2-ThF_4$ MELTS AS EXPRESSED BY Eq. (1)

LiF (mol%)	BeF_2 (mol%)	ThF_4 (mol%)	C_1	$-C_2 \times 10^{-3}$	Solubility at 600°C (mol%)	Solubility at 700°C (mol%)	Ref.
51.7	48.3	0.0	2.77 ± 0.40	2.65 ± 0.34	0.55	1.12	[2]
71.3	28.7	0.0	2.34 ± 1.05	2.13 ± 0.92	0.79	1.40	[2]
68.1	31.9	0.0	3.39 ± 0.32	3.21 ± 0.27	0.52	1.24	[2]
63.0	37.0	0.0	3.14 ± 0.14	3.07 ± 0.13	0.42	0.97	[2]
56.3	43.7	0.0	3.00 ± 0.22	2.94 ± 0.19	0.43	0.95	[2]
66.6	33.4	0.0	3.22 ± 0.20	3.09 ± 0.17	0.48	1.11	[3]
74.0	22.1	3.9	3.55 ± 0.14	2.97 ± 0.14	1.39	3.11	
76.9	17.1	6.0	3.49 ± 0.23	2.82 ± 0.21	1.81	3.88	
75.3	16.7	8.0	3.80 ± 0.05	3.13 ± 0.04	1.64	3.82	
68.2	20.5	11.3	2.98 ± 0.05	2.52 ± 0.05	1.25	2.47	
72.0	16.0	12.0	3.01 ± 0.06	2.41 ± 0.05	1.78	3.42	[4]
71.6	16.2	12.2	2.95 ± 0.07	2.46 ± 0.07	1.36	2.65	
71.3	15.5	13.2	2.62 ± 0.07	2.15 ± 0.06	1.43	2.56	
70.0	14.0	16.0	2.56 ± 0.11	2.06 ± 0.10	1.58	2.76	
75.0	5.0	20.0	2.57 ± 0.14	1.84 ± 0.13	2.88	4.75	
80.0	0.0	20.0	2.62 ± 0.19	1.78 ± 0.19	3.84	6.22	
75.0	0.0	25.0	2.58 ± 0.05	1.76 ± 0.05	3.68	5.92	
70.0	0.0	30.0	2.84 ± 0.07	1.99 ± 0.07	3.60	6.17	
65.0	0.0	35.0	3.01 ± 0.08	2.20 ± 0.08	3.09	5.60	

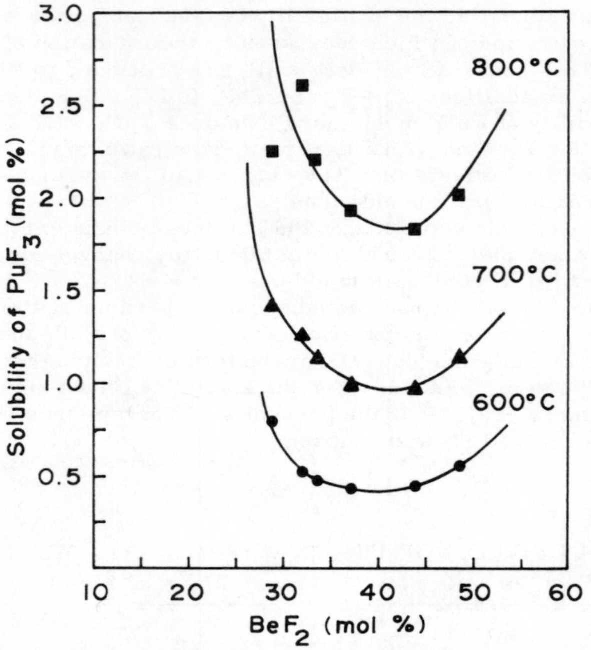

FIG.1. Solubility of PuF_3 in LiF-BeF_2 melts.

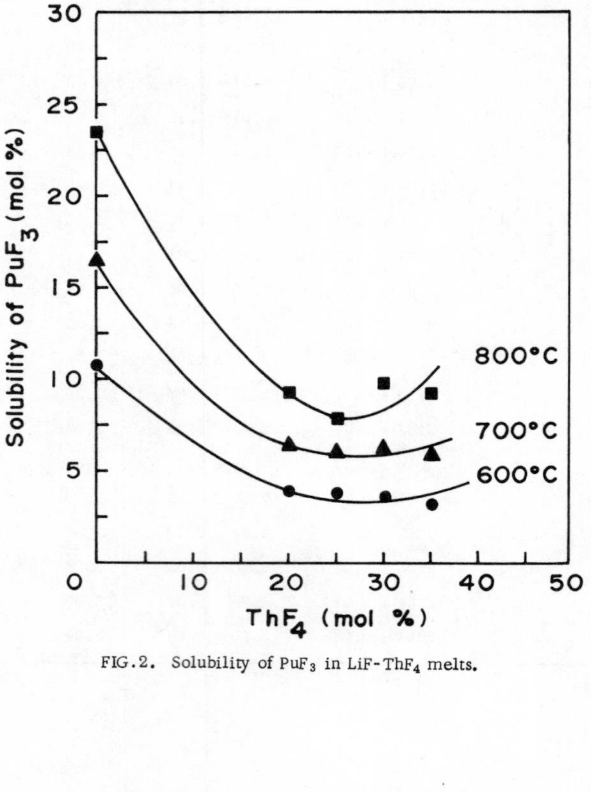

FIG.2. Solubility of PuF_3 in LiF-ThF_4 melts.

$$D_t = \frac{(\text{mol\% ThF}_4)}{(\text{mol\% ThF}_4) + a\cdot(\text{mol\% BeF}_2)} \tag{2}$$

$$E_t = (\text{mol\% LiF}) - b\cdot(\text{mol\% BeF}_2) - c\cdot(\text{mol\% ThF}_4) \tag{3}$$

where a, b and c in these equations are temperature-dependent constants. The parameters D_t and E_t were combined to obtain a single composition parameter X_t as defined by:

$$X_t = D_t + 0.0002\,(E_t^2 + 10E_t) \tag{4}$$

The values for the constants a, b and c were optimized as 1.7, 1.6 and 2.7 for 600°C, and 1.1, 1.0 and 2.7 for 700°C. Using these constants, the values for parameters D_t, E_t and X_t can be calculated for 600°C and 700°C, Denoting X_t by X_6 and X_7 at 600 and 700°C, respectively, the empirical correlations obtained for the solubility of PuF_3 as a function of composition at these temperatures are given by:

$$\text{Solubility}_{600°C}\ (\text{mol\% PuF}_3) = 2.88\,X_6 + 0.44 \tag{5}$$

$$\text{Solubility}_{700°C}\ (\text{mol\% PuF}_3) = 5.69\,X_7 - 0.22 \tag{6}$$

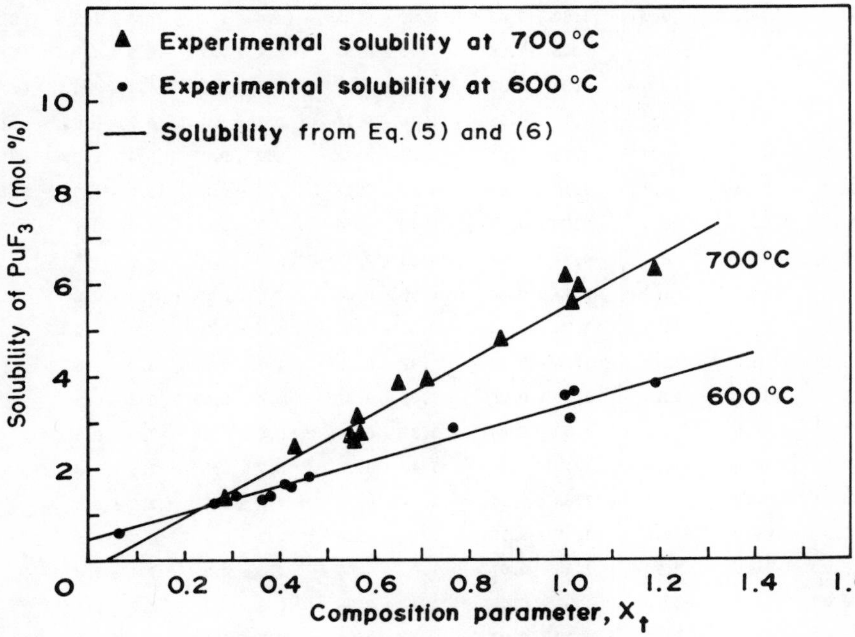

FIG.3. Comparison of PuF_3 solubility data with solubility correlations.

The solubility data from the present investigation and the data of Mailen et al. are compared with the solubility obtained from the correlations in Fig. 3. It is seen that the solubilities calculated from Eqs (5) and (6) agree with the experimental values to within 5 to 10%.

From the solubility data, the activity coefficients of PuF_3 in various fluoride mixtures were calculated, using liquid PuF_3 as the standard state. The activity of PuF_3 at temperature T was calculated from:

$$R \frac{d \ln a_{PuF_3}}{d(1/T)} = -\Delta H_f \tag{7}$$

where a_{PuF_3} is the activity of the saturating phase PuF_3 and ΔH_f is the heat of fusion of PuF_3. On integrating Eq. (7) between limits T and T_m, Eq. (8) is obtained.

$$RT \ln a_{PuF_3} = \Delta H_f^0 \{(T/T_m)-1\} + A_1 \{T_m - T + T \ln (T/T_m)\} + B_1 (T_m - T)^2/2 \tag{8}$$

TABLE II. ACTIVITY COEFFICIENTS OF PuF_3 IN $LiF-BeF_2-ThF_4$ MELTS AS EXPRESSED BY Eq. (9), AND THE VALUES OF THE PARTIAL MOLAL HEATS OF MIXING (H_{PuF_3}) AND EXCESS PARTIAL MOLAL ENTROPIES OF MIXING ($S^E_{PuF_3}$)

LiF (mol%)	BeF_2 (mol%)	ThF_4 (mol%)	A_2	$-B_2 \times 10^{-3}$	$-H_{PuF_3}$ (kcal/mol)	$-S^E_{PuF_3}$ (cal/K·mol)	Ref.
51.7	48.3	0.0	1.60 ± 0.42	0.96 ± 0.35	4.4 ± 1.6	7.3 ± 1.9	[2]
71.3	28.7	0.0	1.95 ± 1.06	1.41 ± 0.93	6.4 ± 4.3	8.9 ± 4.9	[2]
68.1	31.9	0.0	0.91 ± 0.32	0.33 ± 0.27	1.5 ± 1.2	4.1 ± 1.5	[2]
63.0	37.0	0.0	1.14 ± 0.13	0.46 ± 0.11	2.1 ± 0.5	5.2 ± 0.6	[2]
56.3	43.7	0.0	1.28 ± 0.21	0.59 ± 0.18	2.7 ± 0.8	5.9 ± 0.9	[2]
66.6	33.4	0.0	1.10 ± 0.20	0.47 ± 0.17	2.2 ± 0.8	5.0 ± 0.9	[3]
74.0	22.1	3.9	0.66 ± 0.16	0.49 ± 0.15	2.2 ± 0.7	3.0 ± 0.7	
76.9	17.1	6.0	0.71 ± 0.23	0.63 ± 0.21	2.9 ± 1.0	3.2 ± 1.0	
75.3	16.7	8.0	0.43 ± 0.04	0.35 ± 0.04	1.6 ± 0.2	2.0 ± 0.2	
68.2	20.5	11.3	1.22 ± 0.06	0.94 ± 0.05	4.3 ± 0.2	5.6 ± 0.2	
72.0	16.0	12.0	1.16 ± 0.08	1.02 ± 0.06	4.7 ± 0.3	5.3 ± 0.3	[4]
71.6	16.2	12.2	1.27 ± 0.08	1.01 ± 0.07	4.6 ± 0.3	5.8 ± 0.3	
71.3	15.5	13.2	1.59 ± 0.06	1.31 ± 0.06	6.0 ± 0.3	7.3 ± 0.3	
70.0	14.0	16.0	1.72 ± 0.12	1.46 ± 0.10	6.7 ± 0.5	7.9 ± 0.3	
75.0	5.0	20.0	1.66 ± 0.14	1.64 ± 0.13	7.5 ± 0.6	7.6 ± 0.6	
80.0	0.0	20.0	1.47 ± 0.19	1.58 ± 0.18	7.2 ± 0.8	6.7 ± 0.8	
75.0	0.0	25.0	1.55 ± 0.07	1.63 ± 0.06	7.5 ± 0.3	7.1 ± 0.3	
70.0	0.0	30.0	1.32 ± 0.08	1.42 ± 0.08	6.5 ± 0.3	6.0 ± 0.4	
65.0	0.0	35.0	1.10 ± 0.07	1.16 ± 0.07	5.3 ± 0.3	5.0 ± 0.3	

In this equation, T_m is the melting point of PuF_3, and A_1 and B_1 are constants, the values of which depend on the heat capacities of solid and liquid PuF_3. The following literature values were used for the calculation [11].

ΔH_f^0 = 13.0 kcal/mol
T_m = 1699 K
A_1 = -14.3
B_1 = 8.2

The activity data obtained by using Eq. (8) were combined with the solubility data to obtain activity coefficients of PuF_3 in different fluoride melts. The activity coefficient data for a particular salt composition could be represented by an equation of the form:

$$\log f_{PuF_3} = A_2 + B_2/T \qquad (9)$$

From the temperature dependence of the activity coefficients of PuF_3, the partial molal heats of mixing (H_{PuF_3}) and excess partial molal entropies of mixing ($S^E_{PuF_3}$) of liquid PuF_3 with different fluoride salt compositions were calculated. The values of constants A_2 and B_2, the partial molal heats of mixing and excess partial molal entropies of mixing are presented in Table II. It is seen from these data that the partial molal heats of mixing and excess partial molal entropies of mixing of liquid PuF_3 in all the fluoride mixtures are negative, the values ranging from -1.6 to -7.5 kcal/mol for H_{PuF_3} and -2.0 to -8.9 cal/K·mol for $S^E_{PuF_3}$. Negative values for the heat of mixing indicate that the mixing of liquid PuF_3 leads to the formation of additional bonds in the melt. Negative values for the excess entropy of mixing indicate that there is a net increase in order in the system. The H_{PuF_3} and $S^E_{PuF_3}$ values are plotted against the mole-percent BeF_2 or ThF_4 for the binary salt mixtures in Figs 4 and 5, respectively. It is seen that heat and entropy values are low for salt mixtures having a composition close to $2LiF\text{-}BeF_2$ and $2LiF\text{-}ThF_4$. This may be attributed to a higher degree of ordering for these compositions, such that the mixing of liquid PuF_3 in these salts involves minimum bonding and reordering of liquid PuF_3. In the case of ternary salt mixtures, H_{PuF_3} and $S^E_{PuF_3}$ are plotted as a function of

$$D_t = \frac{ThF_4}{ThF_4 + 1.7\ BeF_2}$$

a parameter used for solubility correlations, and are shown in Figs 6 and 7, respectively. It is seen that low heat and entropy effects are observed for salt compositions having a low value for the parameter D_t. The heat and entropy values approach a plateau beyond D_t = 0.5.

4. CONCLUSIONS

The minimum solubility of PuF_3 in $LiF\text{-}BeF_2\text{-}ThF_4$ melts containing 4 to 20 mol% ThF_4 has been observed to be 0.85 mol% at 550°C, which is well above the maximum plutonium concentration of 0.1 to 0.3 mol% estimated by Bauman as being necessary for plutonium start-up molten salt reactors having carrier salts containing 4 to 10 mol% ThF_4 [12]. This indicates that

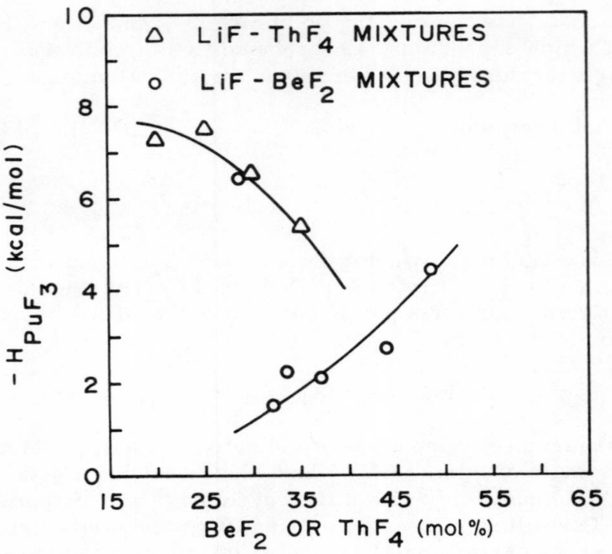

FIG.4. Heat of mixing of liquid PuF_3 in LiF-BeF_2 and LiF-ThF_4 melts.

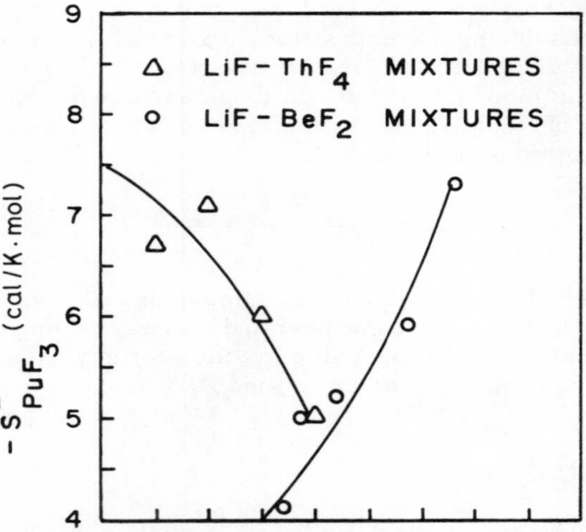

FIG.5. Excess entropy of mixing of liquid PuF_3 in LiF-BeF_2 and LiF-ThF_4 melts.

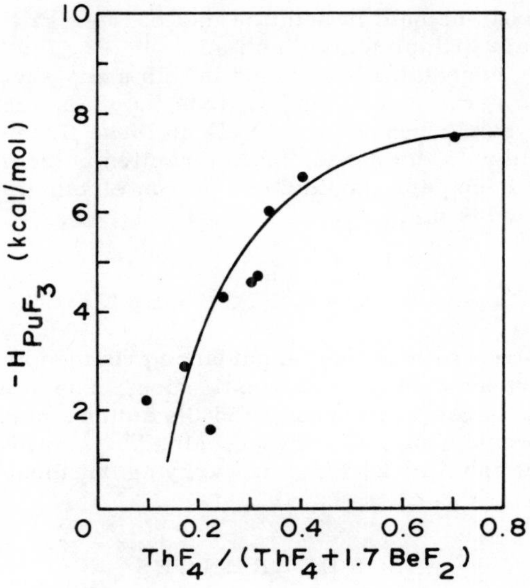

FIG.6. Heat of mixing of liquid PuF_3 with $LiF-BeF_2-ThF_4$ melts.

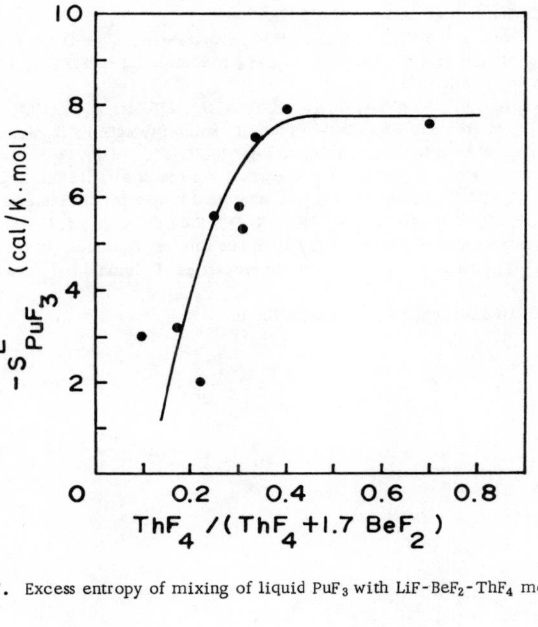

FIG.7. Excess entropy of mixing of liquid PuF_3 with $LiF-BeF_2-ThF_4$ melts.

enough plutonium can be held in solution in a carrier salt to make a molten salt reactor operate with plutonium as fuel.

The solubility correlations obtained in this study should be helpful in evaluating various carrier salt compositions from a solubility point of view.

The thermodynamic behaviour of PuF_3 in these fluoride melts indicated that PuF_3 interacts with the melts, the interaction being higher for less ordered melts. These data should be helpful in elucidating the structure of these complex fluoride melts.

ACKNOWLEDGEMENTS

The authors are grateful for the guidance provided by Dr. M. V. Ramaniah and Mr. N. Srinivasan during this investigation. The work reported is a collective effort of a research group, and the authors particularly wish to acknowledge the work of Mr. R. Prasad, Mr. V.N. Vaidya, Dr. K.N. Roy, Mr. V. Venugopal and Mr. Z. Singh in carrying out these studies.

REFERENCES

[1] ROSENTHAL, M.W., HAUBENREICH, P.N., BRIGGS, R.B., Development Status of Molten Salt Breeder Reactors, USAEC Rep. ORNL-4812 (1972).

[2] BARTON, C.J., Solubility of plutonium-trifluoride in fused alkali fluoride beryllium fluoride mixtures, J. Phys. Chem. 64 (1960) 306.

[3] MAILEN, J.C., SMITH, F.J., FERRIS, L.M., Solubility of PuF_3 in molten $2LiF\text{-}BeF_2$, J. Chem. Eng. Data 16 (1971) 68.

[4] BAMBERGER, C.E., ROSS, R.G., BAES, Jr. C.F., USAEC Rep. ORNL-4622 (1971) 91.

[5] IYER, P.N., PRASAD, R., VAIDYA, V.N., SINGH, Z., SOOD, D.D., RAMANIAH, M.V., Molten Salt Chemistry – Part I. Preparation of CeF_3, PuF_3, ThF_4 and Purification of $LiF\text{-}BeF_2\text{-}ThF_4$ Mixture, Bhabha Atomic Research Centre Rep. BARC-670 (1973).

[6] IYER, P.N., PRASAD, R., VAIDYA, V.N., NAG, K., SINGH, Z., SOOD, D.D., RAMANIAH, M.V., Molten Salt Chemistry – Part II. Solubility of CeF_3 in Molten $LiF\text{-}BeF_2\text{-}ThF_4$ Mixture, Bhabha Atomic Research Centre Rep. BARC-671 (1973).

[7] IYER, P.N., PRASAD, R., VAIDYA, V.N., ROY, K.N., SINGH, Z., SOOD, D.D., RAMANIAH, M.V., SRINIVASAN, N., Molten Salt Chemistry – Part III. Solubility of PuF_3 in Molten Mixtures of $LiF\text{-}BeF_2\text{-}ThF_4$ and $LiF\text{-}ThF_4$, Bhabha Atomic Research Centre Rep. BARC-676 (1973).

[8] THOMA, R.E. (Ed.), Phase Diagrams of Nuclear Reactor Materials, USAEC Rep. ORNL-2548 (1959).

[9] SAMSONOV, G.V. (Ed.), Handbook of Physicochemical Properties of Elements, Plenum Press (1968) 98.

[10] BARTON, C.J., BREDIG, M.A., GILPATRICK, L.O., FREDRICKSEN, J.A., Solubility of cerium trifluoride in molten mixtures of lithium, beryllium and thorium fluorides, Inorg. Chem. 9 (1970) 307.

[11] OETTING, F.L., The chemical thermodynamic properties of plutonium compounds, Chem. Rev. 67 (1967) 261.

[12] BAUMAN, H.F., USAEC Rep. ORNL-4832 (1973) 16.

HIGH-TEMPERATURE VAPORIZATION AND MASS SPECTROMETRIC MEASUREMENT OF THE EXTENT OF NON-STOICHIOMETRY IN OXIDE PHASES*

P.W. GILLES, B.R. CONARD, R.I. SHELDON, J.E. BENNETT
University of Kansas,
Lawrence, Kansas,
United States of America

Abstract

HIGH-TEMPERATURE VAPORIZATION AND MASS SPECTROMETRIC MEASUREMENT OF THE EXTENT OF NON-STOICHIOMETRY IN OXIDE PHASES.

A new method has been devised for establishing the compositions of the lower phase boundary, the congruently vaporizing solution, and the upper phase boundary for high-temperature non-stoichiometric phases having significant ranges of homogeneity. The measurements are made at high temperature rather than on quenched samples. A two-phase sample of known composition is used. One of the phases is the non-stoichiometric one containing the particular composition at which congruent vaporization occurs; the other is the adjacent higher or lower phase. A weighed quantity is vaporized at constant temperature in a mass spectrometer until the congruently vaporizing state has been reached. The intensities of ions are measured during the entire experiment. The loss in mass, the initial composition, the intensities, and the times are used to obtain for the non-stoichiometric phase the compositions of the phase boundary and of the congruently vaporizing solution. The experiment is repeated with another sample whose initial composition is on the other side of the non-stoichiometric phase. The method is illustrated with results for the titanium-oxygen system in which the Ti_3O_5 phase contains the congruently vaporizing solution.

1. INTRODUCTION

The dramatic increase in the range of homogeneity or non-stoichiometry of many oxide phases as the temperature increases is a well known feature of high temperature fuels for nuclear reactors. The purpose of this paper is to describe a new technique for the establishment of the compositions of the phase boundaries and hence the range of homogeneity in oxide phases at temperature by high temperature mass spectrometric vaporization studies. In addition, the composition of the congruently vaporizing solution can be measured.

The chemical potentials of the components in a chemical system at fixed temperature are independent of composition if a sufficient number of phases are present, e.g. three phases in the case of a two-component system, but vary with composition if fewer are present. The chemical potentials of the components can sometimes be established by mass spectrometric measurements of ion intensities. In the new technique these measurements are made during the entire course of a vaporization study; the constancy of intensities or variation of intensities reveals the number of phases present, and the duration of a changing intensity reveals the range of non-stoichiometry.

* This work was supported by the USAEC under Contract AT(11-1)-1140 with the University of Kansas.

Consider for simplicity a two-component system containing one congruently vaporizing solution. As vaporization proceeds, the composition of the condensed system changes until the congruently vaporizing solution has been reached, after which vaporization proceeds without change in composition. Mass spectrometric intensities of ions derived from vaporizing species are monitored continuously during a constant temperature experiment.

Three situations occur: if two condensed phases and a gas phase are present, the intensities remain constant with time; if one condensed phase of variable composition and a gas phase are present, the intensities change with time; if the solution which vaporizes congruently and its gas phase are present, the intensities again remain constant with time. Thus, the moment at which a condensed phase disappears and the system enters a single condensed phase region is revealed by the onset of changing intensities. The moment at which the congruently vaporizing state is reached is revealed by the onset of constant intensities.

Absolute gas pressures at any moment in time can be obtained from intensity measurements with a mass spectrometer that has been calibrated. These pressures, the initial composition of the material, and the initial mass of material can be combined to give the ranges of composition of the single phase regions. If the experiment proceeds until the congruently vaporization state is reached and if the final mass is measured, the experiment itself serves to give the necessary calibration.

In order to approach the congruently vaporizing solution from both sides, the scientist may choose samples whose compositions lie respectively above and below that of the congruently vaporizing solution, and he can observe these systems as they approach the same congruently vaporizing solution.

2. THEORY

For the establishment of the compositions of the lower phase boundary (LPB), the congruently vaporizing solution (CVS), and the upper phase boundary (UPB), the method is to obtain the composition of the sample as a function of time. Two equations containing two unknowns are solved simultaneously. (1) The first equation comes from a mass balance. The mass balance equates the measured mass loss to the integrated rate of mass loss obtained from the intensities. (2) The second equation comes from the congruency conditions for the vaporization of a solution in the single phase field. At congruence, the composition of the gaseous mixture escaping from the crucible must equal the composition of the solid. The gas composition is given by the measured intensities of the gaseous species, and the solid composition is calculated from the initial solid composition, the initial mass, and the intensities integrated over time from the beginning.

The two unknown quantities may be variously defined, but one convenient pair are the quotients Y_1 and Y_2. The first quotient Y_1 is a constant divided by the product of the species-independent instrumental sensitivity times the ionization cross section times the multiplier gain for one species, and the second quotient is the corresponding quantity for the other.

The appropriate equation for the mass balance is,

$$m_b - m_a = \sum_i \int_a^b (dm_i/dt)dt \text{ grams} \tag{1}$$

in which a and b are specific times, and the other symbols have their normal significance. Masses are related to moles by the molecular weight. The relationship between rate of mole loss and pressure is the Knudsen equation,

$$-(dn_i/dt) = p_i aW(2\pi M_i RT)^{-1/2} \tag{2}$$

in which a is the orifice area and W is the clausing factor. The relationship between intensities and pressures is,

$$I_i = S(\gamma_i/T)[p_i\sigma_i + p_j\sigma_j\alpha_{ij}] \tag{3}$$

in which S is the sensitivity of the instrument, regarded as the same for all species, σ_i is the cross section for formation of species i from substance i and p_j is the pressure of a heavier species a fraction of whose ionization α_{ij} produces by fragmentation the ion i.

If fragmentation is negligible, these equations can be combined to give,

$$m_b - m_a = -\sum_i Y_i M_i^{+1/2}(A_i)_{ab} \tag{4}$$

in which the unknown Y_i is,

$$Y_i = aW(2\pi R)^{-1/2}/S\sigma_i\gamma_i \tag{5}$$

and in which the A_i's are integrals over time of the product of intensity times square root of temperature,

$$(A_i)_{ab} = \int_a^b IT^{1/2}dt \tag{6}$$

Of the many possible pairs of times a and b, two are of particular importance. The first pair are the start time s at which the experiment began with a known mass, and the finish time f at which the experiment was terminated and the mass subsequently measured. The second pair are the start time s, and the time at which congruence was achieved c. Identifying by subscripts 1 and 2, two different ions, we get for the masses at these times,

$$m_f - m_s = -Y_1M_1^{+1/2}(A_1)_{sf} - Y_2M_2^{+1/2}(A_2)_{sf} \tag{7}$$

$$m_c - m_s = -Y_1M_1^{+1/2}(A_1)_{sc} - Y_2M_2^{+1/2}(A_2)_{sc} \tag{8}$$

Turn now to the second relationship, the congruency condition. For vaporization to occur without change in composition, the composition of escaping gas and that of the vaporizing solid must be the same. For convenience, we express the composition in terms of the masses of the gaseous products. Elemental compositions can readily be obtained from these. The ratio of rates of mass loss $(dm_2/dt)/(dm_1/dt)$ obtained from the intensity measurements at congruence, $(I_2)_c$ and $(I_1)_c$, is equated to the ratio of the masses $(m_2)_c/(m_1)_c$ of the species present in the solid at congruence,

$$(dm_2/dt)_c/(dm_1/dt)_c = (m_2)_c/(m_1)_c \tag{9}$$

The masses present are obtained from the initial masses and the integrated rate of mass loss. Hence,

$$\frac{M_2^{1/2}(I_2)_c Y_2}{M_1^{1/2}(I_1)_c Y_1} = \frac{(m_2)_s + \int_s^c (dm_2/dt)dt}{(m_1)_s + \int_s^c (dm_1/dt)dt} \tag{10}$$

$$= \frac{(m_2)_s - M_2^{1/2} Y_2 (A_2)_{sc}}{(m_1)_s - M_1^{1/2} Y_1 (A_1)_{sc}} \tag{11}$$

The equations (7) and (11) are solved simultaneously for Y_1 and Y_2. Equation (7) contains the measured mass loss and the measured values of A for the entire experiment, and equation (11) contains the instantaneous intensities at congruence and the values of A at congruence.

Then the mass present at any time is obtained from equation (4). The composition of the solid at any time may be obtained at any time b from the masses of the species present given by the equation,

$$(m_i)_b - (m_i)_s = - Y_i M_i^{+1/2} (A_i)_{sb} \tag{12}$$

The ratio of multiplier gains and ionization cross sections may be obtained from Y_1/Y_2.

Notice that the final composition need not be measured; it is a result. Mass spectrometric samples in high temperature studies are usually small in order that the duration of the experiment is reasonable. Ordinary chemical analysis on the small residues would likely be unreliable. Notice also that the experiment may be stopped any time after congruence has been reached.

The loss in mass can be obtained by weighing, or can be taken as the mass introduced if the experiment is carried out to complete exhaustion of the sample as revealed by the disappearance of the mass spectrometric signals.

3. EXPERIMENTAL

The experiments were performed with a Nuclide high temperature mass spectrometer and an automatic data acquisition and equipment control system at the heart of which is a Hewlett-Packard minicomputer. Intensities were measured through a pulse-counting chain, and the resulting data were stored on a magnetic disc from which data retrieval was easily effected. The mass spectrometer and some of the pulse-counting equipment have been described by Hampson and Gilles [1]. More recently a Hewlett-Packard computer, a 2116B for some experiments and a 2100A for others, with associated magnetic disc has been used for data acquisition and the control of the apparatus.

The titanium-oxygen system contains many phases, the congruently vaporizing one being Ti_3O_5 [2]. Vaporization studies have been accomplished with a metal-rich Ti_3O_5 sample containing a small quantity of Ti_2O_3, and on an oxygen-rich sample, Ti_4O_7. The principal gaseous species in equilibrium with solid samples between Ti_2O_3 and Ti_5O_9 are TiO(g) and TiO_2(g).

The samples used were those described by Hampson and Gilles. The metal-rich sample was PJH-9, $Ti_3O_{4.947\pm0.012}$, whose x-ray pattern revealed a small amount of Ti_2O_3 to be present along with a large proportion of Ti_3O_5. The oxygen-rich sample was PJH-6, $Ti_4O_{7.010\pm0.016}$, whose x-ray pattern showed only one line not assignable to Ti_4O_7.

Six experiments have been performed, three from each side. The initial masses ranged from about 700 micrograms to about 23 milligrams. The durations ranged from about 40 hours to about 350 hours.

4. RESULTS

The data reveal clearly the course of the vaporization of the sample. As shown by the experiment of Hampson and Gilles designated MS-3(K), the metal-rich sample showed apparently for a brief period of a very few hours a constant intensity ratio of $I_{TiO_2^+}/I_{TiO^+}$ indicative of the presence of two solids. For about the next 50 hours at a constant temperature of 1688°C the intensity ratio increased indicating an increase in the oxygen content of a single non-stoichiometric solid phase. Subsequently the ratio remained fixed indicating that the congruent state had been reached. Their runs MS-4 Pt.1(L) and MS-4 Pt.2(M) considered as a single experiment and their run MS-5(N) showed consistent results from the opposite side. In the latter at 1620°C after some erratic measurements early, the ratio remained constant for more than 100 hours, then fell during the next 100 hours and remained constant for the remaining about 100 hours. The sample started with two condensed phases, entered the single phase Ti_3O_5 region of variable composition and then reached the congruent state.

The data indicate that the Ti_3O_5 phase at 1620°C extends from about O/Ti = 1.68 to 1.72 with the congruently vaporizing solution having O/Ti = 1.70. The uncertainties in these compositions are about 0.02. Thus the range of homogeneity is probably 0.04 ± 0.03.

5. DISCUSSION

Very few non-stoichiometric oxide phases of low oxygen potential, the most notable one being UO_{2+x}, have been studied at high temperature to obtain information on the extent of non-stoichiometry. The method described here promises to be very useful.

The experiments are very difficult with the titanium oxides and may be difficult for other materials also. One of the difficulties is that the necessity of measuring the intensities frequently over a period of a week or more taxes human endurance. Too large a sample requires too long a time; too small a sample renders the mass measurements too inaccurate. The temperature and other instrumental conditions are difficult to maintain constant. A difficulty in the titanium oxide case is the slight chemical transport reaction of the tungsten crucible material.

In other cases fragmentation in the mass spectrometer could cause errors, although in the titanium oxide case fragmentation is unimportant.

REFERENCES

[1] HAMPSON, P. J., and GILLES, P. W., J. Chem. Phys. 55 3712 (1971).
[2] GILLES, P. W., CARLSON, K. D., FRANZEN, H. F., and WAHLBECK, P. G., J. Chem. Phys. 46, 2461 (1967).

DISCUSSION

C.B. ALCOCK: It would be very useful to apply this method under circumstances where equilibrium could be maintained throughout the solid phase. To facilitate this it would be advantageous to couple a solid electrolyte cell which could act, through coulometric titration as an external source or sink of oxygen. The temperatures which you indicate for your mass spectrometry study, i.e. around 1650°C, are now possible for such a combination.

P.W. GILLES: Thank you for your suggestion.

A. PATTORET: What were the ionization cross-sections for the gaseous titanium oxides deduced from your measurements?

P.W. GILLES: Our measurements indicate that the $\sigma\gamma$ product for TiO is about three times that for TiO_2.

A. PATTORET: This is further confirmation of the many previous observations that the ionization cross-sections of dioxides are greater than those of monoxides.

P.W. GILLES: Yes, indeed.

A.S. PANOV: In your study of the composition close to Ti_3O_5, did you come across the congruently vaporizing composition by chance, or did you carry out any thermodynamic calculations beforehand?

P.W. GILLES: In earlier work (see Ref.[2] of the paper) the Ti_3O_5 phase was shown to contain the congruently vaporizing solution. This result was confirmed by Hampson and myself as reported in Ref.[1]. I believe that this result was not predicted prior to the experiment.

A.S. PANOV: How does the composition of the congruently vaporizing phase vary with temperature?

P.W. GILLES: Our measurements were made close to one temperature. We do not therefore have any information on the change of the composition of the congruently vaporizing solution with temperature.

Symposium on

THERMODYNAMICS OF NUCLEAR MATERIALS
Vienna, 21-25 October 1974

LIST OF CHAIRMEN

Session I	A. POLLIART	International Atomic Energy Agency
Session II	K.L. KOMAREK	Austria
Session III	H. BLANK	Commission of the European Communities
Session IV	P.E. POTTER	United Kingdom
Session V	R.J. ACKERMANN	United States of America
Session VI	A. PATTORET	France
Session VII	Y. TAKAHASHI	Japan
Session VIII		
morning	E.H.P. CORDFUNKE	The Netherlands
afternoon	O.S. IVANOV	Union of Soviet Socialist Republics

SECRETARIAT OF THE SYMPOSIUM

Scientific Secretaries:	F.L. OETTING	Division of Research and Laboratories, IAEA
	A.A. PUSHKOV	Division of Nuclear Power and Reactors, IAEA
Administrative Secretary:	Gertrude SEILER	Division of External Relations, IAEA
Editor:	E.R.A. BECK	Division of Publications, IAEA
Records Officer:	D. WILSON	Division of Languages, IAEA
Conference Officer:	Claire M. BESNYOE	Division of External Relations, IAEA

LIST OF PARTICIPANTS

ARGENTINA

Santos, E.	Centro Atómico Bariloche (CNEA), San Carlos de Bariloche, Río Negro
Zuzek, E.	Comisión Nacional de Energía Atómica, Av. Libertador 8250, Buenos Aires

AUSTRALIA

Hanna, G.L.	Australian Atomic Energy Commission, Australian Mission to the IAEA, Mattiellistrasse 2-4/III, A-1040 Vienna, Austria

AUSTRIA

Komarek, K.L.	Institute for Inorganic Chemistry of the University of Vienna, Währingerstrasse 42, A-1090 Vienna
Konvicka, H.R.	Österreichische Studiengesellschaft für Atomenergie, Lenaugasse 10, A-1082 Vienna
Kumer, L.	I. Physikalisches Institut, University of Vienna, Kegelgasse 16/9, A-1030 Vienna
Reitsamer, G.F.	Österreichische Studiengesellschaft für Atomenergie, Lenaugasse 10, A-1082 Vienna

BELGIUM

Fuger, J.J.	Institute of Radiochemistry, University of Liège, Liège

CANADA

Alcock, C.B.	Dept. of Metallurgy and Materials Science, University of Toronto, 184 College Street, Toronto M5S 1A4
Tomlinson, M.	Atomic Energy of Canada Ltd., Whiteshell Nuclear Research Establishment, Pinawa, Manitoba R0E 1L0

CZECHOSLOVAKIA

Novotný, M.	Institute of Nuclear Fuel, Prague 5 - Zbraslav

FRANCE

Amblard, M.	Service des Transferts Thermiques, CEA, Centre d'Etudes Nucléaires de Grenoble, B.P. 85, Centre de Tri, F-38041 Grenoble-Cedex
Breton, J.-P.E.	CEA, Centre d'Etudes Nucléaires de Cadarache, B.P. 1, St. Paul-lez-Durance, F-13115 Bouches du Rhône
Cabé, J.L.	CEA, Centre d'Etudes de Valduc, B.P. 14, F-21120 Is-sur-Tille
Calais, D.	CEA, Commissariat à l'Energie Atomique, SEAMA Bat. 19, B.P. 6, F-92260 Fontenay-aux-Roses
Chagrot, M.R.	CEA, Centre d'Etudes Nucléaires de Saclay, B.P. 2, F-91190 Gif-sur-Yvette
Coutures, J.P.	Laboratoire des Ultra Réfractaires, B.P. 5, F-66120 Odeilla Font-Romeu
Déan, G.	Commissariat à l'Energie Atomique, IG/CMN, 33, Rue de la Fédération, F-75015 Paris 15e
de Franco, M.J.	CEA, Centre d'Etudes Nucléaires de Fontenay-aux-Roses, SEAMA, B.P. 6, F-92260 Fontenay-aux-Roses
Fiorese, G.	Commissariat à l'Energie Atomique, Centre de Limeil, 75, Rue de la Fédération, F-75015 Paris
Gatesoupe, J.P.	CEA, Centre d'Etudes Nucléaires de Fontenay-aux-Roses, SEAMA, B.P. 6, F-92260 Fontenay-aux-Roses
Michel, F.A.	Commissariat à l'Energie Atomique, Service des Piles, Avenue des Martyrs, F-38000 Grenoble
Mouchnino, M.	CEA, Centre d'Etudes Nucléaires de Fontenay-aux-Roses, SPuI, B.P. 6, F-92260 Fontenay-aux-Roses
Pattoret, A.	CEA, Centre d'Etudes Nucléaires de Fontenay-aux-Roses, SEAMA, B.P. 6, F-92260 Fontenay-aux-Roses
Pluchery, M.	CEA, Centre d'Etudes Nucléaires de Grenoble, B.P. 85, Centre de Tri, F-38041 Grenoble-Cedex
Schmitz, F.	CEA, Centre d'Etudes Nucléaires de Fontenay-aux-Roses, SEAMA, B.P. 6, F-92260 Fontenay-aux-Roses
Trouvé, J.	CEA, Centre d'Etudes Nucléaires de Cadarache, B.P. 1, St. Paul-lez-Durance, F-13115 Bouches du Rhône
Vierne, J.	Commissariat à l'Energie Atomique, 75, Rue de la Fédération, F-75015 Paris 15e

GERMAN DEMOCRATIC REPUBLIC

Nebel, D.	Zentralinstitut für Kernforschung Rossendorf, Postfach 19, D-8051 Dresden
Schimmel, W.M.	Staatliches Amt für Atomsicherheit und Strahlenschutz der DDR, Waldowallee 117, D-1157 Berlin

GERMANY, FEDERAL REPUBLIC OF

Bober, M.	Institut für Neutronenephysik und Reaktortechnik, Kernforschungszentrum Karlsruhe, Postfach 3640, D-75 Karlsruhe 1
Bogensberger, H.G.	Institut für angewandte Systemtechnik und Reaktorphysik, Kernforschungszentrum Karlsruhe, Postfach 3640, D-75 Karlsruhe 1
Chakraborty, A.K.	Institut für Reaktorsicherheit eV des Technischen Überwachungsvereins, Glockengasse 2, D-5 Köln 1
Fieguth, P.	Rheinisch-Westfälischer Technischer Überwachungsverein eV, Steubenstr. 53, D-43 Essen
Fischer, E.A.	Institut für angewandte Systemtechnik und Reaktorpysik, Kernforschungszentrum Karlsruhe, Postfach 3640, D-75 Karlsruhe 1
Förthmann, R.	Kernforschungsanlage Jülich GmbH, Postfach 365, D-5170 Jülich 1
Grübmeier, H.	Kernforschungsanlage Jülich GmbH, Postfach 365, D-5170 Jülich 1
Herkommer, E.W.	Institut für Reaktorsicherheit eV des Technischen Überwachungsvereins, Glockengasse 2, D-5 Köln 1
Heuvel, H.J.	Interatom, Friedrich-Ebert Str., Bensberg
Hofer, G.	Kraftwerk Union Erlangen, Postfach 3220, D-8520 Erlangen 2
Hofmann, P.	Institut für Material- und Festkörperforschung, Kernforschungszentrum Karlsruhe, Postfach 3640, D-75 Karlsruhe 1
Holleck, H.	Institut für Material- und Festkörperforschung, Kernforschungszentrum Karlsruhe, Postfach 3640, D-75 Karlsruhe 1
Karow, H.-U.	Institut für Neutronenphysik und Reaktortechnik, Kernforschungszentrum Karlsruhe, Postfach 3640, D-75 Karlsruhe 1
Kleykamp, H.	Institut für Material- und Festkörperforschung, Kernforschungszentrum Karlsruhe, Postfach 3640, D-75 Karlsruhe 1

GERMANY, FEDERAL REPUBLIC OF (cont.)

Naoumidis, A. — Kernforschungsanlage Jülich,
Postfach 365, D-5170 Jülich 1

Peehs, M. — Kraftwerk Union,
Hammerbacherstr., D-8520 Erlangen

Rechtlich, R.V. — Technischer Überwachungsverein Baden,
Richard Wagner Str. 2, Postfach 2420, D-68 Mannheim

Robinson, E. — Gesellschaft für Kernenergieverwertung in Schiffbau und Schiffahrt mbH,
Postfach, D-2057 Geesthacht-Tesperhude
(Present address: Institutt for Atomenergi, OECD Halden Project,
P.O. Box 173, N-1751 Halden, Norway)

Schumacher, G. — Institut für Neutronenphysik und Reaktortechnik,
Kernforschungszentrum Karlsruhe,
Postfach 3640, D-75 Karlsruhe 1

Stephan, W. — Kernkraftwerk Obrigheim,
D-6951 Obrigheim/Neckar

Wu, C.-H. — Institut für Nuklearchemie, Kernforschungsanlage Jülich,
Postfach 365, D-5170 Jülich 1

INDIA

Sood, D.D. — Radiochemistry Division, Bhabha Atomic Research Centre,
Trombay, Bombay 400 085

ITALY

Fizzotti, C.F. — CNEN, Laboratorio Tecnologie, Centro Studi Nucleari Casaccia,
C.P. 2400, Rome

Monti, M.A. — AGIP Nucleare SpA,
Corso Porta Romana 68, I-20122 Milano

JAPAN

Asami, N. — Mitsubishi Atomic Power Industries, Inc.,
1-297 Kitabokuro-Machi, Omiya City, Saitama, 330

Furuya, H. — Power Reactor and Nuclear Fuel Development Corporation,
Tokai-mura, Ibaraki-ken

Katsura, M. — Dept. of Nuclear Engineering, Faculty of Engineering,
Osaka University, Osaka

Takahashi, Y. — Dept. of Nuclear Engineering, University of Tokyo,
7-3-1 Hongo, Bankyo-ku, Tokyo

MALAYSIA

Ismail, M.Z. — National University of Malaysia,
Jalan Pantai Barus, Kuala Lumpur

MEXICO

de la Tijera, J.S.	Instituto Nacional de Energía Nuclear, Insurgentes Sur 1079, Mexico D.F.

THE NETHERLANDS

Cordfunke, E.H.P.	Stichting Reactor Centrum Nederland, Petten, N.H.

PAKISTAN

Javed, N.A.	Nuclear Materials Division, Pakistan Institute of Nuclear Science and Technology (PINSTECH), P.O. Nilore, Rawalpindi

POLAND

Moser, Z.	Polish Academy of Sciences, Institute for Metals Research, Reymonta 25, Cracow

ROMANIA

Radulescu, O.	Institut de Physique Atomique, B.P. 35, Bucarest

SWITZERLAND

Peter, J.	Swiss Federal Institute for Reactor Research (EIR), CH-5303 Würenlingen

UNION OF SOVIET SOCIALIST REPUBLICS

Akhachinskij, V.V.	USSR State Committee for Atomic Energy, Moscow
Ivanov, O.S.	Academy of Sciences of the USSR, Institute of Metallurgy, Leninskij prospekt 49, Moscow
Panov, A.S.	Institute of Physical Energetics, Obninsk, Kaluga Region

UNITED KINGDOM

Booth, D.L.	Nuclear Installations Inspectorate, Department of Energy, Thames House South, Millbank, London SW 1
Burton, B.	Central Electricity Generating Board, CEGB Berkeley Nuclear Laboratories, Berkeley, Glos.

UNITED KINGDOM (cont.)

Catlow, C.R.A.	UKAEA, Theoretical Physics Division, Atomic Energy Research Establishment, Harwell, Didcot, OX11 ORA Oxon.
Ewart, F.T.	UKAEA, B.429, Atomic Energy Research Establishment, Harwell, Didcot, OX11 ORA Oxon.
Gillan, M.J.	UKAEA, Atomic Energy Research Establishment, Harwell, Didcot, OX11 ORA Oxon.
Gittus, J.H.	UKAEA, RFL Springfields, Salwick, Preston, Lancs.
Hayns, M.R.	UKAEA, Theoretical Physics Division, Atomic Energy Research Establishment, Harwell, Didcot, OX11 ORA Oxon.
Hoskin, N.E.	Ministry of Defence (PE), AWRE, Aldermaston, Reading RG7 4PR, Berks.
Potter, P.E.	UKAEA, Process Technology, B.220.22, Atomic Energy Research Establishment, Harwell, Didcot, OX11 ORA Oxon.
Rand, M.H.	UKAEA, Process Technology Division, Atomic Energy Research Establishment, Harwell, Didcot, OX11 ORA Oxon.

UNITED STATES OF AMERICA

Ackermann, R.J.	Chemistry Division, Argonne National Laboratory, 9700 S. Cass Ave., Argonne, Ill. 60439
Adamson, M.G.	General Electric Company, Fast Breeder Reactor Department, Vallecitos Nuclear Center, PO Box 846, Pleasanton, Calif. 94566
Barnes, J.F.	Los Alamos Scientific Laboratory, PO Box 1663, Los Alamos, N.Mex. 87544
Davies, J.H.	General Electric Company, Curtner, Ave., San Jose, Calif. 95120
Flotow, H.E.	Chemistry Division, Argonne National Laboratory, 9700 S. Cass Ave., Argonne, Ill. 60439
Gilles, P.	University of Kansas, Dept. of Chemistry, Lawrence, Kansas 66045
Goldenberg, N.	United States Atomic Energy Commission, Washington, D.C. 20545
Hoch, M.	Dept. of Materials Science and Metallurgical Engineering, University of Cincinnati, Cincinnati, Ohio 45221

UNITED STATES OF AMERICA (cont.)

Horton, W.S.	National Bureau of Standards, Materials Building, Rm. B-308, Washington, D.C. 20234
Johnson, C.E.	Chemical Engineering Division, Argonne National Laboratory, 9700 S. Cass Ave., Argonne, Ill. 60439
Johnson, I.	Chemical Engineering Division, Argonne National Laboratory, 9700 S. Cass Ave., Argonne, Ill. 60439
Lindemer, T.B.	Oak Ridge National Laboratory, PO Box X, Oak Ridge, Tenn. 31830
Maroni, V.A.	Chemical Engineering Division, Argonne National Laboratory, 9700 S. Cass Ave., Argonne, Ill. 60439
O'Hare, P.A.	Chemical Engineering Division, Argonne National Laboratory, 9700 S. Cass Ave., Argonne, Ill. 60439
Smith, J.F.	Ames Laboratory USAEC, or Dept. of Metallurgy, Iowa State University, Ames, Iowa 50010
Tetenbaum, M.	Chemical Engineering Division, Argonne National Laboratory, 9700 S. Cass Ave., Argonne, Ill. 60439
Westrum Jr., E.F.	Dept. of Chemistry, University of Michigan, Ann Arbor, Mich.
Wolf, H.	Department of Mechanical Engineering, University of Arkansas, Fayetteville, Arkansas 72701

ORGANIZATIONS

COMMISSION OF THE EUROPEAN COMMUNITIES (CEC)

Barbi, G.B.	Euratom Joint Research Centre, Ispra, Italy
Blank, H.	Commission of the European Communities, European Institute for Transuranium Elements, Postfach 2266, D-75 Karlsruhe, Fed. Rep. of Germany
Drago, A.	Euratom Joint Research Centre, Materials Department, Ispra, Italy
Lindner, H.R.	Commission of the European Communities, European Institute for Transuranium Elements, Postfach 2266, D-75 Karlsruhe, Fed. Rep. of Germany
Matzke, Hj.	Commission of the European Communities, European Institute for Transuranium Elements, Postfach 2266, D-75 Karlsruhe, Fed. Rep. of Germany

COMMISSION OF THE EUROPEAN COMMUNITIES (CEC) (cont.)

Ohse, R.W. — Commission of the European Communities, European Institute for Transuranium Elements, Postfach 2266, D-75 Karlsruhe, Fed. Rep. of Germany

Steven, G.H. — Commission of the European Communities, 29, Rue Aldringen, Luxembourg

FORATOM

Koss, P. — Österreichisches Atomforum, Lenaugasse 10, A-1082 Vienna, Austria

INTERNATIONAL ATOMIC ENERGY AGENCY (IAEA)

Fitts, R.B. — Division of Nuclear Power and Reactors, IAEA, PO Box 590, A-1011 Vienna, Austria

ORGANISATION FOR ECONOMIC CO-OPERATION AND DEVELOPMENT (OECD)

Steemers, T.C. — Dragon Project Office, UKAEA - Winfrith, Dorchester, Dorset, United Kingdom

Wagner-Löffler, M. — Dragon Project Office, UKAEA - Winfrith, Dorchester, Dorset, United Kingdom

AUTHOR INDEX

Roman numerals are volume numbers.
Arabic numerals underlined refer to the first page of a paper by the author concerned.
Further Arabic numerals denote references of other kinds, principally comments and questions in discussions.
Literature references are not indexed.

TRANSLITERATION INDEX

Александрова, Л.Н.	Aleksandrova, L.N.
Алексеева, З.М.	Alekseeva, Z.M.
Андриевский, Р.А.	Andrievskij, R.A.
Ахачинский, В.В.	Akhachinskij, V.V.
Бадаева, Т.А.	Badaeva, T.A.
Бобков, Б.Н.	Bobkov, B.N.
Борисов, Е.В.	Borisov, E.V.
Вамберский, Ю.В.	Vamberskij, Yu.V.
Гусев, В.Н.	Gusev, V.N.
Загрязкин, В.Н.	Zagryaskin, V.N.
Иванов, О.С.	Ivanov, O.S.
Максимов, В.П.	Maksimov, V.P.
Панов, А.С.	Panov, A.S.
Пушков, А.А.	Pushkov, A.A.
Сенчуков, А.Д.	Senchukov, A.D.
Смирнов, Е.А.	Smirnov, E.A.
Спивак, И.И.	Spivak, I.I.
Удовский, А.Л.	Udovskij, A.L.
Ушаков, Б.Ф.	Ushakov, B.F.
Федоров, Г.Б.	Fedorov, G.B.
Федоров, Э.М.	Fedorov, Eh.M.
Фивейский, Е.В.	Fivejskij, E.V.
Хромов, Ю.Ф.	Khromov, Yu.F.
Хромоножкин, В.В.	Khromonozhkin, V.V.
Чирин, Н.А.	Chirin, N.A.
Шлыков, В.И.	Shlykov, V.I.
Янчур, В.П.	Yanchur, V.P.

SUBJECT INDEX

This index of Papers and Discussions is divided into two parts. Part I indexes key phrases and Part II indexes elements, alloy systems and compounds by chemical symbol. Although the two parts do overlap, they are intended to be complementary and there is no guaranteed correspondence between them.

Volume numbers are shown in Roman numerals, and F and T stand for figures and tables, respectively. Semi-colons indicate the end of a reference to a particular Paper or Discussion. *

PART I

* The following examples are given from Parts I and II:

"Entropy of PuO_2: II 26; 486-7, Tl". This means: "Vol. II, page 26; pages 486 to 487 and corresponding Table I".

"Cs-U-O system, phase diagram for: II F1(187)". This means: "Vol. II, Figure 1, this figure being on page 187".

PART II

The following conversion table is provided for the convenience of readers and to encourage the use of SI units.

FACTORS FOR CONVERTING UNITS TO SI SYSTEM EQUIVALENTS *

SI base units are the metre (m), kilogram (kg), second (s), ampere (A), kelvin (K), candela (cd) and mole (mol).
[For further information, see International Standards ISO 1000 (1973), and ISO 31/0 (1974) and its several parts]

Multiply		*by*	*to obtain*
Mass			
pound mass (avoirdupois)	1 lbm	= 4.536×10^{-1}	kg
ounce mass (avoirdupois)	1 ozm	= 2.835×10^{1}	g
ton (long) (= 2240 lbm)	1 ton	= 1.016×10^{3}	kg
ton (short) (= 2000 lbm)	1 short ton	= 9.072×10^{2}	kg
tonne (= metric ton)	1 t	= 1.00×10^{3}	kg
Length			
statute mile	1 mile	= 1.609×10^{0}	km
yard	1 yd	= 9.144×10^{-1}	m
foot	1 ft	= 3.048×10^{-1}	m
inch	1 in	= 2.54×10^{-2}	m
mil (= 10^{-3} in)	1 mil	= 2.54×10^{-2}	mm
Area			
hectare	1 ha	= 1.00×10^{4}	m^2
(statute mile)^2	1 mile^2	= 2.590×10^{0}	km^2
acre	1 acre	= 4.047×10^{3}	m^2
yard^2	1 yd^2	= 8.361×10^{-1}	m^2
foot^2	1 ft^2	= 9.290×10^{-2}	m^2
inch^2	1 in^2	= 6.452×10^{2}	mm^2
Volume			
yard^3	1 yd^3	= 7.646×10^{-1}	m^3
foot^3	1 ft^3	= 2.832×10^{-2}	m^3
inch^3	1 in^3	= 1.639×10^{4}	mm^3
gallon (Brit. or Imp.)	1 gal (Brit)	= 4.546×10^{-3}	m^3
gallon (US liquid)	1 gal (US)	= 3.785×10^{-3}	m^3
litre	1 l	= 1.00×10^{-3}	m^3
Force			
dyne	1 dyn	= 1.00×10^{-5}	N
kilogram force	1 kgf	= 9.807×10^{0}	N
poundal	1 pdl	= 1.383×10^{-1}	N
pound force (avoirdupois)	1 lbf	= 4.448×10^{0}	N
ounce force (avoirdupois)	1 ozf	= 2.780×10^{-1}	N
Power			
British thermal unit/second	1 Btu/s	= 1.054×10^{3}	W
calorie/second	1 cal/s	= 4.184×10^{0}	W
foot-pound force/second	1 ft·lbf/s	= 1.356×10^{0}	W
horsepower (electric)	1 hp	= 7.46×10^{2}	W
horsepower (metric) (= ps)	1 ps	= 7.355×10^{2}	W
horsepower (550 ft·lbf/s)	1 hp	= 7.457×10^{2}	W

* Factors are given exactly or to a maximum of 4 significant figures

Multiply			by	to obtain
Density				
pound mass/inch^3	1 lbm/in^3	=	2.768×10^4	kg/m^3
pound mass/foot^3	1 lbm/ft^3	=	1.602×10^1	kg/m^3
Energy				
British thermal unit	1 Btu	=	1.054×10^3	J
calorie	1 cal	=	4.184×10^0	J
electron-volt	1 eV	≃	1.602×10^{-19}	J
erg	1 erg	=	1.00×10^{-7}	J
foot-pound force	1 ft·lbf	=	1.356×10^0	J
kilowatt-hour	1 kW·h	=	3.60×10^6	J
Pressure				
newtons/metre^2	1 N/m^2	=	1.00	Pa
atmosphere[a]	1 atm	=	1.013×10^5	Pa
bar	1 bar	=	1.00×10^5	Pa
centimetres of mercury (0°C)	1 cmHg	=	1.333×10^3	Pa
dyne/centimetre^2	1 dyn/cm^2	=	1.00×10^{-1}	Pa
feet of water (4°C)	1 ftH_2O	=	2.989×10^3	Pa
inches of mercury (0°C)	1 inHg	=	3.386×10^3	Pa
inches of water (4°C)	1 inH_2O	=	2.491×10^2	Pa
kilogram force/centimetre^2	1 kgf/cm^2	=	9.807×10^4	Pa
pound force/foot^2	1 lbf/ft^2	=	4.788×10^1	Pa
pound force/inch^2 (= psi)[b]	1 lbf/in^2	=	6.895×10^3	Pa
torr (0°C) (= mmHg)	1 torr	=	1.333×10^2	Pa
Velocity, acceleration				
inch/second	1 in/s	=	2.54×10^1	mm/s
foot/second (= fps)	1 ft/s	=	3.048×10^{-1}	m/s
foot/minute	1 ft/min	=	5.08×10^{-3}	m/s
mile/hour (= mph)	1 mile/h	=	4.470×10^{-1}	m/s
			1.609×10^0	km/h
knot	1 knot	=	1.852×10^0	km/h
free fall, standard (= g)		=	9.807×10^0	m/s^2
foot/second^2	1 ft/s^2	=	3.048×10^{-1}	m/s^2
Temperature, thermal conductivity, energy/area·time				
Fahrenheit, degrees −32	°F − 32		$\frac{5}{9}$	°C
Rankine	°R			K
1 Btu·in/ft^2·s·°F		=	5.189×10^2	W/m·K
1 Btu/ft·s·°F		=	6.226×10^1	W/m·K
1 cal/cm·s·°C		=	4.184×10^2	W/m·K
1 Btu/ft^2·s		=	1.135×10^4	W/m^2
1 cal/cm^2·min		=	6.973×10^2	W/m^2
Miscellaneous				
foot^3/second	1 ft^3/s	=	2.832×10^{-2}	m^3/s
foot^3/minute	1 ft^3/min	=	4.719×10^{-4}	m^3/s
rad	rad	=	1.00×10^{-2}	J/kg
roentgen	R	=	2.580×10^{-4}	C/kg
curie	Ci	=	3.70×10^{10}	disintegration/s

[a] atm abs: atmospheres absolute;
atm (g): atmospheres gauge.

[b] lbf/in^2 (g) (= psig): gauge pressure;
lbf/in^2 abs (= psia): absolute pressure.

HOW TO ORDER IAEA PUBLICATIONS

Exclusive sales agents for IAEA publications, to whom all orders and inquiries should be addressed, have been appointed in the following countries:

UNITED KINGDOM	Her Majesty's Stationery Office, P.O. Box 569, London SE 1 9NH
UNITED STATES OF AMERICA	UNIPUB, Inc., P.O. Box 433, Murray Hill Station, New York, N.Y. 10016

In the following countries IAEA publications may be purchased from the sales agents or booksellers listed or through your major local booksellers. Payment can be made in local currency or with UNESCO coupons.

ARGENTINA	Comisión Nacional de Energía Atómica, Avenida del Libertador 8250, Buenos Aires
AUSTRALIA	Hunter Publications, 58 A Gipps Street, Collingwood, Victoria 3066
BELGIUM	Service du Courrier de l'UNESCO, 112, Rue du Trône, B-1050 Brussels
CANADA	Information Canada, 171 Slater Street, Ottawa, Ont. K 1 A OS 9
C.S.S.R.	S.N.T.L., Spálená 51, CS-11000 Prague Alfa, Publishers, Hurbanovo námestie 6, CS-80000 Bratislava
FRANCE	Office International de Documentation et Librairie, 48, rue Gay-Lussac, F-75005 Paris
HUNGARY	Kultura, Hungarian Trading Company for Books and Newspapers, P.O. Box 149, H-1011 Budapest 62
INDIA	Oxford Book and Stationery Comp., 17, Park Street, Calcutta 16
ISRAEL	Heiliger and Co., 3, Nathan Strauss Str., Jerusalem
ITALY	Libreria Scientifica, Dott. de Biasio Lucio "aeiou", Via Meravigli 16, I-20123 Milan
JAPAN	Maruzen Company, Ltd., P.O.Box 5050, 100-31 Tokyo International
NETHERLANDS	Marinus Nijhoff N.V., Lange Voorhout 9-11, P.O. Box 269, The Hague
PAKISTAN	Mirza Book Agency, 65, The Mall, P.O.Box 729, Lahore-3
POLAND	Ars Polona, Centrala Handlu Zagranicznego, Krakowskie Przedmiescie 7, Warsaw
ROMANIA	Cartimex, 3-5 13 Decembrie Street, P.O.Box 134-135, Bucarest
SOUTH AFRICA	Van Schaik's Bookstore, P.O.Box 724, Pretoria Universitas Books (Pty) Ltd., P.O.Box 1557, Pretoria
SPAIN	Nautrónica, S.A., Pérez Ayuso 16, Madrid-2
SWEDEN	C.E. Fritzes Kungl. Hovbokhandel, Fredsgatan 2, S-10307 Stockholm
U.S.S.R.	Mezhdunarodnaya Kniga, Smolenskaya-Sennaya 32-34, Moscow G-200
YUGOSLAVIA	Jugoslovenska Knjiga, Terazije 27, YU-11000 Belgrade

---

Orders from countries where sales agents have not yet been appointed and requests for information should be addressed directly to:

Publishing Section,
International Atomic Energy Agency,
Kärntner Ring 11, P.O.Box 590, A-1011 Vienna, Austria